FOUNDATIONS AND PHILOSOPHY OF SCIENCE AND TECHNOLOGY SERIES

General Editor: MARIO BUNGE

Relativity and Geometry

FOUNDATIONS AND PHILOSOPHY OF SCIENCE AND TECHNOLOGY SERIES

General Editor: MARIO BUNGE
McGill University, Montreal, Canada

Other Titles in the Series

AGASSI, J.
The Philosophy of Technology

ALCOCK, J. E.
Parapsychology—Science or Magic?

ANGEL, R.
Relativity: The Theory and its Philosophy

BUNGE, M.
The Mind–Body Problem

GIEDYMIN, J.
Science and Convention

HATCHER, W.
The Logical Foundations of Mathematics

SIMPSON, G.
Why and How: Some Problems and Methods in Historical Biology

WILDER, R.
Mathematics as a Cultural System

Relativity and Geometry

by

ROBERTO TORRETTI

Department of Philosophy, University of Puerto Rico

PERGAMON PRESS

OXFORD · NEW YORK · TORONTO · SYDNEY · PARIS · FRANKFURT

U.K.	Pergamon Press Ltd., Headington Hill Hall, Oxford OX3 0BW, England
U.S.A.	Pergamon Press Inc., Maxwell House, Fairview Park, Elmsford, New York 10523, U.S.A.
CANADA	Pergamon Press Canada Ltd., Suite 104, 150 Consumers Road, Willowdale, Ontario M2J 1P9, Canada
AUSTRALIA	Pergamon Press (Aust.) Pty. Ltd., P.O. Box 544, Potts Point, N.S.W. 2011, Australia
FRANCE	Pergamon Press SARL, 24 rue des Ecoles, 75240 Paris, Cedex 05, France
FEDERAL REPUBLIC OF GERMANY	Pergamon Press GmbH, Hammerweg 6, D-6242 Kronberg-Taunus, Federal Republic of Germany

First edition 1983

Library of Congress Cataloging in Publication Data
Torretti, Roberto, 1930–
Relativity and geometry.
(Foundations and philosophy of science and technology series)
Includes bibliographical references and index.
1. Relativity (Physics) 2. Geometry, Differential.
3. Geometry—Philosophy. I. Title. II. Series:
Foundations & philosophy of science & technology.
QC173.55.T67 1983 530.1′1 82–9826

British Library Cataloguing in Publication Data
Torretti, Roberto
Relativity and geometry.—(Foundations and philosophy of science and technology series)
1. Relativity 2. Geometry
I. Title II. Series
530.1′1 QC173.55

ISBN 0-08-026773-4

Printed in Great Britain by A. Wheaton & Co. Ltd, Exeter

For Christian

Preface

THE AIM and contents of the book are briefly described in the Introduction. My debt to those who have written on the same subject before me is recorded in the footnotes and in the list of references on pp. 351–379. Here I wish to acknowledge my obligation and to convey my warmest thanks to those who knowingly and willingly have aided me in my work.

Mario Bunge encouraged me to write the book for this series. I also owe him much of my present understanding of fundamental physics as natural philosophy, in the spirit of Aristotle, Newton and Einstein, and in opposition to my earlier instrumentalist leanings.

Carla Cordua stood by me during the long and not always easy time of writing, and gracefully put up with my absentmindedness. She heard much of the book while still half-baked, helping me with her fine philosophical sense and her keen awareness of style to shape my thoughts and to find the words to express them.

John Stachel read the first draft of Chapter 5 and contributed with his vast knowledge and his incisive intelligence to dispel some of my confusions and to correct my errors. He wrote the paragraphs at the end of Section 5.8 (pp. 181 ff.) on the relation between gravitational radiation and the equations of motion in General Relativity, an important subject of current research I did not feel competent to deal with.

Adolf Grünbaum, whose "Geometry, Chronometry and Empiricism" awakened my interest in the philosophical problems of relativity and geometry, has kindly authorized me to quote at length from his writings.

David Malament corrected a serious error in my treatment of causal spaces.

Roger Angel, Mario Bunge, Alberto Coffa, Clark Glymour, Adolf Grünbaum, Peter Havas, Bernulf Kanitscheider, David Malament, Lewis Pyenson, John Stachel and Elie Zahar have sent me copies of some of their latest writings. John Stachel, Kenneth Schaffner and my dear teacher Felix Schwartzmann have called my attention to some papers I would otherwise have overlooked. My son Christian, to whom the book is dedicated, has often verified references and procured me materials to which I did not have direct access.

The Estate of Albert Einstein has kindly authorized me to reproduce the quotations from Einstein's correspondence. Part of them I obtained in the Manuscript Room of the Firestone Library in Princeton University with the expert and polite assistance of its staff. During my short stay in Princeton I enjoyed the pleasant hospitality of the Institute for Advanced Study.

I began to write in August 1979 and finished today. In the second half of 1980 the University of Puerto Rico freed me from my teaching and administrative duties and the John Simon Guggenheim Memorial Foundation granted me a fellowship to pursue my work. I am deeply grateful to both institutions for their renewed trust and their generous support.

Contents

Naturwissenschaft ist der Versuch die Natur durch genaue Begriffe aufzufassen.

BERNHARD RIEMANN

Nach unserer bisherigen Erfahrung sind wir nämlich zu dem Vertrauen berechtigt, dass die Natur die Realisierung des mathematisch denkbar Einfachsten ist. Durch rein mathematische Konstruktion vermögen wir nach meiner Ueberzeugung diejenigen Begriffe und diejenige gesetzliche Verknüpfung zwischen ihnen zu finden, welche den Schlüssel für das Verstehen der Naturerscheinungen liefern.

ALBERT EINSTEIN.*

Introduction

1

Geometry grew in ancient Greece as the science of plane and solid figures. Purists would have had its scope restricted to figures that can be constructed with ruler and compass, but its subject-matter somehow led by itself to the study of curves such as the quadratrix (Hippias, 5th century B.C.), the spiral (Archimedes, 287–212 B.C.), the conchoid (Nicomedes, *ca.* 200 B.C.), the cissoid (Diocles, *ca.* 100 B.C.), which do not fit into that description. The construction of a figure brings out points, which one then naturally regards as pre-existing—at least "potentially"—in a continuous medium that provides, so to speak, the material from which the figures are made. Understandably, geometry came to be conceived of as the study of this medium, the science of points and their relations in space. This idea is in a sense already implicit in the Greek problem of loci—i.e. the search for point-sets satisfying some prescribed condition. It is fully and openly at work in Descartes' *Géométrie* (1637). When space itself, that is the repository of all points required for carrying out the admissible constructions of geometry, was identified with matter (Descartes), or with the locus of the divine presence in which matter is placed (Henry More), the stage was set for the foundation of natural philosophy upon geometrical principles (Newton). Inspired to a considerable extent by Newton's success, Kant developed his view of geometry as a paradigm of our *a priori* knowledge of nature, the surest proof that the order of nature is an outgrowth of human reason.

For all its alluring features, Kant's philosophy of geometry could not in the long run stand its ground before the onrush of the many geometries that flourished in the 19th century. Bold yet entirely plausible variations of the established principles and methods of geometrical thinking gave rise to a wondrous crop of systems, that furnished either generalizations and extensions of, or alternatives to, the geometry of Euclid and Descartes. The almost indecent fertility of reason definitely disqualified it from prescribing the geometrical structure of nature. The choice of a physical geometry, among the many possibilities offered by mathematics, was either a matter of fact, to be resolved by experimental reasoning (Gauss, Lobachevsky), or a mere matter of agreement (Poincaré).

Two widely different proposals were made, in the third quarter of the century, for ordering and unifying the new disciplines of geometry; namely,

Riemann's theory of manifolds, and Klein's Erlangen programme.[1] Riemann sought to classify physical space within the vast genus of structured sets or "manifolds" of which it is an instance. While countenancing the possibility that it might ultimately be discrete, he judged that, at any rate macroscopically, it could be regarded as a three-dimensional continuum (i.e., in current parlance, as a topological space patchwise homeomorphic to $\mathbf{R}^3$). Lines, i.e. one-dimensional continua embedded in physical space, had a definite length independent of the manner of their embedding. The practical success of Euclidean geometry suggested, moreover, that the infinitesimal element of length was equal, at each point of space, to the positive square root of a quadratic form in the coordinate differentials, whose coefficients, however, would in all likelihood—for every choice of a coordinate system—vary continuously with time and place, in utter contrast with the Euclidean case. Riemann's theory of metric manifolds was perfected by Christoffel, Schur, Ricci-Curbastro, etc. and would probably have found an application to classical mechanics if Heinrich Hertz had been granted the time to further develop his ideas on the subject. But it was not taken seriously as the right approach to physical geometry until Albert Einstein based on it his theory of gravitation or, as he preferred to call it, the General Theory of Relativity.

Klein defined a geometry within the purview of his scheme by the following task: Let there be given a set and a group acting on it; to investigate the properties and relations on the set which are not altered by the group transformations.[2] A geometry is thus characterized as the theory of invariants of a definite group of transformations of a given abstract set.[3] The known geometries can then be examined with a view to ascertaining the transformation group under which the properties studied by it are invariant. Since the diverse groups are partially ordered by the relation of inclusion, the geometries associated with each of them form a hierarchical system. Klein's elegant and simple approach won prompt acceptance. It encouraged some of the deepest and most fruitful mathematical research of the late 19th century (Lie on Lie groups; Poincaré on algebraic topology). It stimulated the advent of the conventionalist philosophy of geometry.[4] Its most surprising and perhaps in the long run historically most decisive application was the discovery by Hermann Minkowski that the modified theory of space and time on which Albert Einstein (1905*d*) had founded his relativistic electrodynamics of moving bodies was none other than the theory of invariants of a definite group of linear transformations of $\mathbf{R}^4$, namely, the Lorentz group.[5] It was thus shown that the new non-Newtonian physics—known to us as the Special Theory of Relativity—that was being then systematically developed by Einstein, Planck, Minkowski, Lewis and Tolman, Born, Laue, etc., rested on a new, non-Euclidean, geometry, which incorporated time and space into a unified "chronogeometric" structure. Minkowski's discovery not only furnished a most valuable standpoint for the understanding and fruitful development of

Special Relativity, but was also the key to its subsequent reinterpretation as the local—tangential—approximation to General Relativity. The latter, of course, was conceived from the very beginning as a dynamical theory of spacetime geometry—of a geometry, indeed, of which the Minkowski geometry is a unique and rather implausible special case.

The main purpose of this book is to elucidate the motivation and significance of the changes in physical geometry brought about by Einstein, in both the first and the second phase of Relativity. However, since the geometry is, in either phase, inseparable from the physics, what the book in fact has to offer is a "historico-critical" exposition of the elements of the Special and the General Theory of Relativity. But the emphasis throughout the book is on geometrical ideas, and, although I submit that the standpoint of geometry is particularly congenial to the subject, I also believe that there is plenty of room left for other comparable studies that would regard it from a different, more typically "physical" perspective.[6]

2

The book consists of seven chapters and a mathematical appendix. The first two chapters review some of the historical background to Relativity. **Chapter 1** deals with the principles of Newtonian mechanics—space and time, force and mass—as they apply, in particular, to the foundation of Newtonian kinematics. **Chapter 2** summarily sketches some of the relevant theories and results of pre-Relativity optics and electrodynamics.[7] The next two chapters refer to Special Relativity. **Chapter 3** is mainly devoted to Einstein's first Relativity paper of 1905. It carefully analyses Einstein's definition of time on an inertial frame—which, as I will show, can be seen as a needful improvement of the Neumann–Lange definition of an inertial time scale—and his first derivation of the Lorentz transformation. It also comments on some implications of the Lorentz transformation, such as the relativistic effects usually described as "length contraction" and "time dilation"; on the alternative approach to the Lorentz transformation initiated by W. von Ignatowsky (1910), and on the relationship between Einstein's Special Theory of Relativity and the contemporary, predictively equivalent theory of Lorentz (1904) and Poincaré (1905, 1906). **Chapter 4** presents the Minkowskian formulation of Special Relativity with a view to preparing the reader for the subsequent study of General Relativity. Thus, Sections 4.2 and 4.3 (together with Sections A and B of the Appendix) explain the concept of a Riemannian n-manifold and of a geometric object on an n-manifold, and Section 4.5 introduces von Laue's stress–energy–momentum tensor. Section 4.6 sketches the theory of causal spaces originating with A. A. Robb (1914) and perfected in our generation by E. Kronheimer and R. Penrose (1967). **Chapter 5** deals with Einstein's search for General Relativity from 1907 to 1915. As I am not altogether satisfied with earlier descriptions of this development,[8] I pay here special attention to historical detail and quote extensively from the primary sources. For completeness' sake I have added, in Section 5.8, a brief qualitative *aperçu* of Einstein's later work on the equations of motion of General Relativity. **Chapter 6** studies some aspects and subsequent developments of the theory which I believe are especially significant from a philosophical point of view, such as the Weyl–Ehlers–Pirani–Schild conception of a geometry of light propagation and free fall, the so-called Mach Principle, the standard model of relativistic cosmology and the idea of a spacetime singularity. **Chapter 7** is more explicitly philosophical and deals with the concept of simultaneity, geometric conventionalism, and a few other questions concerning spacetime structure, causality and time. The **Appendix** was designed to furnish a definition of spacetime curvature within the theory of connections in the principal bundle of tetrads. On our way to this definition every concept of differential geometry that is used, but not explained, in the main text is defined as well.

Appended to the main text is a sizable number of notes. They are printed at the end of the book, because I did not want the reader to be continually distracted by their presence at the foot of the page. *He should disregard them until he finishes reading the respective section of the text.* Besides the usual references to sources, the notes contain: (i) definitions of mathematical terms and other supplementary explanations that may be of assistance to some readers; (ii) additional remarks, quotations, proofs and empirical data, that support, illustrate or refine the statements in the main text, but may be omitted without loss of continuity; (iii) suggestions for further reading. Notes of type (i) are usually separate from those of type (ii) and (iii). I am sure that each reader will soon learn to decide at a glance which notes deserve his or her attention.

Equations and formulae are designated by three numbers, separated by points. The first two numbers denote the chapter and section in which the equation or formula occurs. Those contained in the Appendix are designated by a capital A, B, C or D, followed by a number—the letter denoting the pertinent part of the Appendix.

The list of References on pages 351 ff. provides bibliographical information on every book and article mentioned in the main test, the Appendix or the notes. It also contains a few items pertaining to our subject which I have found useful, but which I have not had the opportunity of quoting in the book. Normally references are identified by the author's name followed by the publication year. When I refer to two versions of the same work published in the same year, the second version is identified by an asterisk after the year (e.g. "Lorentz (1899*)"). When several works by the same author appeared in the same year they are distinguished by small italic letters, which do not have any further significance. Please be advised that in such cases references in the text and notes often omit the letter *a* (but not the letters *b*, *c*, etc.). Thus, "Weyl

(1919)" may stand for "Weyl (1919*a*)". In a few instances in which I do not quote a work after the first edition, I identify it by some suitable abbreviation, instead of the publication year (e.g. "Newton (*PNPM*)", instead of "Newton (1972)", which would be incongruous if not misleading).

3

A book like this one would be impossibly long if it tried to explain every notion employed in it. The mathematical concepts more directly relevant to an understanding of Relativity and Geometry belong to differential geometry and will be defined either in the main text (chiefly in Sections 4.2, 4.3, 5.4, 5.5, 6.1 and 6.4) or in the Appendix.[9] Their definition, however, will be in terms of other mathematical concepts. Specifically, I assume that the reader has:

(i) a clear recollection of the fundamental concepts of the calculus (on $\mathbf{R}^n$),[10] even if he is no longer able to work with them;
(ii) a firm grasp of the elements of linear algebra, up to and including the notion of a multilinear mapping of a finite-dimensional module or vector space into another.[11]

In section 5.7 I have tried to reproduce Einstein's reasoning in the three papers of November 1915 in which he proposed three alternative systems of gravitational field equations—culminating with the one that is now named after him. The calculations can be easily followed, using the hints in the footnotes, by someone who has a working knowledge of the elements of the tensor calculus.[12] Since the said Section 5.7 will probably be omitted by readers who have not benefited from a certain measure of formal mathematical training, I took the liberty to use in it some basic ideas and results of the calculus of variations which are not explained—and not otherwise mentioned—in this book.

It is a good deal harder to say precisely what knowledge of physics is presupposed. Presumably no one will turn to Relativity who is not somehow acquainted with classical mechanics and electrodynamics; and Chapters 1 and 2 are meant indeed for readers who have already made this acquaintance. In particular, the critical analysis of Newtonian space, time and inertia in Chapter 1 could be misleading for someone who has yet to learn about them in the ordinary way. All the requisite ideas of Relativity are duly explained; but the treatment of some of the better-known consequences of the Special Theory—particularly in Section 3.5—is perhaps too laconic. However, it is doubtful that anybody would try to read the present book without studying first one of the Relativity primers in circulation, such as Einstein's own "popular exposition", or Bondi's *Relativity and Common Sense*, or, better still, Taylor and Wheeler's *Spacetime Physics*, where those consequences are amply discussed.[13] On the other hand, I do not assume any previous knowledge of General Relativity, but rather seek to give in Chapters 4 and 5 a historical introduction to this theory. However, due to the book's peculiar bias and its almost total neglect of experimental data, it is highly advisable that beginners supplement it with some other introductory book written from a physicist's point of view. At the time of writing my own favourite for this job was Wolfgang Rindler's *Essential Relativity* (second, revised edition, 1977), which also furnishes an introduction to the Special Theory.[14]

4

Throughout the book I have tried to use only standard terminology and notation. However, there are several cases in which the standards are unstable. For the benefit of knowledgeable readers who may wish to consult something in the middle of the book without having to track down the definition of terms they are already familiar with, I shall here draw up a list of such cases, stating my choice among the available alternatives. Readers unacquainted with the ideas involved should omit the rest of this §4 and wait until the relevant definitions and conventions are in due course formally introduced.

The *fibre* of a mapping f over a given value is the set of objects in the domain of f at each of which f takes that value. (The fibre of f over a is $\{x | f(x) = a\}$).

A *curve* in a manifold M is always understood to be a parametrized curve, i.e. a continuous—usually differentiable—mapping of a real interval into M. The curve's range, i.e. the subset of M onto which the said interval is mapped by it, is called a *path*.

A basis of the tangent space at a point of an n-dimensional differentiable manifold is called an *n-ad* (rather than a *frame* or *repère*). A section of the bundle of frames over such a manifold is an *n-ad field* (instead of a *frame field* or a *repère mobile*). If $n = 4$, we speak of *tetrads* and *tetrad fields*.

Beginning on page 23, the *Latin* indices $i, j, k \ldots$ range over $\{0, 1, 2, 3\}$; the *Greek* indices $\alpha, \beta, \gamma \ldots$ range over $\{1, 2, 3\}$.

The *Einstein summation convention* is introduced on page 89 and used thereafter unless otherwise noted.

We speak of *Riemannian* (definite or indefinite) and *proper Riemannian* (positive definite) metrics—meaning what are often called Semi-Riemannian and Riemannian metrics, respectively.

We speak of the *Lorentz group* (including translations) and the *homogeneous Lorentz group*—meaning what are nowadays more frequently called the Poincaré group and the Lorentz group, respectively.

The flat Riemannian metric of Minkowski spacetime is denoted by $\boldsymbol{\eta}$. The matrix of its components relative to an inertial coordinate system with standard time coordinate is $[\eta_{ij}] = \mathrm{diag}\,(1, -1, -1, -1)$. The signature of all relativistic metrics is therefore $(+ - - -)$.

The Lorentz factor $1/\sqrt{1 - v^2/c^2}$ is designated by β_v or, where there is no danger of confusion, simply by β, as in Einstein (1905*d*). (The popular symbol γ does not have the same distinguished pedigree as β. Moreover, γ is here pre-empted to serve as the standard name of a typical curve.)

The prefix *world*- indicates appurtenance to a spacetime. A *worldline* is any curve in spacetime—not necessarily a timelike one.

For brevity's sake, objects are said to move or rest *in* a frame of reference—meaning *relatively to* it.

Beginning on page 89 and unless otherwise noted, the vacuum speed of light c is set equal to 1.

Let me add finally that in this book the classical gravitational potential increases as one recedes from the source (the Newtonian gravitational force is therefore equal to *minus* the gradient of the potential); that relativistic 4-momentum is defined as a vector (not a covector as in current literature); and that the Ricci tensor is formed by contracting the Riemann tensor with respect to the first and *last* (not the *third*) indices.

Many of these conventions do not express the author's personal preference, but arise rather from the necessity of conforming as far as possible with the usage of Einstein and his contemporaries.

5

A final word to the reader. When, in a book of this sort, the author qualifies a statement with such adverbs as "evidently", "obviously", "clearly", what he thereby intends to say is that in order to see that the statement in question is true a small effort of attention and reflection will usually be sufficient. The qualifying adverb is meant to encourage the reader to make this small effort, not to humiliate him because he cannot do without it. Genuinely self-evident statements need not bear a tag to remind one that they are such.

CHAPTER 1

Newtonian Principles

1.1 The Task of Natural Philosophy

In the Preface to the *Principia*, Newton says that the whole task of philosophy appears to consist in this: To find out the forces of nature by studying the phenomena of motion, and then, from those forces, to infer the remaining phenomena. This—he adds—is illustrated in the third Book of his work, where, from celestial phenomena, through propositions mathematically demonstrated in the former Books, he derives the forces of gravity by which bodies tend to the sun and to the several planets, and then, from these forces, with the aid of further mathematical propositions, he deduces the motions of the planets, the comets, the moon, and the sea. He expects that all other phenomena of nature might be derived by the same kind of reasoning, for he surmises, on many grounds, that they all depend upon certain forces by which the particles of bodies, by causes still unknown, are either mutually impelled to one another, and cohere in regular figures, or are repelled and recede from one another.[1]

Newton's programme turns on the concepts of motion and force. Unlike medieval writers, Newton does not understand by *motion—motus*—every kind of quantitative and qualititative change, but only that which was formerly called *motus localis*, that is, locomotion or change of place.[2] *Force—vis*—is "the causal principle of motion and rest".[3] In the early manuscript from which I quote these definitions, Newton neatly distinguishes two kinds of force. It is "either an external principle which, when impressed in a body, generates or destroys or otherwise changes its motion, or an internal principle, by which the motion or rest imparted to a body are conserved and by which any being endeavours to persevere in its state and resists hindrance".[4] These two kinds of force are, respectively, the subject of Definitions IV and III of the *Principia*, where the latter is referred to as "the innate force (*vis insita*) of matter", and the former as "impressed force (*vis impressa*)".[5] Innate force is said to be proportional to the "quantity" of matter—also called by Newton its "body" (*corpus*) or "mass" (*massa*)—and it is said to differ from the natural "laziness" (*inertia*) of matter in our conception only. The innate force must therefore be regarded as an indestructible property of the matter endowed with it. Impressed force, on the other hand, is an action only, and does not remain in the body after the action is over. "For a body perseveres in every new state by its force of inertia alone."[6]

In order to play the fundamental role assigned to them in Newton's mathematical physics, the concepts of motion and force must be made quantitative. The quantity of motion is defined as the product of the velocity and the quantity of matter or mass of the moving body, which, as we saw, is equated with the innate force of inertia. A quantitative relation between the latter and the impressed force can therefore be derived from the Second Law of Motion, according to which the quantity of motion changes with time at a rate proportional to the motive force impressed.[7] It follows that the ratio between the impressed and the innate force is proportional to the time rate of change of velocity. The magnitude of the innate force can in turn be evaluated by comparison with a conventional standard of unit mass, in accordance with a remark that follows the Third Law of Motion. The Third Law says that to every action or impressed force there is always opposed an equal reaction, so that the mutual actions of two bodies upon each other are always equal, and directed to contrary parts. As a consequence of this Law, "if a body impinging upon another body changes by its force in any way the motion of the latter, that body will undergo in its turn the same change, towards the contrary part, in its own motion, by the force of the other. [...] These actions produce equal changes not of the velocities, but of the motions. [...] Hence, because the motions are equally changed, the changes in the velocities made towards contrary parts are inversely proportional to the bodies" or masses, and consequently also to the innate forces.[8] In this way, the Axioms, or Laws of Motion—also called "hypotheses" in Newton's unpublished manuscripts[9]—enable us to trace and even to measure the forces of nature in the light of the observable features of motions, that is, in the first place, of their velocity, and, in the second place, of the time rate at which their velocity is changing in magnitude and direction.

1.2. Absolute Space

In ordinary conversation we indicate the place of a thing by specifying its surroundings. It would seem that Aristotle was trying to make precise the conception that underlies this practice when he wrote that "place (*ho topos*) is the innermost motionless boundary of the surroundings (*tou periekhontos*)".[1] Descartes criticized Aristotle's definition because there is often no such "innermost motionless boundary" and yet things are not said to lack a place, as one can see by considering a boat that lies at anchor in a river on a windy day.[2] Nevertheless he apparently abides by the same ordinary conception, for he characterizes motion or change of place "in the true sense", as "the transport of a part of matter, or a body, from the neighbourhood of those that touch it immediately, and that we regard as being at rest, to the neighbourhood of some others".[3] For Newton, however, "the positions, distances and local motions of bodies are to be referred to the parts of space".[4] Space, in turn, must be conceived absolutely—that is, free from all ties to any bodies placed in it—as a boundless real immaterial medium in which every construction postulated or demonstrated by Euclid can be carried out. We call this entity Newtonian space.

We need not dwell here on the historical and personal antecedents of Newton's revolutionary approach to place and motion.[5] But there are two reasons for it that must be mentioned. In the first place, in order to elicit the true forces of nature from the study of motion one should presumably be able to distinguish true from apparent motions. To achieve this, one cannot just refer place and motion to contiguous bodies "that we regard as being at rest", as required by Descartes' definition, because any surroundings, no matter how stable they might seem, can actually be in the course of being removed from other, vaster surroundings—a possibility vividly brought out by Copernican astronomy. Let us call this first reason for Newton's novel approach "the kinematical requirement". The second reason that we must mention is that Newton's search for the mathematical principles of natural philosophy could only make sense, given the scope of 17th century mathematics, if the full set of points required for every conceivable Euclidean construction was allowed some manner of physical existence.[6] We may call this "the requirement of physical geometry". Since Newton would not grant the infinite extension and divisibility of bodies, he could not admit that the said set of points was actually present in matter and could be recovered from it by abstraction.[7] He therefore located them in Newtonian space,[8] which he regarded as having "its own manner of existence which fits neither substances nor accidents" and consequently falls outside the acknowledged categories of traditional ontology.[9] Having admitted as much in order to satisfy the requirement of physical geometry, it was only natural that Newton should have considered his "absolute"—i.e. unbound or free—space as the perfect answer to the kinematical requirement. For true or "absolute" places can now be defined as parts of absolute space; and the real rest and the real motion of a body as the continuance or variation of the absolute place it occupies.

And yet Newton was wrong on this point. Indeed Newtonian space has been a perpetual source of embarrassment to scientists and philosophers not because it is inherently absurd—after all, it is fully characterized by an admittedly consistent mathematical theory—or because it cannot be defined "operationally"—few serious thinkers would really mind that; but because it is kinematically inoperative. To understand this let us observe first that Newtonian space does not provide any criterion by which even an omniscient demon could determine the position of a body in it. Newtonian space possesses no more structure than is required for it to be a model of Euclidean geometry (completed as indicated in note 6). Hence there are no distinguished points or directions in it. Therefore, a body placed anywhere in Newtonian space cannot be described otherwise than the same body placed elsewhere. Moreover, since the structure of space is preserved by translations, rotations and reflections, it does not of itself permit the identification of points through time, unless four of them, at least, are singled out, say, as the vertices of a rigid tetrahedron. Since motion in Newtonian space can only make a difference in relation to such a four-point system, it is obviously meaningless to speak of the motion or rest of the system itself, or of any system of rigidly connected rigid bodies comprising it, if they are considered absolutely, in complete isolation from any other body.[10] In the *Principia* Newton proposes two *Gedankenexperimente* for testing the absolute motion of an isolated system. In both of them he resorts to deformable bodies: a mass of water, a cord. Their deformations, interpreted in the light of the Laws of Motion, yield the answer to the question regarding the state of motion or immobility of the system. Introducing this passage Newton says that "the causes by which true and relative motions are distinguished one from another are the forces impressed upon bodies to generate motion".[11] A short reflection will persuade us that, if the sole grounds for ascribing *kinematic* predicates are *dynamic*, one cannot discern absolute rest from absolute motion in Newtonian physics. For in it dynamical information can refer only to the magnitude of the innate force or to the magnitude, direction and point of application of the impressed force. A given innate force acts equally in a body at rest or in a body moving with any velocity whatsoever—namely, as a resistance to acceleration; while a given impressed force effects on a rigid body a change of motion depending on the time during which the impressed force acts and on the magnitude of the innate force that resists it, but not on the body's initial state of motion or rest. Hence dynamical data may enable one to ascertain the rate—zero or otherwise—at which a body is accelerating, but not the rate—zero or otherwise—at which it is moving uniformly in a straight line. It follows that Newtonian space cannot fulfil the kinematical requirement; at any rate, not in the context of Newtonian theory. On the other hand, its Euclidean structure constitutes the system of distances and angles on which the quantitative study of motion is based in this theory. However, as we shall see in Section 1.6, there may be found another way of linking that structure with physical reality which does not involve the postulation of absolute Newtonian space.

1.3. Absolute Time

Aristotle characterized time as the measure of motion.[1] He remarked, however, that "we do not only measure the motion by the time but also the time by the motion".[2] Among all the motions that are constantly taking place in the Aristotelian cosmos there is one which is most appropriate for time reckoning, and that is the uniform circular motion (*homales kuklophoria*) of the firmament, "because its number is the best known". This explains why time had been identified (by the Pythagoreans) with the motion of the heavenly sphere, "for the other motions are measured by it".[3] However, Aristotle rejected this identification, because "if there were many heavens, the motion of any of them would likewise be time, so that there would be many times simultaneously (*polloi khronoi hama*)". But this is absurd, for time is "the same everywhere at once (*ho autos pantakhou hama*)".[4]

Aristotle's conception of a single universal time, perpetually scanned by the everlasting motion of the sky, but not to be confused with it, clearly presages Newton's "absolute, true, and mathematical time", every moment of which is everywhere.[5] The latter, it is true, is said to "flow equably in itself and by its own nature, without relation to anything external".[6] But such detachment was to be expected after the new astronomy has shattered to pieces the eternal all-encompassing aethereal dome whose daily rotation measured out Aristotelian time. In a Newtonian world "it is possible that there is no equable motion, whereby time may be accurately measured".[7] But this cannot detract from the reality of time itself, by reference to which the swiftness or slowness of the true motions of bodies can alone be estimated.

Newtonian time, like Newtonian space, is accessible only with the assistance of Newton's dynamical principles. There is one significant difference between space and time however. In Section 1.2 we saw that the dynamical principles enable us to ascertain the absolute acceleration of a body but do not provide any cue as to its absolute velocity. For this reason, we raised serious doubts regarding the physical meaning of the concept of Newtonian space. Newtonian time, on the other hand, is not subject to such doubts, for the First Law of Motion provides the paradigm of a physical process that keeps Newtonian time, and this is enough to ensure that the latter concept is physically meaningful, even if no such process can ever be exactly carried out in the world. According to the First Law, a body moving along a graduated ruler will—if both it and the ruler are free from the action of impressed forces—traverse equal distances in equal times. The flow of Newtonian time can therefore be read directly from the distance marks the body passes by as it moves along the ruler. C. Neumann (1870), in the first serious attempt at a rational reconstruction of Newtonian kinematics, used just this consequence of the First Law for defining a time scale.[8]

The 20th-century reader will be quick to observe that our ideal Newtonian clock keeps time only at the place where the moving body happens to be, and that one is not even free to change that place at will. But this restriction does not seem to have worried anyone before the publication of Einstein's "Electrodynamics of Moving Bodies" (1905). Apparently Neumann and other late 19th-century critics of mechanics tacitly assumed with Aristotle that "time is equally present everywhere and with everything",[9] and understood, with Newton, that "every moment of time is diffused indivisibly throughout all spaces".[10] A more portable kind of Newtonian clock can be conceived in the light of the Principle of the Conservation of Angular Momentum, a familiar consequence of the Laws of Motion.[11] By this Principle a freely rotating rigid sphere of constant mass will indefinitely conserve the same angular velocity. The magnitude of the latter can be measured with a dynamometer attached to the surface of the body at some distance from its axis.[12] We can imagine a set of such contraptions—which we may call spinning clocks—rotating with the same angular velocity. Let some of them be placed at regular intervals along the graduated ruler of the Newtonian clock we described first, and take the moving body to be another one. Let each of the clocks on the ruler be set by the moving clock as it passes by them. (To do this one does not have to tamper with the motion of the clocks; it is enough to adjust their readings.) In this way one could, it seems, "diffuse" the same time over all the universe.

However, this definition of time by standard Newtonian clocks could well be ambiguous. Take two spinning clocks, A and B, rotating with the same angular velocity $\boldsymbol{\omega}$ (as measured by dynamometers on each), and send them along our graduated ruler with constant velocities v_A and v_B (as measured with the clocks on the ruler). Let $v_A \neq v_B$, and assume that A is initially synchronized with B, and also agrees with each clock on the ruler as it passes by it. Do the Laws of Motion constrain B to agree with every spinning clock that it meets along the ruler? Suppose that it does not agree with them, but that, say, it appears to run slower than them. In that case, of course, the angular velocity of B, *measured with the clocks on the ruler*, must be less than $\boldsymbol{\omega}$. However, this does not clash with our assumption that it is exactly $\boldsymbol{\omega}$ *if measured with a dynamometer affixed to B*. To claim that such a discrepancy is absurd would beg the question. The local processes that measure out the time parameter of Newton's Laws of Motion will unambiguously define a universal time only if they agree with each other transitively and stably—i.e. only if the synchronism of any such process A with a process B at one time and with a process C at another time ensures the synchronism of B with C at any time. This strong requirement is *presupposed* by the Laws of Motion as Newton and his followers understood them.

It took the genius of Einstein to discover the problem raised in the preceding paragraph. Before him, the intrinsically local nature of the time kept by ideal and real clocks in a Newtonian world was generally ignored. After it is pointed out, it will seem fairly obvious to anyone who cares to reflect about it. The following remarks may help explain why nobody perceived it sooner. While the uppermost heavenly sphere was regarded as the limit of the world, the time reckoned by its daily rotation had an indisputable claim to universality. By symmetry, the rotating earth reckons the same time. In fact, the earth loses angular momentum through tidal friction, but a Newtonian will readily substitute for it a perfectly spheric and rigid spinning clock. The Copernican earth and its idealized Newtonian surrogate are thus legitimate heirs to the time-keeping role of the Aristotelian firmament, and it is not surprising that the men who first acknowledged them as such should have overlooked the fact that they can only keep time locally, not universally.

1.4. Rigid Frames and Coordinates.

As we saw on page 11, stable and changing positions in a Euclidean space can only be identified if at least four non-coplanar points are singled out in it. A set of four or more distinguished non-coplanar points in a Euclidean space will here be called a *rigid frame.* A rigid frame F is thus uniquely associated with what we may call, after Newton, a "relative space",[1] the relative space of F, hereafter denoted by S_F. In physical applications, where the space must be viewed as lasting through time and as providing a place for things, a rigid frame can be visualized as a rigid body—or a collection of rigidly connected such bodies—on which four or more non-coplanar points are marked. If we assume that material particles do not move discontinuously, it is evident that any particle whose distances from three non-collinear points of a rigid frame F are constant will also preserve its distance from each point of F and may therefore be said to be *at rest in F*. A system of particles is at rest in F if every particle of it is at rest in F; otherwise it is said *to move in F*. If F and F' are two rigid frames such that F is at rest in F', F' is also at rest in F, and any particle at rest in either will be at rest in both. Hence, any two rigid frames such that one is at rest in the other can be regarded as representative parts of one and the same frame. Consequently, if one is given a rigid frame F one may always pick out any four non-coplanar points O, X_1, X_2, and X_3, of its relative space S_F, and let them stand for F. Every point of S_F is then uniquely associated with four non-negative real numbers, namely, its distances to the four points O, X_1, X_2, and X_3. However, it is more usual and more useful to label the points of the space S_F with (negative *or* non-negative) real number triples. This can be done in many different ways. The most familiar and in a sense the most natural way of doing it is by means of Cartesian coordinates. Choose O, X_1, X_2, and X_3 in S_F so that OX_1, OX_2 and OX_3 are mutually perpendicular. A point P in S_F is assigned the real numbers $x^1(P)$, $x^2(P)$ and $x^3(P)$ such that $|x^i(P)|$ is the distance from P to the plane OX_jX_k and $x^i(P)$ is positive if and only if P and X_i lie to the same side of that plane ($i \neq j \neq k \neq i; i, j, k = 1, 2, 3$). The assignment $P \mapsto (x^1(P), x^2(P), x^3(P))$ is an injective (one-to-one) mapping of S_F onto $\mathbf{R}^3$, which we call a *Cartesian chart* of S_F (or a Cartesian coordinate system for the frame F). We denote this chart by x; ($x(P)$, the value of x at P, is then, of course, the number triple $(x^1(P), x^2(P), x^3(P))$.) We say that x is *based* on the tetrahedron $OX_1X_2X_3$. The mapping $P \mapsto x^i(P)$ of S_F onto $\mathbf{R}$ is the ith *coordinate function* of the chart. Note that a Cartesian chart correlates neighbouring points of S_F with neighbouring points of $\mathbf{R}^3$, and vice versa: it is a *homeomorphism* of S_F onto $\mathbf{R}^3$. It is also an *isometry* (an isomorphism of metric spaces) insofar as any two congruent point-pairs in S_F are mapped by it on two equidistant pairs of number triples in $\mathbf{R}^3$.[2] There are many methods of labelling the points of a Euclidean space by real number triples, which are homeomorphisms of the space (or a region of it) into $\mathbf{R}^3$—we call them generically *charts* (or coordinate systems)—but only Cartesian charts have the additional advantage of being isometries. (Later on, we shall consider spaces that do not have any charts with this property.) Let x and y be two charts of S_F, based respectively on $O_xX_1X_2X_3$ and $O_yY_1Y_2Y_3$. For simplicity's sake we assume that the distances that determine the values of x and y are measured in the same units. The *coordinate transformation* $y \cdot x^{-1}$ between chart x and chart y is evidently an isometry of $\mathbf{R}^3$ onto itself. The values of x and y at an arbitrary point P of S_F must therefore be related by a linear equation

$$y(P) = Ax(P) - \mathbf{k} \tag{1.4.1}$$

where A is an orthogonal matrix and $\mathbf{k}$ is a real number triple that can be readily seen to be equal to $Ax(O_y)$ (since $y(O_y) = (0, 0, 0)$). The determinant $|A| = \pm 1$. If $|A| > 0$, x and y are said to be *similarly oriented.*[3] We note in passing that the coordinate transformation $y \cdot x^{-1}$ and its inverse $x \cdot y^{-1}$ are continuous mappings that possess continuous partial derivatives of all orders (mappings of class C^{∞}). The relative space S_F together with the collection of all its Cartesian charts, is therefore a 3-manifold, in the sense defined in the Appendix (pp. 257f.)

In physical applications, in which one must be able to identify what is happening at different places in S_F at different times, one will also associate a time coordinate function with the rigid frame F. The meaning of this move will not become clear until Section 1.6. Let us now say only that this function assigns real numbers to events in S_F, in such a way that simultaneous events are given the same number, and successive events are given successive numbers. We may visualize the assignment as carried out by means of Newtonian clocks at rest in F wherever the events concerned occur, which have been synchronized in accordance with Newton's Laws of Motion. As noted on page 13, Newtonian physics presupposes that such a synchronization can be performed in an essentially unique way, which does not depend on the frame F. If two time coordinate functions t and t' have been constructed in the aforesaid manner, the real numbers $t(E)$ and $t'(E)$ that they assign to a given event E will be related by the linear equation $t'(E) = at(E) + b$, with $a \neq 0$. If $b \neq 0$, the two functions have different origins (that is, they assign the number 0 to different sets of simultaneous events); if $a \neq 1$, they employ different units; if $a < 0$, the coordinate transformation $t' \cdot t^{-1}$ reverses time order. In the rest of this chapter, we always assume that all time coordinate functions associated with rigid frames employ the same time unit and define the same time order, differing at most by their choice of an origin.

1.5. Inertial Frames and Newtonian Relativity.

Consider now a rigid body B moving in a rigid frame F. Since B contains at least four non-coplanar points it is a rigid frame too. Its relative space S_B differs from S_F. Let P be a particle that moves uniformly in a straight line in S_F. Depending on how B moves in S_F, P will be at rest in B or will move in S_B in different ways, uniformly or non-uniformly, describing trajectories of different shape and length. It follows at once that if the Laws of Motion give a true description of the motions of a body in one or more rigid frames they must give a false description of the motions of the same body in other rigid frames. One might remark that this was to be expected, for the Laws of Motion are supposed to govern absolute motions, but a motion in a rigid frame is a relative motion, a change of place in the relative space of that frame. We saw, however, that, even for someone willing to grant absolute space the peculiar manner of existence that Newton claims for it, the concept of absolute motion is completely devoid of meaning. Hence, our next task must be to characterize the rigid frames in which motions are adequately described by Newton's Laws. We shall say that the Laws are *referred* to such frames. If there were only one frame to which the Laws of Motion are referred we could reasonably identify the relative space determined by it with Newton's absolute space. But what we said on page 11 about the kinematical information that can be derived from dynamical data according to those Laws indicates that there must exist an infinite family of frames moving in one another's relative spaces, to which the Newtonian Laws of Motion are referred.

In his inaugural lecture on the Principles of Galileo-Newtonian (G-N) Theory, Carl G. Neumann pointed out that the First Law of Motion was unintelligible as it was stated by Newton, "because a motion which, e.g. is *rectilinear* when viewed from our earth, will appear to be *curvilinear* when viewed from the sun".[1] Hence, in order to give a definite sense to that Law, one must postulate, as the *First Principle of G-N Theory*, that "in some unknown place of the cosmic space there is an unknown body, which is an *absolutely rigid* body, a body whose shape and size are forever unchangeable".[2] Neumann calls it "the Alpha body" (*der Körper Alpha*). He observes that its existence can be asserted with the same right and the same certainty with which his contemporaries asserted the existence of a luminiferous aether, that is, as the fundamental hypothesis of a successful physical theory. For those, however, who might feel that the Alpha body is too far-fetched, he adds that it need not be a material body, but may be constituted, say, by three concurrent geometrical lines, such as the principal axes of inertia of a non-rigid body.[3] The First Law of Motion can now be meaningfully stated in two parts: (*Second Principle of G-N Theory*) A freely moving material particle describes a path that is rectilinear with reference to the Alpha body; (*Third Principle of G-N Theory*) Two such material particles move so that each traverses equal distances while the other traverses equal distances.[4] The last principle evidently presupposes the universality of time, which Neumann, as we observed earlier, simply took for granted. The "Third Principle" yields Neumann's definition of a Newtonian time scale, to which we have already referred: "Equal time intervals can now be defined as those in which a freely moving particle traverses equal distances [in the Alpha body's frame]."[5] We see at once both the analogy and the difference between this definition of a time scale and the postulation of the Alpha body. The former does for absolute time what the latter was meant to do for absolute space, that is, it provides a physically meaningful surrogate, or, as I would rather say, a realization of it. But while the uniqueness of the time scale—tacitly assumed by Neumann and his contemporaries—can be vindicated if the Laws of Motion are true (see Section 3.7), the Alpha body is necessarily *non-unique* unless the Laws of Motion are false. If α is a body that satisfies Neumann's requirements and β is another body which moves uniformly in a straight line without rotation in α's frame, β also satisfies Neumann's requirements and is just as good an Alpha body as α. A freely moving particle will describe a straight line with constant velocity, and, if it is acted on by an impressed force, it will suffer the same change of motion, in the frame of either α or β. Newton himself was well aware of this important feature of his theory.[6] In the 1880s James Thomson and Ludwig Lange based on it their independently arrived at but essentially equivalent rigorous restatements of the Law of Inertia.

We shall follow Lange (1885*b*). Like Neumann, Lange analyses the First Law into two parts concerning, respectively, the shape of the trajectory described by a freely moving particle, and the speed with which such a particle moves along its trajectory. In order to bestow an intelligible meaning on either part of the Law, Lange defines a type of rigid frame in whose relative space the said trajectory is rectilinear, and a type of time scale in terms of which the said speed is constant. Each type is appropriately labelled *inertial*. The full statement of the First Law consists of two definitions and two "theorems". The latter express, in the terminology fixed by the former, the physical content of the First Law. The following is a paraphrase of Lange's text.

> **Definition 1.** A rigid frame F is said to be *inertial* if three freely moving particles projected non-collinearly from a given point in F describe straight lines in S_F.
> **Theorem 1.** Every freely moving particle describes a straight line in an inertial frame.
> **Definition 2.** A time scale t is said to be *inertial* if a single particle moving freely in an inertial frame F travels equal distances in S_F in equal times as measured by t.
> **Theorem 2.** Every freely moving particle travels equal distances (in the relative space of an inertial frame) in equal times (measured by an inertial time scale).[7]

Consider two inertial frames F and F'. F must move uniformly in a straight line without rotation in the relative space $S_{F'}$. (Otherwise, a freely moving particle that satisfies theorems 1 and 2 in F' would not satisfy them in F.) We assume that the same time coordinate function has been associated with F and F'. Each point P of S_F coincides at any given time t with a point of $S_{F'}$ that we shall denote by $f(P, t)$. We make the Euclidean space S_F into a (real, three-dimensional, normed, complete) vector space by choosing in it an arbitrary point $\mathbf{0}$ as zero vector.[8] We denote $f(\mathbf{0}, 0)$ by $\mathbf{0}'$, and choose it as the zero of $S_{F'}$. The mapping $\mathbf{r} \mapsto f(\mathbf{r}, 0)$, $\mathbf{r} \in S_F$, is then clearly an isomorphism of the normed vector space $(S_F, \mathbf{0})$ onto $(S_{F'}, \mathbf{0}')$. We denote this mapping by f_0, and write $f_0(\mathbf{r})$ for $f(\mathbf{r}, 0)$. On the other hand, the mapping $t \mapsto f(\mathbf{0}, t)$, $t \in \mathbf{R}$, is a curve in $S_{F'}$, whose range is the straight line described by $\mathbf{0}$ as F moves in F'. The velocity of $\mathbf{0}$ along this line is the derivative $\partial f(\mathbf{0}, t)/\partial t$, whose constant value may be regarded as a vector in $S_{F'}$.[9] We denote this vector by $\mathbf{v}$. Evidently the constant velocity $\partial f(\mathbf{r}, t)/\partial t$ with which any point $\mathbf{r}$ of S_F moves in $S_{F'}$ is also equal to $\mathbf{v}$. Hence

$$f(\mathbf{r}, t) = f_0(\mathbf{r}) + t\,\mathbf{v} \tag{1.5.1}$$

Consider now a particle α of constant mass m, whose position in S_F at time t is $\mathbf{r}(t)$. Let $\mathbf{r}'(t)$ stand for $f(\mathbf{r}(t), t)$, its simultaneous position in $S_{F'}$. Set $\dot{\mathbf{r}}(t) = \mathrm{d}\mathbf{r}(t)/\mathrm{d}t$ and $\dot{\mathbf{r}}'(t) = \mathrm{d}\mathbf{r}'(t)/\mathrm{d}t$. Recalling that $\mathrm{d}f_0/\mathrm{d}\mathbf{r}$ is everywhere equal to f_0 (because f_0 is a linear mapping), we see that:

$$\begin{aligned}\dot{\mathbf{r}}'(t) &= \frac{\mathrm{d}}{\mathrm{d}t} f(\mathbf{r}(t), t) = \frac{\mathrm{d}}{\mathrm{d}t}(f_0(\mathbf{r}(t)) + t\,\mathbf{v}) \\ &= \left.\frac{\mathrm{d}f_0}{\mathrm{d}\mathbf{r}}\right|_{\mathbf{r}(t)} \cdot \left.\frac{\mathrm{d}\mathbf{r}}{\mathrm{d}t}\right|_t + \mathbf{v} \\ &= f_0(\dot{\mathbf{r}}(t)) + \mathbf{v}.\end{aligned} \tag{1.5.2}$$

In particular, if α moves with constant velocity $\mathbf{w}$ in F, it moves with constant velocity $f_0(\mathbf{w}) + \mathbf{v}$ in F', where $f_0(\mathbf{w})$ is the copy of $\mathbf{w}$ in $S_{F'}$ by the isometric linear mapping f_0. If α is acted on by a force K at time t, and this force is represented in $(S_F, \mathbf{0})$ and $(S_{F'}, \mathbf{0}')$ by the vectors $\mathbf{k}$ and $\mathbf{k}'$, respectively, we have that (writing $\ddot{\mathbf{r}}$ for $\mathrm{d}\dot{\mathbf{r}}/\mathrm{d}t$):

$$\mathbf{k} = \left.\frac{\mathrm{d}(m\dot{\mathbf{r}})}{\mathrm{d}t}\right|_t = m\ddot{\mathbf{r}}(t). \tag{1.5.3}$$

Since m is the same in F and F', and $\mathbf{v}$ is constant,

$$\begin{aligned}\mathbf{k}' = m\ddot{\mathbf{r}}'(t) &= m\frac{\mathrm{d}}{\mathrm{d}t}(f_0(\dot{\mathbf{r}}(t)) + \mathbf{v}) \\ &= m\left.\frac{\mathrm{d}f_0}{\mathrm{d}\mathbf{r}}\right|_{\mathbf{r}=\dot{\mathbf{r}}(t)} \cdot \frac{\mathrm{d}\dot{\mathbf{r}}(t)}{\mathrm{d}t} \\ &= mf_0(\ddot{\mathbf{r}}(t)) = f_0(m\ddot{\mathbf{r}}(t)) \\ &= f_0(\mathbf{k}).\end{aligned} \tag{1.5.4}$$

The vectorial representations of a given force of nature in different inertial frames, according to the Second Law of Motion, are therefore isometric transforms of one another.[10] According to the Third Law, when α is acted on by K, it must instantaneously react on the source of K with a force of equal magnitude and opposite direction. Hence this force will be represented in $(S_F, \mathbf{0})$ by $-\mathbf{k}$ and in $(S_{F'}, \mathbf{0}')$ by $f_0(-\mathbf{k}) = -f_0(\mathbf{k})$. It follows that in Newtonian mechanics all inertial frames are equivalent, and that the Newtonian Laws are referred to such frames. The equivalence of inertial frames for the description of motions and the forces that cause them can appropriately be called the *Principle of (Newtonian) Relativity*. It was stated by Newton himself in Corollary V to the Laws of Motion:

> The motions of bodies included in a given space are the same among themselves, whether that space is at rest, or moves uniformly forwards in a right line without circular motion.[11]

Lange's definition of an inertial frame depends on the concept of a freely moving particle. It might seem that this concept is ambiguous.[12] For let F and G be two rigid frames with the same associated time coordinate function, such that G moves in F in a straight line without rotation and with constant acceleration $\mathbf{a}$. Let $f_0: S_F \to S_G$ assign to each point of S_F the point of S_G with which it coincides at time 0. Let $\mathscr{F}$ and $\mathscr{G}$ be two families of frames, indexed by S_F and S_G, respectively, such that each element $F_\mathbf{v}$ of $\mathscr{F}$ moves in F with constant velocity $\mathbf{v}$, and each element $G_\mathbf{w}$ of $\mathscr{G}$ moves in G with constant velocity $\mathbf{w}$. A particle α at rest in $F_\mathbf{v} \in \mathscr{F}$ moves in $F_\mathbf{u} \in \mathscr{F}$ with a constant velocity represented in S_F by $\mathbf{v} - \mathbf{u}$,[13] but accelerates in G with constant acceleration $-f_0(\mathbf{a})$; while a particle at rest in $G_\mathbf{w} \in \mathscr{G}$ moves in $G_\mathbf{z} \in \mathscr{G}$ with a constant velocity represented in S_G by $\mathbf{w} - \mathbf{z}$, but accelerates in F with constant acceleration $\mathbf{a}$. It seems that we cannot tell which of $\mathscr{F}$ or $\mathscr{G}$ consists of inertial frames, even if we happen to know for certain that one of them does. Indeed, we can look upon $\mathscr{F}$ and $\mathscr{G}$ as members of a family of families, $\mathscr{H}$, indexed by S_F, such that each member $\mathscr{G}_\mathbf{b}$ of $\mathscr{H}$ consists of rigid frames moving uniformly in one another and accelerating with constant acceleration $\mathbf{b}$ in F. (In this indexing, our $\mathscr{F}$ is of course $\mathscr{G}_\mathbf{0}$; our $\mathscr{G}$, $\mathscr{G}_\mathbf{a}$). At first sight, it would seem that Newton's theory gives us full freedom to choose any member of $\mathscr{H}$ to be the family of inertial frames. This tremendous increase in the scope of Newtonian relativity appears to be confirmed by Corollary VI to the Laws of Motion:

> If bodies, moved in any manner among themselves, are urged in the direction of parallel lines by equal accelerative forces, they will all continue to move among themselves, after the same manner as if they had not been urged by those forces.[14]

Such an appearance is deceptive, however. For the Third Law of Motion furnishes a Newtonian physicist with all the needs for distinguishing, in principle, between a particle acted on by a true force of nature and a free particle accelerating in a particular—necessarily non-inertial—frame. If a material particle α of mass m experiences acceleration $\mathbf{a}$ in an inertial frame F, it will instantaneously react with force $-m\mathbf{a}$ on the material source of its acceleration.[15] There must exist therefore a material system β, of mass m/k, whose centre of mass experiences in F the acceleration $-k\mathbf{a}$. On the other hand, if a particle α accelerates in a non-inertial frame, its acceleration must include a component that is not matched by the acceleration of another material system, in direction opposite to the said component, caused by the action of α on that system.[16]

Though Newton's Laws are referred to inertial frames, they can certainly be used for explaining and forecasting the motions of material systems in rigid frames of every kind. The unmatched accelerations exhibited by bodies moving in non-inertial frames are usually ascribed to fictitious forces, computed according to the Second Law. We call such forces, *inertial forces.* They include the *inertial force*—in the ordinary sense of the word—that shows up in frames that move non-uniformly in a straight line (such as a bus or an underground train) and the *centrifugal, Coriolis* and *d'Alembert forces*, that show up in rotating frames.[17] All inertial forces have the following properties: (i) they cannot be traced to material sources on which the particle or body acted on by them is currently reacting; (ii) they act on each material object in proportion to its mass, independently of its state or composition; and (iii) they cannot be shielded off by insulators. All three properties are naturally to be expected in fictitious forces, that are postulated as the putative cause of accelerations arising from the state of motion of the frame in which they occur. But only the first property singles out a force as fictitious. Properties (ii) and (iii) are shared also by the force of gravity, which Newton, while confessing that he was unable to understand it, held to be very real indeed.

1.6 Newtonian Spacetime

As we shall see in Chapter 3, the concept of an inertial frame played an essential role in Einstein's original statement of Special Relativity in 1905. Einstein showed that if, as experience suggests, the propagation of electromagnetic signals *in vacuo* in the relative space of an inertial frame does not depend on the state of motion of their source, the physical equivalence of all inertial frames can only be maintained by ascribing to each such frame a different time, entailing a different partition of physical events into classes of simultaneous events. In 1907, Hermann Minkowski showed that Special Relativity can be restated in a more elegant and perspicuous, and, as proved by the eventual development of Einstein's theory of gravitation, remarkably fruitful manner, by postulating a single, metrically structured, four-dimensional manifold, which Minkowski significantly called "the world", but which is now generally known as *Minkowski spacetime.* The relative space and the relative time of each inertial frame can then be readily obtained, respectively, by orthogonal projection of the manifold onto a hyperplane of a particular kind and onto an axis perpendicular to that hyperplane. Subsequently, Elie Cartan (1923/24/25) and other writers[1] showed how to construct what may appropriately be called *Newtonian spacetimes*, that is, four-dimensional manifolds suitably structured for the reformulation of Newtonian dynamics in a style analogous to that of Minkowski's statement of Special Relativity. Though the main purpose of these constructions was to facilitate the comparison of Newton's gravitational theory with Einstein's, they can be easily adapted to a situation in which matter is so scarce and is so sparsely distributed that gravitational effects can be ignored. As Earman and Friedman (1973) have ably shown, such an empty Newtonian spacetime provides a much more natural setting than Lange's inertial frames for describing and understanding the behaviour of Newtonian test particles moving freely or under the action of impressed forces.

Philosophical attitudes toward the concept of spacetime are often conditioned by the idea that it is unintuitive and hence—it is believed—unnatural. Such a "formal" mathematical artifact may be —it is thought—of great help in calculations, but it cannot be regarded as a proper, even if admittedly simplified and "idealized", representation of a physical reality, in the manner in which Newtonian space, or, at any rate, the relative spaces of Lange's frames, are readily allowed to be. However, unless you unwittingly equate the "intuitive" with the intellectually familiar, you will surely acknowledge that the infinitely extended Euclidean space, with its infinitely articulated net of depthless planes and widthless lines, meeting at dimensionless points, is not a whit more intuitive than, say, Minkowski spacetime.[2] Even if you believe that the idea of a region in space can be obtained by abstraction and idealization from our direct grasp, say, of the interior of a room, you must admit that this idea is one step further removed from reality as experienced than the idea of a *spacetime* region. For we perceive the interior of a room for a lapse of time, no matter how short, during which even the quietest room will exhibit sound patterns—an insect's flutter near the upper right-hand corner in front, the hum of traffic beneath the window to the left—and other manifestations of becoming. If, by an operation of abstractive idealization, we ignore all qualitative differences in the room and form the conception of an homogeneous continuum underlying them, this continuum must be four-dimensional, since it must provide locations where the perceived patterns may not only stand beside each other, but also succeed one another. The conception of the three-dimensional continuum of space can only be arrived at by this method of abstraction through a second operation, by which one ignores even the unqualified possibility of change.

Though some kind of four-dimensional continuity underlies all our representations of physical objects, an exact concept of a spacetime manifold was not achieved before Minkowski. In Minkowski spacetime we must distinguish (i) the underlying four-dimensional differentiable manifold, diffeomorphic with $\mathbf{R}^4$, and (ii) the peculiar affine and metric structure bestowed on that manifold in consonance with Special Relativity. Here we need only busy ourselves with (i), which is shared by Minkowski spacetime and the Newtonian spacetime that is the subject of the present section. Though the notion of an n-manifold (see page 258) was first conceived by Bernhard Riemann in the 1850s, one can now, with the benefit of hindsight, reasonably maintain that spacetime was already handled as a 4-manifold when Galilei plotted spaces against times in his discussion of the motion of projectiles.[3] The development of coordinate geometry by Descartes and his successors furthered the intellectual habits and furnished the conceptual tools that make for this interpretation of Galilei's procedure. We find time explicitly referred to as a fourth dimension of the world in some 18th century texts.[4] However this idea received little or no attention during the 19th century, while n-dimensional geometry became increasingly fashionable among mathematicians.[5] The neglect of the mathematical concept of spacetime was due perhaps to the fact that until late in the 19th century mathematicians did not deal with the topological and differentiable structures of $\mathbf{R}^n$ as with something distinct from and more fundamental than its standard metric structure. For unless one "forgets" the Euclidean metric the analogy of $\mathbf{R}^4$ is more harmful than helpful to the idea of a spacetime manifold.[6]

Perceived continua can be conceived as n-manifolds only if they are suitably idealized.[7] But the concept of spacetime as a 4-manifold does not depend on more extreme idealizations than the standard mathematical representations of physical space and time. If physical locations and durations are idealized, respectively, as points and instants, one can without more ado conceive an instantaneous occurrence at a point, such as the durationless collision of two dimensionless particles. Such ideal occurrences are generally called point-events, or simply *events*. If we do not admit genuine vacua in the world and we regard the mere continuance of a physical state as being in some sense an occurrence, we shall take the collection of all events as the underlying set of the spacetime manifold. However, if in theoretical discussions we wish to countenance an empty or almost empty world, we must conceive the underlying set of our spacetime manifold not as the collection of all physical events but as the collection of all punctual instantaneous locations available for them. This is the set of "possible events" often mentioned by physicists. It should be clear at once that a "possible event" in this peculiar sense is not quite the same sort of thing that is ordinarily designated by this expression. Whoever allows the possible to extend beyond the actual will also grant that several mutually incompatible events might alternatively occur at a given time and place. All these possible events furnish but a single argument to a space-time coordinate system, and must consequently be fused into a single "possible event" in the physicists' special sense. The "possible events" that make up the physicists' spacetime manifold are therefore in effect equivalence classes of (idealized, i.e. instantaneous and punctual) possible events in the ordinary sense. The equivalence that defines them is none other than the identity of spatio-temporal location.

Though the notion of a "possible event" is even more remote from perceived reality than that of a real but instantaneous occurrence at a point, it is already implicit in the kinematical concepts developed in the foregoing sections. Let F and F' be two inertial frames. At each instant t, each point $\mathbf{r}$ of the relative space S_F coincides with a point $\mathbf{r}'$ of the relative space $S_{F'}$. The instantaneous coincidence of $\mathbf{r}$ and $\mathbf{r}'$ specifies a "possible event" in the above sense. It is the locus of a real one if, for instance, there are particles α and α', at rest, respectively, in $\mathbf{r}$ and $\mathbf{r}'$, during a time interval ending at t. Let $\bar{x}$ be a Cartesian coordinate system for F, with coordinate functions $\bar{x}^1$, $\bar{x}^2$ and $\bar{x}^3$. Let $\bar{x}^0$ denote the Newtonian time coordinate function, ranging over all $\mathbf{R}$, associated with F and F'. Hereafter, unless otherwise noted, small Latin indices i, j, k, . . . will range over 0, 1, 2, 3, while small Greek indices, α, β, γ, . . . range over 1, 2, 3. The four functions $\bar{x}^i$ coordinate a different number quadruple with each coincidence of a point of S_F with a point of $S_{F'}$, thereby defining a bijective mapping of the set of all "possible events" onto $\mathbf{R}^4$. Let x denote this mapping. Let $x^i = \pi_i \cdot x$ (where π_i is the i-th projection function of $\mathbf{R}^4$, which sends each real number quadruple to its $(i+1)$-th element). We see at once that the "possible events" mapped by x^α on a given real number are precisely the coincidences involving the points of S_F mapped by $\bar{x}^\alpha$ on the same real number. We call the three mappings x^α the *space coordinate functions* of x. On the other hand, x^0 and $\bar{x}^0$ are in fact the same function, for they are both defined on "possible events" and they map on a given real number precisely the same set of simultaneous events.[8] We call x^0 the *time coordinate function* of x. It is clear that our intuitions of spatiotemporal vicinity will be properly taken care of if the set of "possible events" is endowed with the weakest topology that makes x into a continuous mapping. Let $\mathscr{M}$ denote the set of "possible events", with this topology. x is then a global chart of $\mathscr{M}$ and $\{x\}$ is an atlas of $\mathscr{M}$, in the sense defined on page 257. With the differentiable structure determined by this atlas, $\mathscr{M}$ is a 4-manifold, diffeomorphic with $\mathbf{R}^4$ (see p. 258). $\mathscr{M}$ is indeed the manifold underlying both Newtonian and Minkowski spacetime to which I referred under (i) on page 21. We shall henceforth designate the points of $\mathscr{M}$—or of any other spacetime manifold—by the name of worldpoints, used by Minkowski, which is less clumsy and inaccurate than "possible events". A regular smooth or piecewise smooth curve in $\mathscr{M}$ or the path defined by it (pp. 259, 261) is called a *worldline*.

A chart of $\mathscr{M}$ constructed like x from a Newtonian time coordinate function and a Cartesian coordinate system for a given inertial frame, will be called a *Galilei chart*, *adapted* to that frame. Through the study of Galilei charts and their coordinate transformations we shall arrive at a characterization of the specific structure of Newtonian spacetime. We prove, first of all, that if x and y

are two Galilei charts adapted to the inertial frames, F and G, respectively, the coordinate transformations $x \cdot y^{-1}$ and $y \cdot x^{-1}$ are linear permutations of $\mathbf{R}^4$ (i.e. linear bijections of $\mathbf{R}^4$ onto itself). Since a change of scale amounts to a linear permutation, we can assume, without loss of generality, that the same units of time and length are used in constructing x and y. Let O be the worldpoint such that $x(O) = (0,0,0,0)$, and let $y(O) = (k^0, k^1, k^2, k^3)$. There exists a Galilei chart z, adapted to G, such that

$$z^i = y^i - k^i \tag{1.6.1}$$

It is clear that $z(O) = x(O)$ and that $z^0 = x^0$ on all of $\mathscr{M}$. Consider the worldpoints P_i, such that $x(P_0) = (1,0,0,0)$, $x(P_1) = (0,1,0,0)$, $x(P_2) = (0,0,1,0)$ and $x(P_3) = (0,0,0,1)$, and let $z(P_0) = (1, v_1, v_2, v_3)$ and $z(P_\alpha) = (0, a_{1\alpha}, a_{2\alpha}, a_{3\alpha})$. The meaning of the twelve numbers v_α, $a_{\alpha\beta}$, is not hard to see. Let $\bar{x}$ and $\bar{z}$ denote, as before, the Cartesian systems for F and G from which x and z are built. O can be regarded as the coincidence of a point o in S_F with a point o' in S_G. P_0 is the coincidence of the same point o with a point o'' in S_G. Hence F moves in G with constant velocity in the direction from o' to o''. If $\bar{z}_p(t)$ is the value of $\bar{z}$ at the point of S_G with which the point p of S_F coincides at time t, it is clear that $\mathrm{d}\bar{z}_p/\mathrm{d}z^0 = (v_1, v_2, v_3)$. The v_α are therefore the velocity components of F in the Cartesian system z. The mapping $\bar{x}(p) \mapsto \bar{z}_p(0)$ is a rotation or a rotation-*cum*-parity of $\mathbf{R}^3$, with matrix $(a_{\alpha\beta})$.[9] The matrix $(a_{\alpha\beta})$ is therefore orthogonal and has no more than three independent components.[10] It is now easy to calculate the values of the coordinate functions z^i at an arbitrary worldpoint P if the values of the x^i at P are given. Let P be the coincidence of $p \in S_F$ with $p' \in S_G$. At the time of O, while o coincides with o', p coincides with $p'' \in S_G$. Noting that the time interval between O and P is given by $z^0(P) - z^0(O) = x^0(P)$, it will be seen that $\bar{z}^\alpha(p'') = \bar{z}^\alpha(p') - v_\alpha x^0(P)$. Since $\bar{z}^\alpha(p'') = \Sigma_\beta a_{\alpha\beta} \bar{x}^\beta(p)$, it follows that $\bar{z}^\alpha(p') = \Sigma_\beta a_{\alpha\beta} \bar{x}^\beta(p) + v_\alpha x^0(P)$. Consequently, the coordinate transformation $z \cdot x^{-1}$ is governed by the equations:

$$\begin{aligned} z^0 &= x^0 \\ z^\alpha &= \Sigma_\beta a_{\alpha\beta} x^\beta + v_\alpha x^0 \end{aligned} \tag{1.6.2}$$

Substituting from equations (1.6.1) into (1.6.2) we find the general form of the equations governing the coordinate transformation $y \cdot x^{-1}$, where x and y are any two Galilei charts constructed using the same units of time and length:

$$\begin{aligned} y^0 &= x^0 + k_0 \\ y^\alpha &= \Sigma_\beta a_{\alpha\beta} x^\beta + v_\alpha x^0 + k_\alpha \end{aligned} \tag{1.6.3}$$

$y \cdot x^{-1}$ is therefore a linear permutation of $\mathbf{R}^4$. The same is true of its inverse $x \cdot y^{-1}$.

To avoid useless complications we shall henceforth consider a representative collection of Galilei charts, which we call the Galilei atlas $A_{\mathscr{G}}$. $A_{\mathscr{G}}$ consists of an arbitrarily chosen Galilei chart x and of all Galilei charts y related to x by equations of the form (1.6.3) in which the orthogonal matrix $(a_{\alpha\beta})$ has a positive determinant. This implies that (i) all charts of $A_{\mathscr{G}}$ involve the same units of length and time; (ii) their time coordinate functions increase in the same temporal direction, and (iii) at any given time, the space coordinate functions of two charts in $A_{\mathscr{G}}$, adapted, say, to the frames F and G, can be made to agree on coincident points by subjecting either frame to a translation followed by a rotation in the other frame (no spatial reflection is involved). Let x be any chart in $A_{\mathscr{G}}$, adapted to an inertial frame F. $x(\mathscr{M})$ is then $\mathbf{R}^4$, which we consider to be endowed with the standard Euclidean metric (Sect. 1.4, n. 2). We wish to examine some geometrical features of x. If T is the fibre (note 8) of the time coordinate function x^0 over one of its values, $x(T)$ is a hyperplane in $\mathbf{R}^4$, consisting of all the real number quadruples whose first component is the said value of x^0. This hyperplane is isometric with $\mathbf{R}^3$ and is therefore a copy of Euclidean space. The set of all such hyperplanes, $\pi_0 = t, t \in \mathbf{R}$, is ordered by the parameter t. If s is the collection of all "possible events" in the history of a particle σ, $x(s)$ is a line that meets each hyperplane $\pi_0 = const.$, once and once only. If σ is motionless in F, $x(s)$ is a straight line orthogonal to the hyperplanes $\pi_0 = const.$ If σ moves uniformly in F with speed v, $x(s)$ is a straight line which forms an angle equal to arc $\tan v$ with the direction orthogonal to the hyperplanes $\pi_0 = const.$ Let y be a chart in $A_{\mathscr{G}}$ adapted to a frame G. The mapping $y \cdot x^{-1}$ is a linear permutation of $\mathbf{R}^4$. We call it a *Galilei coordinate transformation*. How does $\mathbf{R}^4$ fáre under such a transformation? A few facts deserve our attention. (1) Each hyperplane $\pi_0 = const.$ is mapped isometrically, without reflection, onto a hyperplane parallel to it.[11] (2) The transformation preserves the order of the hyperplanes $\pi_0 = t, t \in R$, and the distance, in $\mathbf{R}^4$, between any two of them. (3) The family of parallel lines orthogonal to the hyperplanes $\pi_0 = const.$ is mapped onto a family of parallels tilted with respect to the former by an angle α, less than $\pi/2$; $\tan \alpha$ is the speed of F in G. (4) As a consequence of (1) and (3), all parallels are mapped onto parallels: the transformation is affine. It is not difficult to see that any permutation of $\mathbf{R}^4$ that satisfies these four conditions is a Galilei coordinate transformation between some pair of charts in $A_{\mathscr{G}}$. Let $\mathscr{G}_{\mathbf{R}}$ be the set of all such permutations. $\mathscr{G}_{\mathbf{R}}$ includes the identity mapping. If θ belongs to $\mathscr{G}_{\mathbf{R}}$, the inverse permutation θ^{-1} also belongs to $\mathscr{G}_{\mathbf{R}}$. Moreover, if θ and θ' belong to $\mathscr{G}_{\mathbf{R}}$, their product $\theta \cdot \theta' = \theta''$ is affine and satisfies (1) and (2); it also satisfies (3) because, if λ is perpendicular to the hyperplanes $\pi_0 = const.$, the directions of λ and $\theta''(\lambda)$ cannot make a right angle if θ'', which is bijective, satisfies (1), nor an obtuse angle if θ'' satisfies (2) and hence preserves time order. $\mathscr{G}_{\mathbf{R}}$ is therefore a group of permutations of $\mathbf{R}^4$.

The mapping f: $x \cdot y^{-1} \mapsto x^{-1} \cdot y (x, y \in A_{\mathscr{G}})$ assigns to each Galilei coordinate transformation of $\mathbf{R}^4$ a permutation of $\mathscr{M}$, which we call a *Galilei point transformation*. Let $\mathscr{G}_{\mathscr{M}}$ be the range of f. Since f is bijective and maps the product of two elements of $\mathscr{G}_{\mathbf{R}}$ on the product of their images in $\mathscr{G}_{\mathscr{M}}$, $\mathscr{G}_{\mathscr{M}}$ is evidently a group of permutations and f is a group isomorphism. Both $\mathscr{G}_{\mathbf{R}}$ and $\mathscr{G}_{\mathscr{M}}$ are realizations of the same abstract Lie group, known as the *proper orthochronous Galilei group.* We denote it by $\mathscr{G}$.[12] Though we have defined it by means of its realization as a group of permutations of $\mathbf{R}^4$, its structure can be conceived apart from all realizations. Though we have used Lange inertial frames and Newtonian time coordinates as a historically justified prop for explaining the significance of the Galilei group $\mathscr{G}$, we do not need this scaffolding for defining it. This can be better understood if I first introduce the concept of the *action* of a group on a set, which we shall use repeatedly in this book. A group $\mathscr{H}$ is said to act on a set S if there is a surjective mapping $F\colon \mathscr{H} \times S \to S$ such that, for every g, $h \in \mathscr{H}$ and every $s \in S$, $F(gh, s) = F(g, F(h, s))$. F is called the *action* of $\mathscr{H}$ on S. It is said to be *transitive* if for every $s, s' \in S$ there is an $h \in \mathscr{H}$, such that $s' = F(h, s)$. It is said to be *effective* if $F(h, s) = s$, for every $s \in S$, only if h is the neutral element of $\mathscr{H}$. We write hereafter hs instead of $F(h, s)$.[13] The mapping $s \mapsto hs$ $(s \in S)$, for a fixed $h \in \mathscr{H}$, is clearly a permutation or transformation of S. Such transformations form a group, isomorphic with $\mathscr{H}$, the realization of $\mathscr{H}$ by its action F. If $\mathscr{H}$ is a Lie group and S an n-manifold and F is sufficiently smooth (p. 258), we say that $\mathscr{H}$ acts on S as a Lie group of transformations. Each transformation $s \mapsto hs$ is evidently a diffeomorphism (p. 258). By means of the concept of group action we can express the structure of $\mathscr{G}$ in terms of that of other known groups. Since groups are structured sets they can act on each other. The *semidirect product* $\mathscr{H} \times \mathscr{K}$ of a Lie group $\mathscr{H}$ by a Lie group $\mathscr{K}$ on which the former acts is the product manifold $\mathscr{H} \times \mathscr{K}$ endowed with the group product: $(h, k)(h', k') = (hh', khk')$. If $O^+(3)$ denotes the multiplicative group of 3×3 orthogonal matrices with positive determinant, and T_n is the underlying additive group of the vector space $\mathbf{R}^n$, $\mathscr{G}$ is the semidirect product $(O^+(3) \times T_3) \times T_4$—the actions involved being constructed by ordinary matrix multiplication.[14]

Let $\mathscr{H}$ and S be as above. An n-place relation R on S is said to be *invariant under* $\mathscr{H}$, or $\mathscr{H}$*-invariant* if, for every n-tuple $(s_1, \ldots, s_n)$ of points of S and every $h \in \mathscr{H}$, whenever $(s_1, \ldots, s_n)$ have the relation R, $(hs_1, \ldots, hs_n)$ also have it. (View properties as "one-place relations".) This definition is readily extended to mappings of S^n into any set; e.g., $f\colon S^2 \to S'$ is $\mathscr{H}$-invariant if, for any $s_1, s_2 \in S$ and any $h \in \mathscr{H}$, $f(s_1, s_2) = f(hs_1, hs_2)$. In the Erlangen programme, Felix Klein (1872) successfully employed the concept of invariance under a group for unifying in an orderly system most of the seemingly disparate branches of 19th century geometry. In Klein's system, a geometry is induced—or should we say disclosed?—in a manifold by a group that acts on it. The geometric structure is constituted by the group invariants. From this point of view, we define the Newtonian structure—in the absence of matter—of the manifold $\mathscr{M}$ diffeomorphic with $\mathbf{R}^4$, by the properties and relations of $\mathscr{M}$ that are invariant under $\mathscr{G}$—or more precisely, under the action of $\mathscr{G}$ on $\mathscr{M}$

that realizes $\mathscr{G}$ by means of $\mathscr{G}_{\mathscr{M}}$, the group of Galilei point transformations introduced above. A rigorous analytical deduction of the $\mathscr{G}$-invariants is out of the question here, but a quick glance at our earlier description of the action of the Galilei group on $\mathbf{R}^4$ should persuade us that they include the properties, relations and functions listed below. (As usual, x stands for an arbitrary chart of $A_{\mathscr{G}}$, and x^0 for its time coordinate function.)

A. $\mathscr{G}$-invariant relations between worldpoints

(i) Two worldpoints P_1 and P_2 are *simultaneous* if and only if $x^0(P_1) = x^0(P_2)$. (Hence, simultaneity is an equivalence, i.e. a symmetric, reflexive and transitive relation.)
(ii) P_1 *follows* P_2 (and P_2 *precedes* P_1) if and only if $x^0(P_1) > x^0(P_2)$.
(iii) Three worldpoints P_1, P_2 and P_3 are *collinear* if and only if $x(P_1)$, $x(P_2)$ and $x(P_3)$ lie on the same straight line in $\mathbf{R}^4$.
(iv) Four worldpoints $P_1, \ldots, P_4$ are *coplanar* if and only if $x(P_1)$, $\ldots, x(P_4)$ lie on the same plane in $\mathbf{R}^4$.

B. $\mathscr{G}$-invariant properties and relations of worldlines

(i) A worldline λ is *spacelike* if and only if all worldpoints on λ are simultaneous.
(ii) λ is *timelike* if and only if no two worldpoints on λ are simultaneous.
(iii) λ is *straight* if and only if any two worldpoints on λ are collinear with every other worldpoint on λ.
(iv) Two straights λ and μ are *coplanar* if and only if any two worldpoints on λ are coplanar with any two worldpoints on μ.
(v) Two coplanar straights λ and μ are *parallel* if and only if there does not exist a worldpoint P that is collinear with two points of λ and is also collinear with two points of μ.

C. $\mathscr{G}$-invariant functions

(i) The *temporal distance* $\tau(P_1, P_2)$ between two worldpoints P_1 and P_2 is equal to the distance in $\mathbf{R}^4$, between the two hyperplanes $\pi_0 = x^0(P_1)$ and $\pi_0 = x^0(P_2)$.
(ii) The *spatial distance* $\sigma(P_1, P_2)$ between two simultaneous worldpoints P_1 and P_2 is equal to the distance in $\mathbf{R}^4$ between $x(P_1)$ and $x(P_2)$.

The distance function σ evidently satisfies the laws of Euclidean geometry on each set of mutually simultaneous worldpoints. It can be used for defining the length of spacelike straights (viz., as the spatial distance between their endpoints). On the other hand, since $\mathscr{G}_{\mathbf{R}}$ is a group of linear but generally non-isometric transformations of $\mathbf{R}^4$, there is no $\mathscr{G}$-invariant distance function between arbitrary worldpoints.[15] A $\mathscr{G}$-invariant angle measure can only be defined for concurrent spacelike straights. The distance between two spacelike worldlines can be readily defined as in ordinary geometry if their points are simultaneous, but it makes little sense to speak of the distance between non-spacelike worldlines. However, we can define equidistance between two timelike worldlines λ and μ if for for every point P on λ there is a point Q on μ which is simultaneous with P. If this condition is met, we say that λ and μ are *equidistant* if and only if the spatial distance between each pair of simultaneous worldpoints in $\lambda \cup \mu$ is the same. (This constant spatial distance is then the distance between λ and μ.) It is not hard to see that two timelike straights whose union consists of pairwise simultaneous points are equidistant if and only if they are parallel.

Newtonian spacetime, as characterized above, is not a metric space. But it is what we shall call an affine space (p. 91), and its system of straights agrees with that of four-dimensional Euclidean space. We have characterized the $\mathscr{G}$-invariants of Newtonian spacetime by means of an arbitrary Galilei chart, but they can also be given an intrinsic or "coordinate free" definition.[16] The foregoing characterization can then be used for defining a Galilei chart.

The geometry of empty Newtonian spacetime throws much light on the conceptual framework developed in the earlier sections of this chapter. A *congruence of curves* in a manifold S is a set of curves in S such that each point of S lies on the range of one and only one curve of the set. A *frame* in Newtonian spacetime can be defined as a congruence of timelike curves. A frame F is *rigid* if each pair of curves in F are equidistant. A rigid frame is *inertial* if it is a congruence of straights. We see at once how each rigid frame is associated with a copy of Euclidean 3-space. A congruence of curves in a manifold is a partition of the manifold and thereby defines a relation of equivalence between the points of the latter. Let f be the equivalence defined on $\mathscr{M}$ by the rigid frame F. (Two worldpoints are equivalent by f if and only if they lie on the same curve of the congruence F.) S_F, the relative space of F, is none other than the quotient manifold $\mathscr{M}/f$, whose points are the curves of F. Since the distance between two of these curves is the spatial distance between any two simultaneous points of them, the quotient space $\mathscr{M}/f$ is a copy of Euclidean space. There is no longer any mystery about the "mutual penetration" of the relative spaces of two rigid frames, if we conceive them as different quotient manifolds of the same spacetime $\mathscr{M}$ defined by different partitions of its worldpoints. Let s denote simultaneity; s is an equivalence. The quotient manifold $\mathscr{M}/s$ is diffeomorphic with $\mathbf{R}$. We take $\mathscr{M}/s$ to be what we mean by Newtonian time. We make the latter, elusive concept, perspicuous by identifying it with the former, which is clear and distinct. It is immediately clear in what sense Newtonian time is universal: namely, because each instant is an equivalence class of simultaneous "possible events" at all points of the relative space of any frame. And why it is absolute: namely, because the partition of worldpoints into "simultaneity classes" does not depend on the choice of a frame.

The connection between Galilei charts and inertial frames can now be described as follows: If x is a Galilei chart, the parametric lines of the time coordinate function x^0 (i.e. the curves in $\mathscr{M}$ along which x^0 varies while the space coordinate functions x^α take constant values) constitute a congruence of timelike straights, or, in other words, an inertial frame. Call it X. Let $\bar{p}$: $(a, b, c, d) \mapsto (b, c, d)$ be the projection of $\mathbf{R}^4$ onto $\mathbf{R}^3$ defined by "ignoring" the first element of each real number quadruple, and let $\bar{x}$ stand for $\bar{p} \cdot x$. The fibre of $\bar{x}$ over a given real number triple is then a parametric line of x^0, that is, a point of the relative space of the inertial frame X. $\bar{x}$ is evidently a Cartesian coordinate system for X. If y is another Galilei chart, related to x by equations (1.6.3) with $v_1 = v_2 = v_3 = 0$ the parametric lines of y^0 constitute the same inertial frame X. Let us say that two Galilei charts are *frame equivalent* if they thus define the same inertial frame. If x and y are frame inequivalent charts of $A_{\mathscr{G}}$ one can always find a chart z, frame equivalent to y and related to x by equations (1.6.2), with $(a_{\alpha\beta}) = (\delta_{\alpha\beta})$, the 3×3 identity matrix. We call $x \cdot z^{-1}$ a *purely kinematic*(pk) Galilei coordinate transformation. The transformation depends only on the three parametres v_α, which are the velocity components of the frame defined by y and z, in the frame defined by x, along the three directions determined by the parametric lines of the space coordinate functions x^α. The pk transformations constitute a subgroup of $\mathscr{G}$. For if x, y and z are frame inequivalent charts of $A_{\mathscr{G}}$, such that $y \cdot x^{-1}$ and $z \cdot y^{-1}$ are pk Galilei transformations governed by

$$\begin{aligned} y^0 &= x^0 & z^0 &= y^0 \\ y^\alpha &= x^\alpha + v_\alpha x^0 & z^\alpha &= y^\alpha + w_\alpha y^0, \end{aligned} \tag{1.6.4}$$

it follows that $z \cdot x^{-1}$ is also a pk Galilei transformation governed by

$$\begin{aligned} z^0 &= x^0 \\ z^\alpha &= x^\alpha + (v_\alpha + w_\alpha) x^0. \end{aligned} \tag{1.6.5}$$

Writing $\mathbf{r}$ for the number triple (r_1, r_2, r_3) and $\dot{x}_\alpha$ for the coordinate velocity $\mathrm{d}x^\alpha/\mathrm{d}x^0$, it is clear that a particle at rest in the frame defined by x moves with velocity $\dot{\mathbf{y}} = \mathbf{v}$ in the frame defined by y and with velocity $\dot{\mathbf{z}} = \mathbf{v} + \mathbf{w}$ in the frame defined by z. This is the *Galilei Rule for the Transformation of Velocities.*

A material particle is represented in Newtonian spacetime by the set of all "possible events" in its history, that is by a timelike worldline. *A free particle has a straight worldline*: this is the geometric contents of Newton's First Law. That of the Second Law is the following: *Under the influence of an impressed force the worldline of a particle deviates from its direction at each worldpoint in the spacelike direction in which the force is acting, at a time rate directly proportional to that force and inversely proportional to the particle's mass.* The difference between a free or inertial and a non-inertial or forced motion is therefore analogous to that between straightness and crookedness—an analogy that Newton himself suggested in a note of his "Waste Book".[17] As we saw on page 18, the Third Law can teach us, in principle, what particles are free, and hence, what timelike worldlines are straight. It will be noticed that the Newtonian concept of an impressed force, endowed with a direction in space, depends essentially on the existence of an absolute time, that unambiguously determines the simultaneity class of any given worldpoint, and hence the set of spacelike directions at the latter.

The Newtonian Principle of Relativity, which proclaims the dynamical equivalence of all inertial frames, is so to speak built into the structure of Newtonian space-time. Its implications for Newton's programme of natural philosophy and its physical significance become clearer in this perspective. Consider an isolated system of mutually interacting material particles. The history of the system between two instants of time is represented in Newtonian spacetime by a configuration S of timelike worldlines, one for each particle in the system, whose directions are variously changing under the influence of the forces with which the particles act on each other. According to Newton's programme (Section 1.1), the laws that govern the operation and evolution of natural forces should be ascertainable in the light of the phenomena of motion, that is, by studying such configurations as S. If τ is a Galilei point transformation, Newtonian relativity implies that τS, the image of S by τ, is dynamically indistinguishable from S. S and τS are therefore equally good representations of our isolated system, or of physically equivalent isolated systems, and the study of S or of τS must lead to the discovery of exactly the same laws. Consequently, Newtonian relativity demands that the operation and evolution of the forces of nature depend only on the $\mathcal{G}$-invariant features of the spacetime representations of material systems and on intrinsic properties of material objects—such as mass, electric charge, etc.—which must be deemed to be $\mathcal{G}$-invariant as well. Consider now the image $x(S)$ of S by a Galilei chart x. $x(S)$ provides a full description of the shape of S by means of real number quadruples and their relations. Another Galilei chart y yields a different description $y(S)$. But the laws disclosed by the study of S must hold equally well for either description. Suppose that the shape of S is determined by a set of dynamical laws L together with initial and boundary conditions $C(S)$. If $C_x(S)$ denotes the description of the latter in terms of the chart x, there must be a set of equations L_x, involving the coordinate functions of x and the functions representing in $\mathbf{R}^4$ the spacetime distribution of the relevant intrinsic properties of matter, such that the shape of $x(S)$ can be derived from L_x together with $C_x(S)$. Obviously, the equations L_x mus be regarded as the expression of L in terms of x. If the functions involved in L_x are subjected to the Galilei coordinate transformation $y \cdot x^{-1}$, one should obtain another set of equations L_y which, together with the conditions $C_y(S)$, will determine the shape of $y(S)$. Moreover, since the laws L must only concern $\mathcal{G}$-invariant properties and relations, their expressions L_x and L_y in terms of the charts x and y cannot involve traits peculiar to the representation of S by either chart. Hence, one concludes, L_x and L_y must be equations of the same "form", that can be obtained from one another by substitution of variables. These abstract and somewhat imprecise considerations will be illustrated by an important example in Section 1.7. But before passing to it I wish to state one more consequence of the foregoing, which epitomizes the physical meaning of Newtonian relativity. Take again the physical process we have represented by the spacetime configuration S, determined by laws L and conditions $C(S)$. The Galilei point transformation τ maps S onto τS, determined by L and conditions $C(\tau S)$. If $\tau = y^{-1} \cdot x$, where y and x are suitable Galilei charts, $x(S)$ and $y(\tau S)$ provide identical numerical descriptions of S and of τS, respectively. Evidently, the conditions $C_x(S)$, from which, together with L_x, one derives $x(S)$, are the same set of numbers and numerical relations as the conditions $C_y(\tau S)$, from which, together with L_y, one derives $y(\tau S)$. Consequently, if the laws of nature satisfy the requirements of Newtonian relativity, two physical processes, whose determining conditions can be described in the same way in terms of two Galilei charts (of our atlas $A_{\mathcal{G}}$), will receive, through their entire development, the same description in terms of those two charts. As Galilei himself expressed it, with unrivalled clarity:

> Shut yourself up with some friend in the main cabin below decks on some large ship, and have with you there some flies, butterflies, and other small flying animals. Have a large bowl of water with some fish in it; hang up a bottle that empties drop by drop into a wide vessel beneath it. With the ship standing still, observe carefully how the little animals fly with equal speed to all sides of the cabin. The fish swim indifferently in all directions; the drops fall into the vessel beneath; and, in throwing something to your friend, you need throw it no more strongly in one direction than another, the distances being equal; jumping with your feet together, you pass equal spaces in every direction. When you have observed all these things carefully (though there is no doubt that when the ship is standing still everything must happen in this way), have the ship proceed with any speed you like, so long as the motion is uniform and not fluctuating this way and that. You will discover not the least change in all the effects named, nor could you tell from any of them whether the ship was moving or standing still.[18]

1.7 Gravitation

Newton's greatest contribution to the fulfilment of his philosophical programme was the law of universal gravitation. With it he unified, boldly and successfully, two of the most conspicuous regularities of nature, free fall and planetary motion. Although we no longer consider Newton's formula as perfectly accurate and we reject the notion of force that underlies it, today such phenomena can be only conceived of, regardless of their apparent disparateness, as instances of the same general law. (Philosophers who tend to underrate the permanence of scientific achievements would do well to reflect on this fact.)

According to Newton's Law of Gravity, a particle of mass m_1, at a distance r from a particle of mass m_2, suffers the action of an impressed force directed toward the second particle and proportional to $m_1 m_2/r^2$. By the Third Law of Motion, the second particle must then be acted on by a force of the same magnitude and opposite direction. It is clear therefore that the susceptibility of a particle to gravitational force—what we may call its "passive gravitational mass"—must be proportional to its power of exerting such a force—its "active gravitational mass". On the other hand, it is a remarkable coincidence that these properties of matter should be measured by the same quantity that determines matter's "innate force" or resistance to acceleration—what we have hitherto called, with Newton, its "mass" and shall henceforth call its "inertial mass". Newton was well aware that the proportionality between inertial mass and passive gravitational mass, or, as he puts it, between the "quantity of matter" and its "weight", was not a conceptual necessity. He claimed to have established it experimentally, with an accuracy of 1 part in 1000, by means of pendulums "perfectly equal in weight and figure", filled with equal weights of "gold, silver, lead, glass, sand, common salt, wood, water, and wheat".[1]

Let y be a Galilei chart, and let $y_j^\alpha(t)$ be the space coordinates of a particle j, of mass m_j, at time t. In terms of y, the force impressed on j at time t can be represented by a vector in $\mathbf{R}^3$, with components $F^\alpha(t) = m_j \mathrm{d}^2 y_j^\alpha(t)/(\mathrm{d}y^0)^2$. If we use bold face, as on page 29, to denote vectors in $\mathbf{R}^3$, and a dot or two over a function to designate its first or second time derivative, the gravitational force on particle 1, of mass m_1, in the presence of particle 2, of mass m_2, is given by either side of the equation:

$$m_1 \ddot{\mathbf{y}}_1 = -G\frac{m_1 m_2}{|\mathbf{y}_1 - \mathbf{y}_2|^3}(\mathbf{y}_1 - \mathbf{y}_2), \tag{1.7.1}$$

where G, the gravitational constant, is approximately equal to $6.67 \cdot 10^{-11}\ \mathrm{m}^3\,\mathrm{kg}^{-1}\,\mathrm{s}^{-2}$. Let x be another Galilei chart related to y by equations (1.6.3). Let A denote the matrix of the parameters $a_{\alpha\beta}$. Since A is orthogonal, $|A\mathbf{r}| = |\mathbf{r}|$, for any vector $\mathbf{r}$ in $\mathbf{R}^3$. Substituting $\mathbf{x}_1$, $\mathbf{x}_2$ and $\ddot{\mathbf{x}}_1$ in (1.7.1), in accordance with (1.6.3), we have that

$$m_1 A\ddot{\mathbf{x}}_1 = -G\frac{m_1 m_2}{|\mathbf{x}_1 - \mathbf{x}_2|^3}A(\mathbf{x}_1 - \mathbf{x}_2). \tag{1.7.2}$$

Multiplying both sides on the left by A^{-1}, we obtain:

$$m_1 \ddot{\mathbf{x}}_1 = -G\frac{m_1 m_2}{|\mathbf{x}_1 - \mathbf{x}_2|^3}(\mathbf{x}_1 - \mathbf{x}_2). \tag{1.7.3}$$

We have here an example illustrating the general ideas presented on page 30.

The Law of Gravity, expressed in terms of the chart y by equations (1.7.1), is expressed in terms of the chart x by equations (1.7.3). The latter are obtained from the former *via* the Galilei coordinate transformation between x and y, or directly by substitution.[2] Dividing both members of either set of equations by m_1, we find that the gravitational acceleration of a particle does not depend on its mass. This fact, noted in the case of falling bodies by Galilei, is thus explained by Newton as a consequence of the proportionality of inertial and gravitational mass.

Newton's Law of Gravity was ill-received by many of his contemporaries because it postulates instantaneous interaction between distant bodies. As the Earth revolves around the Sun, the opposite and equal forces which they exert on each other adjust continually and without delay to the varying distance between them. In the eyes of Huygens or Leibniz, to allow direct action at a distance was to return to the pre-Cartesian natural philosophy of occult, magic powers. Newton himself rejected the view of gravity as "innate, inherent, and essential to matter". He wrote to Bentley:

> That one body may act upon another at a distance through a *vacuum*, without the mediation of anything else, by and through which their action and force may be conveyed from one to another, is to me so great an absurdity that I believe no man who has in philosophical matters a competent faculty of thinking can ever fall into it.[3]

Gravitational interaction requires therefore a cause besides the mere presence of the interacting bodies. But Newton would frame no hypotheses concerning it. To him it was enough that "gravity does really exist, and act according to the laws" he had stated.[4] However, during the 18th century, the overwhelming success of Newton's theory and the growing awareness that action by contact is not a whit more intelligible than action at a distance, led scientists and philosophers to regard Newtonian gravity as a paradigm of interaction between bodies, and to explain even impenetrability, pressure and collisions by central repulsive—instead of attractive—forces, acting from afar.[5]

New mathematical techniques originating also in the 18th century made it possible to reformulate Newton's Theory of Gravitation as a field theory. A given distribution of matter determines a scalar field, i.e. a Galilei invariant real-valued function, the *gravitational potential*, on all space. The potential Φ determined by a matter distribution of density ρ can be calculated by integrating Poisson's equation:

$$\nabla^2\Phi = 4\pi k\rho \tag{1.7.4}$$

At each point the field exerts on matter a force per unit volume given by:

$$\mathbf{f} = -\rho\nabla\Phi \tag{1.7.5}$$

Equations (1.7.4) and (1.7.5) are the field-theoretic version of Newton's Law of Gravity. Potential theory is a powerful tool for working out the effects of gravity in actual physical situations, but it does not replace the conceptual foundation laid down by Newton. Poisson's equation can be derived from Newton's Law by straightforward mathematical reasoning.[6] And the gravitational potential at any time depends on the distribution of matter at that time and will instantaneously change throughout space when that distribution changes.

The idea of Newtonian spacetime was originally developed as still another reformulation of Newton's Theory of Gravitation. The version of that idea given in Section 1.6 holds only in the limit in which spacetime is practically void. But Newtonian spacetime in the presence of matter cannot be simply explained in terms of group-invariants. It is a manifold of variable curvature and, as such, it lies beyond the scope of Klein's Erlangen programme. We shall refer to it briefly on page 191, in connection with Einstein's theory of gravity.

CHAPTER 2

Electrodynamics and the Aether

2.1 Nineteenth-Century Views on Electromagnetic Action

In 1785 Coulomb verified by means of the torsion balance the law of electric interaction that was proposed by Priestley in 1767. Charged bodies repulse each other—if charged with electricity of the same kind—or attract each other—if charged with electricity of different kinds—with a force directly proportional to their charges and inversely proportional to their distance squared. Coulomb verified also a similar law of magnetic interaction, first published in 1750 by Michell.[1] Naturally enough, Coulomb's Laws—as they are usually called—were greeted as a splendid corroboration of the Newtonian paradigm of action at a distance by central forces. Volta's invention in 1800 of a device capable of generating steady currents opened the way to the tremendous surge of electrical science in the 19th century. In 1820 Oersted discovered the action of a conducting wire on a magnetic needle. Thereupon Ampère conjectured that magnets are but congeries of tiny electric circuits, and that conducting wires must attract or repel each other. Combining clever and painstaking experimentation with brilliant reasoning he developed his "mathematical theory of electrodynamical phenomena inferred from experience alone".[2] Ampère analysed electric circuits into infinitesimal elements of current characterized by their current intensity and their spatial direction. Any two such elements interact, by Ampère's Law, with a force directed along the line joining them and proportional to the product of their current intensities multiplied by a function of their directions and divided by their distance squared. Thus Ampère fulfilled his announced aim of reducing the newly discovered electrodynamic effects "to forces acting between two material particles along the straight line between them and such that the action of one upon the other is equal and opposite to that which the latter has upon the former".[3] His compliance with the Newtonian paradigm must be taken with a grain of salt, however, for his centres of force are in fact line elements, endowed with both position and direction.

Ampère's theory predicted Oersted's effect and the mutual attraction and repulsion between conducting wires that Ampère himself had conjectured and confirmed, but it did not predict the phenomena of electromagnetic induction discovered by Faraday in 1831. In 1847 Wilhelm Weber derived Faraday's Law of Induction and Ampère's Law of Currents from the twofold assumption that electric currents are caused by the symmetric motion in opposite directions of particles charged with equal amounts of positive and negative electricity, and that Coulomb's Law of electric interaction must be supplemented with two additional terms depending on the relative radial velocities and accelerations of the charged bodies. According to Weber's Law, a particle with charge e acts on a particle with charge e', at a distance r, with a force directed along the line joining both particles, of magnitude

$$F = \frac{ee'}{r^2}\left(1 - \frac{\dot{r}^2}{2c^2} + \frac{r\ddot{r}}{c^2}\right) \tag{2.1.1}$$

Here c is a constant, of the dimensions of a velocity, equal to the ratio of the electromagnetic to the electrostatic unit of charge.[4] Weber's electrodynamics found much acceptance in continental Europe, in spite of persistent criticism by Helmholtz.[5] In his great *Treatise* of 1873, Maxwell presents it as the most interesting version of a way of conceiving electromagnetic phenomena—based on "direct action at a distance"—to which his own—based on "action through a medium from one portion to the contiguous portion"—is "completely opposed".[6] Weber's approach withered away when Maxwell's won general approval, after Hertz in the late 1880's experimentally established that electromagnetic actions propagate with a finite speed.[7]

Maxwell avowedly drew his inspiration from Faraday, who conceived the observable behaviour of electrified and magnetized bodies as the manifestation of processes occurring in the surrounding space. Faraday regarded charged particles and magnetic poles as the endpoints of "lines of force", that fill all space, and can be made visible—between magnetic poles—by means of iron filings. Though Faraday's views were scorned by the scientific establishment of the time because he was unable to formulate them in the standard language of mathematical analysis, "his method of conceiving the phenomena—as Maxwell rightly observed—was also a mathematical one, though not exhibited in the conventional form of mathematical symbols".[8] If we define the electric intensity at a point of space at a given time as the force that would act at that time on a test particle of unit charge placed at that point, there is evidently defined, at each instant of Newtonian time, a vector field on all space, viz. the mapping assigning to each point of space the electric intensity vector at that point. An electric line of force is simply an integral curve of this vector field, that is, a curve tangent at each point to the electric intensity vector at that point.[9] A magnetic line of force can be defined analogously. The chief problem of electrodynamics is then to find the laws governing the evolution of the electric and magnetic vector fields through time. Maxwell's "general equations of the electromagnetic field" provide the classical answer to this problem.[10] We have learned to read them as statements about geometric objects in space, and it is worth noting that this reading agrees well with Faraday's conceptions. But Maxwell himself understood them differently. The mechanical action observed between electrified or magnetized bodies involves the transmission of energy. If, as Maxwell assumed, after Faraday, this transmission is not instantaneous, the energy—he believed—must be stored up for some time in a material substance filling the "so-called vacuum" between the bodies. Maxwell regarded his general equations as a phenomenological theory of the behaviour of this substance.

Maxwell's view was all but radical. The existence of a pervasive material medium capable of storing energy on its way from one ordinary body to another was a familiar assumption of contemporary optics, which successfully explained the phenomena of light as the effect of transversal waves in that medium. The medium was called the *aether*—Descartes' name for imponderable matter—and there was no dearth of hypotheses about its structure and dynamics. Maxwell had one very powerful reason for thinking that the opticians' aether was also the carrier of electromagnetic action. His theory predicted the existence of transversal wavelike disturbances of electromagnetic origin, propagating *in vacuo* with a constant speed equal to the ratio c between the electromagnetic and the electrostatic units of charge (see equation 2.1.1 and note 4). This ratio had been measured by Weber and Kohlrausch, and Maxwell observed that it differed from the known value of the speed of light *in vacuo* by less than the allowable experimental error.[11] He concluded that light consisted of the electromagnetic waves predicted by his theory.

In the introductory part of "A Dynamical Theory of the Electromagnetic Field", read before the Royal Society in December 1864, Maxwell summarized his electrical philosophy. He said then:

> It appears therefore that certain phenomena in electricity and magnetism lead to the same conclusion as those of optics, namely, that there is an aethereal medium pervading all bodies, and modified only in degree by their presence; that the parts of this medium are capable of being set in motion by electric currents and magnets; that this motion is communicated from one part of the medium to another by forces arising from the connexions of those parts; that under the action of these forces there is a certain yielding depending on the elasticity of these connexions; and that therefore energy in two different forms may exist in the medium, the one form being the actual energy of motion of its parts, and the other being the potential energy stored up in the connexions, in virtue of their elasticity.
>
> Thus, then, we are led to the conception of a complicated mechanism capable of a vast variety of motion, but at the same time so connected that the motion of one part depends, according to definite relations, on the motion of other parts, these motions being communicated by forces arising from the relative displacement of the connected parts, in virtue of their elasticity. *Such a mechanism must be subject to the general laws of Dynamics.*[12]

In an earlier paper[13] he had "attempted to describe a particular kind of motion and a particular kind of strain, so arranged as to account for the phenomena".[14] But he never repeated such an attempt. He continued to believe that "a complete dynamical theory of electricity" must regard "electrical action [. . .] as the result of known motions of known portions of matter, in which not only the total effects and final results, but the whole intermediate mechanism and detail of the motion, are taken as the objects of study".[15] But he obviated the need for such a complete theory by resorting to the dynamical methods of Lagrange, "which do not require a knowledge of the mechanism of the system".[16] He thus dodged the difficulty that beset all hypotheses concerning the mechanical workings of the aether: the aether must be infinitely rigid, for all disturbances are propagated across it by transversal, never by longitudinal waves; and yet it does not appear to offer the slightest resistance to the motion of the bodies immersed in it. While Maxwell's friend and mentor, William Thomson, Lord Kelvin, devised one unsatisfactory mechanical model of the aether after another,[17] the next generation of electricians was ready to accept the aether governed by Maxwell's Laws as a peculiar mode of physical existence, different from and perhaps more primitive and fundamental than the one exemplified by ponderable matter. In a letter to Hertz of September 13th, 1889, Heaviside wrote:

> It often occurs to me that we may be all wrong in thinking of the ether as a kind of matter (elastic solid for instance) accounting for its properties by those of the matter in bulk with which we are acquainted; and that the true way, could we only see how to do it, is to explain matter in terms of the ether, going from the simpler to the more complex.[18]

2.2 The Relative Motion of the Earth and the Aether

It is often said that Newtonian physics acknowledged the equivalence of inertial frames only with regard to *dynamical* laws and experiments, and that Einstein's peculiar achievement was to declare them equivalent with respect to physical laws and experiments *of every conceivable kind*. This statement, which introduces Einstein as the young genius who boldly and successfully extended a principle that his predecessors had niggardly restricted, is useful in the classroom, where one must try to win over to Einstein's theory the unsympathetic minds of adolescents, steeped in common sense; but it is not fair to Newton or to his 19th century followers. The distinction between dynamical and non-dynamical physical laws and experiments is completely out of place in Newtonian science, which, as we saw in Chapter 1, sought to understand all natural phenomena as effected by the forces and governed by the laws disclosed by the phenomena of motion. And during most of the 19th century this Newtonian programme was faithfully adhered to by the physical establishment—excepting Michael Faraday, who for this very reason was generally spurned as a theoretician and acknowledged only as a great experimentalist. Of course, if we agree to restrict dynamics, by a wanton though viable semantic convention, to the direct interaction of ordinary, tangible matter, we must conclude that optics, according to the Young–Fresnel wave theory of light, and electromagnetism, as conceived by Maxwell and his school, lie beyond its scope. But even then we cannot conclude that they also lie beyond the scope of the Newtonian Principle of Relativity. The gist of this principle is that any two experiments will have the same outcome with regard, respectively, to two inertial frames, if all the physical objects relevant to either experiment are initially in the same state of motion relative to the corresponding frame. Now according to late 19th-century physics the principle will certainly apply to optical and electromagnetic experiments provided that the aether is included among the physical objects whose state of relative motion is taken into account.[1] The proviso is sensible enough. After all, in a "purely dynamical" experiment such as dragging a marlin from a boat nobody would expect the tension on the fishing-line to be the same regardless of the currents in the water.

Maxwell's equations were expressed in terms of a Cartesian coordinate system or in the now familiar compact vectorial notation. In either form they are referred to a definite rigid frame. Are there any restrictions on the choice of the latter? Maxwell does not seem to have paid much attention to this question, but there are signs that he thought that almost any frame one could reasonably care to choose would do, if one was willing to ignore—as indeed one must—experimentally indiscernible differences. In Maxwell's theory, the outcome of electromagnetic experiments depends on the behaviour of (i) a system of ponderable bodies—conductors, magnets, dielectrics—and (ii) the aether. In the *Treatise*, §§ 600 f., Maxwell argues that "the electromotive intensity is expressed by a formula of the same type", whether we refer it to "a fixed system of axes"—that is, I presume, one relatively to which the fixed stars are at rest—or to a system of axes moving with uniform linear and uniform angular velocity with respect to the former. The only change that must be introduced in the said formula as a consequence of such a transformation of coordinates does not alter the value of the electromotive force round a closed circuit. Therefore—Maxwell concludes—"in all phenomena relating to closed circuits and the currents in them, it is indifferent whether the axes to which we refer the system be at rest or in motion".[2] So much for the motion of conductors. That of magnets and dielectrics is not explicitly discussed in the *Treatise*. As to the aether, since the speed c with which electromagnetic action propagates through it is involved in Maxwell's equations, the equations obviously cannot take the same form with respect to any frame, whether the aether is at rest or in motion in it. It would seem, however, that Maxwell believed that any changes in their form arising from this consideration could be safely neglected, at least for the time being, because the differences due to the motion of the aether relative to such frames of reference as our terrestrial physicists can be expected to choose were still beyond the reach of experimental detection.[3]

The appropriate frame of reference for the electromagnetic field equations, or rather, the appropriate form of the latter when referred to one frame or another, was explicitly considered by Heinrich Hertz. His first exposition of Maxwell's theory, which had a tremendous impact on the young physicists of continental Europe, dealt with "the fundamental equations of electromagnetics for bodies at rest".[4] It was followed, in the same year, by a second paper, "On the fundamental equations of electromagnetics for bodies in motion".[5] Hertz remarks here "that whenever in ordinary speech we speak of bodies in motion, we have in mind the motion of ponderable matter alone". However, according to his view of electromagnetic phenomena "the disturbances of the ether, which simultaneously arise, cannot be without effect; and of these we have no knowledge". The question with which the paper is concerned cannot therefore "at present be treated at all without introducing arbitrary assumptions as to the motion of the ether".[6] The fact that we cannot remove the aether from any closed space would indicate indeed that "even in the interior of tangible matter the ether moves independently of it". According to Hertz this would imply that the electromagnetic conditions of the aether and of the ponderable matter at every point of space must be described separately, by means of two distinct pairs of vectors. This complication can be avoided, however,

> if we explicitly content ourselves with representing electromagnetic phenomena in a narrower sense—up to the extent to which they have hitherto been satisfactorily investigated. We may assert that among the phenomena so embraced there is not one which requires the admission of a motion of the ether independently of ponderable matter within the latter; this follows at once from the fact that from this class of phenomena no hint has been obtained as to the magnitude of the relative displacement. At least this class of electric and magnetic phenomena must be compatible with the view that no such displacement occurs, but that the ether which is hypothetically assumed to exist in the interior of ponderable matter only moves with it. [. . .] For the purpose of the present paper we adopt this view.[7]

Having thus temporarily assumed that the aether inside a ponderable body is completely dragged along with the latter, Hertz goes on to propose a set of electromagnetic field equations applicable to bodies in motion in the frame to which the equations are referred. These can be obtained more or less straightforwardly from the Maxwell equations, as given in Hertz's former paper, using the Galilei Rule for the Transformation of Velocities (p. 29).[8] But Hertz's equations for bodies in motion clash with the outcome of some electromagnetic experiments (which, for the most part, had not yet been performed in 1890)[9], and also with some familiar facts of optics, to which we now turn.

In 1728 James Bradley announced "a new discovered Motion of the Fixt Stars": annually, the apparent position of any star describes a small ellipse in the sky. Bradley had been trying to observe the stellar parallax, that is, the variation of the apparent position of a star caused by changes in the actual position of the earth. But the displacement he discovered could not be that which he sought for, because its amount varied precisely with the latitude of the stars observed, while the parallax depends on their presumably random distances to the earth; and also because each star was seen to move at any given time perpendicularly to the direction in which the earth was moving, while the motion associated with parallax must be parallel and opposite to the motion of the earth. The corpuscular theory of light, then in fashion, furnished Bradley with a simple explanation of this phenomenon, subsequently known as the aberration of stars, or the aberration of light. According to that theory, starlight pours on us like rain, in straight lines parallel to the direction in which its source actually lies; however, in order to capture it with our telescope on the moving earth, we must tilt the instrument in the direction of motion, as we tilt our umbrella when we run under the rain. More precisely: if light falls vertically on the objective of a telescope moving with speed v relative to the fixed stars, in order that a light corpuscle may reach the eyepiece, the telescope must be tilted by a small "angle of aberration" α, such that $\tan \alpha = v/c$— provided, of course, that the corpuscle travels inside the telescope with the same vertical velocity c with which it travels *in vacuo.*[10] This result, though in excellent agreement with observation, appeared to clash with the wave theory of light, which in many ways could claim superiority over the corpuscular theory after 1800. For according to the wave theory the angle of aberration will not depend on the velocity of the telescope relative to the source of light, but on its velocity relative to the medium of light propagation, and the latter, as G. G. Stokes remarked, can agree with the former only on "the rather startling hypothesis that the luminiferous ether passes freely through the sides of the telescope and through the earth itself".[11] Stokes, however, by a subtle mathematical argument, explained aberration in accordance with the wave theory on the seemingly natural assumption that the aether is completely carried along by the earth and the bodies on the earth's surface, while its velocity changes as we ascend through the atmosphere, "till, at no great distance, it is at rest in space".[12] Almost thirty years earlier Fresnel (1818) had given a different wave-theoretical explanation of aberration, which, though based on a highly artificial hypothesis, achieved remarkable experimental success. Fresnel's work was prompted by Arago's researches on the refraction of starlight by prisms (e.g. telescope lenses). Arago's measurements, carried out in 1810, while he still adhered to the corpuscular theory, had shown that aberration is not affected by refraction, in spite of the fact that light rays are bent by it to compensate for the change of the speed of light from its value c, in free space, to c/n in the refractive medium (where n is the medium's index of refraction). Arago (1853) concluded that light corpuscles are emitted by stars with different speeds, and that only those whose speed lies within certain narrow limits can be perceived by the eye. But Fresnel thought these conclusions unlikely. He derived Arago's results from the wave theory, on the hypothesis that the aether is partially dragged by refractive bodies. He postulated that the ratio of the aether's density in free space to its density in a medium of refractive index n is $1/n^2$. He assumed moreover that a refractive body in motion drags along with it only the aether it stores in excess of that contained in the same volume of free space. He then showed that, on these assumptions, the velocity of light in a medium of refractive index n, moving with velocity v, was equal, in the direction of the motion, to $c/n + v(1 - 1/n^2)$; c/n being, as we know, the velocity of light in the same medium at rest. Fresnel's theory accounted for the then known facts of aberration and also implied that the angle of aberration would not be affected by using a telescope filled with water, a prediction verified by Airy (1871).[13] The theory achieved a major success when Fizeau (1851) measured the velocity of light in running water and found it to be in good agreement with Fresnel's formula.

In 1881 A. A. Michelson thought that he had confirmed Stokes' theory by his first interferometer experiment, which failed to disclose a "relative motion of the earth and the luminiferous ether".[14] But H. A. Lorentz (1886) showed that Michelson had overrated the effect that such a motion would have on the outcome of the experiment; when corrected, it turned out to lie within the range of experimental error. Lorentz also argued that Stokes' theory is untenable, because it has to postulate that the velocity field of the aether has zero curl and is therefore the gradient of a potential, which is inconsistent with Stokes' assumption that the velocity of the aether is everywhere the same on the earth's surface. In 1886 Michelson and Morley repeated Fizeau's experiment with increased accuracy, providing for the first time a direct quantitative confirmation of Fresnel's "drag coefficient" $(1 - 1/n^2)$. So when in 1887, using an improved interferometer, no longer liable to Lorentz's strictures, Michelson and Morley again found no traces of the relative motion of the earth and the aether, they did not claim to have vindicated Stokes, but only expressed their perplexity, for their result was in open conflict with Fresnel's hypothesis, which they had so splendidly corroborated a year earlier.[15]

Though Hertz's electrodynamics of bodies in motion agreed, on the surface of the earth, with Stokes' theory and Michelson and Morley's experiment of 1887, it was irreconcilable with Fresnel's drag coefficient and Michelson and Morley's result of 1886. Since it also clashed with some new electromagnetic experiments (see note 9) and its author died at 36 in 1894, it was generally abandoned. In the last decade of the 19th century the most influential contributions to the subject were those of H. A. Lorentz, who, deeply impressed by the necessity of explaining the drag coefficient in a less *ad hoc* way than Fresnel's, conceived matter as consisting of charged particles swimming in a stagnant aether that remains unruffled by their motions. In the rest of this Section I shall sketch some ideas of Lorentz that are generally regarded as the immediate background of Einstein's "Electrodynamics of Moving Bodies".[16]

Matter, which Lorentz described as "everything that can be the seat of currents or displacements of electricity, and of electromagnetic motions", comprises "the aether as well as ponderable matter".[17] The latter is "perfectly permeable to the aether" and can move "without communicating the slightest motion to the latter".[18] Lorentz postulates that all ponderable bodies contain a multitude of positively or negatively charged particles—which he called "ions" or "electrons"—and that electrical phenomena are caused by the displacement of these particles. The state of the aether is characterized at each point by the local values of two vector fields, the dielectric displacement **d** and the magnetic force **h**. The position and motion of the electrons governs the fields according to the equations:

$$\nabla \cdot \mathbf{d} = \rho \qquad \nabla \times \mathbf{d} = -\frac{1}{4\pi c^2}\frac{\partial \mathbf{h}}{\partial t}$$
$$\nabla \cdot \mathbf{h} = 0 \qquad \nabla \times \mathbf{h} = 4\pi\left(\rho\mathbf{v} + \frac{\partial \mathbf{d}}{\partial t}\right) \tag{2.2.1}$$

where ρ is the local charge density and $\mathbf{v}$ is the local velocity of matter.[19] At each point the fields exert on matter a force per unit volume given by:

$$\mathbf{f} = \rho(4\pi c^2\mathbf{d} + \mathbf{v} \times \mathbf{h}) \tag{2.2.2}$$

Since the aether is not subject to ponderomotive forces and the electrons do not act instantaneously on one another, Lorentz theory is incompatible with the Third Law of Motion. Lorentz did not shy away from this conclusion, but remarked that, though the Third Law is manifestly satisfied by the observable phenomena, it need not apply to the elementary interactions that cause them.[20] Lorentz (1892*a*) derived the "Lorentz force" (2.2.2) from the "Maxwell–Lorentz" equations (2.2.1) by a variational argument. By integrating over "physically infinitesimal" regions he also obtained from (2.2.1) Maxwell's phenomenological equations for ponderable bodies, relating the four macroscopic fields of electric force **E**, electric displacement **D**, magnetic force **H** and magnetic induction **B**. In the same paper Lorentz derived Fresnel's drag coefficient from his basic equations.[21]

Lorentz naturally assumed that equations (2.2.1) and (2.2.2) are referred to the frame, presumed inertial, in which the aether is permanently at rest. By subjecting them to a Galilei coordinate transformation one can see at once that this is the only frame in which the equations are valid as they stand. However, Lorentz did not identify, as some have said, the aether's rest frame with Newton's absolute space. He was well aware that to speak of "the *absolute* rest of the aether [. . .] would not even make sense".[22]

In the course of his investigations Lorentz (1892*a*) constructs some auxiliary functions that enable him to calculate the state of the aether from data referred to a material system moving in the aether frame with constant velocity $\mathbf{v}$. The differential equation for one of the said functions can be easily solved by suitably changing variables.[23] If the coordinate functions of the aether frame, t, x, y, z, are so chosen that the components of $\mathbf{v}$ with respect to them are $\mathbf{v}_x = v$, $\mathbf{v}_y = \mathbf{v}_z = 0$, a Galilei coordinate system adapted to the moving frame is given by the transformation:

$$\bar{t} = t \qquad \bar{x} = x - vt$$
$$\bar{y} = y \qquad \bar{z} = z \tag{2.2.3}$$

Lorentz introduces the variables

$$x' = \beta\bar{x} \tag{2.2.4a}$$
$$t' = t - \alpha x'/c \tag{2.2.4b}$$

where $\alpha = |(c^2/v^2 - 1)^{-1/2}|$ and $\beta = |(1 - v^2/c^2)^{-1/2}|$. Writing y' for $\bar{y}$, z' for $\bar{z}$, and substituting x for $\bar{x}$ in (2.2.4), in agreement with (2.2.3), we have that[24]

$$t' = \frac{t - (vx/c^2)}{1 - (v^2/c^2)} \qquad x' = \frac{x - vt}{\sqrt{1 - (v^2/c^2)}}$$
$$y' = y \qquad z' = z \tag{2.2.5}$$

Though the substitution (2.2.4) was devised only as an aid to calculation, it involves two ideas for which Lorentz found later a direct physical use. Equation (2.2.4*b*) introduces a time coordinate function that depends on the spatial location of events, a "local time", as Lorentz will call it, to distinguish it from the universal Newtonian time. Equation (2.2.4*a*) introduces a space coordinate function which increases in the direction of motion β times faster than the corresponding Cartesian one, as if it were evaluated by means of a standard measuring rod that had been shortened by a factor of β^{-1}. Let us consider Lorentz's physical application of these two ideas in the stated order.

In his "Inquiry into electrical and optical phenomena in moving bodies" (1895) Lorentz develops a theory that he claims to be correct up to first order terms in $|\mathbf{v}|/c$ ($\mathbf{v}$ being the velocity of the moving system in question). Lorentz proves that the theory satisfies what he calls the Theorem of Corresponding States. Let $\mathbf{r}$ and t be respectively the position vector function and the time coordinate function associated with the aether frame, and let the position vector $\mathbf{r}'$ be given by the standard transformation:

$$\mathbf{r}' = \mathbf{r} - \mathbf{v}t \tag{2.2.6}$$

Introduce the "local time" variable

$$t' = t - \frac{1}{c^2}\mathbf{v} \cdot \mathbf{r}' \tag{2.2.7}$$

The Theorem of Corresponding States can then be stated as follows :

> If there is a system of bodies at rest in the aether frame and a solution of the field equations such that the dielectric displacement, the electric force and the magnetic force in that frame are certain functions of $\mathbf{r}$ and t, there is a solution for the same system of bodies moving in the aether frame

with velocity $\mathbf{v}$, such that the dielectric displacement, the electric force and the magnetic force in the inertial frame in which the system is at rest are exactly the same functions of $\mathbf{r}'$ and t'.[25]

Lorentz concludes that the relative motion of the earth and the aether cannot exert on electromagnetic experiments any effects of the first order in v/c (v being the relative speed of the earth and the aether).[26] Though Lorentz was not quite explicit concerning the physical significance of his "local time" and twenty years later he once described it as "no more than an auxiliary mathematical quantity", it is clear that the Theorem of Corresponding States cannot be experimentally significant unless the local time bears a definite relation to the physical time keeping processes observed in the laboratory.[27] Lorentz, however, did not try to explain how all such "natural clocks" can be affected, and affected in the same way, by carrying them across the aether. He was much more outspoken with regard to the contraction of moving metre rods, the second idea that we saw was implied by a physical interpretation of (2.2.4). To this we now turn.

On August 18th, 1892, Lorentz wrote to Lord Rayleigh that he was "totally at a loss to clear away [the] contradiction" between Fresnel's value for the aether drag coefficient, that agreed so admirably with practically all observed phenomena, and the null result of Michelson's interferometer experiments, which was apparently the only exception to the rule.[28] Soon thereafter it lighted on him that Michelson's result could be understood at once if "the line joining two points of a solid body, if at first parallel to the direction of the earth's motion, does not keep the same length when it is subsequently turned through 90°", so that, if v is the speed of the earth in the aether frame, the first length is $(1-v^2/2c^2)$ times the latter.[29] According to Lorentz this suggestion—which formerly had been made by G. F. Fitzgerald (1889)—is not so implausible as it might seem. For, the size and shape of a solid body being determined by the molecular forces, "any cause which would alter the latter would also influence the shape and dimensions". If electric and magnetic forces are transmitted by the aether "it is not far-fetched to suppose the same to be true of the molecular forces". This hypothesis is untestable, for we are completely ignorant of the nature of the molecular forces; but if the influence of motion across the aether on them is equal to that on electric and magnetic forces, it will cause in any solid body a deformation of precisely the required amount. This conclusion is justified more rigorously in Lorentz's "Inquiry" of 1895. In § 23 of that work Lorentz studies the effect of motion across the aether on the electrostatic interaction between charged partices. Let S be a system of electrons at rest in the aether. If x, y, z are Cartesian coordinates adapted to the aether frame, we denote by F_x, F_y and F_z the components, with respect to those coordinates, of the electric force field F, generated by S. Let S' be a system equal to S, except for the following two conditions: (a) S' moves across the aether with velocity **v** and velocity components, with respect to the said coordinates, $v_x = v$, $v_y = v_z = 0$; (*b*) in S' the dimensions of the electrons and the distances between them are shortened in the direction of motion by the factor $\beta^{-1} = (1 - v^2/c^2)^{1/2}$. Lorentz shows that the components of the electric force field $\mathbf{F}'$ generated by S', with respect to a coordinate system x', y', z', adapted to the rest frame of S', are related to F_x, F_y and F_z by the simple equations

$$F'_{x'} = F_x \qquad F'_{y'} = \beta^{-1}F_y \qquad F'_{z'} = \beta^{-1}F_z \tag{2.2.8}$$

provided that (i) the primed coordinate system is not Cartesian; (ii) if $\bar{x}$, $\bar{y}$, $\bar{z}$ are Cartesian coordinates adapted to the rest frame of S' and so oriented that the $\bar{x}$-axis glides along the x-axis with speed v, the primed system is related to the barred system by the transformation we met already on page 44:

$$x' = \beta\bar{x} \qquad y' = \bar{y} \qquad z' = \bar{z} \tag{2.2.4a}$$

In §§ 89 ff. Lorentz discusses Michelson's results,[30] arguing that they can be reconciled with the hypothesis of the stagnant aether if we assume that the molecular forces that keep together a solid body, such as the arms of the interferometer, are subject to the same transformation law (2.2.8) that is applicable to electrostatic forces. For, on this assumption, the equilibrium of the molecular forces can only be maintained when the body moves across the aether with constant speed v, if its dimensions are shortened in the direction of motion in the ratio of 1 to $(1 - v^2/c^2)^{1/2}$. Lorentz readily admits that "there is certainly no reason why" the said assumption should hold, but he apparently regarded Michelson's result as evidence for it.[31] He also expected that the effect would not be quite so neat in reality, since the molecules of a solid body are not exactly at rest but oscillating relative to each other. However, the interferometer experiments were not so exact that such small effects could modify their results.[32]

In his "Simplified theory of electrical and optical phenomena in moving systems" (1899), Lorentz extends his Theorem of Corresponding States to effects of the second order (i.e. effects of the order of v^2/c^2, where v is the speed of motion across the aether).[33] With this purpose he introduces still one more coordinate system, slightly different from the primed system with respect to which the Theorem is valid up to first order terms. So we now have to do with four coordinate systems:

(i) The unprimed Galilei system (t, x, y, z), adapted to the aether frame.
(ii) A Galilei system that we designate by the barred variables $(\bar{t}, \bar{x}, \bar{y}, \bar{z})$, adapted to the rest frame of a material system moving across the aether with velocity **v**. This system is so chosen that the $\bar{x}$-axis glides along the x-axis with speed $v = |\mathbf{v}|$. It is therefore related to (i) by the Galilei transformation (2.2.3).

(iii) The primed system to which the first order Theorem of Corresponding States is referred in this paper of 1899 (see note 26), and which is related to (ii) by the transformation:

$$t' = \bar{t} - \beta^2 \frac{v\bar{x}}{c^2} \qquad x' = \beta\bar{x}$$
$$y' = \bar{y} \qquad z' = \bar{z} \tag{2.2.9}$$

(iv) A doubly primed system related to (iii) by the transformation:

$$t'' = t'/\beta\varepsilon \qquad x'' = x'/\varepsilon \qquad y'' = y'/\varepsilon \qquad z'' = z'/\varepsilon \tag{2.2.10}$$

where ε is "an indeterminate coefficient differing from unity by a quantity of the second order", and β retains the value we have given it throughout the present Section.[34]

Lorentz also gives transformation formulae for the components of the electromagnetic fields with respect to (iii) and (iv), and goes on to argue that the Theorem of Corresponding States will hold up to terms of second order if we substitute the doubly primed system for the simply primed one in the previous formulation of it, provided that we not only suppose that a system at rest in the aether *may* be changed according to the given transformation into the imaginary doubly primed system, moving across the aether with velocity **v**, "but that, as soon as the translation is given to it, the transformation *really* takes place, of itself, i.e. by the action of the forces acting between the particles of the system, and the aether".[35]

Spoiled as we are by the elegance and clarity that Einstein brought into classical electrodynamics we find Lorentz's method of successive approximations vexingly cautious. And yet it cannot be denied that, though it procured little insight into the structure of things, it did yield some astounding predictions. Thus, the second order theory of Lorentz (1899) implies that the inertial mass of a charged particle must vary with its speed (relative to the aether), and indeed in such a way that it will oppose a different resistance to accelerations parallel and perpendicular to the direction of motion.[36] If matter in motion across the stagnant, ubiquitous aether suffers such impressive changes, there is good reason to look on the aether frame as a standard of rest. It is nevertheless ironic that these effects which single out in theory one inertial frame among all others—and thus make an end of Newtonian Relativity—should have been postulated to explain the impossibility of ascertaining in practice which bodies move and which are at rest in that frame. But it is time now that we address ourselves to Einstein's own work, for the study of which we have gathered, I hope, sufficient background information.

CHAPTER 3

Einstein's 'Electrodynamics of Moving Bodies'

3.1 Motivation

In an undated letter to Conrad Habicht, probably written in the Spring of 1905, Einstein mentions four forthcoming papers of his.[1] The first, which was about to appear, he describes as "very revolutionary". It was the article "On a heuristic point of view about the creation and conversion of light", in which Einstein, boldly generalizing Planck's quantization of energy, postulated that "the energy of light is distributed discontinuously in space" in the form of localized "energy quanta . . . which move without being divided and which can be absorbed or emitted only as a whole".[2] The fourth paper, still a draft, would contain "an electrodynamics of moving bodies, involving modifications of the theory of space and time".[3] This article, submitted in June 1905 to *Annalen der Physik* and published in September, laid down the foundations of Special Relativity. In 1952 Einstein told R. S. Shankland that the subject of this paper "had been his life for over seven years", that is, from 1898 to 1905, while that of the former had busied him for five, that is, since the publication of Planck's trailblazing paper of 1900.[4] Though it would not be easy to guess from either text that they were written by the same man, practically at the same time, the insight leading to the paper on radiation inevitably influenced Einstein's peculiar approach to electrodynamics.

In his Autobiographical Notes Einstein describes the effect of Planck's energy quantization on his mind: "All my attempts to adapt the theoretical foundations of physics to [Planck's approach] failed completely. It was as if the ground had been pulled out from under one's feet, with no firm foundation to be seen anywhere, upon which one could have built".[5] His main problem in those years was to ascertain "what general conclusions can be drawn from [Planck's] radiation formula regarding the structure of radiation and, more generally, regarding the electromagnetic basis of physics".[6] Shortly after 1900 it was already clear to him that classical mechanics could not, except in limiting cases, claim exact validity.[7] The "heuristic viewpoint" on electromagnetic radiation implied that classical electrodynamics was not exactly valid either.[8] Although Einstein had the highest regard for Lorentz's achievements, he could no longer subscribe to the latter's "deep" theory of electron–aether interaction, and it would have been premature to replace it by some new-fangled speculation on the fine structure of matter. On the other hand, the predictive success of Maxwell's equations at the macroscopic level was beyond question. One could therefore incorporate them without qualms in a wide-ranging phenomenological theory.[9] The paradigm of a phenomenological theory was classical thermodynamics, which attained accurate knowledge of thermal phenomena, without any hypotheses as to the ultimate nature of heat, by translating into exact mathematical language and elevating to the rank of universal physical principles two familiar "facts of life", viz. that there is no inexhaustible source of mechanical work, and that a cooling machine cannot run solely on the heat it extracts. The principles of thermodynamics constrain the laws of nature to be such "that it is impossible to construct a *perpetuum mobile* (of the first and second kind)".[10] Einstein was persuaded that only the discovery of a universal principle of this sort could rescue him from the quandary he was in, and lead him to assured results.

The universal principle finally proposed by Einstein in 1905 prescribes that "the laws of physics are invariant under Lorentz transformations (when going from one inertial system to another arbitrarily chosen inertial system)".[11] As we shall see in Section 3.3, this follows from the joint assertion of the thoroughgoing physical equivalence of all inertial frames, proclaimed by the Relativity Principle (RP), and the constancy of the speed of light *in vacuo* regardless of the state of motion of its source, embodied in what we shall hereafter call the Light Principle (LP). The RP was firmly entrenched in the tradition of classical physics and could moreover be said to rest, like the Principles of Thermodynamics, on familiar observations (cf. the quotation from Galilei on page 31). But the LP was traditionally accepted only as a consequence of the wave theory of light, supported by the vast array of facts that corroborated this theory, but not resting on any direct evidence of its own. Einstein's decision to retain the LP, upgrading it from a corollary to a basic principle, at a time when he had lost faith in the exact validity of the wave theory, was a daring but hugely successful gamble.[12]

The LP is indeed too vague, unless one specifies the frame in which light travels with constant speed. According to the wave theory this must be the frame in which the luminiferous medium oscillates about a stationary position. But Einstein had no use for a luminiferous medium and simply referred the LP to an arbitrarily chosen inertial frame. This would clash with the RP if, as hitherto assumed, velocities transform according to the Galilei Rule when going from one inertial frame to another.[13] However the apparent inconsistency is dispelled by Einstein's discovery that the RP, as classically understood, is too vague also, because it does not specify how time is "diffused" throughout an inertial frame. If this is done in the manner prescribed in Einstein (1905*d*), § 1, the conjunction of RP and LP (in their now precise meanings) implies that the Galilei Rule for the Transformation of Velocities is not valid. The rule of velocity transformation that actually follows from Einstein's principles does not lead to inconsistencies.

Before going further into the matters we have just sketched it will be useful to mention some of the reasons that moved Einstein to uphold both RP and LP. In the Autobiographical Notes he speaks of a "paradox" that had haunted him since he was a lad: A man travelling with speed c after a beam of light would perceive the beam as a static field, periodic in space. Einstein deemed the situation unnatural.[14] The joint validity of RP and LP makes it utterly impossible. In the preamble to (1905*d*), Einstein argues persuasively for the RP alone. In the then fashionable version of classical electrodynamics Maxwell's equations were referred, in their standard form, to a particular inertial frame, the aether's rest frame. This leads to startling and altogether unwarranted asymmetries in the description of otherwise indistinguishable phenomena. Take, for instance, the interaction of a magnet and a conductor.

> If the magnet is in motion and the conductor at rest, there arises in the neighbourhood of the magnet an electric field with a definite energy, producing a current at the places where the parts of the conductor are situated. But if the magnet is at rest and the conductor in motion, no electric field arises in the neighbourhood of the magnet. In the conductor, however, there arises an electromotive force, to which no energy corresponds, but which—assuming equality of relative motion in the two cases discussed—generates electric currents of the same path and intensity as those produced by the electric forces in the former case.[15]

Examples of this kind, "together with the unsuccessful attempts to detect a motion of the earth relative to the 'light medium' ",[16] suggest that the Principle of Relativity applies to electrodynamic phenomena and laws, not—as Hertz professed—if due allowance is made for the state of motion of the aether,[17] but if the aether is completely ignored—or if it does not exist at all—and inertial frames are defined exclusively by reference to ponderable bodies. At this point of the preamble Einstein introduces the LP in an informal version similar to the one I gave above. He does not argue for it, but adds only that RP and LP are all that is needed for developing "a simple and consistent electrodynamics of moving bodies, based on Maxwell's equations for bodies at rest".[18] In 1950, however, Einstein told R. S. Shankland of a specific reason for upholding the LP. Before 1905 he had countenanced a ballistic theory of light emission, similar to that developed by Ritz (1908*a*, *b*), but had rejected it because "he was unable to think of any form of differential equation that could have solutions representing waves whose velocity depended on the motion of the source".[19]

3.2 The Definition of Time in an Inertial Frame

It is generally agreed that Einstein's implicit criticism and outright dismissal of Newton's universal time is somehow the key to Special Relativity. Opinions differ, however, as to the true import of the conception of physical time and time order that Einstein introduced instead. While one very influential school maintains that in Einstein's view there can be no natural partition of events into simultaneity classes, so that a universal time order can only be settled by convention, other writers believe that in the world of Special Relativity there are infinitely many such natural partitions, one for each inertial frame.[1] We shall postpone the consideration of these philosophical positions until Section 7.1. At this point we shall only examine Einstein's own discussion of time order in an inertial frame in his (1905*d*), § 1, supplementing it with such remarks as may seem necessary for a clear and consistent reading of the text.

At the beginning of § 1 Einstein bids us choose "a coordinate system in which the equations of Newtonian mechanics hold good". He proposes that, for easy reference, we call it the "stationary" system. He remarks that if a material particle is at rest relatively to this system its position can be defined with regard to it by means of measuring rods and the methods of Euclidean geometry, and can be expressed in Cartesian coordinates. We shall understand therefore that the stationary system is a Cartesian coordinate system for a rigid frame—hereafter called the "stationary frame". It follows from the remaining sections of Einstein's paper that neither the Second nor the Third Newtonian Laws of Motion do really hold good in the said frame. To avoid a blatant inconsistency we must conclude that Einstein expects only the First Law of Motion to be true in the stationary frame. The latter is therefore best described as an arbitrary inertial frame in Lange's sense.[2] We saw on pages 19–20 that in the Newtonian theory the Third Law is quite essential for distinguishing the family of inertial frames from any family of frames travelling past it with the same constant acceleration. But the Third Law is not required for singling out the inertial frames in Einstein's theory. Einstein does not go into the question here perhaps because he thought that to state it clearly he needed the very concept of time which he still was in the course of defining. But in the light of his later writings there can be no doubt that the following is assumed by him: If an inertial and a non-inertial frame move past each other with uniform acceleration, a light-ray emitted through empty space in a direction normal to the mutual acceleration of the frames, describes a straight line in the inertial frame, a curved line in the other.[3] The rectilinear propagation of light *in vacuo* provides therefore an additional criterion for the identification of inertial frames, which, be it noted, does not presuppose a definition of time. Summing up, we may say that Einstein's stationary system is a Cartesian coordinate system defined on the relative space S_F of a rigid frame F, which meets the following conditions:

(I) Three free particles projected non-collinearly from a point in F describe straight lines in S_F.

(II) A light ray transmitted through empty space in any direction from a point in F describes a straight line in S_F.

After making the remark reported above concerning the assignment of coordinates to a particle at rest in the stationary system, Einstein goes on to say:

> If we wish to describe the *motion* of a material point we give the values of its coordinates as functions of the time. Now we must bear in mind that such a mathematical description will only have a physical meaning if we are quite clear beforehand as to what we understand by "time".[4]

He then introduces the distinction that we mentioned in Section 1.3 between time at a point and time throughout space. We saw there that Newton's "true and mathematical time" was relevant to the scientific description of phenomena only insofar as it was embodied in natural processes which, according to the Laws of Motion, do, in ideal circumstances, keep true time. But we also observed that such Newtonian clocks keep time only at the point of space where they are actually located—the point, that is, where the event designated for the role of time counter is taking place. Though Einstein cannot resort any longer to the Newtonian Laws of Motion for the definition of foolproof time-keeping devices, he unhesitatingly relies on the concept of a natural clock, that keeps true time at a point of space. Indeed the possibility of such clocks follows at once from the Law, postulated "in agreement with experience" near the end of § 1, that the round trip speed of light through empty space in the stationary frame is a universal constant. As Pierre Langevin observed, a light-pulse travelling to and fro between two parallel mirrors facing each other at the endpoints of a rigid rod at rest in that frame constitutes such a natural clock. The "Langevin clock" obviously keeps time only at either end of the rod, and one cannot even assert, without further stipulations, that it keeps the same time at both ends. But the time shown by a clock at a point in the stationary frame cannot be used for describing the successive positions of a particle moving in that frame. As the particle goes from one point to another, the time must be read in different clocks. How are they to be set in agreement with one another? As early as 1884 James Thomson had realized that time is not diffused of itself throughout all space, but must be propagated from point to point by some method of signalling. He assumed, however, as a matter of course, that every plausible signalling method—such as transmitting light pulses at appointed times, or sending out messengers furnished with accurate, stable clocks, set by the clock at the origin—would ideally lead to the same universal system of chronometry.[5] Thomson knew that this assumption was not necessarily true. Einstein ventured to think that it was contingently false. He was apparently willing to allow any method for propagating time over space, provided that it met the basic requirement of any physical method, namely, that it be freely reproducible and yield consistent results. But he was intent on defining without ambiguity whatever method might be finally adopted, on stipulating exactly the idealized physical conditions in which two events should be said to happen at the same time. By thus substituting precision for Thomson's looseness, Einstein avoided or diminished the risk of using mutually inconsistent methods, that may seem equivalent in the eyes of common sense.

To emphasize the latitude of choice that we enjoy in this matter Einstein proposes, with tongue in cheek, the following method for defining time on the stationary frame. Let A be a natural clock at rest in the frame. Let P be any point of the frame. An event E at P will be assigned the time coordinate t_0 if A shows time t_0 upon reception of a light signal sent through empty space from P to A when E occurred. Einstein remarks that this method has the disadvantage that it is not independent of the location of A. It is also liable to a stronger objection. If time is defined on the stationary frame by this method a free particle moving in that frame will not travel with constant velocity, but will suffer a sudden loss of speed when its distance from A attains its minimum value. Indeed the requirement that the First Law of Motion be valid on the stationary system implies that the time coordinate function cannot be chosen altogether freely, for it must agree with Lange's definition of an inertial time scale:

(III) A free particle moving in the stationary frame F traverses equal distances in equal times.

But this requirement does not determine the time coordinate function uniquely. Apart from the trivial possibility of changing the origin of time and the time unit, it is clear that, if x, y and z are the space coordinate functions of the stationary system and t is a time coordinate function compatible with (III), the function

$$t' = t + ax + by + cz \tag{3.2.1}$$

(where a, b and c are any three real numbers) is also compatible with (III).

The ambiguity that is still inherent in Lange's stipulations is finally overcome by Einstein's second and, as he says, "more practical" proposal for defining the time coordinate function for the stationary system. Let A and B be two clocks at rest anywhere in the stationary frame. Let L be a light signal transmitted across empty space from A to B, reflected without delay at B, and retransmitted *in vacuo* again from B to A. (I shall eventually refer to signals of this sort as *bouncing* light signals or *radar* signals.) Let t_E and t_R be, respectively, the times of emission and reception of L at A. The clock B can now be set in synchrony with the clock A if we "establish *by definition* that the 'time' required by light to travel from A to B equals the 'time' it requires to travel from B to A".[6] B will then be synchronous with A if and only if the signal L is reflected at B when this clock shows the time $(t_R - t_E)/2$. An event E at the location of B will be assigned the time coordinate t if B shows time t when E occurs.

Einstein's definition of time on the stationary frame F is therefore equivalent to the following stipulation:

(IV) A light pulse transmitted through empty space in any direction from a source at rest in F traverses equal distances in equal times.

Though Einstein here presents (IV) as a matter of definition, he must implicitly assume that it is compatible with (III), or the stationary frame would not be inertial as was prescribed at the outset. He assumes also that all time coordinate functions defined by the foregoing method from any point on the stationary frame will agree with each other up to a change of origin and time unit.[7] These are factual hypotheses that could very well clash with experience. But we need not linger on them for they follow at once from the much stronger factual assumption introduced by Einstein at the beginning of § 2: the Light Principle.

A time scale satisfying conditions (III) and (IV) will be called *Einstein time*. Two events that happen at the same Einstein time are said to be *Einstein simultaneous*.

3.3. The Principles of Special Relativity

Let F be a rigid frame satisfying the conditions (I) and (II) of Section 3.2. We shall say that F is an inertial frame (in Einstein's sense). Let $\bar{x}$, $\bar{y}$ and $\bar{z}$ be the three coordinate functions of a Cartesian system for F. Let the real valued function t assign a "time" to every "possible event" in agreement with conditions (III) and (IV) of Section 3.2. As we did on page 23 we now define the space coordinate functions x, y and z on the set of "possible events" by the following simple method: if the event E occurs at the point P of F, $x(E) = \bar{x}(P)$, $y(E) = \bar{y}(P)$, and $z(E) = \bar{z}(P)$. The four functions t, x, y, z, in that order, define a mapping of "possible events" onto $\mathbf{R}^4$. Any mapping that is constructed in this way, using a given system of stable but otherwise arbitrary units of length and time will be called a *Lorentz chart*, adapted to the frame F.[1] We are now in a position to state the RP and the LP with due precision.

The Principle of Relativity (RP). The laws by which the states of physical systems undergo change are not affected, whether these changes of state be referred to the one or the other of two Lorentz charts.

The Light Principle (LP). A light ray moves in the "stationary" frame defined in Section 3.2 with the constant speed c, whether it be emitted by a stationary or by a moving body. (Here "speed" is the ratio of the distance traversed by the light in the relative space of the frame, to the difference between its times of reception and emission as determined by the time coordinate function of a Lorentz chart adapted to the frame.)[2]

The "laws" mentioned in the RP are not the real relations between classes of events that a philosopher would call by that name, but rather the relations between coordinate functions, and functions of such functions, by which the physicist seeks to express the former. If the RP spoke about real relations of events it would be trivial: obviously, such relations cannot be affected by the choice men make of a coordinate system for describing them. But if it concerns the mathematical expression of such relations in terms of coordinate systems of the kind specified above, the RP is a very powerful criterion for the selection of viable physical laws. As the reader probably knows, it implies that Maxwell's equations do, while Newton's Second and Third Laws of Motion do not, qualify as candidates for that status.

From the "heuristic viewpoint" of Einstein's (1905*b*) the LP may be regarded as the Principle of Inertia applicable to light quanta. (For greater clarity, one ought perhaps to replace in the above statement of the LP, Einstein's expression "a light ray moves", which anyway sounds peculiar, by "a light point moves".) We must not lose sight of the striking analogy between Lange's treatment of the classical Principle of Inertia (IP), that governs—also in Special Relativity—the free motion of material particles, and the manner how Einstein introduces the LP. Lange saw that the IP is meaningless unless we define at least one frame in which it holds good. Lange defines such a frame by considering the behaviour of three free particles in motion. (See Definitions 1 and 2 on page 17, reproduced as conditions (I) and (III) in Section 3.2.) The IP is the factual statement that every other free particle moves in the frame thus defined in the same way as those three, that is, in a straight line, with constant speed (page 17, Theorems 1 and 2). Lange's definition is good enough for the task he set himself, but for the purpose of stating a general principle concerning the propagation of light it is too vague. Even if we narrow it down by adding the condition (II) of Section 3.2 it is still compatible with infinitely many alternative statements about the velocity of light—as one may readily infer from equation (3.2.1). The ambiguity is removed by adding the condition (IV) of 3.2, which Einstein handles as a definition, just as Lange did with conditions (I) and (III). This condition (IV) concerns the behaviour of a single light-pulse, propagating in every direction from a point at rest in a Lange inertial frame. Time is to be defined on the frame in such way that at each instant the light-front lies on a sphere with its centre at the source and its radius proportional to the time elapsed since the light was emitted. Having thus defined the requisite system of reference, Einstein is able to formulate the LP. It is the factual statement that every other light-pulse propagates *in the said frame* in the same way as the former, regardless of the state of motion of its source; i.e. that at each time—*determined in the agreed manner*—the light-front will lie on a sphere with its centre at the point of the frame where the source was placed at the moment of emission, and its radius proportional to the time elapsed since that moment. The LP remains hopelessly imprecise if the system to which it is referred is not defined beforehand. In this the LP does not differ essentially from the IP. It is a sure mark of Lange's and Einstein's genius—yet at the same time a source of seemingly endless philosophical befuddlement—that they so contrived the requisite definitions that the prototype objects—the three free particles, the light-pulse—involved in them were made to follow by convention precisely the same pattern of behaviour which the remaining objects of their kind turn out to have as a matter of fact (to the approximation within which Special Relativity holds good). Such a beautiful consistency of description could not have been achieved in this way if the world were other than it is. But the factual support that makes for the truth of a physical theory need not blind us to the freedom that must be exercised in order to state such a theory intelligibly.

3.4 The Lorentz Transformation. Einstein's Derivation of 1905

Let x and y be Lorentz charts. The mapping $y \cdot x^{-1}$, defined on the range of x in $\mathbf{R}^4$, and with values in $\mathbf{R}^4$, is called a *Lorentz coordinate transformation.* The RP implies that the laws of physics, referred to an arbitrary Lorentz chart, must preserve their form when subjected to a Lorentz transformation. To make Special Relativity into a workable physical theory we therefore need to know the general equations that govern such transformations. In § 2 of his (1905*d*) Einstein inferred those equations from the RP and the LP. There are several other ways of deriving the Lorentz transformation equations, some of which are perhaps simpler, or more elegant, than Einstein's, or proceed from a different set of assumptions. We shall comment on some of them in Sections 3.6 and 4.6. But we shall first give here Einstein's original derivation in full.[1]

We assume that every Lorentz chart agrees with measurements performed with natural clocks and unstressed rods at rest in the frame to which the chart is adapted. We also assume hereafter that the stable system of length and time units employed in Lorentz charts is so chosen that speeds are pure numbers.[2] (This can be achieved by choosing, say, the metre as the unit of length and a convenient multiple of it as the unit of time—a metre of time being the time required by a light signal for traversing a metre of length in an inertial frame in empty space— or by choosing the second as the unit of time and a suitable fraction of the light-second—which is equal to 2.998×10^8 metres—as the unit of length, etc.) c designates the speed of light *in vacuo* in the chosen units, relative to an inertial frame in which a global time has been defined by Einstein's method. If x is a Lorentz chart, we denote its time coordinate function by x^0, its space coordinate functions by x^1, x^2 and x^3. Latin and Greek indices are used as indicated on page 23. As on page 54 we distinguish carefully between the Lorentz space coordinate functions x^α, defined on events, and the corresponding Cartesian coordinate functions $\bar{x}^\alpha$, defined on the relative space S_F of the frame F to which the Lorentz chart x is adapted. The value of $\bar{x}^\alpha$ at a given point of S_F agrees with the value of x^α at any event taking place at that point, but the arguments of $\bar{x}^\alpha$ and x^α are never the same.

If x and y are Lorentz charts adapted to the same frame, it is clear that

$$\begin{aligned} x^0 &= \pm y^0 + k_0 \\ x^\alpha &= \Sigma_\beta a_{\alpha\beta} y^\beta + k_\alpha \end{aligned} \tag{3.4.1}$$

for a suitable set of real numbers k_i, $a_{\alpha\beta}$, such that the matrix $(a_{\alpha\beta})$ is orthogonal. A Lorentz transformation governed by equations (3.4.1) will be said to be non-kinematic (nk). We may obviously, without loss of generality, concentrate our attention on the special case in which x and y are adapted, respectively, to two different inertial frames F and G, such that G moves in F with velocity $\mathbf{v} \neq 0$, and both meet some initial conditions that will simplify our calculations. Obviously, $|\mathbf{v}| \neq c$, for otherwise a light-signal emitted in F in the direction of $\mathbf{v}$ would remain forever at the same point of G, in violation of the LP and the RP. (According to the latter we can always regard G as the "stationary" frame mentioned in the former.) The initial conditions that we shall impose on x and y are the following:

(i) Both x and y assign the quadruple $(0, 0, 0, 0)$ to the same event;
(ii) if σ is a permutation of $\{1, 2, 3\}$ and E is an event such that $x^0(E) = x^{\sigma(1)}(E) = x^{\sigma(2)}(E) = 0$, then $y^{\sigma(1)}(E) = y^{\sigma(2)}(E) = 0$;
(iii) if E is an event such that $y^1(E) = y^2(E) = y^3(E) = 0$, then $x^2(E) = x^3(E) = 0$, and $x^1(E) = vx^0(E)$, where v stands for $|\mathbf{v}|$;
(iv) if E_1 and E_2 are two events such that $x^0(E_1) < x^0(E_2)$ and $x^\alpha(E_1) = x^\alpha(E_2) = 0$ $(\alpha = 1, 2, 3)$, then $y^0(E_1) < y^0(E_2)$.[3]

Two Lorentz charts x and y that satisfy the foregoing conditions will be said to *match* each other. If z and w are two arbitrary Lorentz charts, the transformation $w \cdot z^{-1}$ can always be represented by the composite mapping $(w \cdot y^{-1}) \cdot (y \cdot x^{-1}) \cdot (x \cdot z^{-1})$, where x and y are matching Lorentz charts adapted to the same frames as z and w, respectively, and $w \cdot y^{-1}$ and $x \cdot z^{-1}$ are nk transformations.

Einstein declares that the sought for transformation equations must be *linear* "on account of the properties of homogeneity that we attribute to space and time".[4] We shall discuss this statement in Section 3.6. For the time being, let us regard the linearity of the transformation between the matching charts x and y as a lemma, to be proved later.

Einstein introduces an auxiliary coordinate system which we shall denote by u. It is defined as follows:

$$\begin{aligned} u^0 &= x^0 \qquad & u^1 &= x^1 - vx^0 \\ u^2 &= x^2 \qquad & u^3 &= x^3 \end{aligned} \tag{3.4.2}$$

It is not hard to see that all events that take place at a fixed point of G agree in their u^α-coordinates. Therefore one may say that u is a chart adapted to G. As we shall see it is not a Lorentz chart. Although (3.4.2) is mathematically equivalent to the purely kinematic Galilei transformation (1.6.4), it would also be wrong to describe u as a Galilei coordinate system for G, not only because we have no grounds for identifying $x^0 (= u^0)$ with the universal Newtonian time, but, more importantly, because we do not know as yet whether the coordinate functions u^α agree with distances on G, as Galilei space coordinate functions should. (As a matter of fact they do not, if the principles of Special Relativity hold good.)

We shall now determine y^0 as a function of u. We consider a radar signal transmitted *in vacuo* from a point at rest in G, in a direction perpendicular to the $(\bar{y}^2, \bar{y}^3)$-plane through that point. Let A denote the emission, B the reflection and C the final reception of the signal at its point of emission. To simplify calculations we assume that A is the common origin of x and y (initial condition (i)), and that $x^1(B) > 0$.[5] It follows that $u(A) = (0, 0, 0, 0)$. Since y is the time coordinate function of a Lorentz chart, it is clear that

$$y^0(A) + y^0(C) = 2y^0(B) \tag{3.4.3}$$

According to the LP, the signal propagates in F with speed c. Consequently, $x^1(B) = cx^0(B) = cu^0(B)$. Therefore, $u^1(B) = (c-v)u^0(B)$, and $u(B) = (u^1(B)/(c-v), u^1(B), 0, 0)$. Since C takes place at the same point of G as A, $u^\alpha(C) = 0$. By (3.4.2), $x^1(C) = vx^0(C) = vu^0(C)$. But C is the reception of a signal reflected at B and travelling in the direction of decreasing $\bar{x}^1$ during the time $x^0(C) - x^0(B)$. Hence $x^1(C) = x^1(B) - c(x^0(C) - x^0(B))$. Using (3.4.2) and the foregoing results we infer that $u^0(C) = u^1(B)(1/(c-v) + 1/(c+v))$. Writing $\hat{u}$ for $u^1(B)$ and substituting into (3.4.3) we obtain:

$$y^0 \cdot u^{-1}(0,0,0,0) + y^0 \cdot u^{-1}\left(\left(\frac{\hat{u}}{c-v} + \frac{\hat{u}}{c+v}\right), 0, 0, 0\right) = 2y^0 \cdot u^{-1}\left(\frac{\hat{u}}{c-v}, \hat{u}, 0, 0\right) \tag{3.4.4}$$

(3.4.4) equates two differently constructed real-valued functions of the real parameter $\hat{u}$.[6] Since $y \cdot x^{-1}$ and $x \cdot u^{-1}$ are linear, the functions in question are linear also and have constant first derivatives. Differentiating both sides of (3.4.4) with respect to $\hat{u}$ and writing $\partial y^0/\partial u^i$ for

$$\frac{\partial y^0 \cdot u^{-1}}{\partial u^i \cdot u^{-1}},$$

we have that:

$$\frac{1}{2}\left(\frac{1}{c-v} + \frac{1}{c+v}\right)\frac{\partial y^0}{\partial u^0} = \frac{\partial y^0}{\partial u^1} + \frac{1}{c-v}\frac{\partial y^0}{\partial u^0} \tag{3.4.5}$$

Hence

$$\frac{\partial y^0}{\partial u^1} + \frac{v}{c^2 - v^2}\frac{\partial y^0}{\partial u^0} = 0 \tag{3.4.6a}$$

Since this result does not depend on the choice of A nor on the sign of c, it will not be affected if we lift the restrictions introduced for simplicity's sake before reference number 5. By considering radar signals emitted under similar conditions in directions perpendicular to the $(\bar{y}^1, \bar{y}^2)$-plane and to the $(\bar{y}^1, \bar{y}^3)$-plane, we infer likewise that

$$\frac{\partial y^0}{\partial u^2} = 0; \qquad \frac{\partial y^0}{\partial u^3} = 0 \tag{3.4.6b}$$

Since $y^0 \cdot u^{-1}$ is linear, the system (3.4.6) implies that

$$y^0 = \alpha\left(u^0 - \frac{vu^1}{c^2 - v^2}\right) = \alpha\left(\frac{x^0 - vx^1/c^2}{1 - (v^2/c^2)}\right) \tag{3.4.7}$$

where α is an unknown function of v.

Einstein observes that, according to the RP, the LP applies not only to the frame F, as we assumed above, but also to the frame G.[7] Hence, if σ is a permutation of $\{1, 2, 3\}$ and $B_{\sigma(1)}$ is the reflection of a radar signal emitted at time 0 from the origin of G, at right angles with the $(\bar{y}^{\sigma(2)}, \bar{y}^{\sigma(3)})$-plane, in the direction of increasing $\bar{y}^{\sigma(1)}$,

$$y^{\sigma(1)}(B_{\sigma(1)}) = cy^0(B_{\sigma(1)}) \tag{3.4.8}$$

Einstein's next step is to evaluate $y^\alpha(B_\alpha)$ in terms ot $x(B_\alpha)$ at three events B_1, B_2, and B_3, that satisfy the foregoing description for a suitable permutation σ. Take B_1 first. Substituting from (3.4.7) into (3.4.8), recalling that $u^1(B_1) = (c - v)u^0(B_1)$, and using (3.4.2), we can see that the following relations hold at B_1:

$$y^1 = \alpha c\left(u^0 - \frac{vu^1}{c^2 - v^2}\right) = \frac{\alpha c^2 u^1}{c^2 - v^2} = \alpha\beta^2(x^1 - vx^0) \tag{3.4.9}$$

where β, as on page 44, stands for $|(1 - v^2/c^2)^{-1/2}|$. Turning now to B_2, we note that u^1 must take at it the same value as at the emission of the signal of which B_2 is the reflection, so that $u^1(B_2) = 0$. Hence, by (3.4.2), $x^1(B_2) = vx^0(B_2)$. Looking at the triangle in $\mathbf{R}^3$ whose vertices are the $\bar{x}$-images of the point of emission and the point of reception of the signal, and the perpendicular projection of the latter image on the x^1-axis, we verify that $x^2(B_2) = x^0(B_2)\sqrt{(c^2 - v^2)}$. Consequently, (3.4.7) and (3.4.8) imply that, at B_2,

$$y^2 = \alpha\beta x^2 \tag{3.4.10}$$

By a similar argument one shows that, at B_3,

$$y^3 = \alpha\beta x^3 \tag{3.4.11}$$

Einstein introduces the designation $\varphi(v)$ for $\alpha\beta$, which, like α, is a still unknown function of v, and, seemingly oblivious of the fact that equations (3.4.9)–(3.4.11) have been shown to hold only at certain special events, he collects them, together with (3.4.7), in the following set of transformation equations of purportedly general validity:

$$\begin{aligned} y^0 &= \varphi(v)\beta(x^0 - vx^1/c^2) \qquad & y^1 &= \varphi(v)\beta(x^1 - vx^0) \\ y^2 &= \varphi(v)x^2 & y^3 &= \varphi(v)x^3 \end{aligned} \tag{3.4.12}$$

That these equations do actually hold everywhere follows at once from the fact that $y \cdot x^{-1}$ is a linear transformation of $\mathbf{R}^4$, which is therefore fully determined by its values at four linearly independent points. Now we know that $x^1(B_2) = vx^0(B_2)$, that $x^1(B_3) = vx^0(B_3)$, and that the remaining, hitherto not explicitly given x and y space coordinates of the three events B_α are all zero. From (3.4.7) and the initial condition (iii) on page 57 we infer that if E is an event such that $x(E) = (t, vt, 0, 0)$, then $y(E) = (\alpha t, 0, 0, 0)$. If $t \neq 0$ and, as we assumed at the outset, $|\mathbf{v}| \neq c$, it is clear that $x(E)$, $x(B_1)$, $x(B_2)$ and $x(B_3)$ are four linearly independent points of $\mathbf{R}^4$ at which we know the values of $y \cdot x^{-1}$. Starting from these known values it is not hard to show that (3.4.12) gives the value of $y \cdot x^{-1}$ at an arbitrary point of $\mathbf{R}^4$.[8]

Einstein proves next that the validity of the LP in G—which I shall denote by LP(G)—can be inferred from its validity in F—that is, from LP(F)—and equations (3.4.12). Some writers object that his procedure is circular, for LP(G) had to be assumed in order to prove (3.4.12). But this assumption was made on the strength of the RP and LP(F). Hence, the proof that LP(F) and (3.4.12) imply LP(G), instead of clashing with it, is a welcome test of the consistency of Einstein's principles. Let E and R be the emission and reception of a light signal transmitted through empty space, and write x_E for $x(E)$, etc. Then, the following is a statement of LP(F):

$$\Sigma_\alpha (x_R^\alpha - x_E^\alpha)^2 = (x_R^0 - x_E^0)^2 c^2 \tag{3.4.13}$$

By means of (3.4.12) one infers that

$$\Sigma_\alpha (y_R^\alpha - y_E^\alpha)^2 = (y_R^0 - y_E^0)^2 c^2 \tag{3.4.14}$$

which is a statement of LP(G).

In the last stage of the argument, Einstein succeeds in determining the unknown function $\varphi(v)$. He introduces a new Lorentz chart z, adapted to a frame H (our notation). z fulfils with respect to y the same initial conditions that y fulfils with respect to x, except that, in condition (iii), $-v$ is to be substituted for v. It folows that $z \cdot y^{-1}$ is governed by equations of the same form as (3.4.12), with z, y and $-v$, instead of y, x and v, respectively.

Consequently,

$$\begin{aligned} z^0 &= \varphi(-v)\beta(y^0 + vy^1/c^2) &&= \varphi(v)\varphi(-v)x^0 \\ z^1 &= \varphi(-v)\beta(y^1 + vy^0) &&= \varphi(v)\varphi(-v)x^1 \\ z^2 &= \varphi(-v)y^2 &&= \varphi(v)\varphi(-v)x^2 \\ z^3 &= \varphi(-v)y^3 &&= \varphi(v)\varphi(-v)x^3 \end{aligned} \tag{3.4.15}$$

Einstein observes that, since the relation between the space coordinates z^α and x^α do not depend on x^0, H must be at rest in F. Hence, $z \cdot x^{-1}$ is the identity on $\mathbf{R}^4$,[9] and

$$\varphi(v)\varphi(-v) = 1 \tag{3.4.16}$$

To solve this equation Einstein bids us consider a rod at rest in G, whose endpoints P and Q have, respectively, the Cartesian coordinates $\bar{y}^1(P) = \bar{y}^2(P) = \bar{y}^3(P) = 0$ and $\bar{y}^1(Q) = \bar{y}^3(Q) = 0$, $\bar{y}^2(Q) = \lambda$. The rod moves in F with speed v in a direction perpendicular to its axis. At time $x^0 = t$ its position in F is given by $\bar{x}^1(P_t) = vt$, $\bar{x}^2(P_t) = \bar{x}^3(P_t) = 0$, and $\bar{x}^1(Q_t) = vt$, $\bar{x}^2(Q_t) = \lambda/\varphi(v)$, $\bar{x}^3(Q_t) = 0$. Consequently, the distance between two simultaneous positions of the rod's endpoints in F is $\lambda/\varphi(v)$. Einstein equates this distance with "the length of the rod, measured in [F]", and adds that "from reasons of symmetry it is evident that the length of a given rod moving perpendicular to its axis, measured in the stationary system, must depend only on the velocity and not on the direction and the sense of the motion".[10] Consequently $\lambda/\varphi(v) = \lambda/\varphi(-v)$ and

$$\varphi(v) = \varphi(-v) = 1 \tag{3.4.17}$$

Substituting for φ and β in (3.4.12) we obtain the Lorentz transformation equations for matching charts in their familiar form:

$$\begin{aligned} y^0 &= \frac{x^0 - vx^1/c^2}{\sqrt{1-(v^2/c^2)}} \qquad & y^1 &= \frac{x^1 - vx^0}{\sqrt{1-(v^2/c^2)}} \\ y^2 &= x^2 & y^3 &= x^3. \end{aligned} \tag{3.4.18}$$

This completes Einstein's derivation of 1905. For future reference I shall note down here the Jacobian matrix of the transformation $y \cdot x^{-1}$:

$$\left(\frac{\partial y^i}{\partial x^j}\right) = \begin{bmatrix} \beta & -\beta v c^{-2} & 0 & 0 \\ -\beta v & \beta & 0 & 0 \\ 0 & 0 & 1 & 0 \\ 0 & 0 & 0 & 1 \end{bmatrix}. \tag{3.4.18'}$$

A Lorentz transformation governed by equations (3.4.18) is a *purely kinematic* (pk) transformation—or *boost*—with velocity parameters $(v, 0, 0)$. Pk transformations with velocity parameters $(0, v, 0)$ and $(0, 0, v)$ can be defined by cyclically permuting 1, 2 and 3 in (3.4.18) and in the statement of initial condition (iii) on page 57. It can be shown, after some calculation, that the equations governing the general pk Lorentz transformation with velocity parameters (v_1, v_2, v_3) such that $\Sigma_\alpha v_\alpha^2 < c^2$, are:[11]

$$\begin{aligned} y^0 &= \beta(x^0 - \Sigma_\alpha v_\alpha x^\alpha c^{-2}) \\ y^\mu &= x^\mu - v_\mu\left(\beta x^0 + (1-\beta)\frac{\Sigma_\alpha v_\alpha x^\alpha}{\Sigma_\alpha v_\alpha^2}\right) \\ (\mu &= 1, 2, 3;\ \beta = |(1 - \Sigma_\alpha v_\alpha^2/c^2)^{-1/2}|). \end{aligned} \tag{3.4.19}$$

The typical elements of the Jacobian matrix of this transformation are readily seen to be:

$$\frac{\partial y^0}{\partial x^0} = \beta \quad \frac{\partial y^0}{\partial x^\mu} = \frac{\partial y^\mu}{\partial x^0} = -\beta v c^{-2} \quad \frac{\partial y^\mu}{\partial x^\nu} = \delta_{\mu\nu} + (\beta - 1)\frac{v_\mu v_\nu}{\Sigma_\alpha v_\alpha^2} \tag{3.4.19'}$$

Every Lorentz transformation is the product of a pk transformation governed by equations (3.4.19) and an nk transformation governed by equations (3.4.1). It is not hard to verify that the inverse of a Lorentz transformation and the product of two Lorentz transformations are also Lorentz transformations. The Lorentz coordinate transformations constitute therefore a group of linear transformations of $\mathbf{R}^4$, which is a representation of an abstract group called the *Lorentz group*. (It is increasingly fashionable to call it the *Poincaré group* in honour of its actual discoverer; the name *Lorentz group* is then reserved to the homogeneous Lorentz group defined below.) Since Lorentz charts are global, for each Lorentz coordinate transformation $y \cdot x^{-1}$ between two such charts x and y there is a Lorentz point transformation $y^{-1} \cdot x$, which maps the common domain of all Lorentz charts (the set of "possible events") onto itself. The Lorentz point transformations are also a realization of the Lorentz group.

We can obtain a purely algebraic characterization of the Lorentz group by means of the following simple argument. If x and y are two arbitrary Lorentz charts the transformation $y \cdot x^{-1}$ is defined by a system of linear equations:

$$y^i = \Sigma_j A_{ij} x^j + k^i \tag{3.4.20}$$

The coefficients A_{ij} constitute the Jacobian matrix $(\partial y^i/\partial x^j)$ of the transformation. We denote this matrix by A. Since $y \cdot x^{-1}$ has an inverse $x \cdot y^{-1}$, A is a non-singular matrix (its determinant $\det A \neq 0$). Equations (3.4.20) can be written in matrix notation as follows:

$$y = Ax + k \tag{3.4.21}$$

where $k \in \mathbf{R}^4$. If $k = 0$, equation (3.4.21) defines the general homogeneous Lorentz transformation, between two Lorentz charts that share the same origin. As the product of two homogeneous Lorentz transformations is a homogeneous Lorentz transformation, it is clear that such transformations form a group, the *homogeneous Lorentz group* L (4). The full Lorentz group is the group generated by $L(4)$ and the group of translations (i.e. the additive group underlying the vector space $\mathbf{R}^4$). Thus we shall reach our goal if we succeed in characterizing $L(4)$.

Since the matrix A of a homogeneous Lorentz transformation is non-singular, $L(4)$ is a subgroup of the general linear group $GL(4, \mathbf{R})$, and will be represented by a certain family of 4×4 real non-singular matrices closed under matrix multiplication. We shall now proceed to define this family. Let H_{ij} denote the (i, j)-th element $(0 \leq i, j \leq 3)$ of the matrix

$$H = \begin{bmatrix} c^2 & 0 & 0 & 0 \\ 0 & -1 & 0 & 0 \\ 0 & 0 & -1 & 0 \\ 0 & 0 & 0 & -1 \end{bmatrix} \tag{3.4.22}$$

A short calculation shows that, if the homogeneous Lorentz coordinate transformation $y \cdot x^{-1}$ is governed by equations (3.4.18), then

$$\begin{aligned} \Sigma_i \Sigma_j H_{ij} y^i y^j &= c^2 (y^0)^2 - (y^1)^2 - (y^2)^2 - (y^3)^2 \\ &= c^2 (x^0)^2 - (x^1)^2 - (x^2)^2 - (x^3)^2 \\ &= \Sigma_i \Sigma_j H_{ij} x^i x^j \end{aligned} \tag{3.4.23}$$

On the other hand, if $y \cdot x^{-1}$ is governed by equations (3.4.1), with $k_i = 0$, then $c^2 (y^0)^2 = c^2 (x^0)^2$ and $\Sigma_\alpha \Sigma_\beta H_{\alpha\beta} y^\alpha y^\beta = \Sigma_\alpha \Sigma_\beta H_{\alpha\beta} x^\alpha x^\beta (\alpha, \beta = 1, 2, 3)$, because equations (3.4.1) with $k^0 = 0$ define an isometry of each Euclidean space $\pi_0 =$ const. onto itself. (The projection function π_0 is defined on page 23.) Since every homogeneous Lorentz coordinate transformation is the product of a pk transformation governed by equations (3.4.18) and two nk transformations governed by equations (3.4.1) (page 57), the quadratic form $\Sigma_i \Sigma_j H_{ij} x^i x^j$ is obviously invariant under the transformations of the group $L(4)$. Let $y \cdot x^{-1}$ be an arbitrary homogeneous Lorentz coordinate transformation and let $A = (A_{ij})$ be its Jacobian matrix. We have then that

$$\begin{aligned} \Sigma_i \Sigma_j H_{ij} x^i x^j &= \Sigma_h \Sigma_k H_{hk} y^h y^k \\ &= \Sigma_i \Sigma_j \Sigma_h \Sigma_k H_{hk} A_{hi} x^i A_{kj} x^j \\ (i, j, h, k &= 0, 1, 2, 3) \end{aligned} \tag{3.4.24}$$

Equations (3.4.24) will hold good, as they must, on the entire domain of the chart $x = (x^0, x^1, x^2, x^3)$ if and only if

$$H_{ij} = \Sigma_h \Sigma_k A_{hi} H_{hk} A_{kj} \tag{3.4.25}$$

Or, in other words, if and only if

$$H = A^T H A \tag{3.4.26}$$

where $A^T = (A_{ji})$ is the transpose of the matrix $A = (A_{ij})$. We may therefore characterize the homogeneous Lorentz group $L(4)$ as the subgroup of $GL(4, \mathbf{R})$ represented by all the non-singular 4×4 matrices A that satisfy (3.4.26).[12] It can be shown that this is a Lie group.[13]

By putting $i = j = 0$ in (3.4.25) we promptly verify that

$$(A_{00})^2 = 1 + \Sigma_\alpha (A_{\alpha 0}/c)^2 \geq 1 \qquad (\alpha = 1, 2, 3) \tag{3.4.27}$$

Thus there are two kinds of matrices $A \in L(4)$, those with $A_{00} \geq 1$ and those with $A_{00} \leq 1$; and there is no path in $L(4)$ that joins the former to the latter. (Transformations of the second kind reverse time order.) Recalling that the determinant of the product of several square matrices is equal to the product of the determinants of its factors, we infer from (3.4.26) that

$$\det H = (\det H)(\det A)(\det A^T) \tag{3.4.28}$$

Hence $(\det A)(\det A^T) = 1$, and as $\det A = \det A^T$,

$$(\det A)^2 = 1 \tag{3.4.29}$$

Consequently, if $A \in L(4)$, then either $\det A = 1$ or $\det A = -1$, and again there is no path in $L(4)$ that joins these two classes of matrices. (Transformations of the second class change a right-handed into a left-handed Cartesian system, or vice versa.) Thus the Lie group $L(4)$ splits into four topologically disconnected parts, each of which can be seen to be a connected component:

(i) $\{A \in L(4) | \det A = 1; \quad A_{00} \geq 1\}$;
(ii) $\{A \in L(4) | \det A = 1; \quad A_{00} \leq 1\}$;
(iii) $\{A \in L(4) | \det A = -1; A_{00} \geq 1\}$;
(iv) $\{A \in L(4) | \det A = -1; A_{00} \leq 1\}$.

To clarify the meaning of this fourfold decomposition of $L(4)$ let us note that the set (i) contains the identity matrix $I = (\delta_{ij})$ and constitutes a group in its own right. It is in fact a Lie subgroup of $L(4)$ that we shall designate by $\mathscr{L}_0$. Like any other subgroup, $\mathscr{L}_0$ acts on the left on its parent group $L(4)$ by $(g, h) \mapsto gh$ ($g \in \mathscr{L}_0, h \in L(4)$). The sets (ii), (iii) and (iv) are the orbits, under this action of $\mathscr{L}_0$, of the time reversal transformation T, the parity transformation P (which reflects each Euclidean 3-space $\pi_0 =$ const. about its origin) and their product PT, respectively.[14] We see at once that the union of (i) and (ii) and the union of (i) and (iii) are two subgroups, known as the *proper homogeneous Lorentz group* and the *orthochronous homogeneous Lorentz group*, respectively. Their intersection $\mathscr{L}_0$ is called the *proper orthochronous homogeneous Lorentz group*. The same adjectives are bestowed on the subgroups of the full Lorentz group (or Poincaré group) generated by each of these three subgroups together with the additive group $\mathbf{R}^4$ (the group of translations). The proper orthochronous inhomogeneous Lorentz group, generated by $\mathscr{L}_0$ and $\mathbf{R}^4$, will hereafter be designated by $\mathscr{L}$. One very important remark that we must make before passing to a different subject is that the one-parameter pk transformations governed by equations (3.4.18) *do*, but the general pk transformations governed by equations (3.4.19) *do not* constitute a group on their own right. This is due to the fact that the product of two pk transformations is again a pk transformation only if the two triples of velocity parameters of the former are each a multiple of the other.[15] As the reader will recall, the Galilei group differs in this respect significantly from the Lorentz group (page 29). The full Lorentz group is like the Galilei group a 10-parameter Lie group; each element of the identity component $\mathscr{L}$ can be labelled by the three velocity parameters of the pk transformation, the three parameters of the space rotation and the four parameters of the spacetime translation that it involves.

Einstein's RP of 1905 is a statement about the laws of nature, or rather, about the formulae that express them in terms of coordinate systems of a special kind, viz. Lorentz charts. Such formulae represent the properties and relations of physical events by properties and relations of the real number quadruples assigned to the events by the said charts. The RP says that the formulae remain valid when one such chart is substituted for another one, or, as we may say, when the formulae are subjected to a Lorentz transformation. However, the matter is not quite so simple as it might seem at first sight. One cannot just conclude, for example, that if f is a function on $\mathbf{R}^4$ and h is a Lorentz coordinate transformation, then if $f = 0$ expresses a law of nature, $f \cdot h = 0$ also expresses a law of nature. It would be fitter to say that if $f = 0$ expresses a law, then $h_* f \cdot h$ expresses a law, where h_* is the transformation induced by h in the function space to which f belongs. But, since the meaning of h_* depends on the nature of f, our statement is imprecise and merely suggestive. As we shall see in Chapter 4, one of the great merits of the conceptual framework that Minkowski took over from geometry for the clarification of Einstein's theory, is that it provides the means for a systematic treatment of this question.

The RP is sometimes presented in a different way, as a statement about physical experiments. Thus, we read that, according to the *principle of relativity*, "no physical experiment can distinguish one inertial frame from another", and that "Einstein's special principle of relativity can be stated as follows: *under equal conditions all physical phenomena have the same course of development in any system of inertia.*"[16] The equivalence of both versions of the RP follows from the fact that every Lorentz coordinate transformation—mapping number quadruples on number quadruples—is matched by a Lorentz point transformation—which maps events on events.[17] Hence, the same considerations that, on pp. 30 f., we found applicable to Newtonian relativity, apply *mutatis mutandis* to Einstein's RP. If the laws of nature meet its demands, any two physical processes whose determining conditions can be described in the same way in terms of two Lorentz charts, must be given throughout their development identical descriptions in terms of those charts.[18]

There can hardly be any doubt that in the minds of Einstein and his contemporaries the Lorentz transformations to which the equations of mathematical physics and the physical frames of reference may be subjected without prejudice to the validity of the former or to the expected outcome of experiments in the latter could have been chosen from the full Lorentz group. The fundamental laws of physics, as known to them, were immune to time reversal, and they would not have allowed a purely geometric transformation, such as parity, to have the slightest physical significance. However, in the light of subsequent advances in microphysics, it is advisable to restrict the admissible transformations to the proper orthochronous Lorentz group $\mathscr{L}$.[19] Unless otherwise noted, we shall hereafter proceed on this assumption. We shall also assume that the space coordinate functions of our Lorentz charts always constitute a right-handed system, and that the time coordinate functions never assign a smaller number to a man's death than to his birth.

3.5 The Lorentz Transformation. Some Corollaries and Applications

We cannot dwell here on the many consequences of the transformation equations derived in Section 3.4. But we shall examine a few of them, that have attracted attention in philosophical debates on Relativity, or that will be useful to us later on.

Any mapping partitions its domain into equivalence classes, namely, the fibres of the mapping over each of its values. The fibres of a time coordinate function over each real number in its range are called *simultaneity classes.* Equations (3.4.1) imply that the time coordinate functions of two Lorentz charts adapted to the same frame determine the same partition of events into simultaneity classes. Let x be a Lorentz chart adapted to a frame F. Two events E_1 and E_2 such that $x^0(E_1) = x^0(E_2)$ are said to be *simultaneous in F*. Let y be another Lorentz chart matching x and adapted to a frame G. A glance at the first of equations (3.4.18) will persuade us that if E_1 and E_2 are simultaneous in F they cannot be simultaneous in G unless $x^1(E_1) = x^1(E_2)$. Generally speaking, two events simultaneous in an inertial frame are simultaneous in another inertial frame that moves relatively to the former if and only if the line joining their spatial locations is perpendicular to the direction of motion.

Any real-valued function induces on its domain a relation of partial order, determined by the linear order of **R**. The partial order induced on events by time coordinate functions is called *temporal order*. Temporal order is preserved by nk Lorentz coordinate transformations (governed by equations (3.4.1)). Let x be as above. If $x^0(E_1) < x^0(E_2)$ we say that event E_1 *precedes* event E_2 and that E_2 *follows* E_1 *in* F (abbreviated $E_1 <_F E_2$). Temporal order is not preserved by Lorentz pk transformations (governed by equations (3.4.19)). However, if E_1 and E_2 occur at the same point of some inertial frame F and $E_1 <_F E_2$, E_1 precedes E_2 in every inertial frame. On the other hand, if E_1 and E_2 are simultaneous in some inertial frame F but do not occur at the same point in that frame, there are inertial frames in which E_1 precedes E_2 and others in which E_2 precedes E_1. (Take, for example, any two frames moving relatively to F in opposite senses, in the direction of the line joining the points of occurrence of E_1 and E_2 in F.) Needless to say, in this case E_1 and E_2 do not occur at the same point in any inertial frame.

If simultaneity, as determined by Lorentz charts, were not frame dependent, a light signal could not propagate with the same speed in all inertial frames, and the LP would be incompatible with the RP. For let F and G be two inertial frames in motion relative to one another and consider a light pulse transmitted through empty space from a point O_F at rest in F. Let O_F coincide, at the time of emission, with the point O_G of G. Then, if the RP and the LP are jointly true, the wave front takes up in either frame, at each instant, a sphere centred, respectively, in O_F and in O_G. Since O_F and O_G do not coincide except at the time of emission, the preceding statement would be contradictory if each instantaneous location of the wave front in F were also an instantaneous location of the wave front in G. But due to the frame dependence of simultaneity, the events that make up the wave front at a given time in F are not simultaneous in G, and hence do not constitute the wave front at any time in G. The sphere in F, centred at O_F, that the wave front covers at some instant of F-time, is not a sphere in G, centred at O_G, but neither is it the shape of the wave front at any instant of G-time. The same can be said, of course, with F and G interchanged.

The foregoing considerations throw some light on the phenomenon known as relativistic *length contraction*. Let Q be a solid cube at rest in an inertial frame F_0. Choose three concurrent edges of Q and label them 1, 2, and 3, as one would the axes of a right-handed Cartesian system. The length of edge α in F_0 is, of course, the distance between its endpoints in F_0. We denote it by $\lambda_0(\alpha)$. We assume that Q is not subject to any deforming influence, so that $\lambda_0(\alpha)$ has the same constant value for all three values of α. Let F_v be another inertial frame moving in F_0 with speed v in a direction parallel to edge number 1. The length $\lambda_v(\alpha)$ of edge α in F_v is aptly defined as the distance between two simultaneous positions of its endpoints in F_v.[1] For greater definiteness we introduce a Lorentz chart x, adapted to F_0, such that the meeting point of the three numbered edges of Q lies at the spatial origin of x and that, for any event E_α, taking place at the other end of edge α, $x^\alpha(E_\alpha) = \lambda_0(\alpha)$. Let y be a Lorentz chart matching x and adapted to F_v. Choose four events $E_i((i = 0, 1, 2, 3)$, such that E_0 is mapped by both x and y on $(0, 0, 0, 0)$ and that $E_\alpha(\alpha = 1, 2, 3,)$ is simultaneous with E_0 in F_v and occurs at the other end of edge α. Obviously,

$y^\alpha(E_\alpha) = \lambda_v(\alpha)$. We have that $0 = y^0(E_1) = \beta(x^0(E_1) - x^1(E_1)vc^{-2})$. Hence, $x^0(E_1) = \lambda_0(1)vc^{-2}$, and

$$\begin{aligned} \lambda_v(1) = y^1(E_1) &= \beta(x^1(E_1) - vx^0(E_1)) \\ &= \beta\lambda_0(1)(1 - v^2c^{-2}) \\ &= \beta^{-1}\lambda_0(1) \end{aligned} \tag{3.5.1}$$

$$\lambda_v(2) = y^2(E_2) = x^2(E_2) = \lambda_0(2)$$
$$\lambda_v(3) = y^3(E_3) = x^3(E_3) = \lambda_0(3).$$

Consequently, the space taken up by Q at any given moment in the inertial frame F_v, in which it moves with speed v, is smaller than the space it fills in the inertial frame in which it is at rest, being shorter than it in the direction of motion by a factor of $\beta^{-1} = |(1 - v^2c^{-2})^{1/2}|$. As one would expect, this factor does not depend on the sense of the motion. When considering this phenomenon, one must bear in mind that Q, by hypothesis, is not subject to any action that might deform it. Hence, we are not entitled to say that Q is contracted by motion. On the other hand, it is true that the place Q occupies in its own rest frame differs in both size and shape from the fleeting figure it cuts at each moment in a frame moving past it. But surely there is no more need for this figure to be cubic than for Q's shadows to be all square.

A deeper insight into the changes introduced by Einstein in the received views on space and time is provided by relativistic *time dilation*. Let F_0, F_v, x and y be as in the preceding paragraph. A natural clock C at rest on the spatial origin of x is set to agree with the time coordinate function x^0. Hence, it also agrees with the time coordinate function y^0 as it goes through the spatial origin of y in F_v. But on every other occasion it will not agree with y^0. For let E be any event at C. Then, by (3.4.18),

$$y^0(E) = \beta x^0(E). \tag{3.5.2}$$

Hence, if E_1 and E_2 are two events at C

$$|y^0(E_1) - y^0(E_2)| > |x^0(E_1) - x^0(E_2)|. \tag{3.5.3}$$

If F_v is studded with clocks keeping y^0-time along C's path, C will appear to run slow when checked against those clocks as it passes by them. By symmetry, any clock at rest in F_v and keeping y^0-time must appear to run slow when compared with clocks at rest in F_0 and synchronized with C. (Apply, instead of $y \cdot x^{-1}$, the inverse transformation $x \cdot y^{-1}$.) No contradiction is involved here, for it is impossible to compare two clocks more than once by this method. However, if C is transferred all of a sudden to an inertial frame F'_0 moving relatively to F_v in the opposite sense though with the same speed as F_0, C will again pass by the clocks at rest in F_v with which it was compared earlier. Since β depends on v^2, (3.5.2) holds also for y^0 and a suitably chosen time coordinate function x'^0, adapted to F'_0. If C is not damaged by the sudden bounce it must suffer when changing frames, but continues to behave in F'_0 as the good natural clock that it is, then, when it returns, say, to the spatial origin of y, it will definitely run behind the clock C_y that keeps y^0-time at that place, and with which C agreed as they first passed each other. (It is worth noting that C also runs behind any clocks at rest in F_0 and keeping x^0-time that it meets on its way, although it was in distant synchrony with them until it changed frames.) Since C has rested on two different inertial frames while C_y remained fixed in one, symmetry is broken, and one cannot predict that C runs ahead of C_y.[2] Nevertheless, it has been judged paradoxical that Special Relativity, which is based on the denial of the absolute significance of inertial motion, should associate with it such absolute effects as the retardation of C with respect to C_y.[3] Some authors have sought to explain the different behaviour of the clocks by noting that C is subjected in the course of the experiment to an accelerating force. They apparently believe that this can be held responsible for a physical change in C that mere inertial motion is not supposed to cause. But our argument was predicated on the assumption that C suffered no significant change when we reversed its motion relative to F_v. During its entire trip away from and back to C_y, C is supposed to keep natural time faithfully in agreement with the time coordinate of a Lorentz chart adapted to the frame in which it rests. Moreover, the impulse applied to C when changing frames can be reduced as much as we wish without diminishing the observed time lag, if we correspondingly increase the distance traversed by it in F_v with uniform speed between its two encounters with C_y. Indeed, we can obviate the need for accelerating C if, instead of throwing it from F_0 to F'_0, we set by it a clock C' at rest in F'_0 as it goes past C, and allow C' to carry C's time back to C_y. In this case, no force is exerted, each clock follows undisturbed its own natural course—e.g., if they are Langevin clocks, their respective light signals all obey the LP—and yet, upon meeting C_y, C' lags behind it by exactly the same amount as C did in the former case. I must confess, even at the risk of seeming presumptuous, that the air of paradox which, in the eyes of many, enshrouds relativistic time dilation arises in my opinion from a persistent yet undeclared belief in a unique universal time. If there is but one time encompassing both C and C_y wherever they happen to be, we cannot consistently maintain that they both keep time truly and yet agree on their first meeting and disagree on the second. A cause must be found why the one has counted more seconds while the other has counted less. But if there is no ubiquitous "while" that both clocks are required to measure, it is not surprising that each one, though faithfully keeping time at its own place, should not register the same lapse of time as the other, from one encounter to the next.[4]

Important consequences flow from *Einstein's Rule for the Transformation of Velocities*, to which we now turn. Let A be an object moving with constant velocity in the frame F_v. What is A's velocity in F_0? We use again the Lorentz charts x and y that we introduced above, and let Δx^i and Δy^i denote the increment of the i-th coordinate function of either chart between any two events at A. Clearly, the Δy^i are given as functions of the Δx^i by equations (3.4.18). (Prefix Δ's to all x's and y's.) Write u_α for $\Delta x^\alpha/\Delta x^0$ and w_α for $\Delta y^\alpha/\Delta y^0$, the α-th velocity components of A in terms of x and y. It readily follows from (3.4.18) that

$$u_1 = \frac{w_1 + v}{1 + (vw_1/c^2)} \qquad u_2 = \frac{w_2}{\beta(1 + (vw_1/c^2))} \qquad u_3 = \frac{w_3}{\beta(1 + (vw_1/c^2))} \tag{3.5.4}$$

Setting $\Sigma_\alpha(w_\alpha)^2 = c^2$, one verifies, after a short calculation, that $\Sigma_\alpha(u_\alpha)^2 = c^2$. Hence, if A is a light pulse transmitted *in vacuo*, it moves with speed c in both F_0 and F_v, as indeed it must, if the RP and the LP hold good. But since the same can be said if A is any object whatsoever moving in either frame with that speed, one surmises that it is not so much the nature of light, but rather some structural feature of space and time, that is responsible for the peculiar status of speed c. Suppose, on the other hand, that F_v is the rest frame of a medium of refractive index n moving across the laboratory frame F_0. If A is a light pulse sent across that medium, in the direction of its motion, we have that $w_1 = c/n$, and $w_2 = w_3 = 0$, so that $u_2 = u_3 = 0$, and

$$u_1 = \frac{(c/n) + v}{1 + (v/cn)} = \frac{c}{n} + v\left(1 - \frac{1}{n^2}\right)\left(1 + \frac{v}{cn}\right)^{-1} \tag{3.5.5}$$

This result agrees with Fresnel's formula for the speed of light carried by a moving refractive medium (page 42) if $v \ll c$. Being a direct consequence of the kinematic rule for the transformation of velocities from one inertial frame to another, it need not be explained by physical hypotheses, such as Fresnel's aether drag, or the complex mechanism of matter–light interaction proposed by Lorentz.[5] Einstein's Rule (3.5.4) also yields improved equations for the aberration of starlight and the Doppler effect.

The main task of Relativity physics is to formulate the laws of nature in a manner consonant with Einstein's RP. Classical electrodynamics met this demand splendidly, as Einstein showed without difficulty.[6] But classical mechanics did not, and a new, relativistic mechanics was developed in the next few years by Einstein himself, Planck, Lewis and Tolman, and others.[7] We shall have something to say about it in Chapter 4, after our study of Minkowski spacetime gives us the means to see more clearly into the subject. There is, however, one relativistic law that we ought to mention beforehand, due to its importance for the theory of spacetime. In relativistic mechanics, the energy of a material particle moving in an inertial frame increases beyond all bounds as its speed approaches the speed of light *in vacuo*. Since all sources of power available for accelerating a particle are presumably finite, it follows that no particle moving with speed $v < c$ in an inertial frame can ever attain the speed c in that frame. If we postulate further that every particle must have a definite real-valued energy in each inertial frame, it is impossible that a particle at rest in any such frame should move with speed $v \geq c$ in another one. If we regard inertial frames as being at least potentially the rest frames of material bodies, it follows at once that no two inertial frames can move relatively to each other with a speed greater than that of light. Such a restriction on the relative speeds of inertial frames of reference was inconceivable in classical mechanics because, under the Galilei Rule for the Transformation of Velocities, every bound imposed on relative speeds is surpassed for a suitable choice of frames. But the restriction is in perfect harmony with Einstein's Rule, by which, if F, G and H are three inertial frames such that G moves in F and H moves in G with any speed less than c, H moves in F with a speed less than c.[8] Thus, the fact that the general pk Lorentz transformations (3.4.19) are undefined if $\Sigma_\alpha v_\alpha^2 > c^2$ ($\alpha = 1, 2, 3$) does not impair the usefulness of Lorentz charts.[9] Again, since a rod or clock cannot travel across an inertial frame with the speed of light, there is no paradox involved in the prediction, implicit in (3.5.1) and (3.5.2), that at speed c a rod's length would shrink to zero and a clock would stop completely, in comparison with rods and clocks at rest in the frame. It is worth emphasizing that the speed c, relative to an inertial frame, is an unattainable maximum only for other such frames and for material systems at rest in them. By the Einstein Rule, an object travelling with speed c in one inertial frame, moves with the same speed in all the others. And nothing in Special Relativity precludes the existence of particles moving faster than light in *all* inertial frames. Such particles, known in the literature as *tachyons* (from the Greek word *tachys*, meaning *fast*), have been the subject of lengthy theoretical discussions, but have not hitherto been detected.[10] Within the framework of Relativity physics, they would have to be assigned an imaginary-valued energy. No braking agency could ever slow one of them down to speed c in any inertial frame. Moreover, they could not—lest they be used for deliberately changing the past—act as information carriers, and would therefore interact with the more familiar forms of matter only in a totally random fashion (if at all). The speed of light *in vacuo* thus possesses, according to the Special Theory of Relativity, a fundamental physical significance, which does not depend on any of the conventions leading to the definition of Lorentz charts: it is the maximum signal speed. As a consequence of this, a relativistic world cannot contain a true rigid body, i.e. a material system whose parts are constrained by internal forces to stay at a perfectly constant distance from one another; for, if such a body existed, one could send a signal across it with infinite speed just by pulling or pushing at one of its extremes.

3.6 The Lorentz Transformation. Linearity

As I announced on page 57, we shall now examine Einstein's claim that the coordinate transformation between two Lorentz charts with a common origin must be linear "on account of the properties of homogeneity which we attribute to space and time". To avoid misunderstandings let me point out that the term "Lorentz coordinate transformation" can be used in a physical and a purely mathematical sense. The physical sense was fixed on page 56 by means of the physically defined notion of a Lorentz chart, that is, a coordinate system arising, in the manner described on page 54, from a set of Cartesian coordinates on the relative space of an inertial frame and a time coordinate established on the same frame by Einstein's method (page 52). If x and y are Lorentz charts, the composite mapping $x \cdot y^{-1}$, which sends a real number quadruple to another, is a Lorentz coordinate transformation in the physical sense. The transformation is *orthochronous* if the time coordinates of both charts increase together (i.e. if $\partial x^0 / \partial y^0 > 0$). It is *proper* if the underlying Cartesian systems are both right-handed, or both left-handed. It is *homogeneous* if both charts have a common origin, so that $x \cdot y^{-1}(0) = 0$. In a purely mathematical sense a Lorentz coordinate transformation is simply a member of the (full) Lorentz group of transformations of $\mathbf{R}^4$. It is orthochronous, proper or homogeneous if it belongs to the subgroups of the Lorentz group that bear these names (p. 64). A Lorentz coordinate transformation in the mathematical sense is linear by definition. What Einstein is saying in the passage quoted is that, due to the homogeneity of physical space and time, linearity is also a property of homogeneous Lorentz coordinate transformations in the physical sense.[1] As we saw in Section 3.4 this linearity is an indispensable premise of his proof that both senses of the term are coextensive. We must now try to understand how it can be said to rest on the homogeneity of space and time.

We say that an arbitrary set S is *homogeneous* with respect to a group G if G acts transitively on S.[2] All the elements of S are then equivalent with respect to the action of G. If S has a definite structure, we call it a homogeneous space if it is homogeneous with respect to its group of automorphisms (i.e. the group of structure-preserving permutations of S). Affine spaces are therefore homogeneous *par excellence*, for an affine space is by definition a set transitively and effectively acted on by the additive group underlying some linear space, and its affine structure consists precisely of the properties and relations invariant under the action of this group. The Einstein time and the Euclidean space of an inertial frame are of course affine spaces, transitively and effectively acted on by the additive groups of $\mathbf{R}$ and $\mathbf{R}^3$, respectively; and Einstein was right in calling them homogeneous.[3] But it is not at all clear that the linearity of the Lorentz transformations can follow from this fact alone.

Let me observe, first of all, that the structural properties that a manifold may possess—beyond the one by which, *praecise sumpta*, it is a manifold—do not bestow any peculiar mathematical qualities on the coordinate transformations between its charts. Thus, even if a given manifold M is an affine space, a coordinate transformation between two charts of M might not be linear if one of them is a curvilinear coordinate system (i.e. if its parametric lines are not straight in the affine space M). The structure of the manifold can in fact tell us something about the coordinate transformation only insofar as the charts involved reflect that structure. The following condition, however, is sufficient to ensure the linearity of the coordinate transformations $x \cdot y^{-1}$ and $y \cdot x^{-1}$ between two charts x and y of an n-manifold M: M is an affine space, acted on by $\mathbf{R}^n$, and x and y are linear isomorphisms of the vector spaces $(M, x^{-1}(0))$ and $(M, y^{-1}(0))$, respectively, onto R^n.[4] Do homogeneous Lorentz coordinate transformations automatically meet this condition?

In order to answer this question let us consider two inertial frames F_1 and F_2 and let S_i denote the Euclidean space of F_i $(i = 1, 2)$. We shall designate by T_i the fibres of an Einstein time coordinate function adapted to F_i. (In other words, T_i is the quotient of the set of "possible events" by Einstein simultaneity in F_i.) Let x_i be the Lorentz chart arising from a Cartesian system $\bar{x}_i$ defined on S_i and an Einstein time coordinate x_i^0 defined on T_i. We wish to know whether $x_1 \cdot x_2^{-1}$ is a linear permutation of $\mathbf{R}^4$. As is well-known (see note 4), the mere choice of the charts $\bar{x}_i$ and x_i^0 makes S_i and T_i into vector spaces with their zeroes, respectively, at the origins $O_i^S = \bar{x}_i^{-1}(0)$ and $O_i^T = (x_i^0)^{-1}(0)$ of the said charts. $\bar{x}_i$ and x_i^0 are then linear isomorphisms of (S_i, O_i^S) and (T_i, O_i^T) onto $\mathbf{R}^3$ and $\mathbf{R}$, respectively. But this does not by itself guarantee that the domain of the Lorentz charts x_1 and x_2 is an affine space, nor that $x_1 \cdot x_2^{-1}$ is a linear mapping. For the domain of x_i is neither S_i nor T_i but their Cartesian product $T_i \times S_i$, and, in general, $S_1 \neq S_2$ and $T_1 \neq T_2$. We can indeed view $(T_i \times S_i, x_i^{-1}(0))$, in a natural way, as the direct sum $(T_i, O_i^T) \oplus (S_i, O_i^S)$ of the vector spaces (T_i, O_i^T) and (S_i, O_i^S), whereby the Lorentz chart x_i becomes a linear isomorphism of its domain onto $\mathbf{R}^4$.[5] And we do of course identify the domains of x_1 and x_2 by equating each pair (t_1, P_1) in the former with the pair (t_2, P_2) in the latter, where P_2 is the point in S_2 with which P_1 coincides at time t_1, and t_2 is the time in T_2 at which P_2 coincides with P_1. (It is by virtue of this identification that all Lorentz charts can be said to share the same domain, to affix numerical labels to the same physical world.) If the coordinate systems x_i^0 and $\bar{x}_i$ are so chosen that (O_1^T, O_1^S) is equal in this sense to (O_2^T, O_2^S), the Lorentz charts x_1 and x_2 can be meaningfully said to have the same origin. But it is not self-evident that the sufficient condition of linearity stated above is thereby fulfilled. It is true indeed that each Lorentz chart $x_i (i = 1, 2)$ is defined on an affine space which it maps isomorphically onto $\mathbf{R}^4$. And we have successfully identified the set $T_1 \times S_1$ underlying one of those affine spaces with the set $T_2 \times S_2$ underlying the other. But the correspondence by which the identification was effected did not depend on the affine structures bestowed on those sets, and we have yet to show that it preserves them. In other words, we would still have to prove that the bijective mapping of $T_1 \times S_1$ onto $T_2 \times S_2$ defined above by means of physical coincidences is actually a linear isomorphism of $(T_1, O_1^T) \oplus (S_1, O_1^S)$ onto $(T_2, O_2^T) \oplus (S_2, O_2^S)$. But rather than grapple with the problem of proving this, we shall proceed directly to our main goal of establishing the linearity of the Lorentz coordinate transformation $x_1 \cdot x_2^{-1}$, from which the isomorphic character of the said bijective mapping easily follows. Our goal can be promptly attained if it is true that the additive group of $\mathbf{R}^4 = \mathbf{R} \oplus \mathbf{R}^3$ acts transitively and effectively on the common domain of the Lorentz charts x_1 and x_2, as defined by the above identification; in other words, if not only the Einstein times and the Euclidean spaces belonging to each inertial frame, but also the *common spacetime* constructed from them can be said to be homogeneous. This, however, does not follow from the homogeneity of the spaces S_1 and S_2 and the times T_1 and T_2 alone, but also depends on the way how the inertial frames F_1 and F_2 move in each other.

In order to understand why the linearity of the Lorentz transformations is thus essentially a question of kinematics, and not just one of geometry (in the popular sense) and chronometry, let us consider an example drawn for simplicity's sake from the realm of Newtonian theory. Let F_1 and F_2 stand for two rigid frames, one of which is inertial while the other one rotates in it with constant angular velocity. Let S_i stand as before for the Euclidean space of the frame F_i and set $T_1 = T_2 = T$, the Newtonian time common to all frames. We identify the sets $T_1 \times S_1$ and $T_2 \times S_2$ by the same physical criterion employed above; i.e., we set $(t, P_1) = (t, P_2)$ whenever $P_1 \in S_1$ coincides with $P_2 \in S_2$ at $t \in T$. We choose four points $O_i^T \in T_i$, $O_i^S \in S_i$, such that $(O_1^T, O_1^S) = (O_2^T, O_2^S)$. We construct the linear spaces $(T_i, O_i^T) \oplus (S_i, O_i^S)$. We establish the Cartesian coordinate systems $\bar{y}_i$: $S_i \to \mathbf{R}^3$ and the isometric time coordinate functions y_i^0: $T_i \to \mathbf{R}$, so that $y_i^0(O_i^T) = 0$, $\bar{y}_i(O_i^S) = 0$; and define from them, in an obvious way, the linear isomorphisms y_i: $(T_i, O_T^i) \oplus (S_i, O_T^i) \to \mathbf{R}^4$. Though y_1 and y_2 meet the same conditions explicitly considered in the above discussion of the Lorentz charts x_1 and x_2, it is evident that $y_1 \cdot y_2^{-1}$ cannot be a linear mapping.[6] This is due of course to the kinematic relation between the frames F_1 and F_2, which is quite different from that which obtained in the previous case, when the two frames were assumed to be inertial. Although in each case there are homogeneous times and spaces associated with the frames, we can combine them into a single homogeneous spacetime only if the frames F_1 and F_2 are both inertial. To see this let F_1 alternately rotate and move inertially in the inertial frame F_2, and consider the motion of an arbitrary point $P \in F_1$. If F_1 is rotating the speed of P in F_2 will depend on its distance from the axis of rotation, and the direction of its motion will vary with time. But if F_1 is inertial, the velocity of P in F_2 will be the same at all times, for every $P \in F_1$. Thus *all* points of an *inertial* frame F_1 stand at *all* times in one and the same relation to another *inertial* frame F_2. This is the "homogeneity property" of space and time from which the linearity of the Lorentz transformations actually follows. For, if all points of F_1 always move in F_2 with the same velocity, the homogeneous Lorentz coordinate transformation $x_1 \cdot x_2^{-1}$—in contrast with the non-linear transformation $y_1 \cdot y_2^{-1}$—must be indifferent to the choice of the particular origin $(O_1^T, O_1^S) = (O_2^T, O_2^S)$ shared by the Lorentz charts x_1 and

x_2, and must remain invariant if that origin is changed. Let k and k' stand, respectively, for the values of x_2 and x_1 at the new origin, and, for brevity's sake, set $x_1 \cdot x_2^{-1} = f$. Clearly, $k' = f(k)$. Consequently, the invariance requirement stated above implies that, for any $v \in \mathbf{R}^4$,

$$f(v+k) = f(v) + k' = f(v) + f(k) \tag{3.6.1}$$

This relation holds for any change of origin and thus for any $k \in \mathbf{R}^4$. From (3.6.1) it readily follows that for any $v \in \mathbf{R}^4$ and any rational number λ,

$$f(\lambda v) = \lambda f(v) \tag{3.6.2}$$

If we assume that f is continuous at 0, (3.6.2) holds also for any $\lambda \in \mathbf{R}$.[7] Equations (3.6.1)—for arbitrary v, $k \in \mathbf{R}^4$—and (3.6.2)—for arbitrary $v \in \mathbf{R}^4$, $\lambda \in \mathbf{R}$—say in effect that f is a linear mapping.[8]

The linearity of Lorentz coordinate transformations can also be established on physical grounds, as follows. Let x and y be two arbitrary Lorentz charts. We assume that Einstein's *Relativity Principle* holds good. We have then that:

(A) According to the *Principle of Inertia*, the set of events in the history of any free particle are mapped by both x and y onto a set of linearly dependent real number quadruples.

(B) According to Einstein's *Light Principle*, the images by x and y of the events in the history of a light signal satisfy the equations:

$$c^2(x_0^0 - x^0)^2 - \Sigma_\alpha (x_0^\alpha - x^\alpha)^2 = c^2(y_0^0 - y^0)^2 - \Sigma_\alpha (y_0^\alpha - y^\alpha)^2 = 0 \tag{3.6.3}$$

where (x_0^i) and (y_0^i) are the coordinates in each chart of a fixed event in that history ($i = 0, 1, 2, 3$; $\alpha = 1, 2, 3$).

If we assume moreover that all Lorentz charts are defined on a common domain which they map injectively *onto* $\mathbf{R}^4$, it follows that:

(C) Lorentz coordinate transformations are permutations of $\mathbf{R}^4$.

The linearity of the transformations $x \cdot y^{-1}$ and $y \cdot x^{-1}$ follows from (A) and (C), and also from (B) and (C). Let us now disregard (C) and make allowance for the possibility that the transformations send some points of $\mathbf{R}^4$ to infinity. Then (A) entails that they must be projectivities of the general form

$$y^i = \frac{\Sigma_j a_{ij} x^j + a_i}{\Sigma_j b_j x^j + b} \tag{3.6.4}$$

with $(a_{\alpha\beta})$ a non-singular matrix ($\alpha, \beta = 1, 2, 3$), and the other coefficients arbitrary. (B) implies, on the other hand, that the transformations must belong to the group generated by the transformations of the form

$$y^i = \Sigma_j e_j a_{ij} x^j + a_i \tag{3.6.5}$$

where $e_0 = 1$, $e_1 = e_2 = e_3 = -1$, the a_{ij} satisfy the condition $\Sigma_i e_i a_{ij} a_{ik} = e_j \delta^j_k$, and the a_i are arbitrary; and the so-called Möbius transformations, of the form

$$y^i = \frac{x^i - a_i \Sigma_j e_j (x^j)^2}{1 - 2\Sigma_j e_j a_j x^j + \Sigma_j \Sigma_k e_j a_j^2 e_k (x^k)^2} \tag{3.6.6}$$

where the a_i are arbitrary and the e_i are as in (3.6.5). The intersection of the group of projectivities (3.6.4) and the group generated by the transformations (3.6.5) and (3.6.6) comprises only the special linear transformations (3.6.5) and is in effect none other than the Lorentz group. (The reader ought to satisfy himself that equations (3.6.5) under the conditions stated above agree exactly with equations (3.4.20) subject to conditions (3.4.25).) Thus (A) and (B) jointly entail that the Lorentz coordinate transformations in the physical sense coincide with the Lorentz coordinate transformations in the mathematical sense.[9]

3.7 The Lorentz Transformation. Ignatowsky's Approach

The Russian mathematical physicist W. von Ignatowsky (1910, 1911*a*) claimed that the Lorentz transformation equations can be derived from the Relativity Principle alone, without assuming anything about the velocity of light or any other specific phenomenon. The constant c which occurs in the equations is then demonstrably the least upper bound of the physically possible relative speeds of ordinary matter (i.e. the kind of matter that may be found at rest in an inertial frame). In fact Ignatowsky used several additional explicit and implicit hypotheses besides the Relativity Principle. But they are hypotheses of a purely geometrical or chronogeometrical nature, and do not concern phenomena of a special kind. By relaxing them we can state the general problem of possible relativistic kinematics, first envisaged and fruitfully dealt with by H. Bacry and J. M. Lévy-Leblond (1968). Ignatowsky's work has been repeated and refined by numerous authors, some of whom independently lighted on the same ideas, unaware that they had long been available in well-known journals.[1]

By the Relativity Principle Ignatowsky understands the equivalence of inertial frames. Each such frame F is associated with a Euclidean space S_F, in which it is at rest, and with a universal time T_F, i.e. a unique partition of all events in S_F into simultaneity classes which form a linearly ordered metric continuum. If F and F' are two inertial frames each point in $S_{F'}$ describes a straight line in S_F, traversing equal distances in equal intervals of T_F. As we know, this requirement restricts the choice of the possible partitions T_F but does not fix it uniquely. But Ignatowsky gives no further indications on this matter—no "rules for clock-synchronization"—for he apparently expects that for each inertial frame F the time T_F can be determined by comparison with other inertial frames, in the light of the Relativity Principle. Given any inertial frame F, every physical event can be described at an occurrence in the space S_F at some moment of the time T_F. It is assumed that the laws of nature are invariant under rotations and translations in S_F and under translations of T_F. The equivalence of inertial frames entails that they must also be invariant under appropriate transformations that switch their reference from one inertial frame to another. The aim of Ignatowsky's inquiry is to establish the form of these transformations.

The problem is made mathematically definite by choosing units of time and length and mapping S_F isometrically onto $\mathbf{R}^3$ and T_F isometrically onto $\mathbf{R}$, for every inertial frame F. A coordinate system (t, x, y, z), where (x, y, z) is a Cartesian system on S_F and t is an isometric coordinate function on T_F will be said to be *adapted* to F. (We do not call (t, x, y, z) a Lorentz chart because we do not know as yet whether the time T_F agrees with Einstein time.) Let F and F' be two inertial frames such that F' moves in F with velocity $\mathbf{v}$. Let (t, x, y, z) and (t', x', y', z') be coordinate systems adapted to F and F', respectively. The systems can obviously be so chosen that they both map the same event to 0, that their time coordinates increase together and that the homonymous space coordinate axes point in the same directions at time $t = t' = 0$. A simple spatial rotation of each system is sufficient to ensure that the (x, x')-axis lies along the direction of motion. As on page 57, let us say that the systems (t, x, y, z) and (t', x', y', z') are matched together when they satisfy these conditions. The velocity of F' in F is then equal to its component along the x-axis; we shall designate it henceforth by v. Obviously, the transformation between the matching systems (t, x, y, z) and (t', x', y', z') must be uniquely determined by this velocity v, for, as we observed on page 74, the relation between the two frames is fully expressed by it.[2] Ignatowsky believed that he had proved from the Relativity Principle that the transformation between two matching coordinate systems such as we have described is the Lorentz pk transformation with velocity parameters $(v, 0, 0)$—unless the constant of nature that turned up in his calculations were equal to 0, in which case the transformation would be the Galilei pk transformation with the said velocity parameters. Ignatowsky's proof requires, however, two additional premises, one of which he mistakenly declared to be a corollary of the Relativity Principle. The other he failed to mention, probably because he perceived its denial as physically untenable.

The assumptions made hitherto imply that the transformation is linear. As we saw in Section 3.6, this is a consequence of the structure of the spaces and times of F and F' and the nature of the motion of F' in F.[3] Hence, if the velocities of a set of inertial frames are collinear (i.e. if they point either in the same or in opposite directions) in a given inertial frame, they are collinear also in every inertial frame. Thus if (t, x, y, z), (t', x', y', z') and (t'', x'', y'', z'') are coordinate systems adapted, respectively, to three inertial frames F, F' and F'', and such that the first system is matched with the second and the second is matched with the third, the first system is also matched with the third. Since the transformation from the unprimed to the doubly primed system is obviously the product of the transformation from the unprimed to the simply primed system and the transformation from the latter to the doubly primed one, it is clear that the transformations between matching coordinate systems adapted to inertial frames with collinear velocities constitute a group. We call this group the pk transformation group of inertial frames with collinear velocities, and denote it—until the end of this Section—by G. Evidently, Ignatowsky's claim can be immediately verified or refuted if one knows the structure of the group G. The following remarks will facilitate its study.

Consider the set of all inertial frames moving in either direction of a given straight line in an inertial frame F_0. The set can be indexed by the velocities with which each frame moves in F_0. The index set—i.e. the set of velocities with which an inertial frame can travel in a particular direction in a given such frame—is a set of real numbers which evidently includes 0. We may reasonably assume that if it contains any number $v \neq 0$, it also contains every real number between 0 and v. The index set, which we shall denote by I, is therefore a real interval, possibly identical with $\mathbf{R}$ itself; unless it is equal to $\mathbf{R} \cup \{\infty\}$—i.e. unless there is an inertial frame moving with infinite speed in F_0, a possibility we may just as well countenance for the time being to show that we are open-minded. As we noted earlier the transformation from the first to the second of two matching coordinate systems adapted respectively to F_0 and to another frame F_v in our set is uniquely determined by the velocity $v \in I$. There is therefore a bijective mapping h of I onto G, which assigns to each $v \in I$ the transformation $h(v) \in G$ determined by it. The mapping h induces a group structure in I. If $v, w \in I$, the group product $v * w$ is equal to $h^{-1}(h(w) \cdot h(v))$. We call the group $(I, *)$ the *velocity group* of collinearly moving inertial frames. It is evidently isomorphic with G and thus a realization of G in the set I of possible relative velocities between such frames. By definition $v * w$ is the velocity in F_0 of the frame of our set that moves with velocity w in F_v. Thus the group product $(v, w) \mapsto v * w$ is none other than the Rule of the Transformation of Velocities determined by the group G. Ignatowsky's claim will be fully substantiated if we can show that the group product of the velocity group agrees with the *Einstein Rule*

$$v * w = \frac{v + w}{1 + vwc^{-2}} \tag{3.7.1}$$

degenerating to the *Galilei Rule*

$$v * w = v + w \tag{3.7.2}$$

if the constant c^{-2} that occurs in the right-hand side of (3.7.1) is equal to 0.

We observe that the neutral element of the velocity group $(I, *)$ is 0. Hence (3.7.1) or (3.7.2) can only hold good if the inverse of each $v \in I$ is $-v$. This group-theoretical condition amounts to the following physical

> *Principle of Reciprocity*. If the inertial frame F' moves in the inertial frame F with velocity $\mathbf{v}$, F moves in F' with velocity $-\mathbf{v}$.[4]

Ignatowsky maintains that the Reciprocity Principle is a consequence of the Principle of Relativity. But he is wrong. The Relativity Principle implies that the velocity $\mathbf{v}'$ of F in F' is *the same function* of the velocity $\mathbf{v}$ of F' in F as $\mathbf{v}$ is of $\mathbf{v}'$. This is so if and only if each element v of the velocity group $(I, *)$ is the inverse of its inverse. But this trivial group-theoretical truth cannot by itself teach us anything about the value of v's inverse.[5] Berzi and Gorini (1969) derived the Reciprocity Principle from the following:

> *Principle of Spatial Isotropy*. The transformations of the group G commute with the permutation of $\mathbf{R}^4$ that sends (t, x, y, z) to $(t, -x, y, z)$.

The Isotropy Principle means that if a given $f \in G$ transforms a coordinate system (t, x, y, z) adapted to a frame F into the coordinate system (t', x', y', z') adapted to a frame F', *the same* $f \in G$ transforms the coordinate system obtained from the former by the mapping $x \mapsto -x$ into the coordinate system obtained from the latter by the mapping $x' \mapsto -x'$. At first sight, one may think that the Isotropy Principle follows directly from the twofold fact that the relative spaces of the frames F and F' are Euclidean and that, according to the definition of matching coordinate systems, the x-axis and the x'-axis coincide and are similarly oriented at all times. However, the reader who has learnt the lesson of Section 3.6 will promptly perceive that the Isotropy Principle has a kinematical import which cannot rest exclusively on the spatial geometries of the frames. Though each geometry remains unaffected by spatial reflection, this cannot guarantee the invariance of the kinematic linkage between the frames. After all, a space in which a direction of motion has been singled out is by this very fact kinematically anisotropic. Berzi and Gorini show that the Principle of Spatial Isotropy is equivalent to the following undeniably plausible but manifestly kinematical hypotheses:

(i) If $v \in I$, $-v \in I$.

(ii) If F and F' are two inertial frames such that F' moves with velocity $\mathbf{v}$ in F, then the following quantities depend on the magnitude, but not on the direction of $\mathbf{v}$:

 (*a*) The ratio between the length in F' of a rod at rest in that frame and the distance between two simultaneous positions of its endpoints in F.

 (*b*) The ratio between the duration in F' of a process taking place at a point of that frame and the time interval in F between its beginning and its end.

(c) The ratio between the duration in F of a process taking place at a point of that frame and the time interval in F' between its beginning and its end.[6]

Let v' denote the inverse of v in $(I, *)$. From the Isotropy Principle, Berzi and Gorini easily infer that the mapping that sends each element of I to its inverse is an odd function, i.e. that

$$(-v)' = -v' \tag{3.7.3}$$

By assuming, moreover, that I is entirely contained in $\mathbf{R}$ (i.e. that $\infty \notin I$) and that the mapping $v \mapsto v'$ is continuous, they show that in effect:[7]

$$v' = -v \tag{3.7.4}$$

The Reciprocity Principle (3.7.4) plus Berzi and Gorini's postulate that $I \subset \mathbf{R}$ are the only additional premises that we need in order to prove (3.7.1). (One may surmise that Ignatowsky did not mention the second one because he would not countenance infinite relative velocities between inertial frames.) The typical transformation of G, from a coordinate system (t, x, y, z) adapted to a frame F to a matching coordinate system (t', x', y', z') adapted to a frame F' can be written as follows:

$$\begin{aligned} t' &= a_{00}t + a_{01}x + a_{02}y + a_{03}z \\ x' &= a_{10}t + a_{11}x + a_{12}y + a_{13}z \\ y' &= a_{20}t + a_{21}x + a_{22}y + a_{23}z \\ z' &= a_{30}t + a_{31}x + a_{32}y + a_{33}z \end{aligned} \tag{3.7.5}$$

Since the only significant spatial direction is that of the (x, x')-axis—i.e. the direction of the motion—the transformation must commute with rotations about this axis. This implies that $a_{22} = a_{33}$, $a_{23} = -a_{32}$ and $a_{02} = a_{20} = a_{03} = a_{30} = a_{12} = a_{21} = a_{13} = a_{31} = 0$.[8] Since the y'-axis and the z'-axis lie at time $t = 0$ respectively on the y-axis and on the z-axis, $a_{23} = 0$. The non-zero coefficients depend only on the velocity v with which F' moves in F. We may therefore rewrite (3.7.5) as follows:[9]

$$\begin{aligned} t' &= \mu(v)t + \nu(v)x \qquad & y' &= \lambda(v)y \\ x' &= \alpha(v)t + \beta(v)x \qquad & z' &= \lambda(v)z \end{aligned} \tag{3.7.6a}$$

where μ, ν, α, β and λ are five real-valued functions on I. If F'' is a third frame moving collinearly with the other two, whose velocity in F' is w, the transformation from (t', x', y', z') to the matching system (t'', x'', y'', z'') adapted to F'' is given by:

$$\begin{aligned} t'' &= \mu(w)t' + \nu(w)x' \qquad & y'' &= \lambda(w)y' \\ x'' &= \alpha(w)t' + \beta(w)x' \qquad & z'' &= \lambda(w)z' \end{aligned} \tag{3.7.6b}$$

Let u be the velocity of F'' in F. Then $y'' = \lambda(w)y' = \lambda(w)\lambda(v)y = \lambda(u)y$. Whence $\lambda(u) = \lambda(w)\lambda(v)$. If $w = -v$, then, by the Reciprocity Principle, $u = 0$, and we have that $\lambda(-v)\lambda(v) = \lambda(0) = 1$. By the Relativity Principle, y'' is the same function of (t', x', y', z') as y' is of (t'', x'', y'', z''). Consequently,

$$\lambda(-v) = \lambda(v) = 1 \tag{3.7.7}$$

We also note that, since v is the velocity of F' in F, if $x' = 0$, $x = vt$. Hence $\alpha(v) = -v\beta(v)$. Thus, equations (3.7.6) can be written as follows:

$$\begin{aligned} t' &= \mu(v)t + \nu(v)x & y' &= y \\ x' &= \beta(v)(-vt + x) & z' &= z \end{aligned} \tag{3.7.8a}$$

$$\begin{aligned} t'' &= \mu(w)t' + \nu(w)x' & y'' &= y' \\ x'' &= \beta(w)(-wt' + x') & z'' &= z' \end{aligned} \tag{3.7.8b}$$

Likewise the transformation from the unprimed to the doubly primed system is given by:

$$\begin{aligned} t'' &= \mu(u)t + \nu(u)x & y'' &= y \\ x'' &= \beta(u)(-vt + x) & z'' &= z \end{aligned} \tag{3.7.8c}$$

Equations (3.7.8) are consistent only if

$$\begin{aligned} \begin{bmatrix} \mu(u) & \nu(u) \\ -u\beta(u) & \beta(u) \end{bmatrix} &= \begin{bmatrix} \mu(v) & \nu(v) \\ -v\beta(v) & \beta(v) \end{bmatrix} \begin{bmatrix} \mu(w) & \nu(w) \\ -w\beta(w) & \beta(w) \end{bmatrix} \\ &= \begin{bmatrix} \mu(v)\mu(w) - w\nu(v)\beta(w) & \mu(v)\nu(w) + \nu(v)\beta(w) \\ -v\beta(v)\mu(w) - w\beta(v)\beta(w) & -v\beta(v)\nu(w) + \beta(v)\beta(w) \end{bmatrix} \end{aligned} \tag{3.7.9}$$

If we set $v = -w$, then $u = 0$. Substituting these values into the lower left corner of the first and the last matrix in (3.7.9), we have that $0 = w\beta(-w)\mu(w) - w\beta(-w)\beta(w)$, whence

$$\mu(w) = \beta(w) \tag{3.7.10}$$

Substituting from (3.7.10) into the upper left corner of the matrices in (3.7.9) and equating it with the lower right corner, we verify that

$$\beta(v)\beta(w) - w\nu(v)\beta(w) = -v\beta(v)\nu(w) + \beta(v)\beta(w) \tag{3.7.11}$$

whence obviously

$$-\frac{\nu(v)}{v\beta(v)} = -\frac{\nu(w)}{w\beta(w)} = k \tag{3.7.12}$$

where k is a constant. The lower line of the matrix equation (3.7.9) now reduces to:

$$u\beta(u) = (v + w)\beta(v)\beta(w) \tag{3.7.13a}$$

$$\beta(u) = (1 + vwk)\beta(v)\beta(w) \tag{3.7.13b}$$

Dividing (3.7.13*a*) by (3.7.13*b*) we obtain:

$$u = v * w = \frac{v + w}{1 + vwk} \tag{3.7.14}$$

We see thus that k must have the dimension of the reciprocal value of a velocity squared.[10]

If $k \geq 0$, we can set $k = 1/c^2$, with c a constant with the dimension of a velocity. (3.7.14) becomes then the Einstein Rule:

$$v * w = \frac{v + w}{1 + vwc^{-2}} \tag{3.7.1}$$

degenerating to the Galilei Rule (3.7.2) if $c^{-2} = 0$. Thus the group G is the one-parameter group of Lorentz pk transformations if $k > 0$, and the one-parameter group of Galilei pk transformations if $k = 0$. In either case, the set I of admissible relative velocities of collinearly moving inertial frames is equal to the open interval $(-c, c)$. If $k = 0$, $I = \mathbf{R}$. If $k > 0$, c is finite. It is then the unattainable least upper bound of the speeds with which an inertial frame—and hence an ordinary material particle—can move relatively to another one.[11]

On the other hand, if $k < 0$, $I = \mathbf{R} \cup \{\infty\}$, as the reader can easily verify.[12] The group $(I, *)$ has then the topology of the projective line. The Rule of the Transformation of Velocities (3.7.14) can yield $u = \infty$ even when v and w are finite. It can also yield a negative u for positive v and w. Of course, k cannot be negative if $I \subset \mathbf{R}$, i.e. if no inertial frame moves with infinite speed with respect to another one—as Berzi and Gorini had to assume in order to prove the Reciprocity Principle. Moreover, as Lalan showed,[13] the hypothesis that $k < 0$ is incompatible with the following:

> *Chronology Principle.* If an event E_1 takes place before an event E_2 at a given point of an inertial frame, E_1 takes place before E_2 in every inertial frame.

Denial of this Principle would entail that a travelling mother might be able to see her sedentary son become younger and younger until he is ready to creep back into her womb, while she ages steadily at her usual pace. If we think that this is too far-fetched to be countenanced even as a remote possibility, we must abide by the Chronology Principle and conclude that $k \geq 0$.

I must finally emphasize that if k is positive, the limit speed $c = 1/\sqrt{k}$ is a constant of nature. One is free indeed to establish a coordinate system linked to a given inertial frame F, in terms of which the limit speed of inertial motions relative to F is different in different directions. But such a coordinate system would not reflect the space-time structure implied by the Relativity Principle, the Chronology Principle and the Principle of Spatial Isotropy.

3.8. The "Relativity Theory of Poincaré and Lorentz"

In his *History of the Theories of Aether and Electricity*, Sir Edmund Whittaker describes Einstein's (1905*d*) as "a paper which set forth the relativity theory of Poincaré and Lorentz with some amplifications, and which attracted much attention".[1] Though Whittaker's views on the origin of Special Relativity have been rejected by the great majority of scholars,[2] an examination of their factual grounds can further illuminate the core of Einstein's theory and bring out the true nature of its novelty. Whittaker's case rests on two sorts of evidence. There are some remarks made by Poincaré as early as 1895 or 1898, which definitely anticipate Einstein's conception of the Relativity Principle and may have suggested his criticism of simultaneity, but which do not, by themselves, constitute a physical theory. Then, there is the new theory of "electromagnetic phenomena in a system moving with any velocity less than that of light", proposed by Lorentz in 1904 and perfected by Poincaré in a paper submitted to the Circolo Matematico di Palermo in the summer of 1905.[3] This theory was designed to agree with Poincaré's view of relativity, according to which "optical phenomena depend only on the relative motions of the material bodies—sources of light and optical instruments—involved",[4] and it is, at any rate in Poincaré's streamlined version, experimentally indistinguishable from and mathematically equivalent to Einstein's electrodynamics of moving bodies. However, it gives expression to a natural philosophy very different from Einstein's, for it squarely rests on the assumption of an aether, which acts equably on all material systems that move across it, causing bodies to shrink and natural clocks to go slow by the exact amount required to mask their movement with respect to it.[5]

Poincaré was convinced that experiments can only disclose relations between ordinary material bodies.[6] Already in 1895 he had warned of the impossibility of measuring "the relative motion of ponderable matter with respect to the aether".[7] In 1900 he told the Paris International Congress of Physics that, "in spite of Lorentz", he did not believe "that more precise observations will ever reveal anything but the relative displacements of material bodies".[8] In September 1904, speaking in St Louis, he listed among the fundamental principles of classical physics—together with the two Principles of Thermodynamics, Newton's Third Law of Motion, Mass Conservation and the Principle of Least Action—

> the Principle of Relativity, according to which the laws of physical phenomena should be the same for an observer at rest or for an observer carried along in uniform movement of translation, so that we do not and cannot have any means of determining whether we actually undergo a motion of this kind.[9]

On time, Poincaré writes in Chapter VI of *La Science et l'Hypothèse*:

> There is no absolute time. [. . .] Not only have we no direct intuition of the equality of two time intervals, but we do not even have a direct intuition of the simultaneity of two events occurring at different places.[10]

Hence, simultaneity and time order must be constructed according to agreed rules. Poincaré (1898) discusses several possibilities. He observes, in particular, that astronomers who measure the speed of light

> begin by *admitting* that light has a constant speed, and, in particular, that its speed is the same in every direction. This is a postulate without which no measurement of that speed can be attempted. This postulate cannot be verified by experience, though it could be contradicted by it if the results of different measurements were not consistent. [. . .] The postulate [. . .] has been accepted by everybody. What I wish to retain is that it furnishes a new rule for the determination of simultaneity.[11]

There is no doubt that Einstein could have drawn inspiration and support for his first work on Relativity from the writings of Poincaré.

We saw in Section 2.2 how Lorentz had proposed different explanations for the null results of the first and second order experiments aimed at detecting the relative motion of the earth and the aether. At the 1900 Paris Congress Poincaré spoke disparagingly of the ease with which a hypothesis could always be found, to account for the mutual compensation of the terms of each given order. Isn't it wonderful that a particular circumstance is at work just where it is needed to avoid the detection of the earth's motion by effects of the first order, and that another, altogether different, but no less fitting circumstance, precludes its detection by effects of the second order? "No [said Poincaré]; one must give the same explanation for the former as for the latter; and everything leads us to think that this explanation will be equally valid for all terms of higher order, and that the mutual suppression of such terms will be rigorous and absolute."[12] Lorentz (1904) acknowledges that

> this course of inventing special hypotheses for each new experimental result is somewhat artificial. It would be more satisfactory if it were possible to show by means of certain fundamental assumptions and without neglecting terms of one order of magnitude or another, that many electromagnetic actions are entirely independent of the motion of the system.[13]

Lorentz proposes to do this now, for systems moving with a velocity less than that of light. He considers a material system Σ, travelling across the aether with speed $v < c$. Σ is given a Galilei coordinate system (t, x, y, z), with the positive x-axis pointing in the direction of motion, and an auxiliary coordinate system (t', x', y', z'), related to the former by the equations

$$\begin{aligned} t' &= \beta^{-1} lt - \beta l v c^{-2} x \\ x' &= \beta l x \qquad y' = ly \qquad z' = lz. \end{aligned} \tag{3.8.1}$$

Here $\beta = |(1 - v^2/c^2)^{-1/2}|$, and l is a "function of v whose value is 1 for $v = 0$, and which, for small values of v, differs from unity no more than by a quantity of the second order".[14] By a laborious argument, in the course of which several hypotheses are introduced, Lorentz is able to show that l must be constant, and hence equal to 1.[15] A short calculation will show that, if $l = 1$, the primed system is related to the Galilei coordinate system for the aether rest frame that agrees with the unprimed system at time $t = 0$, by a pk Lorentz transformation with velocity parameters $(v, 0, 0)$, such as the one given by (3.4.18). If the electric and magnetic forces transform according to the equations set forth by Lorentz at the outset, it follows that it is "impossible to detect an influence of the Earth's motion on any optical experiment, made with a terrestrial source of light, in which the geometrical distribution of light and darkness is observed" or "in which the intensities in adjacent parts of the field of view are compared".[16] However, due to a misapprehension of the transformation rules applicable to the electric charge and current densities, Lorentz's new theory does not comply perfectly with the RP, and experiments can still be conceived, involving speeds not much less than c, which, if the theory were correct, would reveal the motion of a material system relative to the aether.[17]

Lorentz's theory of 1904 continues the line of thought pursued by him in the nineties, which it rectifies slightly yet significantly, while maintaining the same basic outlook and methodology.[18] It is hard for me to understand that anyone should have ascribed to Lorentz the authorship of Special Relativity. He, at any rate, never claimed it. In his Columbia lectures of 1906 he unhesitatingly attributes "the principle of relativity" to Einstein,[19] to whom he gives credit "for making us see in the negative result of experiments like those of Michelson, Rayleigh and Brace, *not a fortuitous compensation of opposing effects*, but the manifestation of a general and fundamental principle".[20] Yet he feels that something must be said also for his own approach, for he "cannot but regard the ether, which can be the seat of an electromagnetic field with its energy and vibrations, as endowed with a certain degree of substantiality, however different it may be from all ordinary matter". From this point of view, it seems preferable not to assume, right from the beginning, that "it can never make any difference whether a body moves through the ether or not", and to choose rods and clocks at rest in the aether as the standards of length and time.[21] In a note added to the 1916 edition of the same book he praises "Einstein's theory of relativity" for the simplicity of its treatment of electromagnetic phenomena in moving systems, a simplicity which he, Lorentz, had not been able to achieve. "The chief cause of my failure [he adds] was my clinging to the idea that the variable t only can be considered as the true time and that my local time t' must be regarded as no more than an auxiliary mathematical quantity."[22]

In opposition to Lorentz's cherished opinion, Poincaré (1906) does assume from the outset that "the Relativity Postulate (*Postulat de Relativité*)"—according to which "the absolute motion of the Earth, or rather its motion relative to the aether instead of relative to the other celestial bodies" cannot be demonstrated by experiment—holds good "without restriction". Even if experience, which so far has corroborated the postulate, may well run counter to it in the future, "it is, in any case, of interest to see what consequences follow from it".[23] Poincaré points out that, while the Fitzgerald–Lorentz contraction hypothesis accounts for the null result of Michelson's experiment, "it would nevertheless be inadequate if the relativity postulate were valid in its most general form". However, Lorentz (1904) has succeeded in extending and modifying that hypothesis "so as to bring it into perfect agreement" with the Relativity Postulate.[24] Poincaré's own dynamics of the electron agrees with Lorentz's theory "in all important points", though it "modifies them and completes them in certain details".[25] This preamble suggests that Poincaré in any case embraced the main tenets of Lorentz's natural philosophy; namely, that the aether exists and is the seat of the fields of force governed by the Maxwell equations, that uniform motion across the aether subjects material bodies and systems to strains and stresses that change their shape and delay their internal processes, that real lengths and times can therefore be measured only by rods and clocks at rest in the aether. This suggestion is borne out by several passages in the text,[26] which, on the other hand, bears no trace of Poincaré's having ever countenanced a revision of the fundamental concepts of classical kinematics. Poincaré's electrodynamics of moving bodies definitely does not rest, like Einstein's on a modification of the notions of space and time. For this reason, it does not attain to Special Relativity's universal scope. And yet, in his (1906), Poincaré introduced several ideas of great importance for a correct grasp of the geometrical import of Einstein's theory. He chose the units of length and time so as to make $c = 1$, a practice that throws much light on the symmetries of relativistic spacetime.[27] He proved that Lorentz transformations form a Lie group and investigated its Lie algebra.[28] He characterized the (homogeneous) Lorentz group as the linear group of transformations of $\mathbf{R}^4$ which leave the quadratic form $x^2 + y^2 + z^2 - t^2$ invariant.[29] He further noted that, if one substitutes the complex valued function it for t, so that $(it_\alpha, x_\alpha, y_\alpha, z_\alpha)$ are the coordinates of three points $P_\alpha (\alpha = 1, 2, 3)$ "in four-dimensional space, we see that the [homogeneous] Lorentz transformation is simply a rotation of this space about a fixed origin".[30] He discovered that some physically significant scalars and vectors—e.g. electric charge and current density—can be combined into Lorentz invariant four-component entities (subsequently called "four-vectors"), thus paving the way for the now familiar four-dimensional formulation of Relativity physics.[31]

Perhaps one ought to take Poincaré's acceptance of the aether with a grain of salt. He had written in *La Science et l'Hipothèse*:

> We do not care whether the aether really exists; that is a question for metaphysicians to deal with. For us the essential thing is that everything happens as if it existed and that this hypothesis is convenient [*commode*] for the explanation of phenomena. After all, have we any better reason to believe in the existence of material objects? That again is merely a convenient hypothesis, though one which will never cease to be so, while a day will doubtless come when the aether will be discarded as useless.[32]

His nonchalance on the matter of physical existence may well have moved him to speak in 1906 "as if" Lorentz's aether were real, given that Lorentz's formulae yielded satisfactory predictions. Anyhow, it is surprising that Poincaré, who was so well equipped to discover the chronogeometric key to Special Relativity, should have failed to do so. A. I. Miller blames Poincaré's inductivist philosophy, which led him to treat the Relativity Principle as a contingent generalization from experience, to be ultimately explained by the laws of electrodynamics.[33] But even such a staunch anti-inductivist as Einstein would—one hopes—have rejected the RP, had experience been notoriously hard to reconcile with it. And although the use of the RP as an axiom in Einstein (1905*d*) precludes its derivation from deeper principles, one cannot say that in the more mature formulations of Special Relativity it is left unexplained, for, as we shall see in Chapter 4, it is a natural and obvious consequence of the structure of Minkowski spacetime. I am inclined therefore to attribute Poincaré's failure to another aspect of his philosophy, namely, his conventionalism.[34] Persuaded as he was that physical geometry and chronometry are freely chosen for reasons of convenience and are wholly indifferent to the true nature of things, he could not expect to gain any new knowledge by tampering with Newtonian time or Euclidean space. This particular attitude to the theory of space and time can also help explain why Poincaré did not greet Einstein's paper as the splendid achievement that it was, a decisive breakthrough in the articulation of insights that Poincaré himself had been the first to suggest.[35] Though he recommended Einstein for a chair at the Swiss Federal Polytechnic as "one of the most original thinkers I have ever met",[36] there is no evidence that he ever said a word in praise of Einstein's work on Relativity, and he insistently referred to the RP as "le principe de Relativité de Lorentz" years after Lorentz himself had called it, in the 1908 preface to his Columbia lectures, "Einstein's Principle of Relativity".[37] However, there could be another, at least partial explanation for Poincaré's silence: it may not have been easy, even for a man of his stature, to own that he had lost the glory of founding 20th-century physics to a young Swiss patent clerk.

CHAPTER 4

Minkowski Spacetime

4.1 The Geometry of the Lorentz Group

Hermann Minkowski, professor of mathematics at Göttingen, made several important contributions to number theory and other branches of pure mathematics.[1] However, he is chiefly remembered for having articulated the new theory of space and time that is the core of Special Relativity.[2] Minkowski conducted with Hilbert, Wiechert and Herglotz the 1905 Göttingen seminar on electron theory, based essentially on the work of Lorentz.[3] On November 5th, 1907, he gave a lecture at the Göttingen Mathematical Society about "The Relativity Principle".[4] He introduced the subject as "a complete change in our ideas of space and time", prompted by the electromagnetic theory of light.[5] Mathematicians, he noted, are particularly well prepared to understand the new ideas, which merely specify concepts they have been long familiar with, but "the physicists must now to some extent invent these concepts anew, laboriously carving a path for themselves across a jungle of obscurities, while very close by the mathematicians' highway, excellently laid out long ago, comfortably leads onwards."[6] A month and a half later he submitted to the Göttingen academy a long paper on "The fundamental equations of electromagnetic processes in moving bodies", which gives evidence of the clarity and beauty that accrue to classical electrodynamics through the new ideas of time and space.[7] Finally, on September 21st, 1908, he displayed the new natural philosophy in all its glory before the 80th Congress of German Natural Scientists and Physicians in his lecture, "Space and Time".[8]

Minkowski put his reading of Relativity in a nutshell thus: "The world in space and time is in a certain sense a four-dimensional non-Euclidean manifold."[9] On page 22 we observed that, by their handling of space and time coordinates, the founding fathers of mathematical physics had in effect conceived the changing natural world as a 4-manifold. But Special Relativity goes still further, for, according to Minkowski, it implicitly endows the cosmic 4-manifold with a geometric structure, no less rich than, though distinct from, that of Euclidean 4-space. This is the structure that Minkowski, in the above text, calls non-Euclidean and that I shall now describe.

Before tackling this task I must introduce three conventions which we shall follow from now on, unless otherwise noted: (i) the units of length and time are to be so chosen that $c = 1$; (ii) in each term of a polynomial expression, summation is implied over the entire range of every pair of repeated indices, Latin indices ranging from 0 to 3 and Greek indices from 1 to 3 (*Einstein summation convention*); (iii) η_{ij} is equal to 0 if $i \neq j$, to 1 if $i = j = 0$, and to -1 otherwise. Hence, by (ii) and (iii),

$$\eta_{ij}x^i x^j = (x^0)^2 - (x^1)^2 - (x^2)^2 - (x^3)^2 \tag{4.1.1}$$

We shall continue to regard Lorentz coordinate transformations as permutations of $\mathbf{R}^4$. Consequently a Lorentz chart is a bijective mapping of its domain onto $\mathbf{R}^4$, which assigns a real number quadruple to each "point at an instant", i.e. to each location available for instantaneous punctual events. "Points at an instant" are called "worldpoints" by Minkowski.[10] Let x be a Lorentz chart meeting the conditions stated in the last sentence of Section 3.4 (p. 66), and let $A_{\mathscr{L}}$ denote the set of all Lorentz charts related to x by a transformation of the proper orthochronous Lorentz group $\mathscr{L}$. Since all such transformations are C^∞-differentiable, $A_{\mathscr{L}}$ bestows on the set of worldpoints the structure of a 4-manifold (cf. p. 258). Minkowski calls this manifold, "the world", but to avoid misunderstandings I shall call it *Minkowski's world*. Minkowski's world has the same differentiable structure as the 4-manifold $\mathscr{M}$ underlying our construction of a Newtonian spacetime in Section 1.6, and will therefore be designated by $\mathscr{M}$.

If $a = (a^0, a^1, a^2, a^3)$ ranges over real number quadruples, the real valued function $f_1 : a \mapsto \eta_{ij}a^i a^j$ is a quadratic form on the vector space $\mathbf{R}^4$.[11] If a is mapped on b by a transformation of the homogeneous group $\mathscr{L}_0$, then $f_1(a) = f_1(b)$. f_1 is therefore $\mathscr{L}_0$-invariant.[12] Indeed, the full homogeneous Lorentz group $L(4)$—i.e. the group generated by $\mathscr{L}_0$, plus time reversal and parity—is the largest group of permutations of $\mathbf{R}^4$ that preserves f_1.[13] Since translations preserve coordinate differences, the function $f_2 : \mathbf{R}^4 \times \mathbf{R}^4 \to \mathbf{R}$: $(a, b) \mapsto \eta_{ij}(a^i - b^i)(a^j - b^j)$ is $\mathscr{L}$-invariant. We call f_2 the Minkowski *interval*. The Minkowski interval determines a geometry on $\mathbf{R}^4$ much in the same way that Euclidean distance determines Euclidean geometry. The invariance of $\eta_{ij}a^i a^j$ under the transformations of the homogeneous Lorentz group $L(4)$ captures the essence of the former, just as the invariance of the Pythagorean quadratic form $\delta_{\alpha\beta}a^\alpha a^\beta$ under the orthogonal group $O(3)$ captures that of the latter. The obvious analogy between both characterizations indicates that these are geometries of comparable richness. The fact that $\delta_{\alpha\beta}a^\alpha a^\beta$ is always non-negative, and zero only if $a = 0$, while $\eta_{ij}a^i a^j$ can be positive, negative or even zero for non-zero a, implies that the geometry determined by the Lorentz group is in effect non-Euclidean.

The action of $\mathscr{L}$ on $\mathbf{R}^4$ as a group of coordinate transformations is not too hard to visualize. A proper orthochronous Lorentz transformation is the product of a translation τ and a homogeneous transformation $\sigma \in \mathscr{L}_0$. The effect of τ is very simple: τ maps each real number quadruple (t, x, y, z) on the quadruple $(t+a, x+b, y+c, z+d)$, where (a, b, c, d) is a fixed element of $\mathbf{R}^4$, characteristic of τ. As a result of this, every "point" in $\mathbf{R}^4$ is "transported" a fixed (Euclidean) distance in a fixed direction. The homogeneous transformation σ is, as we know, equal to a pk transformation φ, of the form (3.4.18), preceded and followed by rotations of the form (3.4.1), with $k_i = 0$ and $\det[a_{\alpha\beta}] = 1$. Let ρ denote any of these rotations. ρ maps the linear subspace $\{(t, 0, 0, 0)|t \in \mathbf{R}\}$ identically onto itself. On each hyperplane $\{(k, x, y, z)|x, y, z \in \mathbf{R}; k = \text{const.}\}$, ρ acts like a rotation through a fixed angle θ, characteristic of ρ, about an axis containing the points $(k, 0, 0, 0)$ and (k, r_1, r_2, r_3), where $\mathbf{r} = (r_1, r_2, r_3)$ is an eigenvector of the matrix $(a_{\alpha\beta})$ (in (3.4.1)).[14] It only remains to visualize the effect of the pk transformation φ, with velocity parameters $(v, 0, 0)$. In view of (3.4.18), it is clear that on the plane $\{(0, 0, y, z)|y, z \in \mathbf{R}\}$ φ is the identity, whereas its restriction to each plane $\{(t, x, k_1, k_2)|t, x \in \mathbf{R}; k_1, k_2 \; const.\}$ is a permutation of this plane. φ also maps each hyperboloid $t^2 - x^2 - y^2 - z^2 = const.$ onto itself. Let us consider the effect of φ on the plane $\{(t, x, 0, 0)|t, x \in \mathbf{R}\}$. This plane intersects each of the aforesaid hyperboloids on the hyperbola $t^2 - x^2 = const.$, which φ maps onto itself. φ rotates the axes $T = \{(t, 0, 0, 0)| \; t \in \mathbf{R}\}$ and $X = \{(0, x, 0, 0)| \; x \in \mathbf{R}\}$ about the origin $(0, 0, 0, 0)$ towards one of the two asymptotes $t^2 - x^2 = 0$, through the angles $-\theta$ and θ, respectively, where $\theta = \arctan v$. Since $v < 1$, $-\pi/4 < \theta < \pi/4$. Since φ is linear, the entire hyperplane $H = \{(0, x, y, z)|x, y, z \in \mathbf{R}\}$ is rotated by φ about the fixed plane $\{(0, 0, y, z)| \; y, z \in \mathbf{R}\}$. Let $\langle \; , \; \rangle$ denote the inner product on $\mathbf{R}^4$ obtained by polarizing the quadratic form f_1 (see (4.2.1)). If $t = (t, 0, 0, 0)$ and $h = (0, x, y, z)$ are arbitrary points of T and H, evidently $\langle t, h \rangle = 0$. The $\mathscr{L}_0$-invariance of f_1 implies that $\langle \varphi(t), \varphi(h) \rangle = 0$, as well. We observe finally that, as a linear permutation of $\mathbf{R}^4$, φ preserves parallelism (in the Euclidean sense).

Let $x \in A_{\mathscr{L}}$. We define the *interval* $F(P, Q)$ between two points P and Q of Minkowski's world $\mathscr{M}$ by

$$\begin{aligned} F(P, Q) &= f_2(x(P), x(Q)) \\ &= \eta_{ij}(x^i(P) - x^i(Q))(x^j(P) - x^j(Q)) \end{aligned} \tag{4.1.2}$$

Clearly, F does not depend on the choice of x and is invariant under the action of $\mathscr{L}$ on $\mathscr{M}$ as a group of Lorentz point transformations.[15] By this action, $\mathscr{M}$ is endowed with a geometry that is fully characterized by the $\mathscr{L}$-invariance of the interval function F and is structurally identical with the geometry defined on $\mathbf{R}^4$ by the action of $\mathscr{L}$ as a group of coordinate transformations. *Minkowski spacetime* is Minkowski's world, with this geometry. The comparative ease with which it can be defined should not blind us to its physical significance, which is solidly grounded on the physical hypotheses that go into the definition of Lorentz charts. If, as it turns out, those hypotheses are true only locally and approximately, the physical significance of Minkowski spacetime is correspondingly limited, but by no means suppressed.

4.2 Minkowski Spacetime as an Affine Metric Space and as a Riemannian Manifold

The action of the Lorentz group $\mathscr{L}$ on $\mathbf{R}^4$ illuminates the structure of Minkowski spacetime just as the action of the Galilei group $\mathscr{G}$ on $\mathbf{R}^4$ disclosed us that of Newtonian spacetime (p. 27). Here, however, we shall approach Minkowski spacetime from a different, or rather from two different points of view, that will be useful in the sequel. In the first place, we shall consider it as an affine metric space. This approach is both elegant and natural, and gives an intuitive grasp of its geometry. In the second place, we shall conceive it as a Riemannian manifold. This view of it may seem unnecessarily complex, but it underlies Minkowski's formulation of relativistic mechanics and electrodynamics, and the subsequent development of General Relativity.

An *affine space* consists of a point set, a vector space, and a transitive and effective action of the additive group of the latter on the former. Let H be a mapping of $\mathbf{R}^4 \times \mathscr{M}$ into $\mathscr{M}$ (where $\mathscr{M}$ designates just the set of worldpoints, and we forget its differentiable structure for the time being). Assume that (i) for any $a, b \in \mathbf{R}^4$ and any $P \in \mathscr{M}$, $H(a+b, P) = H(a, H(b, P))$; (ii) for any $P, Q \in \mathscr{M}$ there is some $a \in \mathbf{R}^4$ such that $H(a, P) = Q$; (iii) $H(a, P) = 0$ for every $P \in \mathscr{M}$ if and only if $a = 0$. $\langle \mathscr{M}, \mathbf{R}^4, H \rangle$ is then an affine space. We shall now characterize some subsets of $\mathscr{M}$ which we call *flats*. P and Q denote points of $\mathscr{M}$; u, v and w are linearly independent vectors in $\mathbf{R}^4$; instead of $H(u, P)$, we write uP. The 1-flat or *straight* through P generated by u (under the action H) is the set of all points Q in $\mathscr{M}$ such that, for some real number k, $Q = kuP$. The 2-flat or *plane* through P generated by u and v is the union of all straights through P generated by $k_1 u + k_2 v$, for any real numbers k_1 and k_2. The 3-flat or *hyperplane* through P generated by u, v and w is the union of all straights through P generated by $k_1 u + k_2 v + k_3 w$, for any real numbers k_α $(\alpha = 1, 2, 3)$. Let us say that two n-flats are *parallel* if they are generated by the same set of n linearly independent vectors. The reader should satisfy himself that parallel n-flats either coincide or have no points in common. Hence, the choice of n linearly independent vectors in $\mathbf{R}^4$ $(n < 4)$ induces a partition of $\mathscr{M}$ into parallel n-flats. It is clear, moreover, that if λ is a straight and Q a point of $\mathscr{M}$ outside λ there is one and only one straight through Q parallel to λ, namely, that generated by the same vector that generates λ. $\langle \mathscr{M}, \mathbf{R}^4, H \rangle$ therefore satisfies, like Newtonian spacetime, the Euclidean parallel postulate.

An affine space is an *affine metric space* if an inner product is defined on the vector space. In our case, we define the *Minkowski inner product* on $\mathbf{R}^4$ by polarizing the quadratic form f_1 of Section 4.1.[1] If v and w are two vectors of $\mathbf{R}^4$ their inner product is therefore

$$\langle v, w \rangle = \eta_{ij} v^i w^j \tag{4.2.1}$$

If $\langle v, w \rangle = 0$, v and w are *orthogonal*. Two concurrent straights are orthogonal if they are generated by orthogonal vectors. A straight is orthogonal to a plane (or hyperplane) it meets if the vector that generates the former is orthogonal to those that generate the latter. Since our affine metric space includes some self-orthogonal straights, it is not metrically Euclidean.

We shall now define the *interval* $|P - Q|$ between two points P and Q in $\mathcal{M}$. Since H is transitive and effective, there is one and only one vector v in $\mathbf{R}^4$ such that $Q = vP$. We set

$$|P - vP| = \langle v, v \rangle \tag{4.2.2}$$

The vector v and the interval $|P - vP|$ are said to be *timelike*, *spacelike*, or *null* if $\langle v, v \rangle$ is, respectively, positive, negative, or zero. A null or timelike vector v is said to be *future-directed* if its first component v^0 is greater than zero. The set of null vectors is a hypercone in $\mathbf{R}^4$, which we call the *null cone*.[2] The set of worldpoints Q such that $Q = vP$ for a fixed $P \in \mathcal{M}$ and any null vector v is called the *null cone* at P.

The affine metric space $\langle \mathcal{M}, \mathbf{R}^4, H, f_1 \rangle$ is essentially equal to Minkowski spacetime, as defined in Section 4.1, if the action H meets the physical conditions we shall now state. Let P and Q be two worldpoints such that $Q = vP$. Then, (*a*) if v is a future-directed null vector, P could be the emission and Q the reception of a light signal travelling *in vacuo* between these two events; (*b*) if v is a timelike vector such that $\langle v, v \rangle = 1$, P and Q could be two ticks, separated by a unit of time, of a natural clock at rest in an inertial frame; (*c*) if $\langle v, v \rangle = 1$ and $R = uP = wQ$, where u is null and v and w are orthogonal, there could be an unstressed rod of unit length at rest in an inertial frame, such that P and Q occur at one of its endpoints while R occurs at the other. Any mapping of $\mathbf{R}^4 \times \mathcal{M}$ into $\mathcal{M}$ which satisfies the three mathematical conditions (i)–(iii) on page 91 and the three physical conditions (*a*)–(*c*) listed above will be called an *admissible action*. If H is an admissible action, the mapping $v \mapsto vP$ of $\mathbf{R}^4$ onto $\mathcal{M}$ is the inverse of a Lorentz chart with its origin at P.[3] (Thus $\mathcal{M}$ recovers through an admissible action the differentiable structure that we chose to forget on page 91). If t designates the vector $(1, 0, 0, 0)$, the parallel straights generated by t under the action H agree with the worldlines of particles at rest on a particular inertial frame F (cf. page 29). Let φ be any proper homogeneous Lorentz coordinate transformation. The mapping H_φ defined, for each $v \in \mathbf{R}^4$ and $P \in \mathcal{M}$, by

$$H_\varphi(v, P) = H(\varphi(v), P) \tag{4.2.3}$$

is also an admissible action. (H_φ fulfils the mathematical conditions, for φ is linear; it also fulfils the physical conditions, for φ preserves the inner product on $\mathbf{R}^4$). Again, the parallel straights generated by t under H_φ agree with the worldlines of particles at rest on an inertial frame F_φ. We observe that F_φ differs from F if and only if φ is a pk Lorentz transformation. On the other hand, if G is any inertial frame, there is certainly a $\varphi \in \mathcal{L}_0$ such that the parallel straights generated by $\varphi(t)$ under H, and hence by t under H_φ, agree with the worldlines of particles at rest in G. The mapping $H_\varphi \mapsto F_\varphi$ from admissible actions to inertial frames is therefore surjective. Let $\mathscr{H}$ be the set of admissible actions. The mapping $L: \mathscr{L}_0 \times \mathscr{H} \to \mathscr{H}$ defined by $L(\varphi, H) = H_\varphi$ is an action of $\mathscr{L}_0$ on $\mathscr{H}$.[4] Since L is transitive, under it all elements of $\mathscr{H}$ are in a sense equivalent: Under this action of $\mathscr{L}_0$, $\mathscr{H}$ is a homogeneous space (p. 72). If the stage of physical events actually has the structure of Minkowski spacetime, the equivalence of all inertial frames, proclaimed by the Relativity Principle, follows at once from the equivalence of all points in the homogeneous space $\langle \mathscr{H}, \mathscr{L}_0 \rangle$. The arbitrary choice of a particular action $H \in \mathscr{H}$ for our definition of Minkowski spacetime as an affine metric space mirrors the choice of the "stationary" frame and the directions of three orthogonal spatial axes in Einstein's original presentation of Special Relativity.

Before introducing our second view of Minkowski spacetime let us recall the main features of Riemannian manifolds. To each point P in an n-manifold S there is attached a vector space, the tangent space S_P (p. 260). A *Riemannian metric* $\boldsymbol{\mu}$ on S is a rule that assigns to each $P \in S$ an inner product $\boldsymbol{\mu}_P$ on S_P and satisfies the following requirement of smoothness: if X and Y are any two vector fields on S, and X_P, Y_P are their respective values at P, the mapping $P \mapsto \boldsymbol{\mu}_P(X_P, Y_P)$ is a scalar field, hereafter denoted by $\boldsymbol{\mu}(X,Y)$ (see p. 258). The pair $\langle S, \boldsymbol{\mu} \rangle$ is a *Riemannian n-manifold*. Since $\boldsymbol{\mu}_P$ is a symmetric bilinear function, it depends entirely on its value at the $(n/2)(n+1)$ distinct (unordered) pairs of elements of a basis of S_P. Due to the smoothness of $\boldsymbol{\mu}$, the signature of $\boldsymbol{\mu}_P$ (defined in note 1) cannot change as P ranges over any connected component of S. In particular, if S itself is connected (as indeed it must be if it is a spacetime manifold) $\boldsymbol{\mu}_P$ has the same signature for all $P \in S$, and we may therefore speak of *the signature* $\sigma(\boldsymbol{\mu})$ of the Riemannian metric $\boldsymbol{\mu}$ on S. $\boldsymbol{\mu}$ is *positive definite* if $\sigma(\boldsymbol{\mu}) = n$, *negative definite* if $\sigma(\boldsymbol{\mu}) = -n$. Otherwise $-n < \sigma(\boldsymbol{\mu}) < n$, and $\boldsymbol{\mu}$ is said to be *indefinite*. An indefinite Riemannian metric of signature $2-n$ (on an n-manifold) is called a *Lorentz metric* (cf. page 280).

An n-manifold S is *parallelizable* if there exists a family of n vector fields $V^1, \ldots, V^n$, defined on all S, such that their values at each $P \in S$ form a basis of the tangent space S_P. If $U \subset S$ is the domain of a chart u of S, U is always a parallelizable manifold, for u defines n vector fields $\partial/\partial u^q$ on U $(1 \leq q \leq n)$, such that the value

$$\left.\frac{\partial}{\partial u^q}\right|_P \text{ of } \frac{\partial}{\partial u^q}$$

at a point $P \in U$ is the tangent vector to the q-th parametric line of u through P, and the set

$$\left(\left.\frac{\partial}{\partial u^1}\right|_P, \ldots, \left.\frac{\partial}{\partial u^n}\right|_P\right)$$

is a basis of S_P (p. 259). Hence, a Riemannian metric $\boldsymbol{\mu}$ on S is completely determined on the domain of the chart u by the $(n/2)(n+1)$ scalar fields $\boldsymbol{\mu}(\partial/\partial u^q, \partial/\partial u^r)$ generally known as the *components* of $\boldsymbol{\mu}$ relative to u $(1 \leq q \leq r \leq n)$. Obviously S itself is parallelizable if it admits a global chart. Let $\boldsymbol{\mu}$ be a Riemannian metric on the n-manifold S. If $P \in S$ and v and w belong to S_P, we write $\langle v, w \rangle_P$ for $\boldsymbol{\mu}_P(v, w)$; v and w are orthogonal if $\langle v, w \rangle_P = 0$. We also write $E(v)$ for $\langle v, v \rangle_P$. $E(v)$ is called the "*energy*" of the vector v.[5] With a view to relativistic applications we say that v is *timelike* if $E(v) > 0$, *spacelike* if $E(v) < 0$, and *null* if $E(v) = 0$. (A non-zero null vector is sometimes called *lightlike*.) On the analogy of Euclidean geometry we define the *length* $|v|$ of a vector v as $\sqrt{|E(v)|}$. A *curve* γ is a smooth or piecewise smooth mapping of an interval $I \subset \mathbf{R}$ into S (p. 259). γ is said to *join* any pair of points in its range. As on p. 261, the tangent to γ at the point $\gamma(t) = P(t \in I)$ will be denoted by γ_{*t} or by $\dot{\gamma}_P$. γ is *spacelike* (*timelike*, *null*) if, for all $t \in I$ such that $\gamma_{*t} \neq 0$, γ_{*t} is spacelike (timelike, null). Two curves γ and γ' whose ranges intersect at a point P are orthogonal at P if they have mutually orthogonal non-zero tangent vectors P. This definition can be easily extended to surfaces, etc. If I is the closed interval $[p, q]$, $\gamma(p)$ and $\gamma(q)$ are the *endpoints* of the curve γ, which is said to go from the former to the latter. The integral

$$E(\gamma) = \int_p^q E(\gamma_{*t})\,\mathrm{d}t \tag{4.2.4}$$

is called the "*energy*" of γ. γ is said to be *energy-critical* if it is a critical point of the function E as it ranges over all parametrized curves defined on the same closed interval and having the same endpoints as γ.[6] If I is an arbitrary real interval, the curve $\gamma: I \to S$ is said to be *Riemannian geodesic* of $\langle S, \boldsymbol{\mu} \rangle$ if it is energy-critical on every closed subinterval of I. If $\gamma: [p, q] \to S$ is spacelike or timelike we define the *length* of γ as:

$$L(\gamma) = \int_p^q |\gamma_{*t}|\,\mathrm{d}t \tag{4.2.5}$$

Note that this concept of length only allows for comparison between curves of the same type (spacelike or timelike), and does not apply to null curves. The main significance of the length $L(\gamma)$ lies in the fact that it is invariant under reparametrizations; in other words, if f is any homeomorphism of $\mathbf{R}$ onto itself, $L(\gamma \cdot f) = L(\gamma)$. The length of a parametrized curve may therefore be regarded as a property of the *path* that is its range. If $|\gamma_{*t}|$ is constant, equal to 1, the length of γ's domain (in $\mathbf{R}$) and each of its subintervals is equal to the length of the corresponding image (in S), and γ is naturally said to be "parametrized by arc length". If the metric $\boldsymbol{\mu}$ is either positive or negative definite, $\langle S, \boldsymbol{\mu} \rangle$ is said to be a *proper* Riemannian manifold. In such a manifold all curves are timelike (if $\boldsymbol{\mu}$ is positive definite) or spacelike (if $\boldsymbol{\mu}$ is negative definite) and one may speak of a *length-critical* curve, viz. a curve that is a critical point of the function L on the set of all curves defined on a given interval and sharing a given pair of endpoints. It can be shown that, if γ is a Riemannian geodesic—as defined above—in a proper Riemannian manifold, γ is parametrized by arc length and is length-critical on every closed subinterval of its domain.[7] If $\langle S, \boldsymbol{\mu} \rangle$ is a proper Riemannian manifold, a distance function $d: S \times S \to \mathbf{R}$ can be defined as follows: for any $P, Q \in S, d(P, Q)$ is the infimum or greatest lower bound of the lengths of all the curves that go from P to Q. $\langle S, d \rangle$ is a metric space in the usual sense.[8]

Now we return to the consideration of Minkowski spacetime. Let $\mathcal{M}$ be Minkowski's world, that is, the set of worldpoints, with the differentiable structure defined by the collection of $A_{\mathcal{L}}$ of Lorentz charts. Since any chart $x \in A_{\mathcal{L}}$ is global, $\mathcal{M}$ is parallelizable and a Riemannian metric on $\mathcal{M}$ can be defined by specifying its values on the ten several pairs of vector fields $\partial/\partial x^i$, whose integral curves are the parametric lines of x. In particular, the Minkowski metric $\boldsymbol{\eta}$ is defined by:

$$\boldsymbol{\eta}(\partial/\partial x^i, \partial/\partial x^j) = \eta_{ij} \tag{4.2.6}$$

The η_{ij} have here the values given under convention (iii) on page 89, and must be regarded as constant scalar fields on $\mathcal{M}$. The signature of $\boldsymbol{\eta}$, $\Sigma_i \eta_{ii}$, is equal to -2. The Riemannian 4-manifold $\langle \mathcal{M}, \boldsymbol{\eta} \rangle$ is *Minkowski spacetime*. The name is justified insofar as the Riemannian structure $\langle \mathcal{M}, \boldsymbol{\eta} \rangle$ is determined by the affine metric structure $\langle \mathcal{M}, \mathbf{R}^4, H, f_1 \rangle$ of pp. 91–93, and vice versa. We saw already on p. 92 that the admissible action H defines Lorentz charts on $\mathcal{M}$ whereby it bestows on the latter the same differentiable structure as $A_{\mathcal{L}}$. Let x be one of the Lorentz charts defined by H. The metric $\boldsymbol{\eta}$ on $\mathcal{M}$ is then determined by the simple requirement that, for any pair of tangent vectors v and w, at a worldpoint P, $\boldsymbol{\eta}_P(v, w) = f_1(x_{*P}v, x_{*P}w)$, where x_* is the differential of x (p. 261). Due to the Lorentz invariance of f_1, if the requirement is fulfilled for one such chart x, it is fulfilled for all. On the other hand, if $\langle \mathcal{M}, \boldsymbol{\eta} \rangle$ is given and x is any Lorentz chart, the mapping $H_x: \mathbf{R}^4 \times \mathcal{M} \to \mathcal{M}$, defined by $H_x(r, P) = x^{-1}(x(P) + r)$, is an admissible action in the sense of page 92, as the reader can verify for himself. If y is another Lorentz chart such that $y \cdot x^{-1}$ is a translation, $H_y = H_x$, but if $y \cdot x^{-1} \in \mathcal{L}_0$, H_y is another admissible action related to H_x as H_φ is related to H by equation (4.2.3). The Riemannian and the affine metric structures that characterize Minkowski spacetime do not only determine one another, but are also mutually compatible in the following sense: A curve γ in $\mathcal{M}$ is a geodesic of the said Riemannian structure only if its range lies entirely on a straight of the affine metric structure.[9] If the geodesic γ, with endpoints P and Q, is spacelike or timelike and parametrized by arc length, or if γ is null, the interval $|P - Q|$ defined on p. 92 is equal to the "energy" $E(\gamma)$. If γ is null, P and Q could be, respectively, the emission and reception of a light signal, sent across empty space. If γ is spacelike, there is an inertial frame F in which P and Q are simultaneous, and the length $L(\gamma)$ is the Euclidean distance between their places of occurrence in F. If γ is timelike there is an inertial frame F in which P and Q occur at the same place, and $L(\gamma)$ is the time elapsed between them in F. If π is a massive particle, arbitrarily moving in an inertial frame F, there is a curve γ in $\mathcal{M}$ parametrized by proper time, whose range is the locus of events at π. We call both γ and its range, the *worldline* of π (cf. p. 23). Since π must move in F with a velocity less than that of light (p. 71), γ is timelike. At each instant t the worldline of π is tangent to a timelike geodesic γ^t, which is the worldline of a point at rest in some inertial frame F_t. At the instant t, π moves like F_t. We therefore call F_t the *momentary inertial rest frame* (mirf) of π at that instant. If, but only if, π is a free particle, F_t does not change with t, and the worldline of π is a timelike geodesic.

If $\gamma:[p,q] \to \mathcal{M}$ is a timelike curve, the integral

$$\tau(\gamma) = \int_p^q |\gamma_{*t}|\,dt \tag{4.2.7}$$

is called the *proper time* from event $\gamma(p)$ to event $\gamma(q)$, along the worldline γ. This concept introduced by Minkowski is probably his most important contribution to physics.[10] We have observed already that any time-keeping device or clock keeps time at a place only (e.g., at 0° Lat., 0° Long. on a spherical spinning Newtonian clock, or at one of the endpoints of a Langevin clock, etc.). The worldline of this place is the locus of successive events which mark and measure the time. Since clocks are massive material systems, the said worldline is the range of some timelike curve γ. We now identify the clock with a particle π describing this worldline and we postulate that, if it is accurate, its rate agrees at each instant with that of a Langevin clock at rest on π's mirf. This postulate—sometimes called the *clock hypothesis*[11]—may be viewed as a conventional definition of what we mean by clock accuracy, and hence by physical time. But Special Relativity would doubtless have been rejected or, at any rate, deeply modified if the clock hypothesis were not fulfilled—to a satisfactory approximation—by the timepieces actually used in physical laboratories. The clock hypothesis implies that the time measured by our clock between any two events P and Q is none other than the proper time along the clock's worldline from P to Q. Relativistic clocks are hodometres of timelike worldlines. Had more attention been paid to this fact, much of the effort spent on the so-called Clock Paradox would have been spared. Since the integrand in (4.2.7) is not an exact differential, the proper time must depend on the integration path. Hence, it is not surprising that two accurate clocks which coincide at two events but travel differently between them agree on their first meeting but disagree on the second one (see pp. 68 f.).

The relativistic effects discussed in Section 3.5, pages 67 to 69, can also be explained geometrically, as a consequence of the structure of

Minkowski spacetime. To show it we shall examine the image in $\mathbf{R}^4$ of a neighbourhood of the origin of a Lorentz chart x, adapted to a frame X. Figure 1 is a diagram of the (x^0, x^1)-plane. Vertical lines are the images of the worldlines of points of X. Horizontal lines join the images of events simultaneous in X. The null geodesics described in spacetime by light signals through the origin are mapped by x onto the diagonal lines ω_1 and ω_2. Consider now a Lorentz chart y adapted to a frame Y and related to x by equations (3.4.18). The worldline of the spatial origin of y is mapped by x onto the line μ. All events (on the plane considered) that are simultaneous in Y with the common spatio-temporal origin of both charts are mapped by x on the line ν. The reader ought to verify that, if our diagram is endowed with the restriction to the (x^0, x^1)-plane of the Minkowski inner product (4.2.1), the lines μ and ν are orthogonal, no less than the horizontal and vertical axes. (Cf. also p. 90). Moreover, the null lines ω_1 and ω_2 are not only orthogonal with each other, but each with itself. I have drawn the hyperbolae $(x^0)^2 - (x^1)^2 = 1$ and $(x^0)^2 - (x^1)^2 = -1$. Any spacelike (timelike) geodesic of unit length, drawn from the origin, on the plane represented by our diagram, has its other endpoint mapped by x on a point of the latter (respectively, the former) hyperbola. Thus if the origin is an event E (mapped by x and y on O) at one end of a rod of unit length at rest on frame Y, whose other end points in the direction of increasing y^1, the event that takes place at that other end,

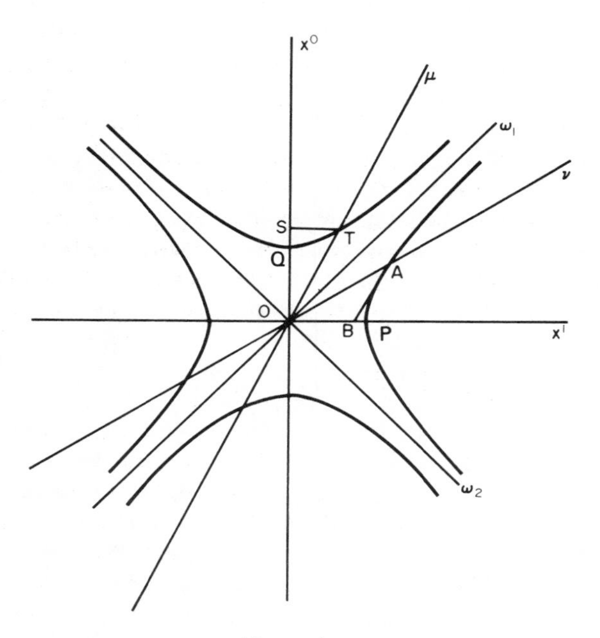

Figure 1.

simultaneously with E in Y, is mapped by x on the point A, where ν meets the right branch of the second hyperbola. The point B where the x^1-axis meets the parallel to μ through A is then the image by x of an event at the same end of the rod, but simultaneous with E in frame X. The distance in X between the two ends of our rod is the length of the spacelike geodesic mapped by x onto OB, which is clearly less than 1. Relativistic length contraction is thus seen to follow from the geometry of Minkowski spacetime.[12] The same can be said of time dilation. Consider the event mapped by x on T, where μ meets the upper branch of the first hyperbola. In frame Y this event occurs exactly one time unit after the event mapped on O. However, the time elapsed between both events in frame X equals the length of the timelike geodesic mapped by x onto OS (where S is the perpendicular projection of T on the x^0-axis), which is of course greater than 1. A process that takes place at a fixed point of Y and lasts unit time by Y's clocks, takes longer than unit time by the clocks of X. The reciprocity of such relativistic effects can be readily verified on the diagram: a rod of unit length at rest in X measures less than 1 in Y (draw the perpendicular to the x^1-axis at P and see where it meets ν); a process that takes unit time in X takes longer in Y (draw the parallel to ν through Q and see where it meets μ). The relativity of simultaneity can also be demonstrated on Fig. 1: each parallel to ν is the image by x of a set of events simultaneous in Y; any two such events are mapped on different horizontal lines; hence, they are not simultaneous in X. Of course, the geometry of events can be read off our diagram only because we initially resolved to read it into it, namely, by introducing the Minkowski inner product in $\mathbf{R}^4$—and hence in the section of $\mathbf{R}^4$ represented in Fig. 1. It is therefore misleading to say that, by means of such diagrams, Minkowski geometry treats time as if it were merely an additional dimension of space. Indeed Fig. 1 is useful only if one succeeds in forgetting its accepted Euclidean metric, and substitutes for it the non-Euclidean metric of Minkowski spacetime. By this act of thought every line that deviates from the vertical direction by less than 45° acquires the formal properties of timelike worldlines. One must therefore learn to "de-spatialize" some of the directions on the page in order to draw on it a non-spacelike section of spacetime.

4.3 Geometrical Objects

If the stage or "arena" of physical events is conceived as a differentiable manifold, physical laws can be expressed as relations between its geometrical objects. Some definitions will be necessary in order to show what a wealth of possibilities this opens. By a *geometrical object* of an n-manifold S we understand here, primarily, a cross-section of a fibre bundle over S or over an open submanifold of S;[1] secondarily, the value of such a cross-section at a point of S (*geometrical object at a point*); and, more loosely, any entity constructed from geometrical objects in the preceding sense, or defined analogously. As explained on p. 262, each point P in an n-manifold S is associated wih an infinite array of vector spaces, namely, the tangent space S_P, its dual space, the cotangent space S_P^* of linear functions on S_P, and the several spaces of multilinear functions on the product spaces formed from S_P and S_P^*, in every conceivable combination. An element of any of these spaces is called a *tensor* at P. A $(0, q)$-tensor (or covariant tensor of order q) at P is a q-linear function on $(S_P)^q$, while a $(q, 0)$-tensor (or contravariant tensor of order q) at P is a q-linear function on $(S_P^*)^q$, and a (q, r)-tensor (mixed tensor of order $q+r$) at P is a $(q+r)$-linear function on $(S_P^*)^q \times (S_P)^r$.[2] Evidently, a $(0, 1)$-tensor at P is an element of S_P^*; it is also called a covariant vector or *covector* at P. On the other hand, in virtue of the canonical identification of a vector space with the dual of its dual, a $(1, 0)$-tensor at P is an element of S_P, and is better known as a contravariant vector or simply as a *vector* at P. We agree to call any scalar (real number) associated with P a $(0, 0)$-tensor at P. A tensor t of order 2 is *symmetric* if, for every pair (u, v) in its domain, $t(u, v) = t(v, u)$; it is *skew-symmetric* if $t(u, v) = -t(v, u)$. The aggregate of all pairs (P, t_P), where $P \in S$ and t_P is a (q, r)-tensor at P, is made into a bundle over S. We call it the *(q, r)-tensor bundle* of S. (The *tangent bundle* of S if $q = 1$ and $r = 0$; the *cotangent bundle* if $q = 0$ and $r = 1$.)[3] A *(q, r)-tensor field* T on S is a cross-section of the (q, r)-tensor bundle of S. The value T(P) of T at $P \in S$ is a pair, whose first member is P itself, and whose second member is a (q, r)-tensor at P, which we denote by T_P. If, for each $P \in S$, T_P is an alternating r-linear function on S_P (that is, a $(0, r)$-tensor which takes the value 0 whenever its argument—a list of r vectors—contains the same vector twice), T is said to be an r-form. (Note that, according to this definition, every covector field is a l-form.) The $(0, 0)$-tensor fields or *scalar fields* on S constitute the algebra $\mathscr{F}(S)$ defined on page 258. The (q, r)-tensor fields on S—for each given pair of non-negative integers q and r—constitute a module over $\mathscr{F}(S)$, which we denote by $\mathscr{V}_r^q(S)$. (In the latter expression we omit the upper or the lower index if it happens to be zero.)[4]

An ordered basis $(e^1, \ldots, e^n)$ of the vector space S_P $(P \in S)$ is called an *n-ad* (*repère*) at P. The ordered basis $(e_1^*, \ldots, e_n^*)$ of the dual space S_P^*, defined by the relations $e_i^*(e^j) = \delta_{ij}$ $(1 \leq i, j \leq n)$, is its dual co-n-ad. The aggregate of all n-ads (co-n-ads) at every $P \in S$ can be made into a principal fibre bundle over S.[5] A cross-section of the bundle of n-ads of S is called an *n-ad field (repère mobile)*. It is obviously an ordered basis of the module $\mathscr{V}^1(S)$. A co-n-ad field is defined analogously and is an ordered basis of $\mathscr{V}_1(S)$. It follows from what we said on page 93 that an n-ad field on S exists if and only if S is parallelizable. If T is a (q, r)-tensor field on S $(q+r \geq 1)$, T is associated on each open submanifold $U \subset S$ with a unique $(q+r)$-linear mapping of $(\mathscr{V}_1(U))^q \times (\mathscr{V}^1(U))^r$ into $\mathscr{F}(U)$, which is also designated by T and is defined as follows: If $\mathrm{V}^1, \ldots, \mathrm{V}^r$ are vector fields on U and $\mathrm{V}_1, \ldots, \mathrm{V}_q$ are covector fields on U, $\mathrm{T}(\mathrm{V}_1, \ldots, \mathrm{V}_q, \mathrm{V}^1, \ldots, \mathrm{V}^r)$ is the scalar field on U whose value at each point $P \in U$ is equal to $\mathrm{T}_P((\mathrm{V}_1)_P, \ldots, (\mathrm{V}_q)_P, \mathrm{V}_P^1, \ldots, \mathrm{V}_P^r)$. If U is parallelizable and

$X = (X^1, \ldots, X^n)$ is an n-ad field on U, and $X^* = (X_1^*, \ldots, X_n^*)$ is the dual co-n-ad field defined by the relations $X_i^*(X^j) = \delta_{ij}$ $(1 \leq i,j \leq n)$, the $(q+r)$-linear mapping T is fully determined by its values on the n^{q+r} possible lists of q elements of X^* followed by r elements of X. These values constitute an array of scalar fields on U, called the *components* of the tensor field T relative to the n-ad field X. If $X = (\partial/\partial x^1, \ldots, \partial/\partial x^n)$, where the $\partial/\partial x^i$ are the vector fields defined on U by a chart x, in the manner explained on page 259, we usually refer to the components of T relative to X as its components relative to the chart x.[6] The (q, r)-tensor field T is therefore uniquely associated, on each parallelizable open submanifold $U \subset S$, with a mapping T of the set Σ_U of n-ad fields on U into $(\mathscr{F}(U))^{n^{q+r}}$, which assigns to each $X \in \Sigma_U$ the list of T's components relative to X.[7] The mutual linear dependence of the n-ad fields of U and their dual co-n-ad fields governs the mutual dependence of the respective components of T, which can therefore be correctly, though somewhat abstrusely described as so many arrays of real-valued functions, which "transform into each other" according to a definite rule. If the manifold S itself is parallelizable one may even identify T with the aggregate of such arrays, structured by the appropriate transformation rule, a view generally taken in the early days of the Tensor Calculus.

Let **g** be a Riemannian metric on S (p. 93). The reader ought to satisfy himself that **g** is in effect a symmetric (0, 2)-tensor field on S. **g** determines a canonic linear isomorphism of $\mathscr{V}^1(S)$ onto $\mathscr{V}_1(S)$, which associates each vector field V with a unique covector field $\overline{\mathrm{V}}$, defined as follows: for any $\mathrm{W} \in \mathscr{V}^1(S)$, $\overline{\mathrm{V}}(\mathrm{W}) = \mathbf{g}(\mathrm{V}, \mathrm{W})$. V and $\overline{\mathrm{V}}$ are usually regarded as the contravariant and the covariant form of the same geometrical object. The said canonic isomorphism induces definite linear isomorphisms between any pair of modules of tensor fields of the same order, on S. Take, for example, the modules $\mathscr{V}^2(S)$, $\mathscr{V}_2(S)$ and $\mathscr{V}_1^1(S)$ of tensor fields of order 2. Each (2, 0)-tensor field T is uniquely associated with a (0, 2)-tensor field $\overline{\mathrm{T}}$ and a (1, 1)-tensor field T′ defined as follows: for every $\mathrm{V}, \mathrm{W} \in \mathscr{V}^1(S)$, $\overline{\mathrm{T}}(\mathrm{V}, \mathrm{W}) = \mathrm{T}'(\overline{\mathrm{V}}, \mathrm{W}) = \mathrm{T}(\overline{\mathrm{V}}, \overline{\mathrm{W}})$. T, $\overline{\mathrm{T}}$ and T′ are regarded as the contravariant, covariant and mixed forms of the same object. Their respective components with respect to a given chart x are usually denoted by T^{ij} $(= \mathrm{T}(\mathrm{d}x^i, \mathrm{d}x^j))$, T_{ij} $(= \overline{\mathrm{T}}(\partial/\partial x^i, \partial/\partial x^j))$, and $\mathrm{T}^i{}_j$ $(= \mathrm{T}'(\mathrm{d}x^i, \partial/\partial x^j))$; whence the habit of saying that such tensor fields are derived from one another by "raising or lowering indices".

On the domain U of a chart x the Riemannian metric **g** can be represented by the matrix of its components relative to x. These are usually denoted by g_{ij} $(= \mathbf{g}(\partial/\partial x^i, \partial/\partial x^j))$. The restriction of **g** to U is an element of $\mathscr{V}_2(U)$ and hence an $\mathscr{F}(U)$-linear combination of the tensor products $\mathrm{d}x^i \otimes \mathrm{d}x^j$ $(1 \leq i,j \leq n)$, which constitute a basis of $\mathscr{V}_2(U)$. (Tensor products are defined on p. 349, note 15; function differentials on p. 261). It will be readily seen that, on U,

$$\mathbf{g} = g_{ij}\,\mathrm{d}x^i \otimes \mathrm{d}x^j \tag{4.3.1a}$$

In classical differential geometry a Riemannian metric **g** was usually expressed in terms of the coordinate differentials associated with a chart x as the *line element* ds, given by

$$ds^2 = g_{ij}dx^i dx^j \tag{4.3.1b}$$

For historical reasons, we shall often use the latter form of expression. But rather than squeeze our brains to penetrate its original meaning, we shall simply regard it as synonymous with (4.3.1a).

If physical realities are represented by tensor fields on the spacetime manifold, physical laws must say how different fields vary together from worldpoint to worldpoint. This involves a comparison between the values of each field at neighbouring worldpoints. Such a comparison is possible in principle, because the several values of a tensor field belong to isomorphic vector spaces—namely, the fibres of the pertinent tensor bundle over the arguments corresponding to those values. However, since there are many different isomorphisms between these spaces, one must be chosen in order to institute a definite method of comparison. Some choices are natural. Thus, as we know, a vector field $\mathrm{V} \in \mathscr{V}^1(S)$ has a maximal integral curve through each $P \in S$, which we shall denote here by γ^P (p. 260). V determines, on a neighbourhood U_P of each $P \in S$, a family $\{F_t | t \in I\}$ of regular mappings of U_P into S (indexed by a real interval I about 0), such that F_t sends each $Q \in U_P$ to $\gamma^Q(t)$. For each $t \in I$, the differential F_{t*} defines the linear isomorphism F_{t*Q} of S_Q onto $S_{F_t(Q)}$ and induces other well-defined isomorphisms between the other vector spaces associated with S_Q and the corresponding vector spaces associated with $S_{F_t(Q)}$. The existence of these isomorphisms makes it possible to define the so-called *Lie derivative* $L_\mathrm{V}\mathrm{T}$ of a (q,r)-tensor field T along the vector field V on an arbitrary n-manifold S. ($L_\mathrm{V}\mathrm{T}$ is a (q,r)-tensor field like T.)[8] The Lie derivative of a vector field along another can be used for characterizing Cartan's *exterior derivative*, a differential operator which, like the Lie derivative, is well defined on any n-manifold S, but acts only on a special kind of tensor fields, namely, r-forms. (The exterior derivative $\mathbf{d}\omega$ of an r-form ω is an $(r+1)$-form.)[9] In Appendix C, I explain how the choice of a "linear connection" on a manifold S determines a linear isomorphism τ^γ_{PQ} of S_P onto S_Q, for each pair of points $P, Q \in S$ and each curve γ joining them, which in turn induces matching isomorphisms between the spaces of (q,r)-tensors at P and Q. (τ^γ_{PQ} is said to "parallel-transport" S_P to S_Q along γ.) These isomorphisms enable us to define the *covariant derivative* $\nabla\mathrm{T}$ of a (q,r)-tensor field T on the n-manifold S, endowed with a given linear connection. ($\nabla\mathrm{T}$ is then a $(q, r+1)$-tensor field on S.)[10] There are indeed infinitely many different ways of bestowing a linear connection on a given n-manifold S, but if S is Riemannian the metric defines a unique linear connection, called the *Levi-Cività connection*, which is in a sense a natural choice and provides a definite and powerful procedure for comparing the values of any given tensor field at nearby points.[11]

Though the theory of tensors on manifolds was created by Ricci in the late 19th century, the conceptual scheme I have just sketched was not developed until much later, after Einstein's use of tensor methods in General Relativity encouraged mathematical research in the field. The first formulations of spacetime physics by Minkowski and others did not represent physical entities by tensor fields in the above sense, but by geometrical objects of a somewhat different sort. Minkowski defined them explicitly as objects *at a point*, but his differential equations express local relations between *fields* that take the objects defined by him as values. Minkowski called his objects *spacetime vectors* of the *first* and the *second* kind, but we shall call them, after A. Sommerfeld (1910), 4-*vectors* and 6-*vectors*, respectively. A 4-vector is defined by Minkowski as an "arbitrary array" of four real numbers (r_0, r_1, r_2, r_3), "together with the rule that", whenever a proper homogeneous Lorentz transformation, governed by the equations $x^i = a_{ij}y^j$, is applied, the array is to be replaced by the quadruple (r'_0, r'_1, r'_2, r'_3), obtained as a solution of the transformation equations when the r_i are substituted for the x_i.[12] In the subsequent discussion it becomes clear that such number quadruples associated with a transformation rule are attached to particular worldpoints. A 4-vector field may therefore be characterized as a correspondence which assigns a 4-vector to each worldpoint and varies smoothly from one worldpoint to another. Such a characterization is achieved simply by repeating the definition of a 4-vector, substituting "array of four smooth real functions (on Minkowski's world)" for "array of four real numbers". Though 4-vector fields do not at first sight seem to have much to do with our earlier concept of a vector field on a 4-manifold they can be easily constructed from the latter. We recall that Minkowski's world $\mathscr{M}$ is parallelizable. Hence a vector field V on $\mathscr{M}$ can be identified with the mapping that assigns to each tetrad field X on $\mathscr{M}$ the components of V relative to X. Let us say that the tetrad field $(\partial/\partial x^0, \partial/\partial x^1, \partial/\partial x^2, \partial/\partial x^3)$ defined on $\mathscr{M}$ by a Lorentz chart x is a *Lorentz tetrad field*. It is not hard to see that the restriction of a vector field on $\mathscr{M}$ (in the sense just explained) to Lorentz tetrad fields is exactly the same as a 4-vector field on $\mathscr{M}$, according to the above characterization. Since each vector v at a worldpoint P can be expressed as a linear combination $v^i\partial/\partial x^i|_P$, of any given Lorentz tetrad at P, v can be described, correctly even if somewhat clumsily, as the set of its components v^i, relative to an arbitrary Lorentz tetrad, plus the rule for calculating from them its components relative to any other Lorentz tetrad. In this way, any vector at a worldpoint may be seen as a 4-vector at a point. This approach to 4-vectors and 4-vector fields not only solves the mystery of their transformation rule, but also enables us to extend Minkowski's conception, almost without effort, to geometrical objects of greater complexity. We simply define a 4-*tensor field* of order q as the restriction to Lorentz tetrad fields of a tensor field of order q on $\mathscr{M}$, conceived as on page 100, i.e. as a mapping of the set $\Sigma_{\mathscr{M}}$ of tetrad fields on $\mathscr{M}$ into the appropriate Cartesian power of the ring

$\mathscr{F}(\mathscr{M})$. In particular, a 6-vector field can be identified with a skew-symmetric 4-tensor field of order 2. (Skew-symmetry implies that it has no more than 6 independent components relative to each Lorentz tetrad field, whence Sommerfeld's name for it.)[13] Again, any tensor at a worldpoint may be viewed as a 4-tensor at that point. Evidently all the concepts of spacetime geometry defined in terms of the Minkowski metric $\boldsymbol{\eta}$ and its operation on ordinary spacetime vectors retain their sense when we view the latter as 4-vectors. We may speak, therefore, of the energy, the orthogonality, etc., of 4-vectors.

4-vectors and 6-vectors are the spacetime analogues of the polar and axial vectors of late 19th-century physics. The latter are geometrical objects of Euclidean space, but not of the ordinary kind defined above for arbitrary manifolds. They belong to the family of so-called Cartesian tensors, which we shall now characterize—as we did with 4-tensors—in terms of ordinary tensors. Let us say that the triad field $(\partial/\partial x^1, \partial/\partial x^2, \partial/\partial x^3)$ determined by a Cartesian chart x on the Euclidean space $\mathscr{E}^3$, is a Cartesian triad field. Since $\mathscr{E}^3$ is parallelizable we can identify tensor fields on $\mathscr{E}^3$ with the corresponding mappings from triad fields to components. A *Cartesian tensor field* of order q on $\mathscr{E}^3$ is the restriction to Cartesian triad fields of an ordinary tensor field of order q on $\mathscr{E}^3$. In particular, a field of polar vectors is a Cartesian tensor field of order 1, or Cartesian vector field, while a field of axial vectors is a skew-symmetric Cartesian tensor field of order 2. (Skew-symmetry implies that it has no more than three independent components relative to each Cartesian triad field.) Again, any tensor at a point of $\mathscr{E}^3$ can be viewed as a Cartesian tensor at that point. As the covariant and contravariant forms of the Pythagorean metric of Euclidean space have the same array of components relative to each Cartesian triad field—namely the 3×3 matrix of constant scalar fields $(\delta_{\alpha\beta})$—there is no point in distinguishing between covariant, contravariant or mixed Cartesian tensors or tensor fields. In physical applications a Cartesian tensor field is defined on the relative space S_F of a given frame of reference F. If, as it is usually the case, the field is supposed to vary with time, it is, strictly speaking, a tensor field on the product manifold $\mathbf{R} \times S_F$. Note, however, that not every tensor field on this 4-dimensional manifold can be equated with a time-varying tensor field on S_F. A tensor field T of order q, on $\mathbf{R} \times S_F$, is, as we shall say, a *time-dependent Cartesian tensor field* on S_F if and only if, for each $t \in \mathbf{R}$, the restriction of T to the 3-dimensional submanifold $\{t\} \times S_F$ is a Cartesian tensor field of order q on this submanifold. Thus, in particular, a vector field V on $\mathbf{R} \times S_F$ is a time-dependent Cartesian vector field on S_F only if the value of V at an arbitrary point $\mathbf{r} \in S_F$ at an arbitrary time t is tangent to the hypersurface $\{t\} \times S_F$ (p. 261). On the other hand, any scalar field f on $\mathbf{R} \times S_F$ is a time-dependent scalar field on S_F, for it assigns a real number $f(t, \mathbf{r})$ to each pair $(t \in \mathbf{R}, \mathbf{r} \in S_F)$, i.e. to each point of space and each time—being of course smooth for fixed t. By the same token, f can also be identified with a scalar field on Minkowski's world. The representation of a physical entity by a time-dependent Cartesian tensor field on the relative space S_F does not depend on the choice of a particular Cartesian chart on S_F, but is tied of course to the reference frame F. The representation of the same physical entity with regard to a different frame F' will be given by a different Cartesian tensor field, defined on the relative space $S_{F'}$. The correspondence between the tensorial representations of the same physical reality with respect to different inertial frames was established in classical physics by means of suitable variations of the Galilei Rule (p. 29). This rule does not follow from the mathematical concept of a Cartesian tensor, but must be regarded, in the present context, as a physical law. It was confirmed by mechanical experiments (involving speeds much less than c), but foundered when applied to electrodynamics. The problem that the Galilei Rule thus failed to solve does not even arise when physical entities are represented by 4-tensors and 4-tensor fields. In contrast with the infinitely many dynamically equivalent relative spaces of Newtonian physics, the physical manifold on which 4-tensors are defined is unique, and the representation of a physical entity by a geometrical object of this manifold is inherently frame-independent. The physical object may therefore in a sense be said to be *conceived as*, not just *represented by* such a geometrical object. (However, we ought not to overlook that the geometrical object with which a given physical object is identified in this way can only be determined up to a choice of units.)

To understand how Minkowski and his successors went about conceiving physical realities as 4-vectors and 6-vectors we must take cognizance of one more geometrical fact. Consider an inertial frame F. A Lorentz chart x adapted to F determines the Lorentz tetrad field $(\partial/\partial x^i)$ on Minkowski spacetime, of which we may also say that it is adapted to F. Let V be a 4-vector field. V is equal to $\mathrm{V}^i\partial/\partial x^i$, where $\mathrm{V}^i = \mathrm{V}x^i$ (see note 6). Relatively to F, V may therefore be regarded as the sum of a field of timelike vectors, $\mathrm{V}^0\partial/\partial x^0$, and a field of spacelike vectors $\mathbf{V} = \mathrm{V}^\alpha\partial/\partial x^\alpha$. This decomposition of V, relative to F, does not depend on the choice of a particular Lorentz chart adapted to F.[14] At each worldpoint P, the local value of $\mathbf{V}$ is orthogonal to that of $\partial/\partial x^0$. This implies that the integral curve of $\mathbf{V}$ through P lies entirely on the hyperplane F_P, of all worldpoints simultaneous with P in F. F_P is a copy of Euclidean space and may be regarded as a representation of the relative space S_F at the time of P (in F). The restriction of $\mathbf{V}$ to F_P can be identified, rather obviously, with a Cartesian vector field on S_F.[15] $\mathbf{V}$ itself can therefore be regarded as a time-dependent Cartesian vector field on S_F. On the other hand, $\mathrm{V}^0\partial/\partial x^0$ can be identified with the scalar field V^0, which may also be viewed as a time-dependent field on S_F. We call $\mathbf{V}$ and V^0, thus understood, the *spacelike* and the *timelike* parts of the 4-vector field V, relative to the frame F.

To illustrate these ideas we shall now define the spacetime counterparts of the basic concepts of classical kinematics. The motion of a particle π in an inertial frame F is represented classically by a curve $\bar{\gamma}$, parametrized by time t, in the relative space S_F. $\bar{\gamma}$ assigns to each real number t in its domain the point of S_F where π is located at time t. The *velocity* of π at time t is the Cartesian vector $\mathbf{u} = \bar{\gamma}_{*t}$, tangent to $\bar{\gamma}$ at $\bar{\gamma}(t)$. Its three components relative to the Cartesian chart $\bar{y}$ (on S_F) are given by

$$u_\alpha = \frac{d\bar{y}^\alpha}{dt} \tag{4.3.2}$$

Spacetime geometry enables us to represent the motion of π, without referring it to a particular frame, by a curve γ, parametrized by proper time τ, in Minkowski spacetime. $\gamma(\tau)$ is the event on π's history at proper time τ (measured along π's worldline). The 4-*velocity* of π at τ is the 4-vector $\mathrm{U} = \gamma_{*\tau}$, tangent to γ at $\gamma(\tau)$. Its components relative to the Lorentz chart y, adapted to F and associated with $\bar{y}$ in the manner described on p. 56, are given by[16]

$$\mathrm{U}^i = \frac{dy^i}{d\tau} \tag{4.3.3}$$

Relativity was born out of the realization that (4.3.2) is meaningless until we make up our minds as to what we understand by the time t. If we assume that t is time measured by natural clocks at rest in F and synchronized by Einstein's method (p. 53) along the path of π, we have that $dt = dy^0$. By the clock hypothesis (p. 96), the proper time τ agrees at each moment with the time measured by natural clocks on the mirf of π. Consequently, if x is a Lorentz chart matching y and adapted to π's mirf at a given moment, then, at that moment, $d\tau = dx^0$. Hence, by (3.5.2),

$$dt = \beta_u d\tau \tag{4.3.4}$$

(where $u = |\mathbf{u}|$ is the speed of π in F at the moment in question, and $\beta_u = 1/\sqrt{1-u^2}$). It follows that

$$\mathrm{U}^0 = \frac{dy^0}{dt}\frac{dt}{d\tau} = \beta_u \qquad \mathrm{U}^\alpha = \frac{dy^\alpha}{dt}\frac{dt}{d\tau} = \beta_u u_\alpha \tag{4.3.5}$$

The spacelike part of the 4-velocity U of π, relative to the inertial frame F, is therefore given by

$$\mathbf{U} = \beta_u \mathbf{u} \tag{4.3.6}$$

while its timelike part U^0 equals the "Lorentz factor" β_u. In particular, if F happens to be the mirf of π, $\mathbf{U} = 0$ and $\mathrm{U}^0 = 1$. Hence the 4-velocity U of any particle π has constant "energy" $E(\mathrm{U}) = 1$.[17] Following the analogy of the classical acceleration $\mathbf{a}$, whose components relative to $\bar{y}$ are $a_\alpha = du_\alpha/dt$, we define the 4-acceleration A, whose components relative to y are

$$\mathrm{A}^i = d\mathrm{U}^i/d\tau \tag{4.3.7}$$

A is always orthogonal to U, for $\langle \mathrm{U}, \mathrm{A} \rangle = \eta_{ij}\mathrm{U}^i \mathrm{dU}^j/\mathrm{d}\tau = \frac{1}{2}\mathrm{d}(\eta_{ij}\mathrm{U}^i\mathrm{U}^j)/\mathrm{d}\tau = \frac{1}{2}\mathrm{d}E(\mathrm{U})/\mathrm{d}\tau = 0$.[18]

The representation of physical realities by 4-tensor fields, instead of ordinary tensors, on Minkowski spacetime $\langle \mathcal{M}, \boldsymbol{\eta} \rangle$ has one considerable practical advantage. If $\mathrm{T}^{\mathbf{a}}_{\mathbf{b}}$ is a component of a (q, r)-tensor field T relative to a Lorentz chart x (where **a** and **b** stand for appropriate lists of indices), the respective component relative to x of the covariant derivative of T along $\partial/\partial x^i$ is simply equal to $\partial \mathrm{T}^{\mathbf{a}}_{\mathbf{b}}/\partial x^i$. Therefore, if T is restricted to Lorentz tetrad fields, one may describe its evolution from worldpoint to worldpoint with the familiar resources of the differential calculus. However this advantage can only be achieved at a price. By definition a 4-tensor field only takes components on Lorentz tetrad fields, defined by Lorentz charts adapted to inertial frames. Hence, if a physical entity is equated with a 4-tensor field one cannot meaningfully assign it components relative to a non-Lorentz tetrad field, such as could be ascertained by measurements performed on a non-inertial frame. The physical entity in question lacks therefore all representation and is as good as non-existent with respect to such a frame. To a philosopher such a price must seem enormous, for, strictly speaking, no terrestrial laboratory is inertial. But the working physicist may well ignore it, just like he ordinarily ignores—or corrects for—the action of extraneous forces on his instruments. As traditional methods of measurement tend to rely heavily on the use of unstressed rods, the results obtained by them only make sense in the ideal limit in which the relevant pieces of the experimental set-up can be regarded as force-free. Thus the use of 4-tensor fields instead of ordinary tensor fields for the representation of physical entities may not be such a big handicap after all. Nevertheless, as soon as one has learnt to understand a 4-tensor field as the restriction of a more natural mapping to a choice part of its domain, one can easily avoid the drawbacks of such a restriction without giving up its advantages. For a 4-tensor field can be extended to an ordinary tensor field in only one way. (If we are given the components of a (q, r)-tensor field T relative to *any* tetrad field on $\mathcal{M}$—such as the one defined by a particular Lorentz chart—its components relative to *any other* such tetrad field are completely determined by the appropriate transformation rule.) Hence there is no difficulty in equating any physical entity suitably represented by a 4-tensor field on $\mathcal{M}$ with the latter's unique extension to an ordinary tensor field on $\mathcal{M}$.[19] However the procedure has been questioned, out of what I can only describe as inductivist squeamishness, by some philosophically minded scientists. As far as I can understand them, their argument appears to be the following: experimental data which, when referred to different inertial frames, transform like the components of a 4-tensor relative to Lorentz charts adapted to those frames, need not, if referred to arbitrary frames, transform like the components of an ordinary tensor relative to suitable charts; hence, one may not conclude that a physical entity which is adequately represented by a 4-tensor field T is also represented adequately by T's extension to an ordinary tensor field, unless one has experimental proof of it.[20] But even if one may not *conclude* it, one can reasonably *conjecture* it, and try and see if experience will prove one wrong. Moreover, if an experimental refutation of the proposed extension of 4-tensor fields to ordinary tensor fields on $\mathscr{M}$ were ever to be found, it would cause the downfall of the entire theoretical outlook on which the representation of physical entities by 4-tensor fields is based. Suppose, for example, that a well-corroborated physical law equates two 4-covector fields on $\mathscr{M}$, ω_1 and ω_2. This means that, for each Lorentz chart x, $\omega_1(\partial/\partial x^i) = \omega_2(\partial/\partial x^i)$. Let $\bar{\omega}_r$ be the ordinary covector field on $\mathscr{M}$ of which ω_r is the restriction to Lorentz tetrad fields ($r = 1, 2$). Suppose that experience somehow shows that $\bar{\omega}_1 \neq \bar{\omega}_2$. This implies that for some arbitrary vector field V on $\mathscr{M}$, or on part of $\mathscr{M}$, $\bar{\omega}_1(\mathrm{V}) \neq \bar{\omega}_2(\mathrm{V})$. Since $\mathrm{V} = \mathrm{V}x^i\partial/\partial x^i$, for any Lorentz chart x, we have that $\bar{\omega}_1(\mathrm{V}) = \mathrm{V}x^i\omega_1(\partial/\partial x^i) = \mathrm{V}x^i\omega_2(\partial/\partial x^i) = \bar{\omega}_2(\mathrm{V})$, a contradiction.

4.4 Concept Mutation at the Birth of Relativistic Dynamics

Let π be a particle moving with velocity $\mathbf{u}$ and acceleration $\mathbf{a}$ in an inertial frame F. Classical dynamics represents π's "quantity of motion" or *momentum* by the vector

$$\mathbf{P} = m\mathbf{u} \tag{4.4.1}$$

(where m is the particle's "quantity of matter" or *mass*). The external *force* acting on π is represented by the vector $\mathbf{k}$, which, by Newton's Second Law of Motion, equals the time rate of change of $\mathbf{p}$:

$$\mathbf{k} = \frac{\mathrm{d}\mathbf{p}}{\mathrm{d}t} = \frac{\mathrm{d}(m\mathbf{u})}{\mathrm{d}t} \tag{4.4.2}$$

As the action of the force on an otherwise isolated particle does not alter m, $\mathrm{d}(m\mathbf{u})/\mathrm{d}t = m\mathrm{d}\mathbf{u}/\mathrm{d}t = m\mathbf{a}$, and the mass equals the ratio of force to acceleration. Mass is thus a measure of the particle's "laziness" or *inertia*, i.e., of its resistance to the accelerative action of a given force. $\mathbf{p}$ and $\mathbf{k}$ are tied, as Cartesian vectors, to a particular Euclidean space, the relative space S_F of frame F. Nevertheless, they display a certain absoluteness, that argues for the Newtonian physicist's belief that they both stand for substantive, frame-independent physical realities. Thus, the force on π is represented in the spaces of any two inertial frames, F and F', by two vectors, say, $\mathbf{k}$ and $\mathbf{k}'$, which have respectively the same components relative to any pair of Cartesian charts on the pertinent relative spaces whose coordinate axes coincide at some time, as F and F' move past each other (p. 19). $\mathbf{k}$ and $\mathbf{k}'$ are thus, in a sense, "the same vector", and, as far as I know, no classical physicist ever bothered to distinguish them. Consider now the momenta $\mathbf{p}_i$, relative to the inertial frame F, of the particles π_i in an isolated material system S. Newton's Third Law implies that the *total momentum* $\Sigma_i \mathbf{p}_i$ of S relative to F is conserved through time. An inertial frame F_0 can always be found, relative to which the total momentum of S is 0. If F moves in F_0 with velocity $\mathbf{u}$, $\Sigma_i \mathbf{p}_i = (\Sigma_i m_i)\mathbf{u}$, where m_i is the mass of π_i and hence $\Sigma_i m_i$ is the *total mass* of S. Consequently, if the inertial frame F' moves in F with velocity $\mathbf{v}$, the total momentum of S relative to F' is

$$\Sigma_i \mathbf{p}'_i = (\Sigma_i m_i)(\mathbf{u} + \mathbf{v}) = \Sigma_i \mathbf{p}_i + (\Sigma_i m_i)\mathbf{v} \tag{4.4.3}$$

These simple and natural relations bespeak the reality of momentum: though its vectorial representation varies from frame to frame, the difference does no more than reflect the change of frame.

The classical definition of momentum (4.4.1) and the Law (4.4.2) that determines the classical representation of force both depend essentially on differentiation with respect to time and are therefore meaningless until the meaning of time is fixed. If time is frame-dependent, (4.4.2) cannot yield a frame-independent representation of force in the manner described above. Moreover, velocities will not transform according to the Galilei Rule, as in the calculation underlying (4.4.3). Classical dynamics is firmly bound to absolute Newtonian time and cannot survive it. Special Relativity had to develop a new dynamics.

In view of the experimental success of the Newtonian theory, relativistic dynamics had to agree well with it when low speeds were involved (i.e, in the case of our particle π, when $u \ll 1 = c$). This condition constrained the problem, but did not yield the answer. A solution could only be found poetically—as Newton had found his—by devising new concepts, and building with them a new framework for judging and describing facts. As on other similar occasions in the history of thought, the new wine was first served in old bottles, the time-honoured words "mass", "momentum", "force", being now used to convey unfamiliar ideas. Stunned readers learned that force was relative and that the mass of a material object would increase dramatically by a mere transformation of coordinates. No wonder Relativity was resisted by some pious ideologues, while others hailed it as bringing about the "de-materialization of matter". We cannot dwell on the historical details of the genesis of relativistic dynamics; but a short discussion will be necessary to motivate our subsequent sketch of the four-dimensional formulation that finally threw light on the meaning of the new conceptual system and on its link with the old.[1]

In his (1905*d*), §6, Einstein assumes that Maxwell's equations for free space hold good in any inertial frame, provided that the coordinate system to which they are referred is a Lorentz chart adapted to the frame in question. The assumption was justified insofar as Maxwell's equations had been suggested and confirmed by experiments referred to laboratory frames moving almost inertially with significantly different velocities relative to the fixed stars. It was also consistent with the form of the equations; indeed, the Lorentz transformation had first turned up in the quest for a "substitution of variables" that would leave the Maxwell equations unchanged in form. In the classical formulation adopted by Einstein, Maxwell's equations for free space equate the time rate of change of the electric field strength **E** with the curl of the magnetic field strength **H**, and vice versa. **E** and **H** are time-dependent, divergence-free Cartesian vector fields in the relative space of an inertial frame F. The time derivative and the curl are taken with respect to the appropriate coordinate functions of a Lorentz chart x, adapted to F, and its associated Cartesian chart $\bar{x}$. We agree that $\varepsilon_{\alpha\beta\gamma}$ is equal to 1 or -1 if $\{\alpha, \beta, \gamma\}$ is, respectively, an even or an odd permutation of $\{1, 2, 3\}$, and is equal to 0 if any two of the indices are equal. If E_α and H_α denote the α-th components of **E** and **H** relative to $\bar{x}$, the equations can be concisely written thus:[2]

$$\frac{\partial E_\alpha}{\partial x^0} = \varepsilon_{\alpha\beta\gamma}\frac{\partial H_\beta}{\partial \bar{x}^\gamma} \qquad -\frac{\partial H_\alpha}{\partial x^0} = \varepsilon_{\alpha\beta\gamma}\frac{\partial E_\beta}{\partial \bar{x}^\gamma} \tag{4.4.4}$$

If y is a Lorentz chart adapted to another frame F' and related to x by the transformation equation (3.4.18), the Maxwell equations

$$\frac{\partial E'_\alpha}{\partial y^0} = \varepsilon_{\alpha\beta\gamma}\frac{\partial H'_\beta}{\partial \bar{y}^\gamma} \qquad -\frac{\partial H'_\alpha}{\partial y^0} = \varepsilon_{\alpha\beta\gamma}\frac{\partial E'_\beta}{\partial \bar{y}^\gamma} \tag{4.4.5}$$

will hold good provided that

$$\begin{aligned} E'_1 &= E_1 & H'_1 &= H_1 \\ E'_2 &= \beta(E_2 - vH_3) & H'_2 &= \beta(H_2 + vE_3) \\ E'_3 &= \beta(E_3 + vH_2) & H'_3 &= \beta(H_3 - vE_2) \end{aligned} \tag{4.4.6}$$

The E'_α and H'_α can be naturally regarded as the components, relative to $\bar{y}$, of two Cartesian vector fields, **E**′ and **H**′, on the relative space of F'.[3] Nothing would mar the beauty of this result if its meaning were clear. It so happens, however, that the electric and magnetic field strengths in Maxwell's equations were conceived until 1905 as fields of (Newtonian) force. Thus, the value of **E** at a point P in S_F was supposed to stand for the force that would be exerted by the field on a particle with unit charge at rest at P.[4] But **E** and **E**′, if related to one another by (4.4.6), cannot be the representations, in the relative spaces of two different inertial frames, of the same Newtonian force field. We may take this as one more indication that the Newtonian concept of force was voided by Relativity. But then the electromagnetic field strengths **E** and **H** no longer can have their original meaning.

Einstein deals expressly with the interpretation of equations (4.4.6) in the second half of § 6.[5] The force on a particle π momentarily at rest in the inertial frame F can be balanced and measured by a dynamometer tied to F. The ratio of its measured magnitude to the magnitude of the acceleration **a** that π experiences in F as the force acts on it, is the inertia of π in its own mirf. We call it the particle's *proper mass* and denote it by m_0. The force on π will be uniquely represented in the relative space S_F by the vector $\mathbf{k}$ of (4.4.2) if t agrees with the time coordinate function of a Lorentz chart adapted to F and we set $m = m_0$. The electrostatic unit of charge can be defined as usual in terms of this concept of force. (Section 2.1, note 4.) The electric field strength $\mathbf{E}$ at a point P in S_F can now be understood to be the force *in the present sense* exerted by the field on a particle with unit charge at rest at P. If the vector field $\mathbf{E}$, defined on S_F, remains constant or changes smoothly in time, equations (4.4.4) determine the vector field $\mathbf{H}$, defined on S_F for each instant.[6] The force exerted by the electromagnetic field $(\mathbf{E}, \mathbf{H})$ on a particle of unit charge moving through S_F is, *by definition*, a vector at the relevant point of S_F, with the same components, relative to matching Lorentz charts,[7] as the force exerted on the particle in its mirf by the electric field obtained from $(\mathbf{E}, \mathbf{H})$ by the appropriate transformation, in agreement with (4.4.6).[8]

Einstein uses the foregoing concepts to calculate the inertia of a charged particle moving in an electromagnetic field. Let q and m_0 be, respectively, the particle's charge and proper mass. Einstein assumes that q is the same in all inertial frames.[9] We denote the particle by π, its mirf by F', and suppose that the electromagnetic field $(\mathbf{E}, \mathbf{H})$ is defined on the relative space S_F of the inertial frame F, in which π moves with velocity $\mathbf{v}$. One can always find two matching Lorentz charts x and y (with associated Cartesian charts $\bar{x}$ and $\bar{y}$), adapted respectively to F and F'. x and y satisfy equations (3.4.18), with $v = |\mathbf{v}|$. The sources that generate the field $(\mathbf{E}, \mathbf{H})$ in S_F, with components E_α, H_α, relative to the chart $\bar{x}$, generate in $S_{F'}$ the field $(\mathbf{E}', \mathbf{H}')$, whose components E'_α, H'_α, relative to $\bar{y}$, are given by (4.4.6). The acceleration $\mathbf{a}'$ of π in its mirf F' satisfies the equation

$$m_0 \mathbf{a}' = q\mathbf{E}' \tag{4.4.7}$$

Consequently, by (3.4.18) and (4.4.6), the components a_α, relative to $\bar{x}$, of π's acceleration in F must satisfy:

$$\begin{aligned} m_0 \beta^3 a_1 &= qE_1 \\ m_0 \beta^2 a_2 &= q\beta(E_2 - vH_3) \\ m_0 \beta^2 a_3 &= q\beta(E_3 + vH_2) \end{aligned} \tag{4.4.8}$$

The quantities on the right-hand side of (4.4.8) are *by definition* the components (relative to $\bar{x}$) of the force with which the field $(\mathbf{E}, \mathbf{H})$ acts on π. The ratio of the force components to the acceleration components is plainly greater than the proper mass m_0. It is $m_0 \beta^3$ for the component parallel to the direction of motion, and $m_0 \beta^2$ for the components perpendicular to that direction. The resistance opposed by a moving particle to the accelerative action of an electromagnetic force depends therefore on the magnitude of the particle's velocity and on the angle between the velocity and the force. Einstein points out that this result applies even to uncharged particles, for it does not depend on the amount of charge.[10] It must apply also to every kind of force: as they are all liable to be balanced by electromagnetic forces, they must have the same effect on matter.

Einstein calculates the kinetic energy T acquired by our particle π, with charge q, accelerated from rest by the field $(\mathbf{E}, \mathbf{H})$ until it moves in F with velocity $\mathbf{v}$. He assumes that π is accelerated slowly and thus loses no energy by radiation. As the force perpendicular to the direction of motion does no work on π,

$$T = \int qE_1 \, d\bar{x}^1 = m_0 \int_0^v \beta^3 v \, dv = m_0 (\beta - 1) \tag{4.4.9}$$

As v approaches 1, T grows beyond all bounds. Hence the particle cannot attain the speed of light. If $v \ll 1$,

$$T \approx \tfrac{1}{2} m_0 v^2 \tag{4.4.10}$$

the classical value of the kinetic energy.

In his (1905*d*), § 8, Einstein shows that if L_0 is the energy content of a train of parallel plane light waves referred to an inertial frame F, its energy content referred to a frame F', moving in F with constant velocity $\mathbf{v}$, at right angles with the direction of propagation of the wave train, is equal to $L_0 \beta$. This result is used by Einstein in the short paper (1905*e*)—published soon after his (1905*d*)—for establishing a most remarkable proposition. Einstein assumes that the Principle of Conservation of Energy holds good in all inertial frames ("according to the Relativity Principle", he says). Then, if a body at rest in an inertial frame F simultaneously emits two trains of plane light waves of energy $L/2$ in opposite directions, it must lose the energy L in F, and the energy $L\beta$ in the frame F' that moves in F with velocity $\mathbf{v}$ at right angles to the direction of emission. In the carefully contrived conditions of this *Gedankenexperiment*, the body will not recoil as it emits the light, but will remain at rest in F, and moving with speed v in F'. Its state of motion remains, so to speak, kinematically unchanged by the emission of radiation, though surely the loss of energy must alter it dynamically. Let t_1 and t_2 be times immediately preceding and following the emission and let Q_r and Q'_r denote the energy content of the body at time t_r in F and F' respectively ($r = 1, 2$). If, with Einstein, we understand by the kinetic energy of a body (with respect to a given inertial frame) the full increment in the body's energy due to motion (in that frame), it is plain that the kinetic energy T_r which our body possesses at time t_r with respect to the frame F' is equal to $(Q'_r - Q_r)$, up to an arbitrary constant C. Since C will not change by the emission of light, it follows that

$$T_1 - T_2 = L(\beta - 1) \tag{4.4.11}$$

We may of course choose v as small as we wish. If $v \ll 1$, the kinetic energy lost by the body in F' as it radiates the energy L in its rest frame is

$$\Delta T \approx \tfrac{1}{2} L v^2 \tag{4.4.12}$$

Comparing (4.4.12) with (4.4.10) we see that when a body at rest loses the energy L its proper mass will decrease by the same amount (viz. if $c = 1$; otherwise $\Delta m_0 = L/c^2$).[11] As it is not essential that the energy be lost by radiation, Einstein draws the general conclusion that "*the mass of a body is a measure of its energy content*".[12] Of course, this conclusion does not follow from his premises. Even if all the energy released by a body has to be subtracted from its mass, the remainder could be more than the body's extant energy store. However, the conversion of colliding particle–antiparticle pairs into light—unsuspected in 1905—lends unexceptionable support to Einstein's dauntless extrapolation. If the energy content of a particle at rest is equal to its proper mass m_0 and its kinetic energy when travelling with speed v is equal to $m_0\beta - m_0$, the *total energy* of the particle relative to an arbitrarily chosen inertial frame F is

$$m(u) = m_0 \beta_u \tag{4.4.13}$$

where u is the speed of the particle in F and $\beta_u = 1/\sqrt{1-u^2}$. For historical reasons that will soon become apparent, $m(u)$ is usually called the *relativistic mass.*[13]

Though Einstein's 1905 definition of force has not been proved inconsistent,[14] and though he relied on it on his way to his most famous discovery, it was soon supplanted by another, in a way more natural definition, introduced by Planck (1906). While Einstein had procured a unique—and hence absolute—representation of the force accelerating a particle in each inertial frame by equating it formally, componentwise, with the force defined physically according to Newton's Second Law, in the particle's mirf, the new definition furnished a physically warranted representation of the force, peculiar to each frame. One simply substitutes $m(u)$, from (4.4.13), for m in (4.4.2) and defines the vector $\mathbf{k}$ as the *relative force* on the particle in the inertial frame to which its velocity $\mathbf{u}$ is referred.

$$\mathbf{p} = m(u)\mathbf{u} \tag{4.4.14}$$

is adopted as the relativistic definition of *momentum.* An argument traceable to Lewis and Tolman (1909) and now included in many textbooks[15] proves that if the Principles of the Conservation of Energy and Momentum hold for an isolated system of freely colliding but otherwise unrelated particles—that is, if $\Sigma_i m(u_i) = const.$ and $\Sigma_i m(u_i)\mathbf{u}_i = const.$, where $\mathbf{u}_i$ is the velocity of the ith particle of the system at any appointed time in any given inertial frame—$m(u)$ must be the function defined by (4.4.13).

One of the most satisfactory features of the concept of relative force is that the time rate of increase of the energy of a particle subject to the force, relative to the frame to which the latter is referred, equals the *power* or rate at which the force does work on the particle. If $\mathbf{k}$ is the relative force and the particle's velocity is $\mathbf{u}$, the power is

$$\begin{aligned}\mathbf{u}\cdot\mathbf{k} &= \mathbf{u}\cdot\frac{\mathrm{d}}{\mathrm{d}t}m(u)\mathbf{u} = m_0\beta_u\mathbf{u}\cdot\frac{\mathrm{d}\mathbf{u}}{\mathrm{d}t}+u^2 m_0\beta_u^3 u\frac{\mathrm{d}u}{\mathrm{d}t}\\ &= m_0\beta_u\left(1+\frac{u^2}{1-u^2}\right)u\frac{\mathrm{d}u}{\mathrm{d}t} = m_0\beta_u^3 u\frac{\mathrm{d}u}{\mathrm{d}t}\\ &= \frac{\mathrm{d}}{\mathrm{d}t}m(u) \end{aligned} \qquad (4.4.15)$$

The result concerning kinetic energy derived in (4.4.9) can now be obtained, by integrating (4.4.15), without appealing to Maxwell's equations. As

$$\mathbf{k} = \frac{\mathrm{d}}{\mathrm{d}t}m(u)\mathbf{u} = m(u)\mathbf{a}+\mathbf{u}(\mathbf{u}\cdot\mathbf{k}) \qquad (4.4.16)$$

it is clear that the relative force $\mathbf{k}$ and the acceleration $\mathbf{a} = \mathrm{d}\mathbf{u}/\mathrm{d}t$ effected by it are linearly dependent if and only if $\mathbf{k}$ is orthogonal to or linearly dependent on the velocity $\mathbf{u}$. The traditional concept of inertia or ratio of force to acceleration has now a definite meaning only in these two cases. Moreover, in each case it has a different value. We must distinguish (i) the *transverse mass* $m_0\beta_u$, or resistance opposed by a particle moving with velocity $\mathbf{u}$ to a force perpendicular to its direction of motion, from (ii) the *longitudinal mass* $m_0\beta_u^3$, or resistance opposed by a particle moving with velocity $\mathbf{u}$ to a force parallel to its direction of motion. Only the former is equal to the energy of the particle, and is often called its "inertial" or "relativistic" mass. (Case (i) includes of course the resistance m_0 offered to any force by a particle at rest: if $\mathbf{u} = 0$, it is orthogonal to every vector.)

The "transverse mass" defined as the ratio of transversal relative force to acceleration differs from the transverse inertia calculated by Einstein from his 1905 concept of force.[16] We must therefore give a different reading of equations (4.4.8): the components of the relative force exerted by the electromagnetic field ($\mathbf{E}$, $\mathbf{H}$), in frame F, on the particle, are qE_1, $q(E_2 - vH_3)$ and $q(E_3 + vH_2)$. The relative force is thus none other than the Lorentz force

$$\mathbf{k} = q(\mathbf{E}+\mathbf{v}\times\mathbf{H}) \qquad (4.4.17)$$

The Lorentz force law (4.4.17) holds in each inertial frame as a corollary of Einstein's RP and Maxwell's equations.

4.5 A Glance at Spacetime Physics[1]

Spacetime physics began with the observation that some of the Cartesian vectors of relativistic mechanics and electrodynamics can be regarded as the spacelike parts, relative to the appropriate frames, of 4-vectors whose timelike parts are familiar scalars, or can be combined in pairs to make a 6-vector. I noted on page 86 that Poincaré had already been aware of this, and I mentioned the current density vector and the charge density scalar as a case in point. Consider a static, continuous charge distribution referred to an inertial frame F. Let $\rho(t, \mathbf{r})$ be the charge density at point $\mathbf{r}$ and time t. ρ is obviously a time-dependent scalar field on S_F and can therefore be identified with a scalar field on $\mathcal{M}$ (p. 103). We define a 4-vector field J by giving its components $(\rho, 0, 0, 0)$, relative to a Lorentz chart x, adapted to F. The spacelike part of J relative to F is 0 and is therefore equal to the current density in F, while the timelike part is of course the charge density ρ. To calculate the components of J relative to a Lorentz chart y, matched with x and adapted to a frame F', moving in F with speed v, we simply multiply $(\rho, 0, 0, 0)$ on the left by the Jacobian matrix (3.4.18′). We obtain: $(\beta\rho, -\beta\rho v, 0, 0)$. By the "length contraction" equations (3.5.1), the charge density in F' equals $\beta\rho$, while $(-\beta\rho v, 0, 0)$ are precisely the components of the current density in F', relative to the Cartesian chart $\bar{y}$, associated with y. The 4-vector J is aptly called the *4-current density*.

An even more striking example is provided by the *4-momentum* P of a particle π of proper mass m_0 and 4-velocity U. Its components, relative to a Lorentz chart x, adapted to a frame F, are defined by

$$\mathrm{P}^i = m_0 \mathrm{U}^i \tag{4.5.1}$$

Hence, by (4.3.5), $\mathrm{P}^0 = m_0\beta_u$ and $\mathrm{P}^\alpha = m_0\beta_u u^\alpha$, where u is the speed of π in F and the u^α are π's velocity components relative to the Cartesian chart $\bar{x}$, associated with x. In other words, the timelike part of P, relative to F, is π's energy $m(u)$, while its spacelike part is π's momentum $m(u)\mathbf{u}$ (4.4.13–14). The Principles of Energy and Momentum Conservation turn out to be no more than the phenomenological, frame-dependent expression of the deeper, absolute Principle of the Conservation of 4-Momentum.

The 4-momentum P, which incorporates π's energy and momentum in a single geometrical object, epitomizes the particle's dynamical state. Hence, if an external force on π is to be measured, as in classical physics, by the rate at which it changes the particle's dynamical state, its components, relative to the Lorentz chart x, must be given by

$$\mathrm{K}^i = \mathrm{dP}^i/\mathrm{d}\tau \tag{4.5.2}$$

It is not hard to show that the K^i are indeed the components of a 4-vector, the *4-force* K, introduced by Minkowski in 1908.[2] Relative to the frame F, in which our particle moves with speed u, the spacelike part $\mathbf{K}$ of K is β_u times the relative force **k** or rate of change of π's momentum, while its timelike part K^0 is β_u times the power $\mathbf{k} \cdot \mathbf{u}$ or rate of change of π's energy. Note that if F happens to be π's mirf, $\mathbf{K} = \mathbf{k}$ and $K^0 = 0$. Thus Minkowski achieves in effect with his 4-force what Einstein in 1905 had sought in vain: a physically motivated, frame-independent representation of the force on a particle, determined by the Newtonian force referred to the particle's own mirf (cf. p. 110).

Let us now consider a continuous distribution of matter. For each worldpoint P there is an inertial frame F_P in which the distribution is practically at rest in a small neighbourhood of P. We call F_P the *local inertial comoving frame* (licf) at P. Let δV^0 denote an infinitesimal "element of volume" in the licf, while δV denotes the volume it takes up in the relative space S_F of an arbitrary inertial frame F. Then, by (3.5.1), if the local speed of the material in F is u, $\delta V^0 = \beta_u \delta V$. The external force on the material is aptly represented, relative to F, by a time-dependent Cartesian vector field on S_F, the force density **f**, which is defined as follows: the relative force acting at time t on the volume element δV about a point **r** in S_F is equal to $\mathbf{f}(t, \mathbf{r})\delta V$. Suppose that the material in δV at the time in question moves in F with velocity **u**. The spacelike part **K**, relative to F, of the 4-force K on δV, is then equal to $\beta_u \mathbf{f}(t, \mathbf{r})\delta V$, while its timelike part K^0 equals $\beta_u \mathbf{u} \cdot \mathbf{f}(t, \mathbf{r})\delta V$. Clearly, then, $(K^0, \mathbf{K}) = (\mathbf{u} \cdot \mathbf{f}, \mathbf{f})\delta V^0$. Consequently, **f** and $\mathbf{u} \cdot \mathbf{f}$ are, respectively, the spacelike and the timelike parts, relative to F, of a 4-vector field, the *4-force density* f. Thus, we have that

$$K = f\delta V^0 \tag{4.5.3}$$

Few chapters in science and philosophy can match the beauty of the four-dimensional formulation of classical electrodynamics. The key to it was Minkowski's discovery that the electric and magnetic field vectors result from the decomposition of a 6-vector relative to an inertial frame. Since we conceive 6-vectors as skew-symmetric 4-tensors (p. 103), we shall refer to the *electromagnetic field tensor*. We denote it by F. The electric and magnetic fields **E** and **H** of classical electrodynamics are none other than its frame-bound facets, whose duality conceals its unity, though at the same time they hint at it by their curious interplay. Let x be a Lorentz chart adapted to the frame on whose space the fields **E** and **H** are defined, and let E_α and H_α be the components of **E** and **H** relative to the associated Cartesian chart $\bar{x}$. The components relative to x of the electromagnetic field tensor F in its contravariant form are then given by:[3]

$$(F^{ij}) = \begin{bmatrix} 0 & E_1 & E_2 & E_3 \\ -E_1 & 0 & H_3 & -H_2 \\ -E_2 & -H_3 & 0 & H_1 \\ -E_3 & H_2 & -H_1 & 0 \end{bmatrix} \tag{4.5.4}$$

An easy calculation shows that the components F'^{ij} of F relative to the Lorentz chart y, matched with x by equations (3.4.18), are obtained by replacing in the above matrix the unprimed values E_α, H_α, by the primed values E'_α, H'_α, obtained from them using equations (4.4.6). The latter are, of course, the components, relative to $\bar{y}$, of the electric and magnetic fields $\mathbf{E}'$ and $\mathbf{H}'$, defined on the space of the frame to which y is adapted.[4] The Maxwell equations for free space, in the presence of convection currents can then be elegantly stated as follows:

$$\varepsilon^{hjki}\frac{\partial \mathrm{F}_{jk}}{\partial x^h} = 0 \qquad \frac{\partial \mathrm{F}^{ij}}{\partial x^j} = 4\pi \mathrm{J}^i \tag{4.5.5}$$

Here the F_{ij} are the components, relative to x, of the covariant form of F (note 3), the J^i are the components of the 4-current density, and we set ε^{hjki} equal to 1, -1 or 0, depending on whether $hjki$ is an even or an odd permutation, or no permutation at all, of $\{0, 1, 2, 3\}$.[5]

Let π be a test particle of charge q and 4-velocity U, whose worldline traverses the electromagnetic field F. We want to calculate the components relative to a Lorentz chart of the 4-force K exerted by the field on π. Consider first a Lorentz chart adapted to π's mirf. Relative to it, $\mathrm{U}^0 = 1, \mathrm{U}^\alpha = 0, \mathrm{K}^0 = 0$, and $\mathrm{K}^\alpha = k_\alpha = qE_\alpha$ (by the definition of E). Hence the components of the 4-force on π, relative to such a chart, satisfy the equations

$$\mathrm{K}^i = q\mathrm{U}^k\mathrm{F}^i_k \tag{4.5.6}$$

(where the F^i_k are the components of the (1, 1) mixed form of F—see note 3). As (4.5.6) equates 4-vector components it remains valid if referred to an arbitrary Lorentz chart (page 305, note 7). A short calculation shows that, for $i = 1$, 2 or 3, (4.5.6) merely restates the Lorentz force law (4.4.17). For $i = 0$, it gives the equation of energy that follows from (4.4.17).

Consider now a time-dependent continuous charge distribution acted on by the field F. As on page 115 we define at each worldpoint P a licf F_P relative to which the charge distribution is practically static in a small neighbourhood of P. If δV^0 denotes the comoving element of volume and ρ is the licf charge density, the amount of charge contained in δV^0 is equal to $\rho\delta V^0$, so that, by (4.5.3) and (4.5.6), the components of the 4-force density f relative to a Lorentz chart adapted to the licf satisfy the equations:

$$\mathrm{f}^i = \mathrm{J}^k\mathrm{F}^i_k \tag{4.5.7}$$

Again, these equations remain valid if the tensorial components on either side are referred to an arbitrary Lorentz chart. By a well-known theorem of classical electrodynamics, the relative force density $\mathbf{f}$ on a charge distribution—i.e. the spatial part of the 4-force density f relative to a given inertial frame—stands in a simple relation to the Poynting vector $\mathbf{S} = (1/4\pi)\mathbf{E}\times\mathbf{H}$ and a symmetric Cartesian tensor $\mathbf{s}$, constructed from the electric and magnetic field strengths **E** and **H**. **s** is known as the Maxwell electromagnetic stress tensor, for Maxwell, who first introduced it, conceived it—on the analogy of the mechanical stress tensor (page 118)—as representing the stresses of the aether, the putative bearer of electromagnetic action.[6] The components of **s**, relative to the Cartesian chart $\bar{x}$, associated with a Lorentz chart x adapted to the inertial frame to which **E** and **H** are referred, are given by

$$s_{\alpha\beta} = -(1/4\pi)(E_\alpha E_\beta + H_\alpha H_\beta - (1/2)(\mathbf{E}^2 + \mathbf{H}^2)\delta_{\alpha\beta}) \tag{4.5.8}$$

(Note that **s** is a symmetric tensor.) The components of the force density **f** relative to $\bar{x}$ are then

$$f_\alpha = -\left(\frac{\partial S_\alpha}{\partial x^0} + \frac{\partial s_{\alpha\beta}}{\partial \bar{x}^\beta}\right) \tag{4.5.9a}$$

By another well-known theorem, if **u** is the velocity field of the charge distribution in the chosen frame, the equation of energy is

$$\mathbf{f}\cdot\mathbf{u} = -\left(\frac{\partial \sigma}{\partial x^0} + \frac{\partial S_\beta}{\partial \bar{x}^\beta}\right) \tag{4.5.9b}$$

where $\sigma = (\mathbf{E}^2 + \mathbf{H}^2)/8\pi$ is the energy density of the electric and magnetic fields. The left-hand sides of equations (4.5.9) are equal to the components, relative to x, of the 4-force density f. Substituting $\partial/\partial x^\beta$ for $\partial/\partial \bar{x}^\beta$ we obtain on the right-hand side the components of a 4-vector. Setting $\mathrm{S}^{00} = \sigma$, $\mathrm{S}^{0\alpha} = \mathrm{S}^{\alpha 0} = S_\alpha$, and $\mathrm{S}^{\alpha\beta} = s_{\alpha\beta}$, the relation between this 4-vector and the 4-force density f can be stated as follows:

$$\mathrm{f}^i = -\frac{\partial \mathrm{S}^{ij}}{\partial x^j} \tag{4.5.10}$$

From (4.5.10), (4.5.7) and the second set of equations (4.5.5), we conclude that

$$\frac{\partial \mathrm{S}^{ij}}{\partial x^j} = -\frac{1}{4\pi}\mathrm{F}^i_{\ k}\frac{\partial \mathrm{F}^{kj}}{\partial x^j} \tag{4.5.11}$$

Equations (4.5.11) can be solved to yield:[7]

$$\mathrm{S}^{ij} = -\frac{1}{4\pi}(\mathrm{F}^{ik}\mathrm{F}^j_{\ k} - \tfrac{1}{4}\eta^{ij}\mathrm{F}_{kh}\mathrm{F}^{kh}) \tag{4.5.12}$$

The S^{ij} are therefore the components, relative to x, of a symmetric contravariant 4-tensor field of order 2, which we shall denote by S.[8] This geometrical object known as the *electromagnetic energy tensor*, splits into three time-dependent fields on the relative space of any given inertial frame, namely, the energy density σ, the Poynting vector **S**, and the Maxwell stress-tensor **s**, which are thus seen to constitute the frame-dependent breakdown of a single physico-mathematical entity.

Following the discovery of the electromagnetic energy tensor,[9] Max von Laue (1911) inferred from the Relativity Principle that "in every domain of physics it must be possible to combine the force density and its power" into a 4-vector field. To each such 4-force density f there are infinitely many 4-tensor fields S that stand to it in the relation expressed by (4.5.10) (see note 7). Laue assumed that "in each domain of physics there is one among such 4-tensors whose components have a physical meaning corresponding to that of the components of the said electromagnetic tensor".[10] Laue postulated moreover that this physically significant solution of (4.5.10) would be, in each case, a symmetric 4-tensor field. As the diverse "domains of physics" interact in a given spacetime region, their respective tensor fields can be added to define a single symmetric tensor field, representing the dynamical state of the whole physical system that fills that region. As an illustration of these ideas, let us define the energy tensor of a material continuum, constructed by Laue from the energy and momentum densities and the mechanical stress tensor.[11] Let Σ denote a continuous distribution of mechanically interacting matter. In the relative space S_F of an inertial frame F the presence of Σ is indicated by a time-dependent scalar field, the energy density μ, while Σ's motion is described by a time-dependent Cartesian vector field, the velocity field $\mathbf{u}$. Each point of Σ is subject to stresses caused by the surrounding material. They are represented by a time-dependent Cartesian tensor field on S_F, the mechanical stress tensor $\mathbf{t}$. Let x be a Lorentz chart adapted to F and let $\bar{x}$ be the associated Cartesian chart on S_F. The nine components $t_{\alpha\beta}$ of $\mathbf{t}$, relative to $\bar{x}$, are defined as follows: Let $Q_{\mathbf{r}}^{\beta}$ be a small square centred at a point $\mathbf{r}$ of S_F and normal to the β-th parametric line of $\bar{x}$ through $\mathbf{r}$; then, the value of $t_{\alpha\beta}$ at $\mathbf{r}$, at time t, is equal to the α-th component, relative to $\bar{x}$, of the force per unit area exerted on $Q_{\mathbf{r}}^{\beta}$, at that time, by the material pressing it from the side on which the coordinate function $\bar{x}^{\beta}$ is smaller. If F happens to be the licf of our matter distribution at worldpoint $(t, \mathbf{r})$, $t_{\alpha\beta}(t, \mathbf{r}) = t_{\beta\alpha}(t, \mathbf{r})$, and the local value of $\mathbf{t}$ is a symmetric tensor.[12] We assume that Σ is subject to no external forces. The force density at each point is then 0, and the energy of the system is conserved. The equations of motion are given by[13]

$$\frac{\partial \mu u_\alpha}{\partial x^0} + \frac{\partial}{\partial \bar{x}^\beta}(t_{\alpha\beta} + \mu u_\alpha u_\beta) = 0 \tag{4.5.13a}$$

while the conservation of energy is expressed by the equation of continuity:

$$\frac{\partial \mu}{\partial x^0} + \frac{\partial \mu u_\beta}{\partial \bar{x}^\beta} = 0 \tag{4.5.13b}$$

The striking similarity of (4.5.13) with (4.5.9) suggests that we substitute x^α for $\bar{x}^\alpha$ and set $\mathbf{M}^{00} = \mu$, $\mathbf{M}^{0\alpha} = \mathbf{M}^{\alpha 0} = \mu u_\alpha$, and $\mathbf{M}^{\alpha\beta} = (t_{\alpha\beta} + \mu u_\alpha u_\beta)$. (4.5.13) can be read after (4.5.10), as a statement about the 4-force density on Σ:

$$\frac{\partial \mathrm{M}^{ij}}{\partial x^j} = 0 \tag{4.5.14}$$

The M^{ij} are the components of a contravariant 4-tensor field of order 2, the *energy tensor* of the material continuum Σ, which we denote by M.[14] The physical significance of equation (4.5.14) can be brought out by rewriting it as $\partial \mathrm{M}^{i0}/\partial x^0 = -\partial \mathrm{M}^{i\alpha}/\partial x^\alpha$, and integrating both sides over an arbitrary volume V bounded by a surface S in the relative space of frame F. In the light of our definitions it is clear that M^{i1}, M^{i2} and M^{i3} are the components, relative to $\bar{x}$, of a Cartesian vector field we shall denote by $\mathbf{M}^i$. The value of $\mathbf{M}^0$ at each point of space equals the local energy flux, while that of $\mathbf{M}^\alpha$ is the local flux of momentum in the direction of increasing $\bar{x}^\alpha$. (Remember that *force* equals *momentum transfer per unit time*.) We have, then, that

$$\frac{\mathrm{d}}{\mathrm{d}x^0}\int_V \mathrm{M}^{i0}\,\mathrm{d}V = \int_V \frac{\partial \mathrm{M}^{i0}}{\partial x^0}\mathrm{d}V = -\int_V \frac{\partial \mathrm{M}^{i\alpha}}{\partial x^\alpha}\mathrm{d}V = -\int_V \nabla\cdot\mathbf{M}^i\mathrm{d}V \tag{4.5.15a}$$

By Gauss's Theorem,

$$-\int_V \nabla\cdot\mathbf{M}^i\mathrm{d}V = -\int_S \mathbf{M}^i\cdot\mathbf{n}\,\mathrm{d}S \tag{4.5.15b}$$

(where $\mathbf{n}$ is an outward directed unit vector normal to the element of surface $\mathrm{d}S$). Equating the left-hand side of (4.5.15*a*) with the right-hand side of (4.5.15*b*) we see that the time rate of increase of each component of 4-momentum in V equals its inward flux across the boundary S. Equation (4.5.14) thus expresses the conservation of 4-momentum for a continuous distribution of mechanically interacting matter. The M^{ij} are the components of a contravariant 4-tensor field of order 2, the *energy tensor* of the material continuum Σ, which we denote by M.[14] Following Laue, we prove that M is symmetric by taking the components of its value at any worldpoint, relative to a chart adapted to the licf. Evidently $\mathrm{M}^{0\alpha} = \mathrm{M}^{\alpha 0} = 0$ (for the velocity field is locally 0 in the comoving frame), while $\mathrm{M}^{\alpha\beta} = t_{\alpha\beta} = t_{\beta\alpha} = \mathrm{M}^{\beta\alpha}$ (for $\mathbf{t}$, as we noted above, is symmetric in the licf). The symmetry of M follows (p. 305, note 7).[15]

As a further illustration of these ideas let us assume that the continuum Σ is a perfect fluid, or in other words that at each worldpoint the pressure is isotropic (i.e. the same in all directions) in the licf. Relative to a Lorentz chart adapted to the licf F_P at a given worldpoint P, the components of the energy tensor M have the local values

$$\bar{\mathrm{M}}^{00} = \rho \qquad \bar{\mathrm{M}}^{0\alpha} = \bar{\mathrm{M}}^{\alpha 0} = 0 \qquad \bar{\mathrm{M}}^{\alpha\beta} = p\delta_{\alpha\beta} \tag{4.5.16}$$

where ρ and p are the local (proper) *energy density* and the local *pressure*, respectively. To calculate the components M^{ij} of M relative to the inertial frame F in which the motion of the fluid Σ is represented by the velocity field **u** we use the fact that, due to the spatial isotropy of the $\overline{M}^{\alpha\beta}$, we may assume, without loss of generality, that F is related to F_P by the pk Lorentz transformation (3.4.19)—with the local value of $-\mathbf{u}$ substituted for the velocity **v**—plus, if need be, a translation which would make no difference to our present purpose. Denoting the typical element of the Jacobian matrix of the transformation by A^i_j (see (3.4.19′), we have that $M^{ij} = A^i_h A^j_k \overline{M}^{hk}$. Consequently:

$$M^{00} = (\rho + p\mathbf{u}^2)\beta_{\mathbf{u}}^2 \qquad M^{\alpha 0} = (\rho + p)u^\alpha \beta_{\mathbf{u}}^2 \qquad M^{\alpha\beta} = p\delta_{\alpha\beta} + (\rho + p)u^\alpha u^\beta \beta_{\mathbf{u}}^2 \tag{4.5.17}$$

where $\beta_{\mathbf{u}} = 1/\sqrt{1-\mathbf{u}^2}$. If we represent the motion of the fluid by the 4-velocity field U, with timelike and spacelike parts, relative to F, given by

$$U^0 = \beta_{\mathbf{u}} \qquad \mathbf{U} = \mathbf{u}\beta_{\mathbf{u}} \tag{4.5.18}$$

we can translate (4.5.17) into the following concise expression for the components of the energy tensor of a perfect fluid:

$$M^{ij} = (\rho + p)U^i U^j - p\eta^{ij} \tag{4.5.17*}$$

Suppose now that Σ is a continuous distribution of charged matter interacting with the electromagnetic field F. The 4-force density on Σ no longer vanishes, but is given, relative to the Lorentz chart x, by (4.5.10). Consequently,

$$\frac{\partial}{\partial x^j}(M^{ij} + S^{ij}) = 0 \tag{4.5.19}$$

But $M^{ij} + S^{ij}$ is precisely the (i, j)-th component, relative to x, of the sum of the tensor fields M and S. If Σ is subject to other external forces, the symmetric tensor fields corresponding to each of them according to Laue's assumption must also be added to M, in order to keep the right-hand side of (4.5.19) equal to 0. Let $T = M + S + \ldots$ denote the sum of all such tensor fields. We call T the total energy tensor. Its components, relative to a Lorentz chart x, satisfy the equations

$$\frac{\partial T^{ij}}{\partial x^j} = 0 \tag{4.5.20}$$

This law is to the physics of continua what 4-momentum conservation is to particle dynamics. It will play a cardinal role in connection with Einstein's field equations of gravitation. The latter, in turn, vindicate the postulated symmetry of the total energy tensor and its summands.[16]

Had we not chosen our units so that $c = 1$, diverse powers of c would turn up as factors of the timelike or the spacelike parts of our 4-vectors, and of some or all the components of our 4-tensors of higher order. What components must be multiplied by what powers of c will depend on the definition of the geometrical objects in question, and some freedom may be exercised in this respect. Such definitions are usually chosen on practical or aesthetical grounds. Our choice can also be fruitfully guided by the defined object's behaviour at the Newtonian limit, $c \to \infty$. W. G. Dixon (1978) consistently takes the latter approach, arriving at some very interesting comparisons.

4.6 The Causal Structure of Minkowski Spacetime

In a Newtonian world, a massive particle can attain any speed relative to a given inertial frame. Moreover, a signal can, in principle at least, be transmitted instantaneously from one spatial point to another, e.g. by poking at the addressee with a rigid rod. Neither possibility is available in the world of Special Relativity, where no signal can travel faster than light *in vacuo*, and massive particles cannot even go as fast as that. We saw in Section 3.7 that this feature of Special Relativity is essentially linked to the Lorentz group. We may therefore expect that it has an important bearing on the geometry of Minkowski spacetime. To see just how important this is we need some definitions.

We say that a worldpoint P *causally precedes* a worldpoint Q (symbolized: $P < Q$) if P and Q are viable locations for, respectively, the emission and the reception of a signal. If the signal in question can be relayed by massive particles, we say that P *chronologically precedes* $Q(P \ll Q)$. Both kinds of precedence are obviously transitive relations. To simplify some statements we stipulate that causal precedence is also reflexive (for every worldpoint P, $P < P$). If P precedes Q causally, but not chronologically, we say that P has the *horismos relation* with $Q(P \to Q)$. P and Q are *causally connectible* if either $P < Q$ or $Q < P$. "Chronologically connectible" and "horismotically connectible" are defined analogously. If P and Q are not causally connectible we say that they are *separate* worldpoints (symbolized: $P \| Q$). P's *causal future*, $J^+(P)$, is the set of all worldpoints that P causally precedes. Its *causal past*, $J^-(P)$, is the set of all worldpoints that causally precede P. P's *chronological future* and *past*, $I^+(P)$ and $I^-(P)$, and its *future* and *past horismoi*, $E^+(P)$ and $E^-(P)$, are defined analogously. Note that, by definition, P lies in the intersection of its causal past and future; also, its future (past) horismos is the complement, relative to its causal future (past), of its chronological future (past). The causal future $J^+(U)$ of a set of worldpoints U is the union of the causal futures of the elements of U. The sets $J^-(U)$, $I^+(U)$, etc. are defined analogously. We denote the union of $J^+(P)$ and $J^-(P)$ by $J(P)$. $I(P)$, $E(P)$, $J(U)$, etc. denotes the unions of the appropriate pasts and futures. The worldpoints separate from P we denote by $\|(P)$.

These concepts apply to Minkowski and Newtonian spacetimes and, more generally, to any point set that may be supposed to furnish locations for physical events. Their properties and relations depend of course on the prevalent physics and can be used for giving a broad characterization of the latter. Thus, for instance, if P is a point in Minkowski spacetime, the set $E(P)$ of all points horismotically connectible with P, is the null cone at P (p. 92). $E(P)$ is the common boundary of two disjoint open sets, namely, $I(P)$ and $\|(P)$. While $\|(P)$ is connected, $I(P)$ consists of two components, $I^+(P)$ and $I^-(P)$, whose boundaries are, respectively, $E^+(P)$ and $E^-(P)$. Since P is the only point in the intersection of the latter, it is the only point on the boundary of both $I^+(P)$ and $I^-(P)$. P lies also on the boundary of $\|(P)$, so that every neighbourhood of P contains a point separate from it. The partition of Minkowski spacetime into the three open sets $I^+(P)$, $I^-(P)$ and $\|(P)$ and the closed set $E(P)$ depends uniquely on P. One the other hand, if P is a point in Newtonian spacetime, $\|(P)$ is empty and $E(P)$ is the common boundary of the disjoint open sets $I^+(P)$ and $I^-(P)$. Any point in $E(P)$ is simultaneous with P and determines the same partition of Newtonian spacetime into the simultaneity class $E(P)$, its chronological future $I^+(E(P)) = I^+(P)$ and its chronological past $I^-(E(P))$. In other words, while each Minkowskian worldpoint has a chronological—and a causal—past and future all to itself, and is separate from—and yet near to—an open, connected region of spacetime, a Newtonian worldpoint is not separate from any other such point and shares its past and future with its contemporaries. This deep and far-reaching ontological difference between Newton's and Einstein's world follows immediately from the properties of the speed of signals and massive particles that we mentioned at the beginning of this Section. By definition, $E(P)$ is the locus of all worldpoints causally, but not chronologically, connectible with P. In Minkowski spacetime $E(P)$ contains, therefore, every location available for the emission or reception of direct light signals travelling *in vacuo* to or from P. Since, apart from P itself, no possible emission site can be a reception site, or vice versa, $E(P)$ naturally falls into two parts, $E^+(P)$ and $E^-(P)$, the two sheets of a null cone, whose only common element is the vertex P. In Newtonian spacetime, $E(P)$ is the locus of possible emissions and receptions of instantaneous signals to and from P and is therefore equal to the simultaneity class of P. Obviously, $E(P) = E^+(P) = E^-(P)$. In the Newtonian world, in which distant interaction demands no time and the speed of massive particles does not have an upper bound, the past and future horismoi of a worldpoint P, i.e. the boundaries of, respectively, its past and future, collapse into a single 3-space, the locus of worldpoints simultaneous with P. We see at once that, while causal precedence defines a partial ordering of Minkowski, but not of Newtonian spacetime, chronological precedence defines a total ordering of the simultaneity classes of the latter.

A. A. Robb (1914) derived the entire geometry of Minkowski spacetime from an axiom system with only two primitives (undefined predicates), namely, "x is an *element*" and "the element x is *after* the element y". Their intended meaning is, respectively, that x is a worldpoint and that y is different from x and causally precedes x.[1] H. Mehlberg (1937) proposed a different axiom system, in which the weaker, symmetrical relation of causal connectibility replaces causal precedence as a primitive.[2] Mehlberg's philosophical remarks tend to foster the claim—of which, however, I do not find any trace in Robb—that such axiom systems effect a reduction of spacetime geometry to causality. To form a proper judgment of the validity, or, at any rate, of the meaning of this claim, one must bear in mind that, in order to yield the theorems of Minkowski geometry, Robb's and Mehlberg's axioms must impose on the primitives much stronger constraints than our ordinary concepts of causality have ever been supposed to entail. The two-place relations we have dubbed "causal precedence" and "causal connectibility" acquire thereby a wealth of connotations that their ordinary language homonyms do not possess. Since, moreover, the relations in question do not, in the intended interpretation, hold between *actual* causes and effects, but between the spacetime locations where such *may* occur, it is plausible to say that Robb's and Mehlberg's systems give, not a reduction of geometric to causal relations, but rather a set of geometric conditions with which the basic forms of physical causality must comply, if the principles of Special Relativity hold good. One may then claim indeed that, in Special Relativity, spacetime is nothing but—and no less than—"the ordering schema governing causal chains",[3] which, so to speak, lays out the permissible avenues of physical action. But a geometric structure which thus *regulates* causality is not properly described as *reducible* to it.

Following E. H. Kronheimer and R. Penrose (1967) we regard Minkowski spacetime, thus conceived, as a paradigm and a special case of a more general mathematical structure, called a *causal space*. Let S be any set, and let $<$ and $\ll$ be two binary relations on S. We define a third binary relation $\rightarrow$ by the condition: for $P, Q \in S$, $P \rightarrow Q$ if and only if $P < Q$ and not $P \ll Q$. We designate $<$, $\ll$, and $\rightarrow$ by the same names as before and define the past and future sets $J^-(P)$, $J^+(P)$, $I^-(P)$, ..., $J^-(U)$, etc. for each $P \in S$, $U \subset S$, as above. We say that $(S, <, \ll)$ is a *causal space* if, for every P, Q, $R \in S$:

(i) $P < P$. (Causal precedence is reflexive.)
(ii) If $P < Q$ and $Q < R$, then $P < R$. (Causal precedence is transitive.)
(iii) If $P \ll Q$, then $P < Q$.
(iv) $P \ll P$ only if, for some $X \neq P$, $P \ll X$ and $X \ll P$.
(v) If $P < Q$ and $Q \ll R$, then $P \ll R$.
(vi) If $P \ll Q$ and $Q < R$, then $P \ll R$.
(vii) If $P < Q$ and $Q < P$, then $P = Q$. (Causal precedence is antisymmetric.)[4]

Let S be the underlying set of a causal space. If S is endowed with a topology it makes sense to speak of causal, chronological or horismotical curves in S. A continuous mapping $\gamma: I \rightarrow S$ (where I is a real interval) is a *future-directed causal curve* in S if $\gamma(t) < \gamma(t')$ whenever $t < t'$ $(t, t' \in I)$; γ is a *past-directed causal curve* if $\gamma(t) < \gamma(t')$ whenever $t' < t$. A curve γ is said to be *closed* if it returns twice to the same point in S (i.e. if there are in I three real numbers $t < t' < t''$, such that $\gamma(t) = \gamma(t'') \neq \gamma(t')$. Evidently, in a causal space there can be no closed causal curves. Chronological and horismotical curves of these several kinds are similarly defined. If S has a differentiable structure the curves under consideration may or may not be smooth. In such a case we shall usually assume that the following condition of *causal smoothness* is fulfilled: any smooth chronological, horismotical, causal or acausal curve in S can be smoothly extended to a curve of the same kind beyond any given point in its range. A vector at a point $P \in S$ will be said to be *causal* if it is, *acausal* if it is not tangent to a smooth causal curve in S. A causal vector is *horismotical* if it is tangent to a horismotical curve, and *chronological* otherwise. Horismotical and chronological vectors are past or future-directed, depending on the nature of the causal curves to which they are tangent. Note that our definitions are so contrived that the zero vector is a horismotical vector of both types. The condition of causal smoothness implies that a vector v at P is future-directed causal, chronological or horismotical if and only if $-v$ is a past-directed vector of the same kind. Consider an arbitrary set $A \subset S$. If no two points of A are joined by a chronological curve, A is said to be *achronal*. If no two points of A are joined by a causal curve, A is said to be *acausal*. A point $P \in S$ is said to lie in the *future domain of dependence* of A (symbolically: $P \in D^+(A)$) if every past-directed maximally extended causal curve through P meets A. Similarly, P belongs to the *past domain of dependence* of $A (P \in D^-(A))$ if every future-directed maximally extended causal curve through P meets A. The *domain of dependence* $D(A)$ of A is the union of its past and its future domains of dependence. If A is acausal and $D(A) = S$, A is said to be a (global) *Cauchy surface* in S.

Let $(S, <, \ll)$ be a causal space. If P. Q, $R \in S$, and $P \ll Q \ll R$, the intersection of $I^+(P)$ and $I^-(R)$ is said to be a *chronological neighbourhood* of Q. The collection of all chronological neighbourhoods of every point $P \in S$ constitutes a base of a topology, the *Alexandrov topology* of S. This is in a sense the natural causal topology of S.[5] Suppose now that S has a topology of its own, for instance, because it was first conceived as a differentiable manifold (e.g. Minkowski's world) and subsequently endowed with a causal structure. If the Alexandrov topology defined by the latter is weaker than (i.e. is included in) the original topology of S we shall say that S is a *topological causal space*. Such is the case if and only if the chronological past and futures of every point of S are open in the original topology. The original topology is, moreover, identical with the Alexandrov topology if and only if the latter is Hausdorff (i.e. if and only if any two distinct points of S lie in mutually disjoint chronological neighbourhoods). For this to be true it is necessary and sufficient that every neighbourhood of each $P \in S$ shall include an open neighbourhood of P into which no causal curve enters more than once.[6] A topological causal space that meets this condition is said to be strongly causal, or simply *strong*.

If the relations of causal and chronological precedence are given the meanings explained on page 121, Minkowski spacetime is indeed a strong topological causal space, satisfying the condition of causal smoothness. On the other hand, Newtonian spacetime, does not meet the condition (vii)— for any neighbourhood of a given worldpoint contains a closed causal curve through that point—and therefore is not a causal space. The three precedence relations in Minkowski spacetime satisfy, besides (i)–(vii), some stringent additional conditions which specify its peculiar causal structure. Robb showed that the same axioms required for defining the unique causal structure of Minkowski spacetime suffice also for characterizing its affine and metric structures (the latter, up to a dilatation). Unfortunately, Robb's elegant construction cannot be summarized in a few pages. In his study of "The Causal Theory of Spacetime", J. A. Winnie (1977) gives fairly simple definitions of Minkowskian straightness and congruence in terms of causal connectibility, but he does not spell out the conditions which his *definiens* must meet in order that the definitions may be satisfactory or even consistent. In the following sketch—which owes much to Winnie—I shall try to be explicit about the assumptions involved, even if, for brevity's sake, I do not analyze them. Let there be given a causal space. I assume, in the first place, that its underlying set of worldpoints is a 4-manifold on its own right, diffeomorphic to $\mathbf{R}^4$, or, in other words, that it is Minkowski's world $\mathcal{M}$.[7] A necessary and sufficient condition for the given structure $\langle \mathcal{M}, <, \ll \rangle$ to agree exactly with the causal structure of Minkowski spacetime is the existence of a global chart x of $\mathcal{M}$ that satisfies the following requirement: for any two points $P, Q \in \mathcal{M}$, such that $x^0(P) \leq x^0(Q)$, the real number $\eta_{ij}(x^i(P) - x^i(Q))(x^j(P) - x^j(Q))$ is, respectively, negative, positive or zero if and only if $P \| Q$, $P \ll Q$ or $P \to Q$.[8] Let $\mathcal{M}_P$ be the tangent space at a point $P \in \mathcal{M}$. If the given causal structure is that of Minkowski spacetime, the future and past-directed horismotical vectors in $\mathcal{M}_P$ lie, respectively, on the two sheets of a hypercone with the zero vector at its vertex, that is the common boundary of the two sets of acausal and chronological vectors in $\mathcal{M}_P$. These two sets are non-empty and open, the former being connected, while the latter has two components, which comprise the future-directed and the past-directed chronological vectors, respectively. Consequently, any pair of linearly independent causal vectors in $\mathcal{M}_P$ spans a subspace containing acausal vectors. On the other hand, a set of up to three linearly independent acausal vectors may span a subspace in which the only causal vector is 0. We say that a vector u is *conjugate* to a vector $v \in \mathcal{M}_P$ if $u + v$ and $u - v$ are horismotical vectors. By the condition of causal smoothness, if $u - v$ is horismotical, $-(u - v) = v - u$ is

RAG - E*

horismotical too. Hence conjugacy is a symmetric relation. The set of all vectors conjugate to a given vector v is its *conjugate class* $C(v)$. The reader ought to satisfy himself that $C(0)$ is the hypercone of horismotical vectors; that if v is horismotical, $C(v)$ is the subspace spanned by v; that if v is acausal, $C(v)$ consists exclusively of chronological vectors, and that if v is chronological, $C(v)$ is a set of acausal vectors which, together with v, span $\mathcal{M}_P$. We shall say that two vectors $u, v \in \mathcal{M}_P$ are *congruent* if and only if (i) there is a vector $w \in \mathcal{M}_P$ which is conjugate to both u and v, or (ii) there is a vector $w \in \mathcal{M}_P$ which is congruent, by (i), to both u and v.[9] The concepts of conjugacy and congruence are all we need for defining, up to a choice of unit, the Minkowski inner product $\langle , \rangle$ in $\mathcal{M}_P$. We pick out a chronological vector $t \in \mathcal{M}_P$, set $\langle t, t \rangle = 1$, and $\langle u, u \rangle = -1$ for any $u \in C(t)$. Since t and $C(t)$ together span $\mathcal{M}_P$, and, by bilinearity, for any $k \in \mathbf{R}$, $u, v \in \mathcal{M}_P$, $\langle ku, ku \rangle = k^2 \langle u, u \rangle$ and $2\langle u, v \rangle = \langle u + v, u + v \rangle - \langle u, u \rangle - \langle v, v \rangle$, our stipulation defines the inner product on all $\mathcal{M}_P$. We see at once that the partition of $\mathcal{M}$ into chronological, horismotical and acausal vectors, due to the causal structure, exactly matches its partition into timelike, null and spacelike vectors, by means of the inner product. Moreover, any otherwise defined Minkowski inner product on $\mathcal{M}_P$ that yields the same matching partition will differ from the above only by a constant factor, arising from the initial choice of a different unit vector. Since the proposed definition depends only on the topological properties of the partition of $\mathcal{M}$ into horismotical, chronological and acausal vectors, the Minkowski inner product can be introduced in the stated manner in the tangent space at any $P \in \mathcal{M}$, provided only that the given causal structure $\langle \mathcal{M}, <, \ll \rangle$ sufficiently ressembles that of Minkowski spacetime on a neighbourhood of P ("sufficiently" meaning "enough to ensure that a partition with the said topological properties obtains in $\mathcal{M}_P$"). But the inner products at each point cannot be welded into a single Minkowski metric field on $\mathcal{M}$ unless the given causal structure meets definite global conditions. Since we have chosen to avoid Robb's painstaking attention to detail I shall put those conditions in a nutshell, thus: The causal structure must be so laid out that, for any $P \in \mathcal{M}$, the additive group of the tangent space $\mathcal{M}_P$ can act transitively and effectively as a group of diffeomorphisms on $\mathcal{M}$ in such a way that the horismotical curves of the causal structure are precisely the straights generated under the action by the horismotical vectors (see p. 91). If a Minkowski inner product $\langle , \rangle_P$ has been defined on $\mathcal{M}_P$, for some $P \in \mathcal{M}$, in the manner indicated above, and H is an action of $\mathcal{M}_P$ on $\mathcal{M}$ which meets the foregoing requirement, H will propagate the inner product on $\mathcal{M}_P$ over all $\mathcal{M}$. Suppose that $Q = H(u, P)$ for some $u \in \mathcal{M}_P$. Let H_* denote the linear mapping of $\mathcal{M}_Q$ onto $\mathcal{M}_P$ by the differential of the diffeomorphism $X \mapsto H(-u, X)$. A Minkowski inner product $\langle , \rangle_Q$ is defined on M_Q by the condition that, for any $v, w \in \mathcal{M}_Q$, $\langle v, w \rangle_Q = \langle H_* v, H_* w \rangle_P$. The mapping $P \mapsto \langle , \rangle_P$ is then a Minkowski metric on $\mathcal{M}$. The horismotical curves of the causal structure are the null geodescis of the chosen metric. Let a and b be two such curves, generated by the same null vector $u \in \mathcal{M}_P$. Suppose that there is a point Q on a and a null vector $v \in \mathcal{M}_P$ such that $H(v, Q)$ lies on b. The 2-flat through Q generated, under H, by u and v is then what Robb called an "acceleration plane". It can be shown that all the straights generated under the action H are the intersections of acceleration planes. In particular, three points $A, B, C \in \mathcal{M}$ such that $A \ll B \ll C$ lie on a straight generated by a timelike vector if and only if there are two distinct acceleration planes containing A, B and C. The same condition is necessary and sufficient for A, B and C to lie on a straight generated by a spacelike vector, if $A \| B \| C \| A$. This provides a causal characterization of the spacelike and timelike straights of Minkowski spacetime.[10]

Perhaps the most striking expression of the link binding the causal and the affine metric structure of Minkowski spacetime is *Zeeman's Theorem*: The group of causal automorphisms of Minkowski spacetime is the group generated by the orthochronous Lorentz group and the dilatations, that is, the group of orthochronous similarities of the Minkowski geometry. A *causal automorphism* of a causal space is a permutation g of its underlying set, such that, for any points P and Q, $g(P) < g(Q)$ if and only if $P < Q$. Chronological and horismotical automorphisms are similarly defined (substitute $\ll$ or $\rightarrow$ for $<$). Clearly the causal automorphisms constitute a group. Let $\langle \mathcal{M}, \mathbf{R}^4, H, f_1 \rangle$ stand, as on page 92, for the Minkowski affine metric structure, with $\mathcal{M}$ an arbitrary set, H a transitive and effective action of $\mathbf{R}^4$ on $\mathcal{M}$ and f_1 the quadratic form on $\mathbf{R}^4$ defined on page 89. Two segments PP' and QQ' are *equivalent* in this structure if, for some $u \in \mathbf{R}^4$, $P = H(u, P')$ and $Q = H(u, Q')$. Two straights a and b are *parallel* if a segment of a is equivalent to a segment of b (and hence also if a and b are generated by the same vector). The Minkowski causal space $\langle \mathcal{M}, <, \ll \rangle$ is defined by the stipulation that, for any $P, Q \in \mathcal{M}$ such that $Q = H(u, P)$ for some $u \in \mathbf{R}^4$, $P \ll Q$ if $f_1(u) > 0 < u_0$ and $P \rightarrow Q$ if $f_1(u) = 0 \leq u_0$. Let G be the group of causal automorphisms of this space. It is not hard to see that every orthochronous similarity of the affine metric structure belongs to G. To prove Zeeman's Theorem we must show moreover that any $g \in G$ is an orthochronous similarity. A. D. Alexandrov and V. V. Ovchinnikova proved in 1953 that a causal automorphism $g \in G$ which is also an affine transformation of Minkowski spacetime is indeed equal to a Lorentz transformation (*modulo* a dilatation).[11] The main effort of Zeeman (1964) was directed at proving that every $g \in G$ is such an affine transformation. His proof rests on five lemmas. By Lemma 1, a permutation of $\mathcal{M}$ is a chronological automorphism of the Minkowski causal space if and only if it is a horismotical automorphism, and hence, if and only if it belongs to G.[12] The remaining four lemmas say that, if $g \in G$, it maps null straights onto null straights (2), preserving parallelism between null straights (3) and equivalence between null segments (5), and that the restriction of g to a null straight is an affine mapping

(4). Turning now to the proof itself, let us recall that the affine space $\langle \mathcal{M}, \mathbf{R}^4, H \rangle$ is made into a vector space by picking out a point $O \in \mathcal{M}$ and stipulating that the mapping of $\mathbf{R}^4$ onto $\mathcal{M}$ by $u \mapsto H(u, O)$ is a linear isomorphism. Let $\langle \mathcal{M}, O \rangle$ denote this vector space and write uO for $H(u, O)$. In $\langle \mathcal{M}, O \rangle$ scalar multiplication and vector addition are defined for any $k \in \mathbf{R}$ and any worldpoints $P = uO$ and $Q = vO$ by the equations $kP = kuO$ and $P + Q = (u + v)O$. A permutation $f: \mathcal{M} \to \mathcal{M}$ is an affine transformation of $\langle \mathcal{M}, \mathbf{R}^4, H \rangle$ if and only if it is a linear mapping of $\langle \mathcal{M}, O \rangle$ onto $\langle \mathcal{M}, f(O) \rangle$. Let $g \in G$. Choose four worldpoints P_1, P_2, P_3, and P_4, such that $P_i = u_i O$ for four linearly independent null vectors $u_i (i = 1, 2, 3, 4)$. Any $X \in \mathcal{M}$ is equal, in $\langle \mathcal{M}, O \rangle$, to a linear combination $\Sigma_i x_i P_i$. The mapping $h: X \mapsto \Sigma_i x_i g(P_i)$ is clearly a linear mapping of $\langle \mathcal{M}, O \rangle$ onto $\langle \mathcal{M}, h(O) \rangle$, and hence an affine transformation of $\langle \mathcal{M}, \mathbf{R}^4, H \rangle$. We shall show that $g = h$. Let M_k be the subspace of $\langle \mathcal{M}, O \rangle$ spanned by the worldpoints $P_j (1 \leq j \leq k \leq 4)$. If $X \in M_1$, $X = x_1 P_1$ for some $x_1 \in \mathbf{R}$. By our choice of P_1, M_1 is a null straight through O, and by Lemma 4, $g(X) = g(x_1 P_1) = x_1 g(P_1) = h(X)$. Hence, in particular, $g(O) = h(O)$ and $\langle \mathcal{M}, h(O) \rangle = \langle \mathcal{M}, g(O) \rangle$. Let $X \in M_k$ and assume that $g = h$ on M_{k-1}. Then, $X = Y + x_k P_k$ for some $Y \in M_{k-1}$. The segment YX is equivalent to the segment joining O to $x_k P_k$. By our choice of P_k the latter segment is null, so that the former is null too, and they both lie on parallel null straights. Hence, by Lemma 5, $g(Y) - g(X) = g(O) - g(x_k P_k)$. Recalling that h is linear, that by our inductive assumption $g(Y) = h(Y)$, and that by Lemma 4 g is linear on the subspace of $\langle \mathcal{M}, O \rangle$ spanned by P_k, we infer that, in $\langle \mathcal{M}, g(O) \rangle$, $g(X) = g(Y) + g(x_k P_k) = g(Y) + x_k g(P_k) = h(Y) + h(x_k P_k) = h(Y + x_k P_k) = h(X)$. (No summation over repeated indices!) Hence, $g = h$ on all $\mathcal{M}$ and is therefore an affine transformation of Minkowski spacetime.[13]

The proof of Zeeman's Theorem does not depend on any topological assumptions. The Theorem does imply indeed that the causal automorphisms of Minkowski spacetime are homeomorphisms of the standard manifold topology of Minkowski's world $\mathcal{M}$ (the topology of $\mathbf{R}^4$). However, though this topology is identical with the Alexandrov topology defined by the causal structure, Zeeman contends that it is "wrong" to attribute it to Minkowski spacetime, because it does not mark any distinction between spacelike and timelike curves and admits many homeomorphisms of no physical significance.[14] Zeeman (1967) proposes a different topology for Minkowski spacetime, which we may call the *Zeeman topology*. It is defined as the strongest topology that induces the 1-dimensional Euclidean topology (the topology of $\mathbf{R}$) on each timelike straight and the 3-dimensional Euclidean topology (the topology of $\mathbf{R}^3$) on each 3-flat generated by spacelike vectors.[15] The manifold topology of $\mathcal{M}$ also has these properties and is therefore weaker—in fact, strictly weaker—than the Zeeman topology. Zeeman proves that the group of homeomorphisms of his topology is precisely the full group of similarities of the Minkowski geometry (i.e. the group generated by the causal automorphisms and time reversal). The Zeeman topology has some curious properties. It induces the discrete topology on each null straight (i.e. the topology in which each singleton is an open set). Moreover, the range of any chronological curve, continuous in the Zeeman topology, consists of a series of straight timelike segments, like the worldline of a particle that moves freely between collisions. Hence, the worldline of a particle accelerated by a field of force cannot be the range of a continuous curve in a spacetime endowed with the Zeeman topology —a limitation that greatly impairs the latter's physical significance.[16] S. W. Hawking, A. R. King and P. J. McCarthy (1976) criticize the Zeeman topology on this and other grounds and propose another topology for causal spaces endowed with any standard 4-manifold topology of their own. (This includes all causal spaces of interest for General Relativity.) They call it the *path topology* and define it as the strongest topology that agrees on every continuous chronological curve with the topology induced in it by the standard 4-manifold topology of the space.[17] D. Malament (1977*c*) has shown that any permutation of such a space which preserves the path topology is also a causal automorphism, so that the causal structure is, so to speak, "encoded" in the path topology. One may object to such proposals that the new-fangled topologies put forward actually presuppose the standard manifold topology they are meant to supplant. Thus, the Zeeman topology is defined in terms of flats of the Minkowski affine metric structure, and the latter, as we know, generates linear isomorphisms of $\mathbf{R}^4$ onto the set of worldpoints, which naturally induce on it the $\mathbf{R}^4$ topology (page 92). Philosophical realists may argue that the concepts involved in such a definition are what enable *us* to grasp the defined notion from a position of ignorance, but need not belong to the essential nature of the real entity which the said notion purports to represent. However, it seems to me that a distinction and opposition between the order and connection of ideas and the order and connection of things is particularly out of place when we are dealing with mathematical structures. If we postulate the existence of an object of this kind we commit ourselves *eo ipso* to accepting the reality of whatever is required to make it intelligible.

CHAPTER 5

Einstein's Quest for a Theory of Gravitation

5.1 Gravitation and Relativity

At the turn of the century Newton's theory of gravitation was generally regarded as the most successful theory of physics. There were indeed a few unsolved anomalies, i.e. discrepancies between some observed quantities and their predicted values, the best known being the secular acceleration of the mean motion of the Moon and the secular advance of Mercury's perihelion.[1] Slight modifications of Newton's Law (1.7.1) had been suggested every now and then to explain the anomalies away, but the majority of physicists were confident that they would eventually be solved, without tampering with the Law's marvelous simplicity, by taking into account circumstances previously ignored or as yet unknown. Einstein expressed this general feeling when he said, at the 1913 Congress of Natural Scientists in Vienna, that

> [Newton's] laws have revealed themselves to be so exactly right that, from the standpoint of experience, there is no decisive ground for doubting their strict validity.[2]

A stronger motive for revising Newton's theory was the uneasiness felt by the less accommodating scientists and philosophers on account of its inherent notion of instantaneous action at a distance. A change in the position of any body relative to others immediately changes, according to Newton's Law, the gravitational pull experienced by the latter. The gravitational potential (p. 33) adjusts itself everywhere without delay to the slightest alteration in the density of matter at any point. The unlikelihood of this manner of action, so utterly at variance with our ordinary experience, was further enhanced when Hertz verified Maxwell's claim that the seemingly instantaneous effects of electromagnetic action propagated with finite speed. As Einstein told the 1913 Congress:

> After the untenability of the theory of action at a distance had thus been proved in the domain of electrodynamics, confidence in the correctness of Newton's action-at-a-distance theory of gravitation was shaken. One had to believe that Newton's Law of Gravity could not embrace the phenomena of gravitation in their entirety, any more than Coulomb's Law of electrostatics and magnetostatics embraced the totality of electromagnetic processes.[3]

A desire to develop "a theory of gravity resembling the electromagnetic theories as much as possible" and to show "that gravity can be attributed to actions that do not propagate faster than light", was the driving force behind Lorentz's "Considerations on Gravity" (1900).[4] He assumed that all ponderable matter consists of positively and negatively charged particles (electrons), and that a density of positive charge ρ moving with local velocity $\mathbf{v}$ generates the electromagnetic field $(\mathbf{d}, \mathbf{h})$, according to equations (2.2.1), on page 43, whereas a density of negative charge ρ' moving with local velocity $\mathbf{v}'$ generates the electromagnetic field $(\mathbf{d}', \mathbf{h}')$, according to the equations obtained by priming the relevant variables in (2.2.1). He further postulated that $(\mathbf{d}, \mathbf{h})$ exerts the Lorentz force $\mathbf{k}_1 = \alpha e(4\pi c^2\mathbf{d} + \mathbf{u} \times \mathbf{h})$ on a particle of positive charge e and velocity $\mathbf{u}$, and the Lorentz force $\mathbf{k}_3 = \beta e'(4\pi c^2\mathbf{d} + \mathbf{u}' \times \mathbf{h})$ on a particle of negative charge e' and velocity $\mathbf{u}'$ (where α and β are two constants such that α is slightly smaller than β), while $(\mathbf{d}', \mathbf{h}')$ exerts the forces $\mathbf{k}_2 = \beta e(4\pi c^2\mathbf{d}' + \mathbf{u} \times \mathbf{h}')$ on the former and $\mathbf{k}_4 = \alpha e'(4\pi c^2\mathbf{d}' + \mathbf{u}' \times \mathbf{h}')$. In a first approximation we may regard ordinary, electrically neutral bodies, as consisting of coupled and mutually interpenetrating positive and negative electrons, so that $\rho' = \rho$, everywhere. If we assume moreover that the local velocity of positively and negatively charged matter is the same it follows that $\mathbf{d}' = -\mathbf{d}$ and $\mathbf{h}' = -\mathbf{h}$. Two test particles with positive charge e and negative charge $e' = -e$, moving with the same velocity $\mathbf{u}$, will then experience, respectively, the forces $\mathbf{k}_1 + \mathbf{k}_2 = \mathbf{k}_1(1 - \beta/\alpha)$ and $\mathbf{k}_3 + \mathbf{k}_4 = \mathbf{k}_1(1 - \beta/\alpha)$. Since the force on each particle is the same, there is, under the circumstances, no electric field, but there is a resultant force on the two-particle system equal to $2\mathbf{k}_1(1 - \beta/\alpha)$. Lorentz expected that this force would suffice to explain the phenomena of gravity. It agrees with the Newtonian force for masses at rest. But if a body A moves with varying speed w about another body B, that moves inertially with speed v, the Newtonian attraction between A and B must be supplemented, in Lorentz's theory, with four force terms depending on v or w, of the second order relatively to "the very small quantities" v/c and w/c.[5] The assumption that gravitational action propagates with the speed of light c does not therefore clash with experience—as it would if the supplementary force terms were of the first order in v/c.[6] Lorentz calculated the precession of Mercury's perihelion according to his theory. The value he obtained was very close to the Newtonian prediction and too small to account for the anomaly.[7]

The old wish to get rid of instantaneous action at a distance became an urgent need with the advent of Special Relativity. As we know, the principles of this theory imply that information cannot travel faster than light, and hence, that nothing can react instantaneously, according to a deterministic law, to a change in the position of a distant body. The incompatibility of Newtonian gravitational theory with Special Relativity can be inferred from the fact that the formulae (1.7.1) and (1.7.4) on pages 32 and 33 do not preserve their form when subjected to a Lorentz transformation. It is also made evident by the following consideration: in a Newtonian two-body system the acceleration of each body at a given time depends on a simultaneous position of the other body; in Special Relativity, however, the latter is not univocally defined, but depends on the chosen inertial frame of reference—and what is simultaneous in one such frame is generally not simultaneous in another. We recall moreover that Newtonian gravity is a function of the inertial masses of the interacting bodies which, in Special Relativity, are also frame-dependent. Given the pervasiveness of gravitational phenomena and the Newtonian theory's success in predicting them, the situation was fraught with difficulty for Special Relativity.

The problem was first noted and tackled by Henri Poincaré, in the final section of his paper "Sur la dynamique de l'électron". Though his motivating statement is couched in the language of the Lorentz-Poincaré aether theory (Section 3.7), his treatment of the subject is well adjusted to the chronogeometric approach proper to Einsteinian Relativity. Indeed, it was for dealing with this question that Poincaré devised the mathematical methods subsequently taken up and developed by Minkowski (p. 86). Poincaré points out that the Lorentz–Poincaré theory would fully account for the impossibility of detecting absolute motion if all forces had an electromagnetic origin. But some forces, such as gravity, do not have that origin. One must therefore assume that every force "is affected by a translation (or, if you prefer, by the Lorentz transformation) in the same way as the electromagnetic forces".[9] It follows that gravitational attraction must depend not only on the positions, but also on the velocities, of the interacting bodies. Moreover, while it is natural to assume that the force acting on the attracted body A at a given moment depends on A's position and velocity at this moment, it shall depend on the position and velocity of the attracting body B at an earlier moment, "as if gravitation took some time to propagate". Also, since "astronomical observations do not seem to exhibit any significant deviation from Newton's law",[10] our Lorentz invariant formulae must agree completely with the latter if A and B are both at rest and must differ from it as little as possible if they are moving slowly. Choose a Lorentz chart x and let x_A^0 and x_B^0 denote, respectively, the times, in this chart, at which A suffers and B exerts the gravitational attraction that we wish to calculate. Let x_A^α be the space coordinates of A at time x_A^0 and x_B^α those of B at x_B^0. We further denote by U_A and U_B the 4-velocities of A (at the worldpoint mapped by x on x_A) and B (at the worldpoint mapped by x on x_B), and by F the gravitational 4-force exerted by B on A (as each goes through the said worldpoints). U_A^i, U_B^i and F^i are the components of these 4-vectors relative to x. We let R^i stand for the coordinate difference $x_B^i - x_A^i$. The R^i are plainly the components, relative to x, of a 4-vector which we shall denote by R. Poincaré seeks for Lorentz invariant laws stating the dependence of the time lag R^0 on the position and velocity components R^α, U_A^α and U_B^α, and of the 4-force F on R, U_A and U_B. For the former he proposes two alternatives: either $\langle \mathrm{R}, \mathrm{R} \rangle = \eta_{ij}\mathrm{R}^i\mathrm{R}^j = 0$, or $\langle \mathrm{R}, \mathrm{U}_B \rangle = \eta_{ij}\mathrm{R}^i\mathrm{U}_B^j = 0$. In the latter case, however, R^0 might turn out to be positive, implying that $x_B^0 > x_A^0$, a possibility that

Poincaré will not countenance, for "if one conceives that the effect of gravitation requires some time for propagating, it would be harder to understand how that effect could depend on a position not yet attained by the attracting body".[11] He assumes therefore that R^0 equals the negative square root of $R^\alpha R^\alpha$, so that gravitation propagates *in vacuo* with the speed of light (equal to 1 in our—and in Poincaré's—units).[12] As a 4-vector, F will depend on R, U_A and U_B linearly, or not at all. We must therefore have $F = \alpha R + \beta U_A + \gamma U_B$, where α, β, and γ are scalar functions of R, U_A and U_B. Tacitly assuming that the Newtonian gravitational constant times the "gravity charges" (the gravitational masses) of *A* and *B* equals 1, Poincaré concludes that the simplest solution that converges to the Newtonian Law as *A* and *B* approach rest is obtained by setting $\beta = 0$, $\gamma = \langle R, U_A \rangle / \alpha \langle U_A, U_B \rangle$ and $\alpha = \langle R, U_B \rangle^{-3}$. Consequently, in the special case considered:[13]

$$F = \langle R, U_B \rangle^{-3} \left(R - \frac{\langle R, U_A \rangle}{\langle U_A, U_B \rangle} U_B \right). \tag{5.1.1}$$

Poincaré points out that other, less simple solutions will also meet the prescribed requirements. He is well aware that the discussion has been sketchy and tentative, but he believes that it would be premature to carry it any further.

Two points are worth noting before we turn to Einstein's work on gravity. *In the first place*, Poincaré's force F is not the gradient of a potential. Poincaré's theory, if true, would explain gravitational phenomena by direct—though delayed—action-at-a-distance of one body on another, without mediation of an intervening field. Only in the Newtonian limit does it become equivalent to a field theory, based on Poisson's equation (1.7.4). Einstein's aim, on the other hand, was always a gravitational field theory, governed by differential equations that in the Newtonian limit converge to Poisson's, and thus, through the equivalence of the latter with Newton's Law, simulate, *in that ideal limit*, the magic of direct action-at-a-distance. *In the second place*, Poincaré, does not touch on the question, critical to any relativistic reform of Newton's theory of gravity, regarding the concept, or concepts, of mass (p. 32). In Lorentz invariant dynamics the inertial mass depends on velocity. Does the same hold for the gravitational mass, i.e. for the quantity, characteristic of each body, that determines the magnitude of the gravitational action it is capable of suffering or exerting? In the special case considered by Poincaré the gravitational masses of the attracting and the attracted body did not have to be mentioned by name. Nor did he need to worry about the inertial mass, as he only sought a Lorentz invariant expression of the force of gravity, not of the gravitating body's law of motion.[14]

5.2 The Principle of Equivalence.

At the begining of his article "Zur Dynamik bewegter Systeme"—submitted to the Berlin Academy on June 13th, 1907—Max Planck remarked that a series of ideas and laws which until then had been usually regarded as steadfast and almost self-evident presuppositions of all theoretical speculations in physics had to be radically revised in the light of recent research, and that some of the simplest and most important among them could no longer claim exact validity, even if they might still be useful as excellent approximations.[1] The last of the three examples with which Planck illustrates his remark is presented as follows:

> The thermic radiation in a perfectly void cavity surrounded by reflecting walls certainly possesses inertial mass. Does it also have ponderable mass? If, as it seems most likely, this question is to be answered in the negative, the generally assumed identity of inertial and ponderable mass, confirmed hitherto by all experiments, is evidently destroyed.[2]

A few months later, in the long review article on Relativity commissioned by the editor of the *Jahrbuch der Radioaktivität* and delivered on December 4th, 1907, Einstein made this identity, that Planck so readily called in question, into a matter of principle, which he thenceforth consistently treated as the key to the problem of gravitation. In § 11, "On the dependence of mass on energy", after arguing for the thesis that "all inertial mass is to be thought of as a store of energy",[3] he discusses the possibility—already suggested at the end of Einstein (1905*e*)—of comparing the mass loss suffered by a radioactive substance with the amount of energy radiated by it. Since the mass loss can only be measured by weighing the substance, the experiment "tacitly presupposes [. . .] that the inertia and the weight of a system are exactly proportional under all circumstances. Hence we must also assume, for instance, that the radiation enclosed in a void cavity not only possesses inertia, but weight as well".[4] After recalling that the proportionality between inertial and gravitational mass holds—with the hitherto achieved experimental accuracy—for all bodies without exception,[5] Einstein adds that in the final section of the paper he will propose a new argument in support of the preceding assumption. The argument rests on an alleged extension of the Relativity Principle, presented in § 17, which I shall now quote in full.

> Until now we have applied the Principle of Relativity, i.e. the assumption that natural laws are independent of the state of motion of the reference frame, only to *unaccelerated* frames. Is it conceivable that the Principle of Relativity also holds for frames that are accelerated relative to each other?
>
> This is not the place for a thoroughgoing discussion of this question. But since it cannot be avoided by anyone who has followed the applications of the Relativity Principle to this point, I shall here take a stance to the question.
>
> We consider two moving systems Σ_1 and Σ_2. Let Σ_1 be accelerated in the direction of its x-axis, and let γ be the (constant) magnitude of its acceleration. Let Σ_2 be at rest in a homogeneous gravitational field which imparts to all objects the acceleration $-\gamma$ in the direction of the x-axis.
>
> As far as we know, the physical laws referred to Σ_1 do not differ from the laws referred to Σ_2; this is due to the fact that all bodies undergo the same acceleration in the gravitational field. Hence, in the present state of our experience, we have no inducement [*Anlass*] to assume that the systems Σ_1 and Σ_2 differ in any respect. We agree therefore to assume from

now on the complete physical equivalence of a gravitational field and the corresponding acceleration of the reference frame.

This assumption extends the Principle of Relativity to the case of the uniformly accelerated translation of the reference system. The heuristic value of the assumption lies in the fact that it enables us to substitute a uniformly accelerated reference frame—which can, up to a point, be dealt with theoretically—for any homogeneous gravitational field.[6]

The assumption introduced here as an extension of the Relativity Principle to uniformly accelerated frames will later be called by Einstein the *Principle of Equivalence.* We note that the only empirical evidence adduced for it consists of the fact "that all bodies undergo the same acceleration in the gravitational field", or, in other words, that the inertial and the gravitational mass are proportional. The statement that they are always exactly so has been dubbed by R. H. Dicke the Weak Equivalence Principle. It implies that ordinary mechanical experiments with ponderable bodies will follow the same course in a uniformly accelerated laboratory and in a laboratory at rest in a matching homogeneous gravitational field. But Einstein, encouraged perhaps by the analogy with his RP—normally understood as extending Newtonian relativity from mechanics to the whole of physics—did not hesitate to proclaim that two such laboratories are *completely* equivalent, so that it is impossible to distinguish between them by means of physical experiments of any kind. Dicke calls this statement the Strong Equivalence Principle.[7] No analogue of the Michelson–Morley experiment was at hand to support the extension of the Equivalence Principle to radiation,[8] but, as we have seen, Einstein was content to point out that there was no evidence against it. In the same paper he inferred from the Strong Principle two effects of gravitation on radiant energy, which have been verified later, though with nothing like the precision attained in the verification of the Weak Principle: the bending of light-rays in a strong gravitational field and the frequency shift suffered by radiation as it goes up or down a gravitational gradient.[9] These effects can count, to some extent, as evidence for the Strong Principle. However, the most powerful inducement for embracing it is that it explains the equality of inertial and gravitational mass, which the Weak Principle merely asserts as a fact: if the systems Σ_1 and Σ_2 are physically equivalent, the same property of a body that manifests itself as *weight*—a propensity to fall—in Σ_2 must show up as *inertia*—a propensity to lag behind—in Σ_1.[10]

While Einstein's description of his strong Principle of Equivalence (hereafter EP) as "a natural extrapolation of one of the most universal empirical statements of physics"[11] is not unjustified, his original presentation of it as a generalization of the Relativity Principle is completely misleading. The physical equivalence between reference frames at rest in a homogeneous gravitational field and uniformly accelerated frames does not obliterate the physical inequivalence between the latter and inertial frames. It simply entails that a reference frame at rest in a gravitational field cannot be inertial. Thus, for instance, a light-ray emitted in Σ_1 or Σ_2 parallel to the y-axis does not keep this direction but inclines towards the decreasing x's; a light-pulse sent by a source at rest in Σ_1 or Σ_2 in the direction of increasing x (i.e. up the gravitational potential slope in Σ_2) will appear redder in the eyes of an observer placed in either system, the farther he lies from the source.[12] Indeed, if the system Σ_1, moving relative to the inertial frames with constant acceleration γ, is equivalent to the system Σ_2, at rest in a homogeneous field G in which all bodies gravitate with acceleration $-\gamma$, a frame immersed in G can behave as if it were inertial only if it is falling freely. Just as the inertial force that apparently acts on bodies referred to Σ_1 will disappear as soon as we refer them to an inertial frame, the gravitational pull they seem to suffer in Σ_2 can be made to vanish by referring them to a freely falling frame. The force of gravity appears thus as a kinematic—or chronogeometric—effect, that can be dispelled by a suitable coordinate transformation. The real gravitational field is of course inhomogeneous and none of the above need apply to it. However, one can always choose, around any worldpoint Q, a spacetime region U_Q in which the gravitational field is homogeneous to some agreed approximation. Let A be a body falling freely and without rotation, on whose worldpath lies Q. A chart x defined on U_Q, according to the standard rules for the construction of Lorentz charts (p. 54), by means of natural clocks and unstressed rods at rest in A, is called a *local Lorentz chart at Q*. The EP, as understood by Einstein, implies that no effect of gravity will be disclosed—within the agreed margin of precision—by any description of natural phenomena in terms of x; and that the laws of nature take the same form in U_Q, when referred to x, as they would when referred to an ordinary Lorentz chart in a spacetime region where gravity is absent.[13] Since gravity-free regions, as far as we can tell, are confined to Utopia, the EP implies that Special Relativity can in effect hold good only in the domains of local Lorentz charts, within the margin of accuracy to which the gravitational field is homogeneous there. Far from generalizing the Principle of Relativity, the EP drastically alters its meaning and restricts its scope. For let y be a local Lorentz chart at a point Q', distinct from Q. Then $x \cdot y^{-1}$ is generally *not* a Lorentz coordinate transformation because the domains of x and y do not usually overlap, and, even if they do, their respective frames generally move with non-constant velocity past each other. On the other hand, x and y can always be so chosen that their ranges coincide, at least in part. The mapping $x^{-1} \cdot y$—or a suitable restriction of it—is then indeed a bijection of a neighbourhood of Q' onto a neighbourhood of Q, which we may call a local Lorentz point transformation, and use for comparing experiments performed near Q and Q'. Conjoined with the EP, the RP implies that two experiments whose initial conditions read alike in terms of x and y will also have the same outcome in terms of these charts. (The modified RP can be expressed intuitively by rewriting the Galilei text on page 31 as a comparison between two non-rotating skylabs in orbit around any two celestial bodies, instead of two ship-cabins respectively at rest and in uniform motion; needless to say, all the empirical evidence for the RP is local and supports precisely its modified version.)

The domains of local Lorentz charts may cover indeed all spacetime, and thus the natural laws that hold in each can be said to hold everywhere. But see with what limitations: as referred to such local charts, the laws do not encompass the universe, but only some tiny, ephemerous fragment of it. The aggregate of such fragments, in each of which Special Relativity is approximately valid, can only be thought of as a cosmos in the light of another, more powerful and truly universal theory. This theory would have to weld into a single, continuous, though highly inhomogeneous gravitational field the mosaic of nearly homogeneous fields into which spacetime can be broken down. Einstein's quest for such a theory lasted eight years, until November 1915. But it probably did not begin in earnest before 1912, when he publicly acknowledged that the Theory of Relativity, in the original form, was valid only in "the important limiting case of spatiotemporal becoming in the presence of a constant gravitational potential"—i.e., in the absence of gravity—while its new version, enriched by the EP, could only be applied to "infinitesimal spaces".[14]

5.3 Gravitation and Geometry, circa 1912

Einstein returned to the subject of gravitation in 1911, in a paper "Concerning the influence of gravity on the propagation of light". He no longer presents here the Equivalence Principle as an extended Principle of Relativity. In §1 he argues that the EP was already part and parcel of Newtonian dynamics, though nobody had acknowledged its role in the foundations of natural philosophy.[1] Einstein grants that we are assured of the principle solely in its weak form, but contends that it attains a deeper significance only in the strong version, i.e., as the statement that "the laws of nature referred to [a coordinate system at rest in a homogeneous gravitational field with gravitational acceleration $-\gamma$] and those referred to [a coordinate system moving with uniform acceleration γ in the absence of gravity] are perfectly coincident".[2] In §2 he shows by means of one of his clever thought-experiments that radiation must gravitate if the weight of a body is proportional to its inertial mass, and, in agreement with Special Relativity, the latter increases when the body absorbs radiant energy. For if radiation were not affected by gravity and could be lifted at no cost, say, from the ground level A to a point B at a higher gravitational potential, we would be able to create energy by radiating it from A to B, and returning it from B to A stored in a falling body. After a full cycle the energy E initially radiated would increase to $E(1+\gamma hc^{-2})$ where γ is the magnitude of the gravitational acceleration, h the height of B above A and c ($=1$, in our units) is the speed of light. The conservation of energy requires therefore that the energy absorbed at B be less than that radiated from A. In § 3 Einstein uses the EP to prove, in the manner sketched in note 12 to Section 5.2, that radiation must also lose frequency as it travels from A to B.[3] As in 1907, he interprets this to imply that gravity has a chronometric significance. Radiant energy can be transferred from one place to another by an essentially stationary periodic process. The number of periods per unit of time can then be less at the receiving point than at the point of emission only if the time unit itself is longer at the latter than at the former. If we assume, as usual, that our equations predict the measurements that will be obtained with relativistic clocks obeying the clock hypothesis (p. 96), we must conclude that such clocks slow down as the gravitational potential decreases. Gravitation must therefore be linked to proper time—the quantity measured by relativistic clocks. As the speed of light *in vacuo* is supposed to be constant when measured with clocks free from the influence of gravity, Einstein infers that when measured with local clocks it must also vary with the gravitational potential. If c_φ denotes the speed of light, thus measured, at a point where the gravitational potential is φ, we must have that

$$c_\varphi = c_0(1 + \varphi c_\varphi^{-2}) \tag{5.3.1}$$

Hence, "according to this theory, the Principle of the Constancy of the Speed of Light does not hold in the formulation in which it usually underlies the ordinary [*gewöhnliche*] Theory of Relativity".[4] Einstein postulates, without adducing further evidence, that the equations derived from the EP for homogeneous gravitational fields also hold for inhomogeneous fields.[5] This enables him to calculate the redshift suffered by light coming to the Earth from the surface of the Sun as 2 Hz per million,[6] and the deviation toward the Sun of a light-ray grazing it as 0.83″.[7]

Except for the latter—erroneous—prediction, Einstein's paper of 1911 does not add much to what he had already said on the question of gravity in the review article of 1907, but he was now much more successful in drawing to it the attention of his fellow scientists. The year 1912 witnessed a veritable flowering of relativistic and post-relativistic theories of gravitation. Max Abraham incorporated Einstein's new conception of the dependence of the speed of light on gravity into a full-fledged field theory of gravitation, which used the Minkowskian idiom to convey a non-relativistic message.[8] Günnar Nordström proposed a different, no less comprehensive theory, based on inconditional adherence to the constancy of the speed of light.[9] Gustav Mie, in a long three-part article on the "Foundations of a Theory of Matter" put forward a strictly relativistic field theory which supposedly accounted for both electromagnetism and gravitation and was expected to throw light on the unsolved problems of microphysics.[10] Einstein himself published two papers dealing only with static gravitational fields, in which—as in Abraham's theory—the local speed of light is a measure of the potential.[11] But he must have continued to ponder over the subject in a broader, more realistic perspective, for in the following year he published, in collaboration with Marcel Grossmann, the outline of a radically new approach to gravity, together with a general theory, in partial agreement with it.[12] Though the new approach did not come to fruition until almost 3 years later, when Einstein discovered the field equations of General Relativity in November 1915, the Einstein–Grossmann paper of 1913 already subscribes to the following closely interrelated propositions, that set it apart from all earlier contributions to the subject.

A. It is assumed a priori that spacetime is a 4-manifold endowed with a Riemannian metric of signature -2 (i.e. a Lorentz metric, as defined on page 93) but the further determination of this metric is left to experience. Indeed, the assumption that it has the same signature as the Minkowski metric $\boldsymbol{\eta}$ is itself empirically motivated, insofar as it entails that every tangent space of the manifold is isometric with Minkowski spacetime, and thus accounts for the local success of Special Relativity.
B. The gravitational potential is represented by a symmetric, nowhere zero, (0, 2)-tensor field on the spacetime manifold. We denote it by **g**. Observe that, by definition, **g** is a Riemannian metric (pp. 93, 100). Naturally, **g** must depend on the actual distribution of matter.
C. The tensor field **g** is precisely the spacetime metric referred to under A. Natural clocks measure the length, determined by **g**, of their respective worldlines (p. 94, eqn. 4.2.5). Moreover, it is postulated that a freely falling particle describes a timelike Riemannian geodesic of the metric **g**, and that the worldline of a light-pulse is everywhere tangent to a null vector.[13] (This, in turn, implies that light-pulses describe null geodesics.)[14]

Before trying to ascertain the reasons that led Einstein to these momentous innovations, it will be useful to comment on the underlying spacetime geometry and the mathematical representation of the potential in the rival theories of gravity mentioned above.

Abraham's geometry is perplexing. He represents all physical magnitudes by 4-tensors and 4-tensor fields on spacetime and states the laws of nature as equations involving the components of these geometrical objects relative to a typical chart with space coordinates x^1, x^2 and x^3, and time coordinate $x^4 = ict$, where c is the speed of light and t a chronometric function which Abraham wastes no time in defining. The gravitational potential Φ is a scalar field satisfying the equation

$$\frac{\partial^2 \Phi}{\partial x^i \partial x^i} = 4\pi\gamma\nu \qquad (5.3.2)$$

(where γ is the gravitational constant, ν is the "density of energy at rest", and the index i ranges over $\{1, 2, 3, 4\}$). The 4-force F exerted by the gravitational field on a particle of unit mass has components

$$F^i = -\frac{\partial\Phi}{\partial x^i} \tag{5.3.3}$$

It follows at once that the "inner product" $F^i dx^i/d\tau$ of the 4-force by the particle's 4-velocity equals $-d\Phi/d\tau$, and must therefore be different from 0 if, as one would normally assume, the potential varies along the particle's worldline. Abraham unquestioningly equates the 4-force F with the particle's 4-acceleration A (components: $d^2x^i/d\tau^2$) and naturally concludes that the speed of light c cannot be a universal constant but must satisfy the equation $cdc = d\Phi$. Hence, the chart employed is certainly not a Lorentz chart. Abraham gives the line element squared as $ds^2 = dx^\alpha dx^\alpha - c^2dt^2$.[15] Consequently, his spacetime is indeed a Riemannian 4-manifold with a Lorentz metric, but the metric is not Minkowski's.[16] We may well ask what do Abraham's 4-tensor fields, their magnitudes and components mean? For, as we noted on p. 106, such geometrical objects only take components relative to Lorentz charts. And what is the sense of the differential operators that he allows to act on them? For he can hardly expect to obtain any physically significant quantities by differentiating tensor components with respect to the coordinates on an arbitrary Riemannian manifold. At first Abraham claimed that his charts were mutually related by Lorentz transformations in each infinitesimal region,[17] but Einstein had no difficulty in proving that he was mistaken.[18] Abraham accepted Einstein's argument but emphatically denied the implication that his own "conception of time and space [was] untenable even from a purely formal mathematical point of view".[19] According to him, Einstein had merely proved the untenability of "every relativistic conception of spacetime"; but he, Abraham, was willing to return to an absolutist spacetime theory and to single out the reference frame in which the gravitational field is static or quasi-static as the one in which bodies rest or move "absolutely".[20] Soon thereafter he brazenly announced that he preferred "to develop the new theory of gravitation without going into the space–time problem".[21] Yet he continued to express the laws of nature in the chronogeometrical idiom he had learnt from Minkowski and Sommerfeld. No wonder Einstein felt that Abraham's theory was "the monstruous child of perplexity".[22]

Gunnar Nordström (1912*a*) accepts equations (5.3.2) and (5.3.3), but aptly observes that they can be asserted together with the constancy of the speed of light c, if, as he will assume, mass is a function of the gravitational potential Φ. If m is the rest mass of a particle moving with 4-velocity U across the potential Φ, the equations of motion will be:

$$-m\frac{\partial\Phi}{\partial x^i} = m\frac{dU^i}{d\tau} + U^i\frac{dm}{d\tau} \tag{5.3.4}$$

We have that $U^iU^i = -c^2$ and $U^i\partial\Phi/\partial x^i = d\Phi/d\tau$. If c is constant, $U^i dU^i/d\tau = 0$. Multiplying both sides of (5.3.4) by U^i and summing over i, we obtain:

$$m\frac{d\Phi}{d\tau} = c^2\frac{dm}{d\tau} \tag{5.3.5}$$

Hence, $m = m_0 \exp(\Phi/c^2)$, where m_0 denotes the rest mass of the particle at potential $\Phi = 0$.[23] Another innovation subsequently introduced by Nordström was to regard the gravitational factor γ in (5.3.2) not as a constant, but as a function of (i) the internal state of the gravitating body, and (ii) the gravitational potential.[24] By assuming (i) one is able to define mass with considerable freedom. Thus Nordström (1913) proposes to equate ν, the "density of energy at rest", in the source term of equation (5.3.2), with the trace T^{ii} of the components, relative to the chosen chart, of Laue's energy tensor (p. 120).[25] This stipulation, plus the hypothesis (ii) that γ depends on Φ, will ensure, according to Nordström, that the Weak Equivalence Principle holds, at least in an important family of cases.[26]

Contrary to what one might expect, the universal constancy of c in Nordström's theory does not guarantee that the charts employed are Lorentzian or that spacetime has the Minkowski metric. For, as it turns out, the gravitational potential has an effect on rods and clocks that muddles up the physical meaning of the coordinates. Nordström squarely faces the issue. After proving that according to his theory the rest-length of a body decreases with the potential, he remarks that

> since the speed of light is constant, it is clear that the time in which a light-signal propagates from one end of a rod to the other increases in the same proportion as the rod's length. This time is therefore inversely proportional to the gravitational potential.[27]

Hence it is necessary to take the gravitational potential into account in the definition of the units of time and length. However, the units defined, say, at the potential of the Earth's surface can be used, according to Nordström, everywhere, if measurements are performed with telescopes and by exchange of light signals, "for light-rays propagate along straight lines with constant speed c".[28] In the last sentence there is an ambiguity that must be dispelled. If by velocity we mean coordinate velocity—i.e. the derivative of the position coordinates of the moving object with respect to the time coordinate—the constancy of c implies that the charts we use will map the worldline of any light-ray onto a straight segment in $\mathbf{R}^4$, but it does not tell us anything about the shape of the worldline itself, unless we already know the chronogeometric nature of the charts (e.g. that they are isometries). If, on the other hand, the speed of light is conceived—as in Nordström's discussion of the effects of gravity on rods and clocks—as a ratio of locally measured distance to local time, its constancy may indeed teach us something about the actual shape of light-rays in spacetime, but only if those local measurements are related to the global metric (e.g. by postulating that a good clock measures the length of its worldline, etc.). In either case we cannot adjudge the shape of light-rays or carry out measurements in the manner suggested by Nordström unless the spacetime metric is defined beforehand.

Einstein and Fokker (1914) threw light on the geometry underlying Nordström's theory a few months after the Einstein–Grossmann theory was made public. Their main purpose was to show that Nordström's equations can be derived from propositions A, B, and C on page 139, together with a single additional hypothesis; namely, that spacetime admits a family of global charts, with Cartesian space coordinates, in terms of which the coordinate speed of light is a universal constant. Let us call them Nordström charts. For simplicity's sake we set the said constant equal to 1. As on p. 139, we denote the spacetime metric and gravitational potential tensor by **g**. Let g_{ij} be its components relative to a Nordström chart x. It can then be proved that $-g_{00} = g_{11} = g_{22} = g_{33}$, and that $g_{ij} = 0$ if $i \neq j$.[29] Hence **g** stands to the Minkowski metric $\boldsymbol{\eta}$ in the simple relation $\mathbf{g} = \Phi^2\eta$, where Φ is a scalar field equal, at each worldpoint, to the local value of $\sqrt{(-g_{00})}$.[30] Einstein and Fokker discuss the mathematically possible and physically plausible relations between Φ and the energy tensor density by which they represent—as in the Einstein–Grossmann theory (p. 158)—the distribution of matter. Their argument leads to a field equation equivalent to Nordström's final version of (5.3.2), with the trace of the energy tensor density as source term, and Φ, of course, as the gravitational potential. The identification of Nordström's scalar potential Φ with the only independent component—relative to a Nordström chart—of the Einstein–Fokker metric tensor potential **g**, is consistent with the effect of Φ on natural clocks and rods and contributes to make it intelligible. Since the coordinate differentials $\mathrm{d}x^i$ of a Nordström chart x and the line element $\mathrm{d}s$ satisfy the equations $\mathrm{d}s^2 = -\Phi^2\mathrm{d}x^\alpha\mathrm{d}x^\alpha$ along any spacelike worldline on which $x^0 = \text{const.}$, and $\mathrm{d}s^2 = \Phi^2(\mathrm{d}x^0)^2$ along any parametric line of the time coordinate x^0, all "naturally measured times and lengths must be multiplied by the factor $1/\Phi$ in order to yield coordinate times and coordinate lengths".[31]

Let us now turn briefly to the third theory of gravity developed in 1912. This is part of the broad scheme of Mie's Theory of Matter, which was expressly designed to comply with Einstein's Relativity Principle, as originally conceived. Mie emphasizes that his field equations "are the same in every coordinate system that can be reached by means of a Lorentz transformation".[32] We may therefore presume—even though Mie does not discuss geometry—that his spacetime is Minkowskian and that his coordinate systems are Lorentz charts. The gravitational field of a material particle is represented by a 4-vector field with timelike part u and spacelike part **g**, related to a scalar field, the gravitational potential ω, by the equations

$$u = -\frac{\partial\omega}{\partial x^0} \qquad g_\alpha = \frac{\partial\omega}{\partial x^\alpha} \tag{5.3.6}$$

If the particle has gravitational rest mass m and moves with speed v in the direction of increasing x^1, the value of ω at a point P, reached from the particle by the radius vector $\mathbf{r}$, and with which the x^1 axis makes an angle θ, is given by:

$$\omega_P = \frac{\gamma m}{4\pi p|\mathbf{r}|} \tag{5.3.7}$$

where γ is a constant and $p = |(1 + v^2 \cos^2 \theta \, (1 - v^2)^{-1})^{1/2}|$.[33] Einstein showed considerably less interest in this theory than in Nordström's or even in Abraham's, because it openly disavows the Equivalence Principle, even in its weak form.[34] In Mie's theory the gravitational and the inertial mass of a particle are equal if the latter is at rest, but not if it is moving. Hence they differ also in any body that contains thermal energy. Since the difference will then depend on the specific heat of the body, it can in principle be detected by an Eötvös-type experiment. Mie reckons that such an experiment must be accurate to one part in 10^{11} or 10^{12} in order to test the predicted difference at room temperature.[35]

5.4 Departure from Flatness

Einstein and Grossmann's "Outline of a Generalized Theory of Relativity and a Theory of Gravitation" (1913) sets the stage for physical becoming in a Riemannian 4-manifold with a Lorentz metric which is equated to the gravitational potential. Since the latter depends on the presence and distribution of matter, the theory, when completed, should achieve Riemann's aim of explaining geometry by physical interactions.[1] Einstein learnt about Riemann's work and its formal development by Christoffel, Ricci and Levi-Cività, from his friend, the mathematician Marcel Grossmann, after joining the faculty of the Swiss Federal Polytechnic in Zürich in August 1912. There is evidence, however, that before benefiting from Grossmann's help he had already lighted on the key ideas of the new approach to gravity. We may conjecture, therefore, that Grossmann, upon hearing of Einstein's tentative thoughts, late in the Summer of 1912, saw that they could be made precise by couching them in the language of Riemannian geometry, and directed his friend's attention to the intellectual resources that this comparatively little known branch of mathematics placed at his disposal.

In this book, the general notion of an n-manifold (Riemann's "n-fold extended continuous manifold") is explained, in agreement with current mathematical usage, in Appendix A. The more specific notion of a Riemannian manifold was defined, together with the concepts of Riemannian metric and Riemannian geodesic, on pages 93–94. In order to understand Einstein's new approach to gravity we also need Riemann's concept of curvature. This is rigorously constructed in Appendix C. However, it will be useful to give here some informal, partly historical indications about it.

Riemannian curvature extends to n-dimensional manifolds the more intuitive and familiar concept of the Gaussian curvature of a surface. Let S be a smooth surface in Euclidean space, diffeomorphic to $\mathbf{R}^2$. Let P be any point of S. Every tangent to a curve drawn from P on S lies on a plane, the tangent plane at P, which we denote by S_P. If S is part of a cone or a cylinder, it can be mapped isometrically onto a neighbourhood of P in S_P. But this is not possible in general. In order to fit S snugly onto a region of the plane S_P one usually has to stretch it or shrink it at one place or another. A surface isometric to a region of the plane or pieced together from such surfaces is said to be *flat*, even if, like a cylinder, it looks perfectly round. The Gaussian curvature of our surface S measures, so to speak, its departure from flatness—in this sense of the word. The Gaussian curvature is a smooth real-valued function, i.e. a scalar field, on S. We cannot go here into Gauss's deep but difficult definition of this function.[2] I shall merely give, by way of illustration, one of Gauss's methods for calculating it. The reader is presumably familiar with the concept of the curvature of a plane curve.[3] A plane through the point P and orthogonal to S_P intersects S at a curve, called a normal section through P. The curvatures of all normal sections through P from a set of real numbers bounded above and below. It has therefore an infimum and a supremum, called the principal curvatures of S at P. The Gaussian curvature of S at P is equal to the product of the two principal curvatures. Hence, it can only be negative if the principal curvatures have different signs, i.e. if some of the normal sections through P lie to one side and some to to the other side of S_P. Otherwise it is always positive, unless one of the normal sections lies on S_P and is therefore straight, in which case the Gaussian curvature at P is 0. Observe that, as one would expect, it is 0 everywhere on a flat surface, and is constant on a sphere; but on arbitrary S—e.g. on the inner surface of a sea-shell—it varies smoothly from point to point. Gauss showed also how to calculate the Gaussian curvature solely in terms of the functions E, F and G, by means of which he specifies the metrical relations on a surface "intrinsically", without regard to the manner how it lies in space.[4] The mere existence of such an "intrinsic" formula for curvature proves Gauss's *Theorema Egregium*: "If a curved surface is developed upon any other surface whatever, the [Gaussian] curvature in each point remains unchanged."[5]

Let us now consider a point P in an n-manifold M with Riemannian metric $\boldsymbol{\mu}$. P has a neighbourhood U, diffeomorphic to $\mathbf{R}^n$, such that each $Q \in U$ is joined to P by a unique geodetic path wholly in U.[6] We shall define a diffeomorphism f of U onto a neighbourhood of the zero vector in the tangent space M_P. For each $Q \in U$ there is a parametrization γ of the geodetic path that joins Q to P such that γ is a geodesic, $\gamma(0) = P$ and $\gamma(1) = Q$. We set $f(Q)$ equal to γ_{*P}, the vector tangent to γ at P. Let v and w be two linearly independent vectors in M_P. They span a two-dimensional subspace $M_P(v, w) \subset M_P$. The intersection of $M_P(v, w)$ with $f(U)$ is the image by f of a two-dimensional submanifold of U which we shall denote by $S_{(v, w)}$. If M, as Riemann assumed, is a proper Riemannian manifold (p. 94), $S_{(v, w)}$, with the metric induced by $\boldsymbol{\mu}$, is isometric to a smooth surface in Euclidean space and consequently has a Gaussian curvature $k_{(v,w)}$ at P. We call $k_{(v,w)}$ the *sectional curvature* of M at P, determined by the tangent vectors v and w. If all sectional curvatures of M, at every $P \in M$, have the same value, M is a *Riemannian manifold of constant curvature.*[7] If, in particular, that constant value happens to be zero, the manifold M and its metric μ are said to be *flat*.

The foregoing discussion agrees with Riemann's treatment of curvature in his inaugural lecture of 1854. By 1861 he had found a (0, 4)-tensor field, strictly dependent on the metric μ of a Riemannian manifold M, which determines and is determined by the sectional curvatures defined above. We call it the *Riemann tensor*, and denote it by R.[8] It possesses the following interesting symmetries: for any vector fields X, Y, Z, W on M,

$$\mathrm{R}(\mathrm{X},\mathrm{Y},\mathrm{Z},\mathrm{W}) = -\mathrm{R}(\mathrm{Y},\mathrm{X},\mathrm{Z},\mathrm{W}) = -\mathrm{R}(\mathrm{X},\mathrm{Y},\mathrm{W},\mathrm{Z}) = \mathrm{R}(\mathrm{Z},\mathrm{W},\mathrm{X},\mathrm{Y}) \tag{5.4.1a}$$

$$\mathrm{R}(\mathrm{X},\mathrm{Y},\mathrm{Z},\mathrm{W}) + \mathrm{R}(\mathrm{X},\mathrm{Z},\mathrm{W},\mathrm{Y}) + \mathrm{R}(\mathrm{X},\mathrm{W},\mathrm{Y},\mathrm{Z}) = 0 \tag{5.4.1b}$$

It can be shown that the sectional curvature of M at P, determined by the vectors v and w, satisfies the equation:

$$k_{(v,w)} = \alpha \mathrm{R}_P(v, w, v, w) \tag{5.4.2}$$

where the factor α depends on the length of v and w and the angle between them.[9] On the other hand, the symmetries (5.4.1) imply that the sectional curvatures at P, determined by every pair of linearly independent vectors in M_P, determine the value of the Riemann tensor at P.[10] Hence, in particular, M is flat if and only if R is everywhere zero. Though we have defined sectional curvatures only for proper Riemannian manifolds, whose metric is positive definite, the Riemann tensor can be constructed from a Riemannian metric also if the latter is indefinite.[11] In this case, we simply use (5.4.2) as our *definition* of sectional curvature, setting $\alpha = 1$ when v and w are any pair of orthogonal unit vectors. The concepts of *flatness* and *constant curvature* are thereby extended to arbitrary Riemannian manifolds. The components of the Riemann tensor R relative to a chart x of the manifold M depend only on the first and second derivatives, $\partial\mu_{ij}/\partial x^h$, $\partial^2\mu_{ij}/\partial x^h \partial x^k$, of the components of the metric μ relative to the chart x.[12] Consequently, if M has a global chart relative to which the components of μ satisfy the relations $\mu_{ij} = \pm\delta_{ij}$, R is identically 0 and M is flat. Hence, according to our definitions, *Minkowski spacetime is a flat Riemannian manifold.* On the other hand, it can be shown that a Riemannian 4-manifold with a Lorentz metric is flat only if each point of it has a neighbourhood isometric with a region of Minkowski spacetime. A manifold with this property is said to be *locally Minkowskian.* Such locally Minkowskian manifolds are, of course, an exception among Riemannian 4-manifolds with a Lorentz metric which, in general, do not even have a constant curvature.[13] Hereafter any Riemannian 4-manifold with a Lorentz metric (i.e. a metric of signature -2) will be called a *relativistic spacetime*, or simply a *spacetime*. We refer to its underlying manifold as the *spacetime manifold*. The global topology of a 4-manifold is only minimally specified by the requirement that it should admit a Lorentz metric. The spacetime manifold need not be homeomorphic to $\mathbf{R}^4$ and can therefore differ considerably from Minkowski's world $\mathscr{M}$. For obvious physical reasons we shall always assume, however, that it is path-connected, i.e. that any two points of it can be joined by a curve.[14] When speaking of a given spacetime manifold we shall often denote it by $\mathscr{W}$ (for *world*). A point of $\mathscr{W}$ is a worldpoint, a path in $\mathscr{W}$ a worldline.

Why did Einstein give up the secure definiteness of Minkowski spacetime and decided to build the theory of gravity on the moving sands of general Riemannian geometry? As we look for an answer to this question we ought to recall that Einstein must have almost made up his mind on this matter before Grossmann introduced him to Riemann's theory. Otherwise, he would have shown little or no interest in this arduous and, at that time, esoteric subject.[15] The grounds for Einstein's decision must therefore have been such as might have led him to invent Riemannian geometry, *proprio Marte*, had it not been already available. In the Preface to the Czech edition of his popular exposition of Relativity, Einstein wrote:

> The decisive idea [*den entscheidenden Gedanken*] of the analogy between the mathematical problem associated with the theory and Gauss's theory of surfaces first occurred to me in 1912, after my return to Zürich, while I still did not know Riemann's and Ricci's or Levi-Cività's investigations.[16]

Which was this analogy that Einstein termed *entscheidend* (literally: "deciding", "effecting a decision")? It must have been striking enough to be perceived by someone who lacked a full grasp of one of its terms, and yet sufficiently deep to precipitate a revolution in natural philosophy. The parallel I shall now draw meets both requirements. And even though Einstein might well not have seen it until much later,[17] it will help us understand the less straightforward analogies that probably struck him first.

Take a smooth surface *S*. *S* can be entirely covered by charts that map a neighbourhood of each point diffeomorphically onto a region of the plane, just like a topographical survey covers a continent by charting it piecewise onto small flat papers. If the neighbourhoods are small enough, the diffeomorphisms can be so chosen that they are nearly isometric, their accuracy improving as the Gaussian curvature on their respective domains approaches 0. We may for certain purposes ignore the curvature, as we do when we read distances with a ruler on a road map. But it cannot be neglected in the long run; thus, the curvature of the Earth must certainly be reckoned with in intercontinental flights. Now consider spacetime, in the presence of ordinary gravitational fields, as we described it, in the light of the EP, on pages 136 f. It can be entirely covered by local Lorentz charts that map a small neighbourhood of each worldpoint diffeomorphically onto a region of $\mathbf{R}^4$. If $\mathbf{R}^4$ is endowed with the Minkowski inner product (p. 91), the local Lorentz charts are nearly isometric, their accuracy improving as the gravitational field in their respective domains approaches homogeneity. We can sometimes ignore the inhomogeneities of the field, e.g. in a local test of Special Relativity. But we shall not rise above a narrow parochial worldview unless we take them into account. Spacetime can be likened to an aggregate of small contiguous Minkowskian regions, just as a surface can be approximated by a flat-faced polyhedron; but gravity "warps" the former as curvature does the latter.

Though the foregoing analogy is both deep and striking there is no documentary evidence that in 1912 Einstein was thinking along such lines. Indeed, the link between gravity and the Riemann tensor, which the above comparison with Gaussian curvature obviously suggests, apparently dawned on him only in 1914.[18] It seems that at first he did not pay much attention to local Lorentz charts, perhaps because, for lack of a tangible example such as we have in our orbiting spaceships, he would not fancy the idea of referring the laws of physics to a freely falling frame. On the other hand, he was much impressed by the impossibility of constructing a metrically meaningful spacetime chart by means of rods and clocks at rest in a gravitational field. This fact, which may have reminded him of the non-isometric parametrizations by which curved surfaces were represented in the Gaussian tradition, is a direct consequence of the EP: a chart adapted to a uniformly accelerated frame cannot be metrically meaningful, for it maps the worldline of a particle at rest in the frame onto a straight segment in $\mathbf{R}^4$, though such a particle does not describe a geodesic in Minkowski spacetime (p. 96). Einstein however did not readily accept this consequence, for—as he remarks in his "Autobiographical Notes"—"it is not easy to free oneself from the idea that the coordinates must have an immediate metrical meaning".[19]

There is strong evidence that Einstein, enlightened by his reflections on uniformly rotating frames of reference, did free himself from this idea before moving to Zürich in August 1912.[20] The behaviour, in Special Relativity, of a uniformly rotating rigid body was the subject of conversations and letters between Einstein, Born and Sommerfeld after Born proposed his definition of rigid motion at a congress in Salzburg in September 1909.[21] On September 29th Einstein wrote to Sommerfeld that the problem interested him on account of a possible extension of the Relativity Principle to uniformly rotating frames, corresponding to its extension to uniformly accelerated frames in the final section of the survey paper of 1907 (cf. p. 134).[22] Though Einstein did not publish anything on the subject in the next few years, he must have pursued it privately, for in the first paper on the static gravitational field, submitted on February 26th, 1912, he cites a novel conclusion regarding rotating frames in support of an important remark. After postulating that the lengths measured with standard rods at rest in a homogeneous gravitational field or on a uniformly accelerated reference frame obey the laws of Euclidean geometry, he observes that this stipulation is not self-evident, and involves physical hypotheses that could eventually turn out to be false. Thus, for instance,

> in all likelihood, they do not hold in a uniformly rotating system, in which, due to the Lorentz contraction, the ratio of the circumference of a circle to its diameter would have to differ from π, if our definition of length is applied.[23]

Einstein did not spell out the grounds for this statement until much later, in 1916. In § 3 of "The Foundations of the General Theory of Relativity", he analyses the geometry of a uniformly rotating plane in order to show that the familiar description of spatial relations by means of isometric (Cartesian) coordinates defined by means of measuring rods "must be dropped and replaced by a more general one [. . .] if the Special Theory of Relativity holds in the limit in which there is no gravitational field".[24] Einstein's analysis can be rephrased as follows. We consider a region of the world in which no effects of gravity are felt. Let F_1 be an inertial frame and let the frame F_2 rotate with constant angular velocity ω about a fixed point O in F_1. (We may imagine each frame F_r as consisting of three mutually perpendicular rods, joined at one of their endpoints.) We denote by S_r the relative space of F_r, i.e. the set of spatial locations that, relatively to F_r, are permanently at rest. As usual, all metrical relations in S_r are to be ascertained by means of measuring rods at rest relative to F_r. It is assumed that every rod always agrees with a standard rod at rest in the former's momentary inertial rest frame (mirf). This entails that the rods at rest in F_2 are unaffected by acceleration, or that the effects of the latter—e.g. the longitudinal stress on rods issuing at right angles from the axis of rotation—have been duly corrected for.[25] The geometry set up by such means in the relative space S_r will be called the standard rod geometry G_r. If Special Relativity holds good in our gravitation-free region, G_1 must be Euclidean. In that case, however, as we shall now see, G_2 is not Euclidean. Let C_1 be a circle in S_1, centred at O and orthogonal to the axis of rotation of F_2. Let the radius of C_1 be R times the length of a standard rod, where $R|\omega|$ is less than c, the speed of light in F_1 (expressed in standard rod lengths per time unit), and yet R is so large that $2\pi R$ standard rods will fit snugly, end to end, on the circumference of C_1. Let C_2 be the locus of all points of S_2 that lie on C_1 at a given time. By reasons of symmetry, C_2 must be a circle, the same at all times (so that we do not need to worry about the definition of time in F_2). We shall now calculate the ratio of C_2's circumference to its radius in the standard rod geometry G_2. A standard rod at rest in F_2 along a radius of C_2 has zero longitudinal velocity relative to F_1. Hence the rod's length in F_1—i.e the G_1-distance between simultaneous positions of its endpoints in S_1—is equal to the length of a standard rod at rest in F_1. Consequently, the number of rods at rest in F_2 that can be placed end to end along a radius of C_2 is R. Consider now a standard rod at rest in F_2 along the circumference of C_2. At any time the rod's mirf moves relative to F_1, along a tangent to C_1, with speed $v = R|\omega|$. Hence the rod's length in F_1 is $\sqrt{(1 - v^2c^{-2})}$ times the length of a standard rod at rest in F_1. How many standard rods at rest in F_2 can fit snugly, end to end, along the circumference of C_2? Evidently, as many as can leave their distinct imprints simultaneously (in F_1-time), end to end, on the circumference of C_1, namely, $2\pi R/\sqrt{(1 - v^2c^{-2})}$. Hence the ratio of the circumference of C_2 to its radius is greater than 2π. The standard rod geometry G_2 is therefore Non-Euclidean. It follows that S_2—with this geometry—cannot be mapped isometrically onto $\mathbf{R}^3$ or be described by means of Cartesian coordinates.[26]

By the beginning of the Summer of 1912 Einstein was thoroughly convinced that physics could no longer rely on the availability of isometric coordinate systems and that the laws and observations involving the effects of gravity had to be referred to spacetime charts whose precise metrical significance was not fixed *a priori*. In his reply to Abraham (1912*d*), submitted on July 4th, he emphasized that even in the "very special case of the gravitation of masses at rest, the spacetime coordinates will forfeit their simple physical interpretation".[27] The use of non-isometric coordinate systems was indeed a well-known feature of Gauss's theory of surfaces. Such systems can be defined on any surface diffeomorphic to $\mathbf{R}^2$.[28] They are unavoidable, however, if the surface under consideration is not flat—and hence not isometric with $\mathbf{R}^2$. At some point in the course of his reflections Einstein must have been struck by the analogy between this familiar fact and the situation emerging in the relativistic theory of spacetime. Here again arbitrary charts can be introduced at will on a neighbourhood of any worldpoint. But if spacetime is Minkowskian they can always be replaced by a metrically meaningful Lorentz chart. When he wrote the reply to Abraham, Einstein had already understood, however, that Lorentz charts cannot be defined—not even by means of freely falling rods and clocks—on the domain of a real gravitational field, that changes with place and time.[29] Thus gravity would prevent a metrically significant charting of the world, much like Gaussian curvature precludes the definition of Cartesian coordinates on a surface that is not flat. Though the analogy is clear enough, it must have seemed amazing, for Einstein did not yield to its lure until he found a reason that made it comprehensible. This was provided by the hypothesis—included in Proposition C on page 139 that a freely falling particle describes a geodesic of the spacetime metric constituted by the gravitational potential. In his "Notes on the Origin of the General Theory of Relativity" (1934), Einstein explains how this hypothesis enabled him at last, in 1912, to see what coordinates really meant in physics.[30] I shall now try to reconstruct the considerations that probably led him to it.[31]

According to the Law of Inertia a free particle π describes a timelike geodesic in Minkowski spacetime (p. 96). Let γ be the worldline of π, parametrized by proper time τ. γ is then "length-critical" on every closed subinterval of its domain (pp. 94–95). This property of γ is expressed in the classical notation of the Calculus of Variations by the variational equation

$$\delta \int d\tau = 0 \tag{5.4.3}$$

which says that γ's length, $\int d\tau$, remains stationary in every variation of γ (see note 6, on p. 304). In terms of a Lorentz chart x, the element of proper time $d\tau$ is given by

$$d\tau^2 = \eta_{ij} dx^i dx^j \tag{5.4.4}$$

and γ satisfies the equations:

$$\frac{d^2 x^i \cdot \gamma}{d\tau^2} = 0 \tag{5.4.5}$$

In terms of an arbitrary chart y, defined on a neighbourhood of π's worldline,

$$d\tau^2 = g_{ij} dy^i dy^j \tag{5.4.6a}$$

where the coefficients g_{ij} are the components $\boldsymbol{\eta}(\partial/\partial y^i, \partial/\partial y^j)$ of the Minkowski metric $\boldsymbol{\eta}$ relative to y. (Note that, according to this definition, the right-hand sides of (5.4.4) and (5.4.6a) are equal; we have replaced the Lorentz chart by arbitrary coordinates but have retained the same flat metric.) Writing g^{ij} for the components, relative to y, of the contravariant form of $\boldsymbol{\eta}$ (p. 100), we introduce the functions[32]

$$\Gamma^i_{jk} = \tfrac{1}{2} g^{ih} \left(\frac{\partial g_{hj}}{\partial y^k} + \frac{\partial g_{kh}}{\partial y^j} - \frac{\partial g_{jk}}{\partial y^h} \right) \tag{5.4.6b}$$

Equations (5.4.3) and (5.4.6) imply that γ also satisfies the system:

$$\frac{d^2 y^i \cdot \gamma}{d\tau^2} = -\Gamma^i_{jk} \cdot \gamma \frac{dy^j \cdot \gamma}{d\tau} \frac{dy^k \cdot \gamma}{d\tau} \tag{5.4.7}$$

To an observer who, for any reason, views y as metrically significant, the right-hand sides of equations (5.4.7) will appear as the components of a 4-force that acts on π as it traverses the field represented by the Γ^i_{jk}. However, the 4-force and its supporting field can be made to vanish—like the inertial forces of classical mechanics—by a coordinate transformation, namely, $x \cdot y^{-1}$, where x, a Lorentz chart, truly discloses the metric.

Suppose now that the particle π is falling freely in a real, inhomogeneous gravitational field. Its worldline will successively pass through the domains U_1, U_2, . . . of diverse local Lorentz charts $x_1, x_2, \ldots$ On each such domain U, a natural clock satisfying the clock hypothesis (p. 96) measures along its worldline the proper time $\int d\tau$ of a Minkowski metric on U within the margin of precision to which the field is homogeneous on U. But the several flat metrics which thus approximately hold good in the domain of each local

Lorentz chart cannot be regarded as the restriction to their respective domains of a global Minkowski metric, any more than the Euclidean metrics that show up, say, in the street plans of Mannheim and Manhattan are the restrictions to these boroughs of a Euclidean metric defined on the entire Earth. On each domain U the worldline of π satisfies, within the said margin of error, the variational law (5.4.3); yet the law does not make any sense beyond the boundary of U. However, the following conjecture is near at hand. Just as a tape-line measures, at any place of the Earth, the line elements of both the local Euclidean and the global "geoidal" metric, so a natural clock might measure, at any worldpoint, the element of proper time of the local Minkowski metric and the global, still to be determined, metric of spacetime. All that is required is that the global metric—which we henceforth denote by $\mathbf{g}$—be approximated, on a small neighbourhood of each worldpoint, by the local flat metric $\boldsymbol{\eta}$. To ensure it we must only demand (i) that $\mathbf{g}$ has the Lorentz signature and (ii) that the local $\boldsymbol{\eta}$ agrees with $\mathbf{g}$ at the origin of the pertinent local Lorentz chart.[33] If this is allowed, we can postulate further that, if $\int d\tau$ is now made to stand for the length of π's worldline as determined by $\mathbf{g}$—i.e. for the proper time measured along it by a natural clock at π—the variational principle

$$\delta \int d\tau = 0 \tag{5.4.3*}$$

is obeyed by the freely falling particle π *between any two events in its history*. We call this postulate the Geodesic Law of Motion, or simply the *Geodesic Law*. Einstein has referred to it as "a natural" extension of the Principle of Equivalence to arbitrary gravitational fields,[34] and also as "the core of the 'Hypothesis of Equivalence'".[35] But it is certainly not a corollary of the EP, as the typographical identity of (5.4.3) and (5.4.3*) might suggest. The line element of the local flat metric $\boldsymbol{\eta}$ which occurs in (5.4.3) is not to be confused with the line element of the metric $\mathbf{g}$ which occurs in (5.4.3*), even if, by hypothesis, they are both measured at each worldpoint with the same instrument.[36] If the g_{ij} and g^{ij} in (5.4.6) denote the components of $\mathbf{g}$—not $\boldsymbol{\eta}$—relative to an arbitrary chart y, then, in the domain of y, the worldline γ of the freely falling π must satisfy the equations:

$$\frac{d^2 y^i \cdot \gamma}{d\tau^2} = -\Gamma^i_{jk} \cdot \gamma \frac{dy^j \cdot \gamma}{d\tau} \frac{dy^k \cdot \gamma}{d\tau} \tag{5.4.7*}$$

The functions Γ^i_{jk} appear here in the guise of "gravitational field strengths", derived from the "potentials" g_{ij}. Since in the presence of inhomogeneous gravity a coordinate transformation cannot reduce the g_{ij} to the constant functions η_{ij} over a finite region, the spacetime metric $\mathbf{g}$ is not flat. This explains why the coordinate systems of physics, which chart the world into the flat "number manifold" $\mathbf{R}^4$, do not allow a direct metrical interpretation. As Einstein expressed it in London in 1921:

> The metrically real is now given only through the combination of the space-time coordinates with the mathematical quantities which describe the gravitational field.[37]

Though the foregoing ideas could not become perspicuous until Einstein had learnt the essentials of Riemannian geometry with Grossmann, they were hovering in his mind several months before he joined his friend in Zürich. At the end of his second paper on the static gravitational field he observes that, if we set $H = -m\sqrt{(c^2 - v^2)}$, where m and v are, respectively, the rest mass and the speed of a particle, and c denotes the local speed of light (which in that work is a function of the gravitational potential—p. 138), the particle's free fall in a static field obeys the variational principle "$\delta \int H \mathrm{d}t = 0$, or $\delta \int \sqrt{(c^2 \mathrm{d}t^2 - \mathrm{d}x^2 - \mathrm{d}y^2 - \mathrm{d}z^2)} = 0$. He then remarks: "The second Hamiltonian equation suggests [*lässt ahnen*] how the equations of motion of a particle in a dynamical gravitational field are built."[38]

5.5 General Covariance and the Einstein–Grossmann Theory.

A particular relativistic spacetime may admit a chart or a family of charts specially suited to its structure. Thus, the Lorentz charts peculiar to flat spacetime are admirably fit for describing it. A Lorentz chart x partitions flat spacetime into equidistant hypersurfaces $x^0 = const.$, each of which is a flat spacelike 3-manifold, isometric to Euclidean space. x is an isometry on each such hypersurface and also on each parametric line of the time coordinate function x^0. Moreover, x is a global isometry of spacetime onto $\mathbf{R}^4$, if the latter is endowed with the Minkowskian—instead of the usual Euclidean—inner product. On the other hand, x can only be defined if spacetime is flat. Since Special Relativity assumed from the outset that Lorentz charts were readily definable in the real world, it naturally assigned them a preferred status and represented all physical quantities and expressed all natural laws in terms of such charts. But if nothing is known a priori about the actual geometry of spacetime and the specific charts that it admits, the representation of physical quantities and the formulation of natural laws must be referred, in principle, to arbitrary charts. (Unless, of course, one resorts to the chartless or "coordinate-free" mode of expression, an alternative that was not yet available in 1913, due to lack of clarity regarding the nature of geometrical objects.[1])

The new approach to gravity proposed in Einstein and Grossmann's "Outline" of 1913 therefore involves the demand that all physically significant quantities and the equations between them be referred indifferently to any conceivable spacetime chart. In 1916 Einstein stated this demand as the Postulate of General Covariance:

> The universal laws of nature must be expressed by equations which hold good for all coordinate systems, that is, are covariant with respect to arbitrary transformations (generally covariant).[2]

Compliance with this postulate obviously implies the following broadened version of Einstein's Relativity Principle of 1905:

> **The Principle of General Relativity** (GRP). The laws by which the states of physical systems undergo change are not affected, whether these changes of state be referred to one or the other of two arbitrary spacetime charts.

By deliberately imitating the wording of the 1905 RP (p. 54) I hope to make clear why Einstein persistently maintained that the new approach to gravity involves a generalization of Relativity. This view, embodied in the name "General Relativity" by which we still designate the theory of gravity in which the new approach finally proved its worth, is certainly legitimate. But we must not let it obscure the considerable differences of meaning and motivation which separate the first or "special" RP from the GRP proposed above. Though the former may seem to be, like the latter, a purely methodological rule for the statement of laws, it yields immediately—through the one–one correspondence between Lorentz coordinate and point transformations—a genuine factual principle concerning the observable course of phenomena (p. 65). Indeed it was the factual principle, corroborated by an increasing store of data that motivated the adoption of the rule of method: the laws of nature must take the same form with respect to all Lorentz charts because the inertial frames to which such charts are adapted are observationally equivalent. On the other hand, the GRP cannot rest on such grounds. In the first place, there is no one–one correspondence between arbitrary coordinate transformations and point transformations of spacetime. If x and y are two arbitrary charts such that $x \cdot y^{-1}$ is defined on some open subset of $\mathbf{R}^4$, $x^{-1} \cdot y$ may just map the empty set onto itself, for the ranges of x and y need not overlap. (This is in sharp contrast with the case of Lorentz charts, all of which share the same range—viz., $\mathbf{R}^4$—and the same domain—flat spacetime.) In the second place, even if $x^{-1} \cdot y$ does map a region of spacetime onto another, this does not entail the equivalence of the frames to which x and y are adapted, for there might not be any such frames at all. Suppose however that x and y are each linked to something that we may reasonably call a frame of reference. Suppose, for example, that each chart has a clear-cut time coordinate function. Such is the nature of, say, x^0 if the fibre of x^0 over each of its values is a spacelike 3-manifold and its parametric lines (i.e. the integral curves of $\partial/\partial x^0$) constitute a congruence of timelike worldlines in the chart's domain. Since a congruence of timelike worldlines is the geometrical equivalent of a frame (p. 28), we may say that our charts x and y are adapted to the frames—i.e. the congruences—F_x and F_y defined by their respective time coordinates. (Since the lines of each congruence need not be equidistant, such reference frames are not rigid—they are more like a creeping mollusc than a standing grille.) However, the fact that the laws of nature take the same form when referred to x or y does not imply that F_x and F_y are physically equivalent frames. They evidently are not equivalent if one of them is and the other is not a congruence of geodesics, or if both are congruences of geodesics but F_x covers a strongly curved region of spacetime and F_y an almost flat one.[3] Indeed such results were to be expected, given the motivation for the GRP. Acceptance of this principle is based not on physical knowledge but on geometrical ignorance; not on experiments purporting to show that under equal conditions all physical phenomena have the same course of development in any frame of reference, but rather on our inability to decide a priori the structure of spacetime and the metrical significance of our coordinate systems.[4] The 1905 RP can translate a fundamental physical fact into a formal requirement concerning the invariance, under a particular group of coordinate transformations, of laws referred to charts of a particular family, because those charts have, by assumption, a fixed metrical meaning, and that group is, by definition, a representation of the group of motions of the underlying flat spacetime. The arbitrary charts and coordinate transformations mentioned in the Postulate of General Covariance do not fulfil such conditions. The spacetime coordinates have now "sunk down to the status of meaningless freely choosable auxiliary variables"[5] while the set of coordinate transformations relating them is not the realization of a group,[6] let alone a group representation—the transformations are generally non-linear—and its structure betokens only the differentiable structure of the spacetime manifold, but has nothing to say about the spacetime metric or its group of motions (which, if the Riemann tensor varies freely from point to point, is probably none other than the trivial group, consisting of the identity alone).[7]

In Sections 4.3 and 4.5 we studied the method used by Minkowski—after Poincaré—for complying with the demands of the RP. By identifying physical quantities with suitable 4-tensor fields on spacetime and expressing the laws of nature as equations between the field components, he ensured that the laws took the same form with respect to any Lorentz chart. We also noted the method's great limitation: by definition, 4-tensor fields only have components relative to Lorentz charts, so that the equations of physics, in the Minkowskian version, become not just more or less false but meaningless when referred to non-inertial frames of reference. Since the EP entails that all laboratory frames on earth are non-inertial, it was imperative to find a less restrictive statement of the laws of nature. On page 106 I suggested how to achieve it within the framework of Special Relativity. All one has to do is to read the differential equations that link the components of 4-tensor fields relative to a Lorentz chart x, as equations between the components, relative to x, of the corresponding *ordinary* tensor fields on flat spacetime. The same equations will hold between the components of the said tensor fields relative to an *arbitrary* chart, provided that we replace the ordinary differential operators which occur in their original formulation (referred to the Lorentz chart x) by the corresponding covariant differential operators.[8] To explain what this means and how it works let me introduce some new notation that we shall be using from now on. If T is a (q, r)-tensor field on a relativistic spacetime $(\mathscr{W}, \mathbf{g})$ and x is any given chart of $\mathscr{W}$, we denote the $(^{\mathbf{a}}_{\mathbf{b}})$-th component of T relative to x by $\mathrm{T}^{\mathbf{a}}_{\mathbf{b}}$ (where $\mathbf{a} = (a_1, \ldots, a_q)$ and $\mathbf{b} = (b_1, \ldots, b_r)$ are lists of indices drawn from $\{0, 1, 2, 3\}$). The covariant derivative $\nabla \mathrm{T}$ is a $(q, r+1)$-tensor field on $\mathscr{W}$.[9] We write $\mathrm{T}^{\mathbf{a}}_{\mathbf{b};k}$ for $(\nabla \mathrm{T})^{\mathbf{a}}_{\mathbf{b}k}$, the $(^{\mathbf{a}}_{\mathbf{b}k})$-th component of $\nabla \mathrm{T}$ relative to x (where $\mathbf{b}k$ stands for the list of $r+1$ indices $(b_1, \ldots, b_r, k)$). We write $\nabla_k \mathrm{T}$ for $\nabla_{\partial/\partial x^k}\mathrm{T}$, the covariant derivative of T along $\partial/\partial x^k$, which is a (q, r)-tensor field, like T. It can be shown that $\mathrm{T}^{\mathbf{a}}_{\mathbf{b};k} = (\nabla_k \mathrm{T})^{\mathbf{a}}_{\mathbf{b}}$.[10] We shall represent partial differentiation with respect to the k-th coordinate function by a comma followed by k. Thus, $\mathrm{T}^{\mathbf{a}}_{\mathbf{b},k} = \partial \mathrm{T}^{\mathbf{a}}_{\mathbf{b}}/\partial x^k$. We denote by $\mathbf{a}(m/a_i)$ the list obtained by substituting m for a_i in the list $\mathbf{a}$. We have that

$$\mathrm{T}^{\mathbf{a}}_{\mathbf{b};k} = \mathrm{T}^{\mathbf{a}}_{\mathbf{b},k} + \sum_{i=1}^{q} \Gamma^{a_i}_{mk} \mathrm{T}^{\mathbf{a}(m/a_i)}_{\mathbf{b}} - \sum_{i=1}^{r} \Gamma^{m}_{b_i k} \mathrm{T}^{\mathbf{a}}_{\mathbf{b}(m/b_i)} \tag{5.5.1}$$

where the Γ^i_{jk} are the scalar fields defined by (5.4.6*b*) in terms of the components of the metric $\mathbf{g}$ relative to x. If $\mathbf{g}$ is the Minkowski metric $\boldsymbol{\eta}$ and x happens to be a Lorentz chart, the Γ^i_{jk} are of course identically 0 and

$$\mathrm{T}^{\mathbf{a}}_{\mathbf{b};k} = \mathrm{T}^{\mathbf{a}}_{\mathbf{b},k} \tag{5.5.2}$$

Every 4-tensor field T′ is the restriction to Lorentz tetrad fields of an ordinary tensor field T on Minkowski spacetime. A functional relation $\Phi'(\mathrm{T}'_1, \mathrm{T}'_2, \ldots) = 0$ between 4-tensor fields is equivalent to a relation $\Phi(\mathrm{T}_1, \mathrm{T}_2, \ldots) = 0$ between the corresponding ordinary fields. (5.5.2) tells us how to obtain a generally covariant formulation of the latter from a Lorentz covariant formulation of the former. Let $\Phi' = 0$ be expressed as a system Σ' of differential equations involving the components $\mathrm{T}'_{1}{}^{\mathbf{a}}_{\mathbf{b}}$, $\mathrm{T}'_{2}{}^{\mathbf{c}}_{\mathbf{d}}, \ldots$ of T'_1, $\mathrm{T}'_2, \ldots$ relative to a Lorentz chart x, and the partial derivatives $\mathrm{T}'_{1}{}^{\mathbf{a}}_{\mathbf{b},k}$, $\mathrm{T}'_{1}{}^{\mathbf{a}}_{\mathbf{b},kj}, \ldots$ of those components with respect to the coordinate functions of x. Then, by (5.5.2), the same system Σ' expresses $\Phi = 0$ relative to x. To obtain a system Σ expressing $\Phi = 0$ in terms of an arbitrary chart y, replace in Σ' (i) the components $\mathrm{T}'_{1}{}^{\mathbf{a}}_{\mathbf{b}}$, $\mathrm{T}'_{2}{}^{\mathbf{c}}_{\mathbf{d}}, \ldots$ of T_1, $\mathrm{T}_2, \ldots$ relative to x, by their components $\mathrm{T}_{1}{}^{\mathbf{a}}_{\mathbf{b}}$, $\mathrm{T}_{2}{}^{\mathbf{c}}_{\mathbf{d}}, \ldots$ relative to y; (ii) the partial derivatives $\mathrm{T}'_{1}{}^{\mathbf{a}}_{\mathbf{b},k}$, $\mathrm{T}'_{1}{}^{\mathbf{a}}_{\mathbf{b},kj}, \ldots$ of the former components with respect to the coordinate functions of x, by the components $\mathrm{T}_{1}{}^{\mathbf{a}}_{\mathbf{b};k}$, $\mathrm{T}_{1}{}^{\mathbf{a}}_{\mathbf{b};kj}, \ldots$ relative to y, of the covariant derivatives $\nabla \mathrm{T}_1$, $\nabla\nabla \mathrm{T}_1, \ldots$ Thus, within the bounds of Special Relativity, general covariance can be achieved by "dotting the commas and unpriming the T's".

In the curved spacetime of Einstein and Grossmann's "Outline" the Lorentz covariant equations of Special Relativity hold good approximately with respect to each local Lorentz chart. By the typographical procedure sketched above one can obtain from them generally covariant equations with a *prima facie* claim to global validity. Although this case differs significantly from the former, Einstein had no qualms about using the same procedure to figure out the laws of Relativity physics in the presence of irreducible gravitational fields. In 1914 he wrote with Fokker:

> To each quantity and operation of vector analysis on the [flat] manifold there is a corresponding more general quantity and operation of vector analysis on the manifold characterized by an arbitrary line element. Hence, the laws for physical phenomena of the original Theory of Relativity can be set in correspondence with matching laws of the generalized Theory of Relativity. The laws thus obtained, which are generally covariant, include the influence of the gravitational field on the physical processes.[11]

The procedure certainly yields the simplest conceivable generalizations of the local laws, and it is reasonable to follow it while there is no inducement to search for something more elaborate (and hardly any indication of how to find it). However, we must bear in mind that its justification depends on a stronger hypothesis and a far less straightforward argument if spacetime is curved than if it is assumed to be flat. If T is a tensor field on the curved spacetime $(\mathscr{W}, \mathbf{g})$ and ${}_P\mathrm{T}'$ is a 4-tensor field defined in terms of local Lorentz charts near a worldpoint P, T cannot be obtained simply by lifting the restriction of ${}_P\mathrm{T}'$ to the local Lorentz tetrad fields. For on the one hand we expect T to be defined on all spacetime, not just on a narrow neighbourhood of P. And on the other hand ${}_P\mathrm{T}'$, as a 4-tensor field, cannot be properly defined even on a small region of the curved spacetime $(\mathscr{W}, \mathbf{g})$. To see in what sense ${}_P\mathrm{T}'$ can exist, after all, near $P \in \mathscr{W}$, let us recall that a neighbourhood U_P of P is mapped onto a neighbourhood of zero in the tangent space $\mathscr{W}_P$ by the diffeomorphism f described on p. 144. If, as we assume, $\mathbf{g}$ is a Lorentz metric $\mathscr{W}_P$ is a 4-manifold endowed with the flat Minkowski metric $\boldsymbol{\eta}$. The metric $f^*\boldsymbol{\eta}$ on U_P is also flat[12] and the 4-tensor field ${}_P\mathrm{T}'$ can certainly be defined on $(U_P, f^*\boldsymbol{\eta})$ and extended—by "dotting the commas"—to an ordinary tensor field ${}_P\mathrm{T}$ on this flat manifold. But $(U_P, f^*\boldsymbol{\eta})$ is not a sub-space of $(\mathscr{W}, \mathbf{g})$, for $f^*\boldsymbol{\eta}$ does not agree with the restriction of $\mathbf{g}$ to U_P. Moreover, if $\bar{f}: U_Q \to \mathscr{W}_Q$ is defined like $f: U_P \to \mathscr{W}_P$ on a suitable neighbourhood U_Q of a point $Q \in U_P$, $\bar{f}^*\boldsymbol{\eta}$ will not generally agree with $f^*\boldsymbol{\eta}$ on $U_P \cap U_Q$.[13] The metrics $\mathbf{g}$, $f^*\boldsymbol{\eta}$ and $\bar{f}^*\boldsymbol{\eta}$ determine distinct covariant differential operators on $U_P \cap U_Q$. Consequently, "dotting the commas" does not mean the same with respect to the three metrics and there is no compelling reason why ${}_P\mathrm{T}$ should agree on $U_P \cap U_Q$ with a tensor field ${}_Q\mathrm{T}$ constructed by the same method on $(U_Q, \bar{f}^*\boldsymbol{\eta})$. However, on a sufficiently small neighbourhood of P (or Q), $\mathbf{g}$ very nearly agrees with $f^*\boldsymbol{\eta}$ (or $\bar{f}^*\boldsymbol{\eta}$).[14] One may therefore reasonably expect that if $\{U_P, U_Q, \ldots\}$ is a covering of $(\mathscr{W}, \mathbf{g})$ by sufficiently small neighbourhoods of suitable worldpoints $P, Q, \ldots$, a collection of geometrical objects ${}_P\mathrm{T}$, ${}_Q\mathrm{T}, \ldots$, defined on each neighbourhood by equivalent physical methods, following the above procedure, will merge almost smoothly into a single global object T on $\mathscr{W}$. (Note that in the definition of T one must ignore—or even up—the small disagreements between the differential operators of the local flat metrics that go into the definition of $_P\mathrm{T}$, $_Q\mathrm{T}, \ldots$, and the homologous differential operators of the global metric **g**.) Such global objects can be constructed from all the tensor fields of the locally valid Special Relativity physics. Einstein and Fokker's assertion, in the passage quoted above, that the laws of the original Theory of Relativity can be matched by corresponding laws of the new generalized theory amounts to the bold but not implausible hypothesis that the generally covariant equations which, on each small neighbourhood, link the local fields to one another, hold also in the same form between the respective global fields. Some such hypothesis is needed to unify the many equivalent yet unrelated Special Relativities in a system of the universe (cf. p. 137).

Though the procedure of "dotting the commas" is automatically justified by the said hypothesis, it must be administered with caution, for it may occasionally turn equations into inequalities when derivatives of the second or higher order are involved. The reason why this can happen is that, although $\partial^2/\partial x^j \partial x^k$ is always equal to $\partial^2/\partial x^k \partial x^j$, on curved spacetime $\nabla_j \nabla_k$ generally differs from $\nabla_k \nabla_j$.[15] Consequently, if T is the generalization to curved spacetime of the local tensor field $_P\mathrm{T}$, obtained in its turn by the standard procedure from the local 4-tensor field $_P\mathrm{T}'$, it is possible that $\mathrm{T}^{\mathbf{a}}_{\mathbf{b};jk} \neq \mathrm{T}^{\mathbf{a}}_{\mathbf{b};kj}$, even though $_P\mathrm{T}'^{\mathbf{a}}_{\mathbf{b},jk} = {_P\mathrm{T}'^{\mathbf{a}}_{\mathbf{b},kj}}$. (This and other related difficulties are discussed by A. Trautman (1965), pp. 123–7.)

By judiciously substituting covariant for ordinary derivatives wherever the latter occur in the definitions, we can extend to a curved spacetime all the physical concepts employed in the four-dimensional formulation of Special Relativity. The generalizations of Minkowskian 4-tensors thus obtained will be called *world-tensors* (e.g., world-velocity, world-momentum, world-force, etc.). As an illustration that will be useful later, let us figure out the components A^i, relative to a chart x, of the world-acceleration of a particle describing a timelike worldline, parametrized by proper time τ, in an arbitrary relativistic spacetime. By (4.3.3) and (4.3.7), the i-th component of the 4-acceleration relative to a suitable local Lorentz chart y is $\mathrm{d}^2y^i/\mathrm{d}\tau^2 = (\mathrm{d}y^i/\mathrm{d}\tau)_{,j}\mathrm{d}y^j/\mathrm{d}\tau$. "Dotting the comma" and writing U^i for $\mathrm{d}x^i/\mathrm{d}\tau$ we have therefore that:

$$\begin{aligned}
\mathrm{A}^i &= \mathrm{U}^i{}_{;j}\mathrm{U}^j \\
&= (\mathrm{U}^i{}_{,j} + \Gamma^i_{kj}\mathrm{U}^k)\mathrm{U}^j \\
&= \frac{\partial \mathrm{U}^i}{\partial x^j}\frac{\mathrm{d}x^j}{\mathrm{d}\tau} + \Gamma^i_{kj}\mathrm{U}^k\mathrm{U}^j \\
&= \frac{\mathrm{d}^2 x^i}{\mathrm{d}\tau^2} + \Gamma^i_{kj}\frac{\mathrm{d}x^k}{\mathrm{d}\tau}\frac{\mathrm{d}x^j}{\mathrm{d}\tau} \qquad (5.5.3)
\end{aligned}$$

Comparing the above with (5.4.7*) we see at once that the particle describes a geodesic if and only if its world-acceleration is always equal to zero.

According to the Einstein–Grossmann "Outline" the spacetime metric **g** must depend on the distribution of matter (proposition B, p. 139). Special Relativity represented the distribution of matter by means of Laue's energy tensor T (pp. 118 ff.). We must now regard this representation as local and construct from it a global tensor field, in the manner indicated above. We shall denote this global energy tensor by T also. Like its local counterpart, it is a symmetric tensor field of order 2. T represents the total energy of bodies and fields, *except the gravitational field*. The generally covariant law that matches the conservation principle (4.5.20) can be stated as:

$$\mathrm{T}^{ij}{}_{;j} = 0 \tag{5.5.4a}$$

where the T^{ij} are the components of T (in its contravariant form), relative to an arbitrary spacetime chart x and the four functions $\mathrm{T}^{ij}{}_{;j}$ are the components relative to the same chart of a vector field known as the *covariant divergence* of T. Einstein describes (5.5.4*a*) as "the most universal law of theoretical physics".[16] It can be spelt out as follows:

$$(\sqrt{-g}\, g_{ik}\mathrm{T}^{ij})_{,j} - \tfrac{1}{2}\sqrt{-g}\,\mathrm{T}^{ij}g_{ij,k} = 0 \tag{5.5.4b}$$

where the g_{ij} are, as usual, the components relative to x of the metric **g**, and g denotes the determinant of their matrix.[17] $g_{ik}\mathrm{T}^{ij}$ is equal to $\mathrm{T}^{j}{}_{k}$, the $(^{j}_{k})$-th component of the mixed form of T. If we set $\mathrm{H}^{i}_{jk} = \frac{1}{2}g^{ih}g_{hj,k}$ and $\mathfrak{T}^{j}_{k} = \mathrm{T}^{j}_{k}\sqrt{-g}$,[18] equations (5.5.4) take the simpler form:

$$\mathfrak{T}^{j}_{k,j} = \mathrm{H}^{i}_{jk}\mathfrak{T}^{j}_{i} \tag{5.5.5}$$

Comparing (5.5.5) with (4.5.14) on page 119, and recalling the argument that led to equations (4.5.15), we see at once that (5.5.5) cannot express a conservation law unless the right-hand side vanishes. Einstein remarks that it will vanish indeed "if no gravitational field is present", but that otherwise equations (5.5.5) do not have "the form of a pure conservation principle".[19] This, he adds, is quite understandable from a physical point of view, for in the presence of a gravitational field matter cannot fulfil the conservation principles all by itself, because it exchanges energy and momentum with the field. The interaction is expressed by the right-hand side of (5.5.5). On account of it, Einstein regarded for some time the functions H^{i}_{jk} as "the natural expression for the components of the gravitational field".[20] The conservation of energy and momentum can only be fulfilled by matter and the field together. True to von Laue's scheme (page 118), Einstein proposed to find a matrix (t^{j}_{k}), satisfying the equations

$$\frac{\partial}{\partial x^{j}}(\mathfrak{T}^{j}_{k} + \mathrm{t}^{j}_{k}) = 0 \tag{5.5.6}$$

and representing the dynamical state of the gravitational field. (In order to play this role, the t_k^j had to be constructed only from the components of the metric and their derivatives.)

The core of a theory of gravity that meets the requirements of the 1913 "Outline" must be a law binding the metric to the distribution of matter. Evidently this law cannot be obtained by generalization of some special-relativistic law, assumed valid on local flat spacetimes, but must be found anew. Indeed, by saying how the spacetime geometry is determined by the existent matter, the law in question will furnish the necessary ground for building a cosmos out of local experiences. If the distribution of matter is represented by the energy tensor T, the law must be given by differential equations of the form

$$\mathrm{G} = \kappa \mathrm{T} \tag{5.5.7}$$

with G a tensor field of order 2, constructed from the metric **g**, and κ a factor of proportionality, the "gravitational constant" of the theory. Einstein writes (5.5.7) componentwise, in terms of a chart x:[21]

$$\mathrm{G}^{ij} = \kappa \mathrm{T}^{ij} \tag{5.5.8}$$

Equations (5.5.8) must meet the following requirements:

(i) In the limit of weak fields and low velocities they must tend to agree with Poisson's equation (1.7.4), the well-attested field equation of Newtonian theory.
(ii) G must be symmetric, like T; hence, $\mathrm{G}^{ij} = \mathrm{G}^{ji}$.
(iii) In agreement with (5.5.4*a*) the covariant divergence of G must vanish everywhere; i.e., $\mathrm{G}^{ij}{}_{;j} = 0$.
(iv) Since (i) implies that the equations (5.5.8) must be at least of the second order, but it would be premature, in the present state of our knowledge of the gravitational field, to countenance equations of a higher order, the G^{ij} should depend exclusively on the components of the metric **g** and their first and second derivatives with respect to the coordinate functions.[22]

Grossmann explicitly says that the only tensor field on a Riemannian n-manifold whose components depend exclusively on the components of the metric and their first and second derivatives is the Riemann tensor R.[23] He surely knew that there is essentially only one way of deriving from it a tensor of order 2, namely, by contracting R with respect to the first and the last indices (p. 349, note 15) to form the *Ricci tensor*,[24] given componentwise by

$$\mathrm{R}_{ij} = \mathrm{R}^{\;k}_{ijk} \tag{5.5.9}$$

This implies that every tensor field on the spacetime $(\mathscr{W}, \mathbf{g})$ that meets requirement (iv) must be an $\mathscr{F}(\mathscr{W})$-linear combination of the Ricci tensor and the metric **g**, whose coefficients are constants, or scalar fields constructed from

R and **g**. One such combination, which I shall call the *proper Einstein tensor*, stands on the left-hand side of the field equations of General Relativity, when they are written after the scheme (5.5.8):

$$\mathrm{R}^{ij} - \tfrac{1}{2}\mathrm{R}^{hk} g_{hk} g^{ij} = -\kappa \mathrm{T}^{ij} \tag{5.5.10}$$

(where the R^{ij} are the components of the Ricci tensor in its contravariant form and we have assumed that $\kappa > 0$). Under a suitable physical interpretation, equations (5.5.10) fulfil condition (i).[25] They also satisfy condition (ii), for the Einstein tensor is obviously symmetric. It can be shown, moreover, that they always meet requirement (iii), independently of the particular values of the metric components, for the covariant divergence of the Einstein tensor vanishes identically.[26] Equations (5.5.10) are only a particular member of a one-parameter family of equation systems that satisfy conditions (i)–(iv). The typical member of the family is:

$$\mathrm{R}^{ij} - (\tfrac{1}{2}\mathrm{R}^{hk} g_{hk} - \lambda) g^{ij} = -\kappa \mathrm{T}^{ij} \tag{5.5.11}$$

Equations (5.5.11) are the modified field equations of General Relativity proposed by Einstein in 1917 (and withdrawn in 1931).[27] I shall therefore refer to the typical tensor field represented on their left-hand side as the *modified Einstein tensor*. David Lovelock (1971, 1972) has proved that, in an arbitrary Riemannian 4-manifold, the modified Einstein tensor is the only tensor field constructed in accordance with condition (iv) whose covariant divergence vanishes identically. Since it automatically fulfils condition (ii)—for both the metric and the Ricci tensor are symmetric—it turns out that the modified Einstein tensor constitutes a natural solution for G in equations (5.5.7) and that the field equations (5.5.11) furnish a mathematically determinate answer to the problem posed by the "Outline" of 1913. (The choice of a particular value for λ, like the assignment of a definite sign to the factor κ, must rest of course on physical grounds.) Had Lovelock's theorem been discovered sixty years earlier, there can hardly be any doubt that Einstein would have filled in the left-hand side of equations (5.5.8) with the components of the Einstein tensor.[28] He would then have established the field equations of General Relativity by a purely mathematical argument as soon as he designed a clear blueprint for the theory of gravity. But in actual fact the Einstein tensor was first constructed by Einstein himself—who finally found equations (5.5.10) guided by physical considerations—and it is unlikely that, without his prompting, the mathematicians would have paid any attention to such a special object.

Although the Einstein tensor was unknown in 1913 it is somehow surprising that Einstein did not produce it at once, for it is a fairly simple combination of the metric and the Ricci tensor, and the inexorable vanishing of its covariant divergence is surely striking. It should be noted, however, that the fundamental law expressed by equations (5.5.4*a*) apparently did not play an important heuristic role in Einstein's search for the precise form of G in (5.5.7)—except indirectly, through its conjectural, non-covariant corollary (5.5.6). At any rate, in the papers of 1913–15 Einstein never lists condition (iii) on p. 000 among the requirements that G must meet. He may have omitted it because it was far too obvious; but also, perhaps, because he expected it to be fulfilled as a matter of fact, and not as a matter of mathematical necessity—in which case, it would impose a constraint on the metric **g**, but not on the form of G. But the main obstacle to the early discovery of the Einstein tensor was doubtless Grossmann's verdict that if the left-hand side of equations (5.5.8) were built from the Ricci tensor they could not possibly agree in the Newtonian limit with Poisson's equation. From this somewhat rash judgment, Grossmann rightly inferred that it would not be possible to find suitable generally covariant second order equations of the form (5.5.8), and Einstein accepted his expert opinion.[29]

Einstein noted that "since the definitive, exact gravitational equations might well be of an order higher than the second [. . .], there remains the possibility that the perfect exact differential equations of gravitation could be covariant with respect to *arbitrary* coordinate transformations". But while the utility of speculating about such higher order equations remained severely limited by our inability to test them and perhaps even to integrate them, the theory of gravity would have to rest on a not generally covariant system of second order equations. Einstein postulated that they should be covariant at least under arbitrary *linear* transformations. After observing that, if the components of a $(q, 0)$-tensor field are denoted by $\mathrm{T}^{i_1 \cdots i_q}$, the functions $(g^{rs}\mathrm{T}^{i_1 \cdots i_q}{}_{,s})_{,r}$ behave like tensor field components under linear transformations of the coordinates, he claimed that the operator that maps the former quantities to the latter is a generalization of the Laplacian. Hence—he concluded—if equations (5.5.8) are to agree in the Newtonian limit with Poisson's equation, the G^{ij} must be equal, on a first approximation, to $(g^{rs}g^{ij}{}_{,s})_{,r}$. Substituting $\kappa^{-1}\mathrm{G}^{ij}$ for T^{ij} in (5.5.4*b*) we see that the second term, that is,

$$\frac{1}{2\kappa}\sqrt{-g}\, g_{ij,k}\mathrm{G}^{ij} \approx \frac{1}{2\kappa}\sqrt{-g}\, g_{ij,k}(g^{rs}g^{ij}{}_{,s})_{,r} \tag{5.5.11}$$

must be equal to a sum of partial derivatives. From these heuristic assumptions, and neglecting the higher order terms omitted on the right-hand side of (5.5.11), Einstein derives, after some lengthy but straightforward analytic manipulation, the following system of field equations:

$$(\sqrt{-g}\, g_{ik}g^{rs}g^{ij}{}_{,s})_{,r} = \kappa(\mathfrak{T}^j{}_k + \mathrm{t}^j{}_k) \tag{5.5.12a}$$

where the gravitational "energy matrix" (t^j_k) is given by:

$$\mathrm{t}^j_k = \frac{2\sqrt{-g}}{\kappa}\left(\tfrac{1}{2}\delta^j_k g^{ab} H^r_{sa} H^s_{rb} - g^{js} H^a_{bk} H^b_{as}\right) \tag{5.5.12b}$$

in striking analogy with the energy tensor of the electromagnetic field (4.5.12).[30] The theory of gravity based on equations (5.5.12) is known as the *Einstein–Grossmann* theory. It is generally acknowledged that all its basic ideas originated with Einstein.

5.6 Einstein's Arguments against General Covariance: 1913–14[1]

The method described on pp. 155 ff. enabled Einstein, in principle, to describe the influence of the gravitational field on any known physical process by means of generally covariant equations. In the Vienna lecture of September 1913 he remarked that

> the whole problem of gravitation would therefore be solved if one succeeded in finding a set of equations, covariant under arbitrary coordinate transformations, which is satisfied by the quantities g_{ij} that determine the gravitational field.[2]

Einstein and Grossmann had not found such a set, and the field equations (5.5.12) proposed by them were not generally covariant. In the main text of the "Outline" Einstein attributed their failure to the fact that they had—for good reasons—restricted their search to second-order equations.[3] But obviously he was not happy about this state of affairs. In a letter to H. A. Lorentz of August 14th, 1913, in reply to the latter's "warm" reception of the "Outline", Einstein confessed that his own "confidence in the admissibility of the theory was still shaky". The action of the gravitational field on other physical processes was satisfactorily dealt with, for it had been subjected, by means of the absolute differential calculus, to generally covariant laws. Thereby the gravitational field g_{ij} showed itself "so to speak as the framework that supports everything [*das Gerippe an dem alle hängt*]".

> *But the gravitational equations themselves unfortunately do not have the property of general covariance.* Only their covariance with respect to *linear* transformations is assured. All our confidence in the theory rests however on the belief that acceleration of the reference frame is equivalent to a gravitational field. Hence, unless all the equation systems of the theory, including equations (18), admit other transformations beside the linear ones, the theory rebuts its own starting point and does not stand on solid ground.[4]

Two days later, however, on August 16th, Einstein wrote to Lorentz:

> To my great joy I found yesterday that the reservations regarding the theory of gravity stated in my last letter and in the paper itself are not warranted. The question can apparently be solved as follows: The expression of the energy principle for matter and the gravitational field together is an equation of the form (19), i.e., of the form $\Sigma \partial T_{ij}/\partial x^j = 0$. Starting from this assumption I set up equations (19).[5] Examination of the general differential operators of the absolute differential calculus shows however that an equation of this structure can never be absolutely covariant.[6]

By postulating the said equations we therefore implicitly specialize the available selection of reference systems: we restrict ourselves to those in which the principle of energy and momentum conservation holds good "in this form".

Summing up: by postulating the conservation principle we are led to a highly determined choice of the reference system and of the admissible transformations.[7]

This is precisely the explanation given by Einstein in the Vienna lecture of September 1913 for his and Grossmann's inability to provide generally covariant gravitational field equations.[8] However, in a footnote to the printed version of the lecture—published on December 15th, 1913—he announced the discovery of a much more cogent explanation.[9] This must have been found before November 2nd, 1913, for on that day Einstein wrote to Ludwig Hopf:

> I am now quite happy [*sehr zufrieden*] with the gravitation theory. The fact that the gravitational equations are not generally covariant, which sometime ago still worried me inordinately [*welche mich vor einiger Zeit noch ungemein störte*] has turned out to be inevitable. It is easily proved that a theory with generally covariant equations cannot exist if we demand that the field be mathematically completely determined by matter.[10]

The newly found proof was presented in one of the notes appended to the "Outline" when it finally came out in the *Zeitschrift fur Mathematik und Physik.*[11] The following paraphrase uses our terminology and notation.

Let the spacetime ($\mathscr{W}$, **g**) include an open subset H on which the energy tensor T vanishes. (H is a spacetime region in which no "material process" occurs, i.e. a hole in the universe.) Let the closure of H be included in the domain of a chart z. Let g^{ij} and T^{ij} denote the (i, j)-th components relative to z of the contravariant forms of g and T, respectively. We assume that the T^{ij}, given outside H, completely determine the g^{ij} on the entire domain of z. Let y be another chart defined on the same domain as z, and such that (i) outside H, $y = z$, but (ii) on H, $y \neq z$ and the matrix of the components $g'^{ij} = \mathbf{g}(\mathrm{d}y^i, \mathrm{d}y^j)$ is different from the matrix of the $g^{ij} = \mathbf{g}(\mathrm{d}z^i, \mathrm{d}z^j)$. Evidently, y and the g'^{ij} can be chosen in several ways. Let T'^{ij} denote the (i, j)-th component of T relative to y. Since $\mathrm{T} = 0$ on H and $z = y$ outside H, $\mathrm{T}'^{ij} = \mathrm{T}^{ij}$ everywhere, on the common domain of z and y. Consequently, in the particular case under consideration, a given matrix (T^{ij}) of components of the energy tensor corresponds to several distinct matrices (g^{ij}), (g'^{ij}), . . . of components of the metric tensor. Contrary to our assumption, the T^{ij} do not determine the g^{ij}.

The argument, as it stands, is impeccable, and certainly proves that a particular array of metric tensor components g^{ij} cannot be the sole solution of a generally covariant system of differential equations of the form

$$\mathrm{G}^{ij} = \kappa \mathrm{T}^{ij} \tag{5.5.8}$$

—where the G^{ij} may depend on the g^{ij} and their derivatives to an arbitrarily high order—even if the values of the T^{ij} are specified everywhere. However, an objection will immediately suggest itself, if we have learned to regard the metric **g** and the energy tensor T as coordinate-free geometric objects. From this point of view, they may be considered as mappings which assign a matrix of scalar fields (components) to each spacetime chart—or to the corresponding tetrad field.[12] Einstein's argument shows that **g** can take different values at two charts z and y at which T takes the same value. This does not entail, however, that **g** is not uniquely determined by T. For a mapping can be determined by another even if its value varies on parts of its domain on which the latter is constant.[13]

I have no evidence that in 1913–14 anyone raised the foregoing objection.[14] Nevertheless, in two papers published in 1914 Einstein modified his argument in a way that seems designed to forestall it. Instead of a *coordinate* transformation that changes the components of the metric while preserving those of the energy tensor, he now considers a *point* transformation, by which all tensor fields are naturally "dragged along", but which is so contrived that the metric is mapped on a different metric, while the energy tensor is mapped on itself. In this form the argument is no longer liable to our earlier stricture. Einstein's clearest formulation of the modified argument is to be found in the second paper.[15] The two charts that agree outside the hole H but disagree inside it are here called K and K'. The matrices of the components of the covariant metric tensor relative to K and K' are denoted by (g_{ij}) and $g'_{ij})$; their typical values, by $G(x)$ and $G'(x')$, where x and x' presumably stand for the real number quadruples assigned, respectively, by K and K' to an arbitrary worldpoint P. Einstein grants that $G(x)$ and $G'(x')$ are merely two different representations of the same gravitational field, referred to two coordinate systems. However, if we substitute "the coordinates x_k for x'_k in the functions g'_{ij}" to form $G'(x)$, we obtain a description, relative to K, of a different gravitational field, which does not agree with the one originally given, relative to the same coordinate system, by $G(x)$. Einstein proceeds:

> If we now assume that the differential equations of the gravitational field are generally covariant, they are satisfied by $G'(x')$ (relative to K') if they are satisfied by $G(x)$ relative to K. Hence they are also satisfied, relative to K, by $G'(x)$. There are therefore two distinct solutions, $G(x)$ and $G'(x)$, relative to K, although both solutions agree on the boundary [of the hole H]. In other words, *the course of events in the gravitational field cannot be fixed univocally by generally covariant differential equations.*[16]

The meaning of this text will become clearer if we reintroduce our notation. Let z and y denote Einstein's coordinate systems K and K'. $G(x)$ stands then for the value of the matrix of the components $g_{ij} = \mathbf{g}(\partial/\partial z^i, \partial/\partial z^j)$ at an arbitrary point P mapped by z on the number quadruple $z(P)$ $(= x$, in Einstein's notation); while $G'(x')$ denotes the value of the matrix of the components $g'_{ij} = \mathbf{g}(\partial/\partial y^i, \partial/\partial y^j)$ at the same point P, which is mapped by y on $y(P)$ $(= x'$, in Einstein's notation). How are we to understand Einstein's prompting to "substitute the coordinates x_k for x'_k in the functions g'_{ij}", so as to form $G'(x)$? I can think of two alternative interpretations. Either (i) we are asked hereby to evaluate the matrix $(g'_{ij}) = (\mathbf{g}(\partial/\partial y^i, \partial/\partial y^j))$ at the point mapped by y on $x = z(P)$, i.e. at the point $y^{-1} \cdot z(P)$; or (ii), less naturally but more probably, we are required to substitute not only the unprimed for the primed coordinate *values* (i.e., $z(P)$ for $y(P)$), but also the unprimed for the primed coordinate *functions* (i.e., z^k for y^k), and hence to evaluate the matrix $(\mathbf{g}(\partial/\partial z^i, \partial/\partial z^j))$ at $y^{-1}\cdot z(P)$.[17] In either alternative, Einstein's move shifts our attention from the coordinate transformation $y\cdot z^{-1}$, which could only change the chart-dependent representation of the given metric, to the point transformation $y^{-1}\cdot z$, which evidently can induce a change of the metric itself. Set $f = y^{-1}\cdot z$. f is a diffeomorphic permutation of an open submanifold $U \subset \mathscr{W}$, which contains the closure of H.[18] The conditions prescribed for y and z imply that f is the identity on $U - H$, but is arbitrary on the interior of H. Hence on H the originally given metric **g** will generally differ from its "pull-back" $f^*\mathbf{g}$.[19] It can be readily shown than Einstein's $G'(x)$, in our second, preferred interpretation, is equal to the value at P of the matrix of the components of $f^*\mathbf{g}$ relative to z.[20] Does Einstein's argument prove that, under its assumptions, $f^*\mathbf{g}$ will satisfy any set of equations of the form (5.5.8) which is satisfied by **g**? To show that it must indeed be so let us first define the "drag-along" $h_*\omega$ of a covariant tensor field ω on an n-manifold M, by an arbitrary diffeomorphism $h: M \to M'$. We set $h_*\omega$ equal to $(h^{-1})^*\omega$, the "pull-back" of ω by the inverse diffeomorphism h^{-1}. $h_*\omega$ is therefore a covariant tensor field on M', of the same order as ω.[21] Returning now to the metric **g** and the energy tensor T on U, and the diffeomorphism f of U onto itself, we see at once that, while $f_*\mathrm{T} = \mathrm{T}$ on all U, $f_*\mathbf{g}$ generally differs from **g** in the interior of the hole H. According to our assumptions, U is mapped by the charts z and y onto the same region of $\mathbf{R}^4$ (see note 18). We denote this region by V. We have also assumed that, if P lies in H, $z(P)$ generally differs from $y(P)$. On the other hand, since $f = y^{-1}\cdot z$, y assigns to $f(P)$ the same real number quadruple which z assigns to P. Hence the spacetime region U is portrayed by z in V in exactly the same way as its deformed image $f(U)$ is portrayed in V by y. In particular, the "drag-along" $z_*\mathbf{g}$ of the metric **g** by z is exactly the same tensor field on V as the "drag-along" $y_*f_*\mathbf{g}$ of the metric $f_*\mathbf{g}$ by y.[22] The same holds, of course, *mutatis mutandis*, for the energy tensor T. Consequently, any system of differential equations that ties $z_*\mathbf{g}$ **to** $z_*\mathrm{T}$ and yields **g** as the metric on U determined by T will, by the same token, tie $y_*f_*\mathbf{g}$ to $y_*f_*\mathrm{T}$ and yield $f_*\mathbf{g}$ as the metric on U determined by $f_*\mathrm{T}$. Since $f_*\,\mathrm{T} = \mathrm{T}$, it turns out that a given energy tensor is compatible with more than one metric.

As we shall see in the next section, in November 1915 Einstein proposed with unconcealed joy a generally covariant system of field equations, through which, he believed, the spacetime metric responsible for the gravitational field was uniquely determined by the energy tensor. How did he dispose of his mathematical proof that no such thing was possible? His paper of November 18th, 1915, on the perihelion advance of Mercury, contains a paragraph that expresses his new view on the matter. After setting up a system of gravitational field equations in the absence of matter ($\mathrm{T} = 0$), covariant with respect to coordinate transformations with Jacobian determinant equal to 1, he remarks that the metric in the vacuum surrounding the sun might not be uniquely determined by these equations and the given solar mass.

> This follows from the fact that these equations are covariant with respect to arbitrary transformations with determinant 1. However, it might be justified to assume that all the solutions can be reduced to one another by means of such transformations, so that (for a given set of boundary conditions) they differ from each other only formally, but not physically.[23]

Einstein's newly acquired perception that the alternative solutions of a system of covariant equations may in fact be indistinguishable from a physical point of view motivates his rejection of the "hole" argument against general covariance. He reexamined this argument in two letters sent, respectively, to Paul Ehrenfest on December 26th, 1915, and to Michele Besso on January 3rd, 1916. The letter to Ehrenfest explicitly refers to the section of Einstein (1914*b*) that we discussed above:

> Everything is correct in Section 12 of my paper of last year (the first three paragraphs) up to the italicized part at the end of the third paragraph [quoted above, on p. 164—R.T.]. Absolutely no contradiction with the uniqueness of events follows from the fact that both systems $G(x)$ and $G'(x)$, referred to the same system of reference, satisfy the conditions for the gravitational field. The apparent cogency of this argument disappears at once if one considers that 1) the system of reference signifies nothing real, 2) that the (simultaneous) realisation of two different g-systems (better said: of two different gravitational fields) in the same region of the continuum is impossible by the nature of the theory.
>
> The following consideration should take the place of Section 12. The physically real in what happens in the world (as opposed to what depends on the choice of the system of reference) consists of *spatiotemporal coincidences*. For example, the point of intersection of two worldlines, or the assertion that they *do not* intersect, is real. Such assertions referring to the physically real are not lost through any (single-valued) coordinate transformation. If two systems of g_{ij} (or generally of any variables used for the description of the world) are so constituted that the second can be obtained from the first by a pure space-time transformation, then they are fully equivalent. For they have all spatiotemporal point coincidences in common, that is, everything observable.
>
> This argument shows at the same time how natural is the demand for general covariance.[24]

The letter to Besso repeats the same ideas more concisely and perhaps more elegantly.[25]

To understand what Einstein had in mind when he wrote these letters let us again consider the spacetime region U, containing a hole H. The gravitational field in H is probed, in principle, by means of test particles describing timelike geodesics of the Riemannian 4-manifold $(U, \mathbf{g})$. Denote one such timelike geodesic by γ. The gravitational field determined by $\mathbf{g}$ at each worldpoint $P \in \mathbf{g}$ is disclosed by γ's behaviour *vis-à-vis* nearby timelike geodesics on a small neighbourhood of P.[26] In order to perform its probing mission, the particle along γ must be somehow linked to instruments placed in the matter-filled region outside H. Thus, for example, the exact location of a worldpoint $P \in \gamma \subset H$ can be ascertained by means of three bouncing signals, emitted at Q_1, Q_2, Q_3 from suitable devices in $U - H$, and recovered at Q'_1, Q'_2, Q'_3. Let σ_r

denote the worldline of the signal from Q_r through P to Q_r' $(r = 1, 2, 3)$. Let f be, as before, a diffeomorphism of U onto itself, that agrees with the identity on $U - H$. f maps the Riemannian manifold $(U, \mathbf{g})$ isometrically onto $(U, f_*\mathbf{g})$.[27] $f \cdot \gamma$ is therefore a timelike geodesic of $(U, f_*\mathbf{g})$. The gravitational field determined by the metric $f_*\mathbf{g}$ at $f(P) \in f \cdot \gamma$ is disclosed by $f \cdot \gamma$'s behaviour *vis-à-vis* nearby geodesics—of $(U, f_*\mathbf{g})$—on a small neighbourhood of $f(P)$. The location of $f(P)$ can be ascertained by three bouncing signals, of the same kind used for locating P, emitted at $f(Q_r) = Q_r$ in $U - H$ and recovered at $f(Q_r')$ $= Q_r'$. The worldline of each such signal in $(U, f_*\mathbf{g})$ can be no other than $f \cdot \sigma_r$, which is identical with σ_r in $U - H$ and hence in the vicinity of both Q_r and Q_r'. The length of σ_r from Q_r through P to Q_r' in $(U, \mathbf{g})$ is of course the same as the length of $f \cdot \sigma_r$ from Q_r through $f(P)$ to Q_r' in $(U, f_*\mathbf{g})$. There is evidently no way of distinguishing between an event P on the worldline γ of a test particle traversing the hole H in $(U, \mathbf{g})$, and the matching event $f(P)$ on the worldline $f \cdot \gamma$ in $(U, f_*\mathbf{g})$. The 4-manifold U, endowed with the metric $\mathbf{g}$ and the energy tensor T, and the same manifold U, endowed with $f_*\mathbf{g} \neq \mathbf{g}$ and with $f_*\mathrm{T} = \mathrm{T}$, do represent but one physical situation.

John Stachel, who—as far as I know—was the first to understand the significance of Einstein's reappraisal of the "hole" argument in the letters to Ehrenfest and Besso,[28] conjectures that Einstein's self-criticism in those letters rests on the tacit assumption that the points of spacetime where no matter is present cannot be physically distinguished except by the properties and relations induced by the spacetime metric. This assumption would certainly account for Einstein's remark (in the letter to Ehrenfest) that the simultaneous realisation of two different metrics in the same region of spacetime "is impossible by the nature of the theory", and for his even stronger statement (in the letter to Besso) that to attribute two different metrics to the same spacetime manifold is nonsense. The idea that spacetime points are what they are only by virtue of the metric structure to which they belong agrees well with the thesis, common to Leibniz *and* Newton, that "it is only by their mutual order and position that the parts of time and space are understood to be the very same which in truth they are", for "they do not possess any principle of individuation apart from this order and these positions."[29] The assumption that Stachel conjecturally attributes to Einstein has some very important consequences. In the light of it, it is obviously meaningless to speak in General Relativity of a spacetime point at which the metric is not defined. (This view is anyhow prevalent in the literature on the so-called singularities of relativistic spacetimes—see Section 6.4) It also makes havoc of the controversial philosophical doctrine (examined in Section 7.2) according to which the metric of a relativistic spacetime is not a matter of fact, but of mere convention. Finally, if conjoined with a realist conception of spacetime as the aggregate of its points, it entails that those points are subject to the philosophically alluring but notoriously difficult principle of individuation by form.[30] Such a conclusion is clearly at variance with the fashionable semantic theory that conceives of proper names as "rigid designators", denoting the same individual in many alternative diversely structured "possible worlds".[31] Proper names cannot function in this way if the very individuals which are their referents owe their identity to the structure in which they are enmeshed.

Although I am not addicted to this brand of semantics, I do not believe that it clashes with the existence of generally covariant field equations of gravity, as the foregoing considerations apparently imply. I shall therefore propose a different way of dealing with Einstein's "hole" argument against general covariance, which does not assume that spacetime points can only be physically distinguished by means of their metric properties and relations. I must emphasize, however, that I do not claim that Einstein ever countenanced what I am about to say, and that I do not dispute the historical accuracy of Stachel's reading of the letters to Ehrenfest and Besso. I submit that two physical objects or states of affairs can be distinguished from each other either (i) *empirically*, in the blunt, not further analyzable way in which two mint-fresh dimes or two blank 3×5 cards are regarded as unquestionably different, even if they look alike in every respect; or (ii) *rationally*, if, on any grounds, they are equated to or represented by structurally unequal conceptual systems. In the case in point, the physical situations represented respectively by the manifold U with metric $\mathbf{g}$ and energy tensor T, and by the same manifold U with metric $f_*\mathbf{g}$ and energy tensor $f_*\mathrm{T}$, are, as we have just seen, observationally indistinguishable. Since the structures $\langle U, \mathbf{g}, \mathrm{T} \rangle$ and $\langle U, f_*\mathbf{g}, f_*\mathrm{T} \rangle$ are isomorphic, the situations they stand for are, moreover, conceptually indistinguishable—at any rate in terms of this representation. Hence, the said situations are, as far as our assumptions go, perfectly indiscernible, and therefore must be regarded as identical.[32] In the view I have just put forward, the onus of individuating the points of spacetime does not rest with the metric, which is a structural feature of the world. The role of structure is not to individuate, but to specify; and of course it cannot perform this role beyond what its own specific identity will permit, that is, "up to isomorphism". It is only on non-conceptual, extra-structural grounds that two isomorphic structures can be held to represent two really different things.[33]

5.7 Einstein's Papers of November 1915

In March 1914 Einstein wrote to Besso about a "novelty" concerning gravity. Let the Einstein–Grossmann equations (5.5.12) be referred to a chart x. If we differentiate the left-hand side of (5.5.12*a*) with respect to x^j and sum over j we obtain four functions:

$$B_k = (\sqrt{-g}\, g_{ik} g^{rs} g^{ij}{}_{,s})_{,rj} \tag{5.7.1}$$

Equations (5.5.12*a*) and (5.5.6) entail that

$$B_k = 0 \tag{5.7.2}$$

Since the B_k are not the components, relative to x, of a vector field, equations (5.7.2) are not generally covariant and must be regarded as a set of conditions restricting the choice of the chart x. The Einstein–Grossmann equations evidently cannot hold good if referred to a chart which does not satisfy the conditions (5.7.2). The good news that Einstein reported to Besso was that he had succeeded in proving "by a *simple* calculation, that the *gravitational equations* [5.5.12] *are valid for every reference system* which fulfills those conditions".[1] Einstein took this to mean that the covariance group of the Einstein–Grossmann equations was as large as one could wish.[2] The proof was given by Einstein and Grossmann (1914), and again, at greater length, by Einstein (1914*b*) alone. We shall not go into it. Its main interest lies in the fact that in the course of it the Einstein–Grossmann equations are derived from a variational principle. Credit for this methodological innovation is given to Paul Bernays.[3] Unfortunately, the derivation is not rigorous. But it did set an important historical precedent.

After the meeting of the Prussian Academy of October 29th, 1914, at which Einstein submitted his paper (1914*b*), he did not publish any further contributions to the theory of gravity for a whole year—though he did not remain entirely silent on the subject, which he discussed, for instance, at Göttingen in the Summer of 1915, being "extremely pleased that everything was understood in detail".[4] Then, on the four Thursdays of November 1915, he presented to the Prussian Academy four papers—"On the General Theory of Relativity", "On the General Theory of Relativity (Addendum)", "Explanation of the Perihelion Motion of Mercury by the General Theory of Relativity" and "The Field Equations of Gravitation"—in which he discarded the Einstein–Grossmann equations and proposed instead, in quick succession, three alternative systems of gravitational field equations, the last two of which were generally covariant.[5] Let us take a look at this wondrous development.

In a letter to Arnold Sommerfeld of November 28th, 1915, after declaring that he had just gone through "one of the most exciting, most strenuous times of my life, also one of the most rewarding", Einstein explains why he had to admit that his "former gravitational field equations [i.e. eqns. (5.5.12)] were completely unfounded". The equations yield less than half the observed value of Mercury's secular perihelion advance (18″, instead of 45″) and, in violation of the Equivalence Principle, they are not satisfied by the inertial force field on a uniformly rotating system. Moreover, the covariance requirement of the 1914 papers does not determine the Lagrangian of the variational principle from which the field equations were derived, but allows the choice of an arbitrary Lagrangian.[6]

In the first paper of the series Einstein had said that the indeterminacy of the

Lagrangian had caused him to lose his confidence in his earlier field equations and to look for "a way of restricting the possibilities in a natural fashion". He was thus led back to require a "more general [*allgemeinere*] covariance of the field equations", a requirement which three years before he had given up "with a heavy heart".[7] He added:

> Just as the Special Theory of Relativity is based on the postulate that its equations must be covariant under linear orthogonal transformations, the theory to be presented here rests on the postulate of *the covariance of all equation systems under transformations whose Jacobian determinant is equal to* 1.[8]

Let us say that an atlas of an n-manifold is a J-atlas if the coordinate transformations between any two of its charts has a Jacobian determinant equal to 1. Einstein's new theory tacitly presupposes that the maximal atlas of the real spacetime manifold $(\mathscr{W}, \mathbf{g})$ contains at least one J-subatlas.[9] Let A_J denote a maximal such subatlas, i.e. one that is not contained in another J-subatlas. The assignment to each chart of A_J of an ordered set of 4^{q+r} smooth functions on its respective domain will be said to define a J-tensor field of type (q, r) on spacetime, if such ordered sets of functions are mutually related like the components of an ordinary (q, r)-tensor field relative to the same charts (i.e. if they "transform" like the latter under the coordinate transformations of A_J). The members of each such ordered set of functions are then the components of the J-tensor field relative to the pertinent chart. Thus, for instance, the determinant g of the metric components g_{ij} relative to any given chart of A_J defines a J-tensor field of type (0, 0) or J-scalar field, for it is invariant under coordinate transformations whose Jacobian equals 1 (see p. 317, note 18). Evidently, if the $\mathrm{T}^{\mathfrak{a}}_{\mathfrak{b}}$ in equations (5.5.1) are the components, relative to a chart $x \in A_J$, of a J-tensor field T, of type (q, r), the $\mathrm{T}^{\mathfrak{a}}_{\mathfrak{b};k}$ given by those equations are the components of a J-tensor field of type $(q, r+1)$, which we may call the J-covariant derivative of T.

The rationale for Einstein's new theory and its curious covariance postulate will flow from the following considerations. Einstein recalls that "all covariants formed exclusively from the g_{ij} and their derivatives" can be obtained from the Riemann tensor R. The problem of gravity directs our attention to tensors of order 2, constructed from R and the metric **g**. The symmetries of R imply that "such a construction can be effected in only *one* way", yielding the Ricci tensor.[10] The matrix of the Ricci tensor's components relative to an arbitrary chart x can be regarded as the sum of two matrices:[11]

$$\mathrm{R}_{ij} = \mathrm{U}_{ij} + \mathrm{S}_{ij} \tag{5.7.3a}$$

with

$$\mathrm{U}_{ij} = -\Gamma^{k}_{ij,k} + \Gamma^{k}_{mj}\Gamma^{m}_{ik} \tag{5.7.3b}$$

and

$$\begin{aligned}\mathrm{S}_{ij} &= \Gamma^{k}_{ik,j} - \Gamma^{k}_{mk}\Gamma^{m}_{ij} \\ &= (\log\sqrt{-g})_{,ij} - \Gamma^{m}_{ij}(\log\sqrt{-g})_{,m}\end{aligned} \tag{5.7.3c}$$

Suppose that x belongs to the J-atlas A_J. Then, $\log \sqrt{-g}$ defines a J-scalar field. The components, relative to x, of its first J-covariant derivative are then its four ordinary derivatives with respect to the x^j, $(\log \sqrt{-g})_{,j}$. The components, relative to x, of its second J-covariant derivative are therefore the S_{ij}. Hence, the matrices (S_{ij}) and (R_{ij}) are arrays of J-tensor field components, relative to x, and their difference, the matrix (U_{ij}), is likewise such an array. Let T_{ij} denote the components of the covariant energy tensor relative to the arbitrary chart $x \in A_J$. The field equations of Einstein's new theory of gravity are:

$$\mathrm{U}_{ij} = -\kappa \mathrm{T}_{ij} \tag{5.7.4}$$

As a system of equations between matching J-tensor field components, (5.7.4) is obviously covariant under coordinate transformations whose Jacobian is equal to 1. Einstein motivates his choice of this system of field equations by deriving them from the variational principle

$$\delta \int (L - \kappa g^{hk} \mathrm{T}_{hk}) \mathrm{d}^4 x = 0 \tag{5.7.5a}$$

where the variation affects the metric **g** but not the energy tensor T, and

$$L = g^{hk} \Gamma^i_{hj} \Gamma^j_{ki} \tag{5.7.5b}$$

The variational equation (5.7.5*a*) will hold only if

$$\frac{\partial}{\partial x^k}\left(\frac{\partial L}{\partial g^{ij}{}_{,k}}\right) - \frac{\partial L}{\partial g^{ij}} = -\kappa \mathrm{T}_{ij} \tag{5.7.6}$$

Noting that $\partial L/\partial g^{ij} = -\Gamma^k_{mj}\Gamma^m_{ik}$ and that $\partial L/\partial g^{ij}{}_{,k} = -\Gamma^k_{ij}$,[12] (5.7.6) can be rewritten as:

$$-\Gamma^k_{ij,k} + \Gamma^k_{mj}\Gamma^m_{ik} = -\kappa \mathrm{T}_{ij} \tag{5.7.6*}$$

which is of course the same as (5.7.4). The meaning of Einstein's variational argument is obscured, however, by the fact that L is not a J-scalar field.

Einstein shows next that the new theory satisfies the "conservation principle"

$$(\mathrm{T}^j_k + \mathrm{t}^j_k)_{,j} = 0 \tag{5.7.7}$$

if the energy matrix of the gravitational field, (t^j_k), is suitably redefined. He sets

$$\kappa \mathrm{t}^j_k = \tfrac{1}{2}\delta^j_k g^{ab}\Gamma^r_{as}\Gamma^s_{br} - g^{ab}\Gamma^r_{ak}\Gamma^j_{br} \tag{5.7.8}$$

an expression reminiscent of the earlier definition (5.5.12*b*).[13] Multiplying both sides of (5.7.8) by δ^k_j and summing over k we verify that $\kappa \mathrm{t}^j_j = g^{ab}\Gamma^r_{as}\Gamma^s_{br} = L$. The Lagrangian in (5.7.5a) is therefore equal to $\kappa(\mathrm{t}^j_j - \mathrm{T}^j_j)$.

Multiplying both sides of (5.7.6*) by g^{ih} and summing over i, we obtain:[14]

$$-(g^{ih}\Gamma^k_{ij})_{,k} - g^{ab}\Gamma^r_{aj}\Gamma^h_{br} = -\kappa \mathrm{T}^h_j \tag{5.7.9}$$

If $\kappa \mathrm{t}^h_j$ is subtracted from each side, (5.7.9) becomes:

$$-(g^{ih}\Gamma^k_{ij})_{,k} - \tfrac{1}{2}\kappa \mathrm{t}^k_k \delta^h_j = -\kappa(\mathrm{T}^h_j + \mathrm{t}^h_j) \tag{5.7.10}$$

In view of (5.7.7), equations (5.7.10) imply that:[15]

$$(g^{ih}_{\ \ ,ih} - \kappa \mathrm{t}^k_k)_{,j} = 0 \tag{5.7.11}$$

In other words, $(g^{ih}_{\ \ ,ih} - \kappa \mathrm{t}^k_k) = \text{const.}$ Einstein sets this constant equal to 0.

Let us now multiply both sides of the field equations (5.7.6*) by g^{ij} and sum over i and j. We obtain:[16]

$$(g^{ij}(\log\sqrt{-g})_{,j})_{,i} + g^{ij}_{\ \ ,ij} - \kappa \mathrm{t}^j_j = -\kappa \mathrm{T}^j_j \tag{5.7.12}$$

Since $(g^{ij}_{\ \ ,ij} - \kappa \mathrm{t}^j_j)$ has been set equal to 0, (5.7.12) amounts to

$$(g^{ij}(\log\sqrt{-g})_{,j})_{,i} = -\kappa \mathrm{T}^j_j \tag{5.7.13}$$

Equation (5.7.13) implies that if g is constant, the invariant trace T^j_j of the energy tensor vanishes. This is hard to accept. For instance, the energy tensor of a continuous distribution of incoherent matter (pressureless dust) is given by

$$\mathrm{T}^{ij} = \rho \mathrm{U}^i \mathrm{U}^j \tag{5.7.14}$$

where ρ is the proper energy density and the U^i are the components of the world-velocity field (pp. 119, 120, 157). Since $g_{ij}\mathrm{U}^i\mathrm{U}^j = 1$, the trace T^j_j is equal to ρ, and can only vanish in a perfect vacuum. Einstein concludes therefore that "it is impossible to choose the coordinate system so that $\sqrt{-g}$ equals 1".[17]

Einstein must have felt uneasy about this highly artificial restriction for a week after submitting his (1915*a*) he came up with a physical hypothesis designed to make the restriction unnecessary. He conjectures that the trace of the energy tensor might in fact vanish everywhere if all the gravitational energy hidden in every chunk of matter is deducted. The electromagnetic energy tensor of classical electrodynamics does have a vanishing trace,[18] and Einstein seeks to justify his bold conjecture by recalling that macroscopic matter is not something "primitively given, physically simple", and that quite a few physicists "hope to reduce matter to purely electromagnetic processes". Although such processes would doubtless obey the laws of a new, non-Maxwellian electrodynamics, we may just as well assume that "in such a perfected electrodynamics the trace of the energy tensor would also vanish".[19] This hypothesis can be reconciled with the fact that the energy tensor of dust (5.7.14) does not have a vanishing trace, because

> gravitational fields are liable to be an essential ingredient of the "matter" to which the above expression refers. Thus T^j_j can be *apparently* positive for the entire system, while in reality only $\mathrm{T}^j_j + \mathrm{t}^j_j$ is positive, while T^j_j vanishes everywhere.[20]

By assuming that "in fact the condition $T^j_j = 0$ is universally fulfilled", we are at last in a position to formulate the gravitational field equations in generally covariant form. Einstein writes them as follows:

$$R_{ij} = -\kappa T_{ij} \tag{5.7.15}$$

where the R_{ij} are the components of the Ricci tensor, "the only tensor available for setting up generally covariant equations of gravity".[21] The theory based on equations (5.7.15) places no restrictions on the choice of spacetime charts. In particular, if we choose a chart x relative to which $\sqrt{-g} = 1$, the S_{ij} defined as in (5.7.3*c*) in terms of x will all vanish. Therefore, the new field equations (5.7.15), when referred to such a chart, are equivalent to the field equations (5.7.4) of Einstein (1915*a*), and all the consequences derived in that paper, including the validity relative to a suitable reference system of the "conservation principle" (5.5.6), still hold good in the theory of Einstein (1915*b*).

In the absence of matter, $T_{ij} = 0$, and equations (5.7.15) amount to

$$R_{ij} = 0 \tag{5.7.16}$$

Einstein's third paper of November 1915 gives an approximate solution of (5.7.16) for a spherically symmetric static metric that becomes flat at infinity.[22] This is a reasonable model of the gravitational field surrounding the sun. After noting that his results agree with his earlier prediction of the gravitational redshift of signals coming from the sun to the earth, Einstein uses them for calculating (i) the deviation that will be experienced by a light-ray grazing the surface of the sun, and (ii) the secular advance of Mercury's perihelion in the absence of perturbing factors. The value of (i) turns out to be 1.75″, or twice the amount predicted by Einstein in 1911. The new figure was confirmed, to 3 parts in 10, by the British Equatorial Expedition of 1919.[23] For (ii), Einstein obtains 43″, which, he says, is "in complete agreement" with the anomaly of $45'' \pm 5''$ per century, unexplained by the Newtonian theory.[24]

The theory of gravity of Einstein (1915*b*) rests on a groundless hypothesis about the fine structure of matter. It is as if Einstein's usual wariness of microphysical speculation had temporarily subsided in the excitement over the discovery of genuine generally covariant field equations. However, in spite of the theory's admirable success in explaining the anomaly of Mercury, Einstein did not rest content with it, but very soon proposed still another system of generally covariant gravitational field equations, namely:

$$R_{ij} = -\kappa(T_{ij} - \tfrac{1}{2}g_{ij}T^h_h) \tag{5.7.17}$$

Equations (5.7.17) are known as *the* Einstein field equations or the field equations of General Relativity. In the absence of matter they are obviously equivalent to (5.7.16), so that, as Einstein is quick to point out, the predictions and post-dictions of his (1915*c*) are also entailed by them. They are equivalent, too, to the system (5.5.10), on page 160, which equates the energy tensor to the proper Einstein tensor.[25] Equations (5.7.17) therefore *imply*, by the sheer force of algebra, that

$$\mathrm{T}_{ij;j} = 0 \tag{5.7.18}$$

However, Einstein does not mention this remarkable fact in his (1915*d*). On the other hand, he devotes almost one third of this brief paper to showing that his latest theory satisfies, no less than its predecessors, the non-covariant conservation law

$$(\mathrm{T}^j_k + \mathrm{t}^j_k)_{,j} = 0 \tag{5.7.19}$$

if "the 'energy tensor' of the gravitational field",[26] t^j_k, is given by equations (5.7.8), as in his (1915*a*). Since the t^j_k are not tensor components, any argument involving them must be referred to a definite chart. As in his (1915*b*), Einstein chooses a chart x, relative to which the determinant g of the metric components is equal to -1. He uses the same special chart to explain why he added the new term $(1/2)g_{ij}\mathrm{T}^h_h$ to the right-hand side of (5.7.15) to form the new field equations (5.7.17). The explanation is as follows. When referred to the said chart, Γ^k_{jk} vanishes and (5.7.17) becomes:[27]

$$-\Gamma^k_{ij,k} + \Gamma^k_{mj}\Gamma^m_{ik} = -\kappa(\mathrm{T}_{ij} - \tfrac{1}{2}g_{ij}\mathrm{T}^k_k) \tag{5.7.20}$$

Einstein subjects (5.7.20) to the same operations applied in his (1915*a*) to (5.7.6∗) to obtain (5.7.11) and (5.7.12).[28] Multiplying both sides of (5.7.20) by g^{ih}, summing over h and subtracting t^h_j from the resulting expressions, we obtain:

$$-(g^{ih}\Gamma^k_{ij})_{,k} = -\kappa((\mathrm{T}^h_j + \mathrm{t}^h_j) - \tfrac{1}{2}\delta^h_j(\mathrm{T}^k_k + \mathrm{t}^k_k)) \tag{5.7.21}$$

This is the analogue, in the present theory, of equations (5.7.10).[29] The fact that the energy tensor of matter and the gravitational energy matrix play identical roles in (5.7.21) was seen by Einstein as a considerable improvement with respect to the earlier system. In view of (5.7.19), equations (5.7.21) imply that:

$$(g^{ih}{}_{,ih} - \kappa(\mathrm{T}^k_k + \mathrm{t}^k_k))_{,j} = 0 \tag{5.7.22}$$

This corresponds to (5.7.11), the equations that placed Einstein in the alternative of having to forbid coordinate systems relative to which $g = -1$ or to postulate that the trace of the energy tensor is a universal constant. The dilemma does not arise in his latest theory, for the operations that transformed (5.7.6*) into (5.7.12) have a very different effect on (5.7.20). If we multiply both sides of the latter by g^{ij} and sum over i and j we are led to:

$$g^{ij}{}_{,ij} - \kappa(\mathrm{T}^k_k + \mathrm{t}^k_k) = 0 \tag{5.7.23}$$

which evidently *entails* equations (5.7.22). The latter, therefore, do not add anything new to (5.7.23), and "it is not necessary to make any further assumptions regarding the energy tensor of matter, except that it satisfies the energy-momentum principle [5.7.18]".[30]

In the long and important paper on "The Foundation of the General Theory of Relativity", submitted on March 20th, 1916, Einstein retraces, with the benefit of hindsight, the main line of thought of his November papers. Because the Ricci tensor is the only tensor of order 2 "which is formed from the g_{ij} and their derivatives, contains no derivatives of order higher than the second, and is linear in the second derivatives",[31] he deems it justified to postulate the vacuum field equations (5.7.16). Once again he refers them to a special chart x relative to which $\sqrt{-g} = 1$. (5.7.16) then become (5.7.20), with 0 on the right-hand side. These equations he derives from the variational principle (5.7.5), which, in the absence of matter, reads $\delta \int L d^4 x = 0$. By the same procedure that carried us from (5.7.20) to (5.7.21), Einstein infers that, in a vacuum,

$$-(g^{ih}\Gamma^k_{ij})_{,k} = -\kappa(\mathrm{t}^h_j - \tfrac{1}{2}\delta^h_j \mathrm{t}^k_k) \tag{5.7.24}$$

He then considers the gravitational action of a continuous matter distribution, represented by the energy tensor T. He introduces a physical argument not found in the November papers, but clearly traceable to a fundamental idea of the 1913 "Outline" (see p. 158).

> If we consider a complete system (e.g. the solar system), the total mass of the system, and hence also its total gravitational action, will depend on the total energy of the system, and therefore on the ponderable and the gravitational energies taken together. This can be expressed by replacing in [5.7.24] the energy components t^h_j of the gravitational field alone by the sums $\mathrm{T}^h_j + \mathrm{t}^h_j$ of the energy components of matter and the gravitational field.[32]

The substitution proposed by Einstein yields (5.7.21), from which equations (5.7.20) can be recovered in their above form. From them Einstein derives (5.7.19) and, for the first time, the energy-momentum principle (5.7.18).[33]

The variational principle invoked by Einstein in his (1916*a*) is liable to the same strictures as its homologue in Einstein (1915*a*), insofar as the Lagrangian L given by (5.7.5*b*) is not a scalar density and consequently is not an appropriate integrand for a generally covariant law. But if we set $L = \sqrt{(-g)}\mathrm{R}^k_k$, the variational equation $\delta \int L d^4 x = 0$ is generally covariant and entails the vanishing of the Einstein tensor and hence the Einstein field equations in the absence of matter.[34] The Einstein field equations in the presence of matter follow from $\delta \int L d^4 x = 0$ if $L = \sqrt{(-g)}\mathrm{R}^k_k + \kappa g_{ij}\mathfrak{T}^{ij}$, and the variation affects the metric **g** but not the energy tensor density $\mathfrak{T}$.[35] A scalar density Lagrangian and a generally covariant variational principle were first used in the theory of gravity by David Hilbert in a paper on "The Foundations of Physics", submitted to the Göttingen Gesellschaft der Wissenschaften on November 20th, 1915, five days before Einstein presented "The Field Equations of Gravitation" in Berlin.[36] Hilbert combines Mie's grand design for an electromagnetic theory of matter with Einstein's outline of a geometric theory of gravity. Hilbert's variational argument yields in fact the Einstein field equations (5.5.10) if the energy tensor in the right-hand side is identified with the electromagnetic energy tensor of Mie's theory. This is indeed an unwarranted "specializing assumption" about the constitution of matter, as Einstein suggests in his (1916*f*). But Hilbert's method of proof does not depend on the particular form of the energy tensor or on the laws that govern its behaviour.

5.8 Einstein's Field Equations and the Geodesic Law of Motion.

With the discovery of field equations (5.7.17) Einstein finally achieved the goal he had set himself in the first paper with Grossmann. He had produced a field theory consonant with the local validity of Special Relativity, in which the gravitational action of matter on matter was mediated by the spacetime geometry. Einstein's theory agreed with Newton's in all the familiar cases in which the latter was well confirmed, and solved the vexing anomaly of Mercury. The new idea of gravitation as geometry dispelled the mysteries attendant on Newtonian attraction, explaining at one stroke why it was ubiquitous and there was no shielding against it, why it acted equally on all kinds of bodies and behaved like the force of inertia over small spacetime regions. Persuaded of the theory's strengths and seduced by its beauty, some of the best minds in Europe, young and old, Felix Klein and Hendrik Anton Lorentz, David Hilbert and Elie Cartan, Karl Schwarzschild, Willem de Sitter, Tullio Levi-Civitȧ, Hermann Weyl, Arthur Eddington, Erwin Schrödinger, sought to master it and to further its development. We shall come across a few of their contributions in Chapter 6. But the relation between the Einstein field equations and the law of the motion of matter in a gravitational field belongs to the foundations of the theory and it is therefore appropriate that we deal with it in this chapter, even though, due to the notorious difficulty of the subject, I can barely hint at it.[1]

The classical field theories of gravity and electricity rest on two sets of laws: the *field equations* state how a given distribution of material sources (masses, charges) generates a field of force on all space, while the *equations of motion* describe how a given field affects the movements of the material objects (masses, charges) liable to its influence. (See equations (1.7.4) and (1.7.5), (2.2.1) and (2.2.2).) Because the field equations are linear, the contributions of different sources to the total field are mutually independent and can be added vectorially. It would seem, therefore, that any aggregate of sources will yield an admissible solution of the field equations and that the latter will not constrain the movement of the sources in any way. If such is the case, the equations of motion do not depend on the field equations, but must be postulated separately.[2] When Einstein, who understood the classical theories in this way, put forth his plan for a new field theory of gravity, he postulated the Geodesic Law of motion (5.4.7*) from the outset, apparently assuming that any suitable system of field equations of the form (5.5.7) would be consistent with it. But the system (5.7.17) which finally won his assent is non-linear, so that the above analysis does not apply to it. We cannot expect that every distribution of material sources moving in any arbitrary way is a possible realization of such a system. At any rate, in the case of a closed material system, not exposed to freely specifiable outside influences, there is every reason to suppose that the worldlines of its elements will be severely constrained by their non-linear gravitational interaction under the Einstein field equations.[3] It is then obviously worth asking whether those constraints are or are not compatible with the Geodesic Law.

While an exact general solution to the problem of motion in General Relativity is not yet known and may even be impossible, many authors have dealt with it by approximation methods in a variety of cases. Under their hypotheses, which necessarily include restrictions on the form of the matter tensor, the gravitational field equations do actually determine, and not just constrain, the motion of freely falling matter. For example, in the absence of non-gravitational fields, matter consisting of gravity monopoles describes timelike geodesics of the metric determined by Einstein's field equations. In the cases considered and within the limitations inherent in the chosen methods of proof the Geodesic Law is therefore a consequence of the Einstein equations. This remarkable result is linked to the fact, noted on pages 160 and 174, that the field equations *entail* the vanishing of the covariant divergence of the energy tensor of matter, or in other words, that equations (5.5.4) are a mathematical corollary of equations (5.5.10).

In the 1913 "Outline" Einstein had already observed that the Geodesic Law of motion follows from equations (5.5.4) in the case of a continuous distribution of incoherent matter or dust.[4] W. Rindler offers the following proof of Einstein's remark.[5] The energy tensor of dust satisfies equations (5.7.14), whence, by (5.5.4)

$$(\rho \mathrm{U}^i \mathrm{U}^j)_{;j} = \mathrm{U}^i (\rho \mathrm{U}^j)_{;j} + \rho \mathrm{U}^i{}_{;j} \mathrm{U}^j = 0 \tag{5.8.1}$$

Multiplying both sides of the second set of equations by $\mathrm{U}_i (= g_{ij}\mathrm{U}^j)$ and summing over i, we obtain:

$$(\rho \mathrm{U}^j)_{;j} + \rho \mathrm{U}_i \mathrm{U}^i{}_{;j} \mathrm{U}^j = 0 \tag{5.8.2}$$

Let us calculate the value of the second term of (5.8.2) at any given worldpoint P. It is $\rho(P)$ times the inner product of the local world velocity U_P (with components $\mathrm{U}^k|_P$) and the local world acceleration A_P (with components $\mathrm{A}^i|_P = \mathrm{U}^i{}_{;j}\mathrm{U}^j|_P$—see p. 157). But the world velocity is everywhere orthogonal to the world acceleration (p. 106), whence the second term of (5.8.2) vanishes at P. Since P is arbitrary,

$$(\rho \mathrm{U}^j)_{;j} = 0 \tag{5.8.3}$$

Substituting from (5.8.3) into (5.8.1) we conclude that

$$\mathrm{U}^i{}_{;j} \mathrm{U}^j = 0 \tag{5.8.4}$$

This means that the world acceleration field of a cloud of freely falling dust is everywhere zero. Consequently, by (5.5.3), each speck of dust describes a timelike geodesic. This exact and fairly simple derivation is especially significant because, as we shall see in Chapter 6, pressureless dust is generally regarded as a reasonable representation of matter in the very large during the present cosmic epoch. The Geodesic Law can be taken for granted in relativistic cosmology at least insofar as the dust world model holds good.

In the fifth edition of *Space–Time–Matter*, published in 1923, Hermann Weyl attempted a derivation of the Geodesic Law of motion for an isolated system of material particles, without taking into consideration the form of the energy tensor in their interior.[6] A few years later, in a paper coauthored by J. Grommer, Einstein introduced a novel way of grappling with the problem.[7] Convinced that in the absence of a satisfactory theory of matter every proposed specification of the energy tensor would be arbitrary, he treated the massive particles whose law of motion was the subject of inquiry as "moving singularities" in a gravitational field governed by the vacuum field equations (5.7.16). As in the classical field theories, a singularity is to be thought of as a worldpoint—a "moving singularity" as a worldline—where the functions characteristic of the field—in the present case, the metric components g_{ij}—are undefined or unsmooth. Einstein and Grommer intend to show that "the law of motion of singularities is fully determined by the field equations and the character of the singularities".[8] Now the notion of a singularity of the metric field is beset with difficulties of its own and cannot be simply taken for granted as a counterpart of the concept of a singularity of the classical gravitational and electromagnetic fields. Still guided by this analogy, Einstein himself wrote with Rosen in 1935 that "a singularity brings so much arbitrariness into the theory that it actually nullifies its laws", and that "every field theory, in our opinion, must therefore adhere to the principle that singularities of the field are to be excluded".[9] But never, to my knowledge, did he pronounce the one objection to singularities in General Relativity that is truly damaging to his treatment of particles in the Einstein–Grommer paper, namely, that a relativistic spacetime cannot include a point where the metric or the Riemann tensor are undefined without ceasing to be *eo ipso* a relativistic spacetime.[10] In fact, Einstein persisted in treating particles as singularities in all his later work on the equations of motion, in collaboration with L. Infeld and B. Hoffmann (1938) and with L. Infeld alone (1940, 1949).[11] In these papers, the two-body problem of General Relativity is solved by approximation methods under the restrictive hypothesis that the metric is almost flat, motion is slow, and the gravitational field is weak. Under such circumstances and with the margin of error allowed by the approximation employed, two particles of finite mass moving *in vacuo* across the gravitational field generated by them would each describe a timelike geodesic. The same basic approach is followed by Infeld and Schild (1949) in their much simpler derivation of the Geodesic Law for test particles of negligible mass. The results thus obtained are somehow perplexing, for a geodesic maps its parameter onto a spacetime path along which the geometry is well-defined and cannot therefore be the locus of a "moving singularity". To cope with this difficulty Infeld and Schild propose the following construction:

> A time-like world line $L_{(0)}$ in a Riemannian 4-space $R_{(0)}$ with coordinates x^k and metric tensor $g_{(0)ij}(x^k)$, which is *analytic* at all points of $L_{(0)}$, represents a *test particle* if the following criterion applies: There exists a sequence of Riemannian 4-spaces $R_{(N)}$ ($N = 1, 2, \ldots, \infty$) with coordinates x^k, with a world line $L_{(N)}$ in each, and with a metric tensor $g_{(N)ij}(x^k)$ which, along $L_{(N)}$, has a singularity of the type representing a particle [described elsewhere in Infeld and Schild's paper—R.T.], such that $\text{Lim}_{N \to \infty} g_{(N)ij}(x^k) = g_{(0)ij}(x^k)$ for all values of x^k which in $R_{(0)}$ represent a point not on $L_{(0)}$, and such that all points of $L_{(N)}$ do not tend to infinity as $N \to \infty$. It follows from this criterion that $\text{Lim}_{N \to \infty} L_{(N)} = L_{(0)}$; this last relation to be interpreted by means of the point–point correspondence which exists between the spaces $R_{(N)}$, $R_{(0)}$ by virtue of their common coordinate system x^k. The geodesic postulate requires that any $L_{(0)}$ representing a test particle be a geodesic in $R_{(0)}$. This statement, whether true or false, is meaningful, since $g_{(0)ij}$ is analytic along $L_{(0)}$.[12]

Yet surely the worldline $L_{(N)}$, along which the $g_{(N)ij}$ are singular, does not lie in the *Riemannian* 4-manifold $R_{(N)}$, but rather in an extension of it. The authors apparently believe that if $R_{(N)}$ and $R_{(0)} - L_{(0)}$ are mapped by the "common coordinate system x^k" onto the same subset of $\mathbf{R}^4$, there is a unique extension of $R_{(N)}$ to a 4-manifold diffeomorphic with $R_{(0)}$. But this belief is illusory. Such an extension can be contrived in many ways, in some of which $R_{(N)}$ is enriched by more than just a one-dimensional manifold diffeomorphic with $L_{(0)}$. It therefore makes little sense to speak about a definite sequence of worldlines $L_{(1)}, L_{(2)}, \ldots$ converging to $L_{(0)}$.

A different approach to the problem of motion in General Relativity was tried by V. A. Fock.[13] A particle is represented by a continuous matter distribution filling the interior of a very narrow tube in spacetime. The energy tensor vanishes, by hypothesis, outside such tubes, whereas inside them it takes a definite agreed form, e.g. that of the Laue tensor for a perfect fluid.[14] A particle's trajectory is represented by an "axis" of its respective tube, i.e. by a one-dimensional submanifold of the tube's interior which meets every spacelike 3-dimensional section of the tube once and once only. It is shown by approximation methods that, on the strength of Einstein's field equations, a non-spinning particle has a geodesic trajectory.[15] But Fock's approach is not exempt from difficulties and ambiguities and has not won general approval.[16]

P. Havas and J. N. Goldberg (1962) distinguish the *laws of motion* which relate the variables describing the trajectory of a particle to some *unspecified* force from the *equations of motion* which specify such a force in terms of fields or of the variables describing the particles that exert it. Typical of the former is Newton's Second Law, $\mathbf{F} = m\mathbf{a}$; of the latter, the Newtonian Law of Gravity (1.7.1) or the Lorentz Force Law (2.2.2). In the spirit of this distinction Havas and Goldberg derive the Geodesic Law as a "law of motion" for an arbitrary isolated system of n material particles gravitationally interacting in an *unspecified* metric field generated by them in agreement with the Einstein equations. The material system in question must consist of simple poles of the gravitational fields.[17] Einstein's field equations enter into the argument only insofar as they imply the vanishing of the covariant divergence of the energy tensor (equations (5.5.4)). Thus, they entail the Geodesic Law of motion both in their original version (5.7.17) and in their modified form (D.6)—on page 281. Indeed any other geometric theory of the gravitational field that implies the energy conservation principle (5.5.4) would entail the same law of motion. Havas and Goldberg's derivation presupposes that the unspecified metric is everywhere defined. As Havas has recently remarked, "no entirely rigorous proof has been given thus far that it is indeed possible to formulate an exact theory for interacting mass points for which [the metric components are] finite on the worldlines".[18] However, in their subsequent derivation of "equations of motion", whereby they seek to specify the metric in successive approximations (with the Minkowski metric as the zeroth-order approximation), Havas and Goldberg (1962) and Havas (1979) manage to keep the metric components finite at any given stage of the procedure.

In the last decade some important work has been done by W. G. Dixon and other researchers on the motion of extended bodies in General Relativity.[19] They have shown that for any body represented by a timelike worldtube B, inside which the energy tensor of matter satisfies some definite positivity-of-energy requirement, there exists a unique center-of-mass worldline λ contained in the convex hull of B. Having covariantly defined on λ, in terms of the metric and the energy tensor, a timelike linear momentum worldvector, an angular momentum bivector and a sequence of multipole moments, they prove that these objects together with the variables descriptive of λ constitute a set of dynamic "body variables" satisfying a finite system of ordinary tensor differential equations. This, and the fact that, if the metric is specified, an energy tensor of matter satisfying the local conservation principle (5.5.4) can be fully recovered from the body variables, indicates according to J. Ehlers "that the body variables describe the body *completely*, and that the equations [between them] contain the same information as the local equation [5.5.4]".[20] Since Dixon's argument depends only on the energy principle (5.5.4) and does not otherwise appeal to Einstein's field equations, the metric is left unspecified. Hence, says Ehlers, the equations for the body variables may be described, in Havas and Goldberg's terminology, as "laws of motion" for extended bodies. "They relate the evolution of the body variables to the gravitational potential [**g**] due (at least partly) to these bodies themselves."[21] "If **g** could be expressed as a functional of the body variables as a consequence of [the Einstein field equations] and boundary conditions and if these functionals were used to transform [the said "laws of motion"] into equations for the body variables alone, the latter would be 'equations of motion'."[22] However, Ehlers suspects that exact "equations of motion" in this sense do not exist in General Relativity and that they can be found as approximations only.

Some advances in the theory of motion in General Relativity might be attained through the study of the temporal development of geometry from initial data, as pursued in connection with the Cauchy problem and the so-called dynamic formulation of the theory. If a spacetime chart x is specified and the respective metric components g_{ij} and their time derivatives $\partial g_{ij}/\partial x^0$ are given on a spacelike hypersurface S—say, on the 3-manifold $x^0 = 0$—the Einstein field equations determine the values of the g_{ij} on the entire domain of dependence $D(S)$ of the said hypersurface.[23] But if the g_{ij} are known at every point of an open submanifold of spacetime the local values of the energy tensor components T_{ij} can be figured out directly from the field equations. Consequently, the evolution of the geometry to and from the instantaneous spacetime slice S prescribes the motion of matter in the interior of $D(S)$. It is doubtful, however, that this method can be used for predicting the outcome of actual physical experiments.[24]

In the discussion following Einstein's Vienna lecture of 1913, Max Born asked him how fast gravitational effects would propagate according to the Einstein–Grossmann theory. Einstein replied:

> It is extraordinarily simple to write the equations for the case in which the disturbances introduced into the field are infinitely small. In that case, the [potentials g_{ij}] differ infinitely little from what they would be without such disturbances. The disturbances propagate then with the same speed as light.[25]

Einstein took this approach to "the important question as to how the propagation of gravitational fields take place"[26] when he tackled it, a few years later, in the context of General Relativity. Assuming an almost flat metric, linked by the field equations (5.7.17) to an infinitely weak energy tensor, he was able to show that a small local change in the distribution of matter would bring about a change in the metric propagating throughout the universe with the speed of light.[27] However, it was not until much later that the question of gravitational radiation in General Relativity was satisfactorily dealt with under less restrictive assumptions.[28] Although the subject is fraught with conceptual, mathematical and experimental difficulties and does not seem to be promising from a technological point of view, it has aroused considerable interest, for, as Steven Weinberg points out, "the theory of gravitational radiation provides a crucial link between general relativity and the microscopic frontier of physics".[29]

The problem of the equations of motion in General Relativity is closely connected with the problem of gravitational radiation.* This may be seen by analogy with the situation in electromagnetic theory. In that theory, conservation of the combined energy tensor for the field and its material sources

* The rest of this Section, until page 185, was especially written by Professor John Stachel, for insertion in this place. (R.T.)

(equations (4.5.19)) must be postulated in addition to the Maxwell equations for the field; while in General Relativity the generally covariant conservation laws for the energy tensor follow from the field equations. But in both cases, these conservation laws imply some connection between the behaviour of the field far from a system of sources and the motion of these sources. In the electromagnetic case, for example, one may derive the force law for a charged monopole particle from the conservation laws by a variety of techniques, such as those based on the work of Dirac (1938). With the assumption of retarded potential interactions between the sources, the force equation takes the form of the Lorentz force law plus a radiation reaction term. As a consequence of the radiation reaction term an isolated system of charges will lose energy as a result of their electromagnetic interaction. On the other hand, integration of the electromagnetic energy-momentum flux over the "sphere at infinity" indicates that electromagnetic radiation is carrying off energy to an extent that just balances the energy lost by the moving sources. These general results are supported by the existence of some exact solutions for radiating electromagnetic fields including their sources. The earliest and best known is the Hertz oscillating dipole solution. In this case, the system is not closed, and what one actually verifies is that the energy fed into the system to keep the oscillator in a steady state just balances the energy radiated away by the field produced by the oscillator. What one would like in General Relativity are some analogous results for gravitating systems. But so far no one has been able to prove any comparable exact results, and even the approximate calculations that have been done are not yet fully satisfactory to all workers in the field.[30]

First of all, no exact solution is known representing an isolated gravitating system in a state of motion that would produce gravitational radiation. The problem in the case of gravity is much more difficult than in that of electromagnetism. In the latter case, non-electromagnetic forces may be used to produce and sustain the motion of charged particles which results in electromagnetic radiation. Thus, open electromagnetic systems are perfectly acceptable. But in the case of gravity, anything non-gravitational that is capable of affecting the motion of the sources must itself have an energy tensor and should therefore properly be included in the total energy tensor producing the gravitational field in question. So only closed systems could provide really satisfactory models of sources plus radiation fields. This means that no steady-state radiative solution can be satisfactory.

An additional problem in General Relativity concerns the definition of energy and momentum for the gravitational field. In a Lorentz-invariant theory, such as electrodynamics, the role of energy and momentum as generators of temporal and spatial translations serves to define the components of an energy-momentum 4-vector, which is an integral of the appropriate components of the energy tensor over a spacelike hypersurface. Since the typical relativistic spacetime has no symmetries, (p. 154), no such procedure may be used. For asymptotically flat spacetimes corresponding to the gravitational fields produced by isolated sources, asymptotic symmetries exist allowing the definition of global quantities which can be identified as the total energy and momentum of the system. In the past, such quantities were usually defined in terms of integrals over spacelike hypersurfaces of various non-tensorial quantities—the so-called gravitational energy pseudotensors (like the energy matrix defined by (5.7.8)). For asymptotically stationary spacetimes the resulting global quantities could be shown to be invariant under a sufficiently restricted class of coordinate transformations. But the analysis of such solutions contributed nothing to the radiation problem, since the energy so defined was constant in time. The use of such gravitational energy pseudotensors in the treatment of radiation problems remained therefore somewhat dubious.

However, great progress has been made in the last 15 years in the analysis of the exact asymptotic structure of the gravitational fields of isolated sources. (Asymptotic structure here means the structure of the field far from its sources.) This concept can be made precise, and indeed by the addition of a boundary to the spacetime manifold it has been possible to define past (and future) null infinities for such systems. That is, the ideal points, defined by past (future) null geodesics of manifolds corresponding to suitably well-behaved solutions of the vacuum Einstein equations (5.7.16), themselves form 3-manifolds which may be attached as boundaries to the 4-manifolds on which the field equations are respectively satisfied.[31] In a suitably defined conformal sense, the said 3-manifolds are null hypersurfaces of the manifold-with-boundary.[32] This technique enables one to define a pair of so-called "news functions" on future null infinity, whose non-vanishing is a criterion for the existence of gravitational radiation. A mass-loss formula then relates these news functions to a "mass function", whose integral over the sphere at infinity is monotonically decreasing whenever the news functions are non-vanishing. (The two news functions correspond to the two degrees of freedom of the gravitational field, comparable to the two independent states of polarization of the electromagnetic field.) The great remaining problem is to connect these exact but asymptotic results with some aspects of the "near field" behaviour of the presumed sources of those "far fields". That there is some connection seems clear. For static spacetimes, for example, the integral of the mass function corresponds to the well-defined "Schwarzschild mass" of the solution.[33] Some linearized approximation results also exist which seem to indicate that the news functions are connected with the changing multipole structure of the field sources in a way reminiscent of the corresponding electromagnetic results.[34]

The linearized expansion technique is based upon the assumption that the gravitational field may be taken to differ only slightly from flat, Minkowski spacetime.[35] The first-order linearized field already gives an excellent approximation to the behaviour of the gravitational radiation field far from its sources.

The major problem with any attempt to approach the problem of the relation between gravitational radiation and motion in terms of the linearized approximation technique is that the linearized gravitational field equations themselves imply an integrability condition on their sources, namely, the special-relativistic conservation law (4.5.20) for the energy tensor. In the absence of any non-gravitational interactions, this means that monopole sources, for example, must move along straight lines in Minkowski spacetime without interacting. If there is any linearized gravitational radiation present, it cannot have been produced by free gravitational motion of linearized sources; nor can the radiation even interact with linearized matter in a linearized theory. Thus, linearized gravitational theory cannot be consistently applied to the problem at hand; at best it might form the starting point for a complicated higher-order approximation procedure. Attempts to carry out such approximations based on the linearized theory have so far not yielded any conclusive results.

The other major approximation technique which has been applied to this problem is the so-called new approximation or EIH method (after Einstein, Infeld and Hoffmann (1938), although it actually goes back to Lorentz and Droste). As noted on page 178, the EIH technique is based upon the assumption that all motions of sources are slow and all gravitational fields are weak. Starting from the Newtonian gravitational field as the lowest order approximation, it gives an excellent description of post-Newtonian and even post-post-Newtonian corrections to the motion of such slow, weak sources of the gravitational field. No effects of gravitational radiation occur up to this order of approximation. Attempts to apply the technique directly to the radiation problem result in horrendous calculations, since radiation effects do not arise until such a very high order in the approximation scheme. Indeed, the very assumption of slow motions and weak fields makes the method ill-suited to the radiation problem. Attempts to use this method have accordingly also failed to produce any conclusive results.

However, a method has been developed which enables one to combine the advantages of the EIH technique for treating the motion of the sources with those of the linearized technique for treating the distant radiation field.[36] The EIH technique is used for the near zone, where the sources are located, resulting in (elliptic) Poisson-type equations relating the metric components to the components of the energy tensor of the sources. The linearized technique is applied in the far zone, many wavelengths away from the sources, resulting in (hyperbolic) wave equations for the propagation of the metric field components describing the radiation field. Relating the solutions in the two zones requires the use of singular perturbation techniques. The method of matched asymptotic expansions is used to match the two solutions in an intermediate zone. This enables the radiation loss in the far zone, for an assumed outgoing wave solution, to be matched with secular "radiation reaction" effects on the motion of the sources in the near zone. The method is known to give reasonable results in the electromagnetic case, where some problems can also be independently solved by other methods, thus allowing for comparison. In the gravitational case, no such independent procedures for evaluating the results are available; nor, indeed, is there any proof that the "approximate solutions" obtained really approximate anything like an exact solution. However, with these cautions, the method of matched asymptotic expansions may be said to furnish the most reliable approximation results so far obtained for this important problem.

CHAPTER 6

Gravitational Geometry

6.1 Structures of Spacetime

The development of General Relativity has been one of the great intellectual adventures of our time. The story of the theory's reception and growth, its lasting successes and apparent failures would fit well in this place, but would also lengthen the book considerably and go beyond its intended scope. Moreover, the story cannot be told properly until much more detailed scholarly research has been carried out.[1] Therefore I shall not try to follow here the same method of historical reconstruction as in the preceding chapter. Instead I shall elucidate a few matters that I consider especially significant from a philosophical point of view. I begin with some mathematical ideas which greatly clarify the meaning of the hypothesis that physical becoming is staged in a relativistic spacetime.

Let us recall our twofold approach to Minkowski spacetime in Section 4.2. We regarded it as the 4-manifold $\mathscr{M}$—diffeomorphic with $\mathbf{R}^4$—endowed with the flat Minkowski metric $\boldsymbol{\eta}$, and also as the affine metric space arising from the transitive and effective action of the additive group of $\mathbf{R}^4$—endowed with the Minkowski inner product $\langle a, b \rangle = (a_0 b_0 - a_\alpha b_\alpha)$—on the *unstructured* set of worldpoints. We saw that both views were equivalent inasmuch as the differentiable structure of $\mathscr{M}$ could be obtained from the action, the straights of the affine space were the ranges of Riemannian geodesics of $(\mathscr{M}, \boldsymbol{\eta})$ and the interval assigned to each pair of worldpoints by the affine metric was measured by the energy of the Riemannian geodesic joining them (p. 95). That the affine metric induced by the Minkowski inner product should agree with $\boldsymbol{\eta}$ is not surprising, but the perfect agreement of the affine straights with the geodesics of $(\mathscr{M}, \boldsymbol{\eta})$ is quite remarkable. For the concept of a Riemannian geodesic was defined in terms of the metric (p. 94), while the straights of an affine space are generated by the action of the associated vector space independently of any inner product that could be defined on the latter (p. 91). This suggests that we should be able to produce a non-metrical characterization of the geodesics of $(\mathscr{M}, \boldsymbol{\eta})$. In order to do so, let us look on Minkowski's world $\mathscr{M}$ as the 4-manifold generated by the action of $\mathbf{R}^4$ on the set of worldpoints (p. 92). Relative to this manifold structure the action of $\mathbf{R}^4$ is smooth, and the additive group of $\mathbf{R}^4$ acts on $\mathscr{M}$ as a Lie group of transformations (p. 26). A manifold constructed in this way from an affine space may be called an affine manifold. The action of $\mathbf{R}^4$ on $\mathcal{M}$ induces a partition of the tangent bundle of $\mathcal{M}$ into equivalence classes, as follows. Each $t \in \mathbf{R}^4$ determines a translation of $\mathcal{M}$ by $P \mapsto tP$. We say that two vectors tangent to $\mathcal{M}$ are *parallel* if one is the image of the other by the differential of a translation.[2] The reader will easily verify that vector parallelism, thus defined, is indeed an equivalence. Let us now bestow the Minkowski metric $\boldsymbol{\eta}$ on the affine manifold $\mathcal{M}$ in such a way that each translation turns out to be an isometry (i.e., so that $t^*\boldsymbol{\eta} = \boldsymbol{\eta}$, for each translation t). The metric is then said to be consistent with the action of $\mathbf{R}^4$. A curve γ in $\mathcal{M}$ is a Riemannian geodesic of $(\mathcal{M}, \boldsymbol{\eta})$ if and only if all its tangents (p. 259) are mutually parallel. We can thus give the following non-metric characterization of geodesics: a curve γ in the affine manifold $\mathcal{M}$ is an (affine) geodesic if and only if γ is an integral curve (p. 260) of a field of parallel vectors. The affine geodesics of the affine manifold become the Riemannian geodesics of the Riemannian manifold if the metric is so defined that translations become isometries.

Let us turn now to the more or less arbitrary relativistic spacetime $(\mathcal{W}, \mathbf{g})$ countenanced by General Relativity. $\mathbf{R}^4$ can act transitively on $\mathcal{W}$ as a Lie group of translations if and only if $\mathcal{W}$ admits a global chart, which of course would entail that $\mathcal{W}$ is a copy of Minkowski's world.[3] But even in the exceptional case in which this is so, no such action of $\mathbf{R}^4$ on $\mathcal{W}$ can be consistent with the metric $\mathbf{g}$, in the sense explained above, unless the latter is flat. Consequently our concept of vector parallelism is generally inapplicable to $(\mathcal{W}, \mathbf{g})$. Nevertheless, by a masterly extension of this concept, Tullio Levi-Cività (1917) provided a non-metric characterization of the geodesics of any Riemannian manifold.[4]

Levi-Cività's idea is beautifully simple, as the following consideration—made in his treatise on *The Absolute Differential Calculus*—will show.[5] Let S be a smooth surface embedded in Euclidean 3-space. (Any 2-manifold can be constructed by pasting together a collection of such surfaces.) If S is developable onto a plane, like a cylinder or a cone, there is an isometric map f of S into $\mathbf{R}^2$. Since $\mathbf{R}^2$ is an affine manifold in an obvious way, our earlier concept of vector parallel is applicable to S. (Two vectors v and w are parallel in S if $f_* v$ and $f_* w$ are parallel in $\mathbf{R}^2$.) Assume therefore that S is non-developable and let $t \mapsto \gamma t$ be a smooth curve in S, joining two not necessarily distinct points P and Q. As t ranges over the interval on which our curve is defined, the plane tangent to S at γt generally changes. We thus obtain a family of planes $\{\pi_t\}$, indexed by the parametre t. The family is enveloped by a smooth developable surface S' which touches S along the range of γ.[6] Each vector tangent to S at a point of γ can be naturally regarded as a vector tangent to S'. We say that two vectors v and w, tangent to S at points of γ, are *parallel along* γ in S if v and w are parallel in S'. Vector parallelism in non-developable surfaces is essentially relative to the curve—or rather to the path of the curve—along which it is said to hold. If

P and Q are any points of S joined by two curves γ_1 and γ_2, a vector at Q parallel to a given vector at P along γ_1 is generally not parallel to it along γ_2. This statement is true even if $P = Q$ and γ_1 and γ_2 are two loops.

Levi-Civitả defines vector parallelism along a curve in an arbitrary Riemannian n-manifold by a straightforward generalization to n dimensions of the foregoing notion of parallelism in a surface. He uses the fact that any such n-manifold is locally isometrically embeddable into a *flat* m-manifold for sufficiently large m.[7] But, as he points out, the concept defined is intrinsic, that is, independent of any such embedding. Levi-Civitả establishes the following analytic criterion of parallelism. Let us say that a vector field V on the Riemannian n-manifold $(M, \mathbf{g})$ is parallel along a curve γ in M if all values of V at points of γ are mutually parallel along γ. Let x be a chart defined on a neighbourhood of the points of γ and let n^3 scalar fields Γ^i_{jk} be defined on the domain of x, in terms of the components of $\mathbf{g}$ relative to x, by equations (D.1) on page 281. The vector field $\mathrm{V} = \mathrm{V}^i\partial/\partial x^i$ is parallel along γ if and only if

$$\frac{\mathrm{d}\mathrm{V}^i \cdot \gamma}{\mathrm{d}t} + (\Gamma^i_{jk} \cdot \gamma)\,(\mathrm{V}^k \cdot \gamma)\frac{\mathrm{d}x^j \cdot \gamma}{\mathrm{d}t} = 0 \tag{6.1.1}$$

Since the Γ^i_{jk} depend only on the metric $\mathbf{g}$ and not on a particular embedding of M into a manifold of higher dimension number, equations (6.1.1) constitute an intrinsic definition of parallelism along γ. Moreover, they will provide a *non-metric* definition of parallelism along γ if, following Hermann Weyl (1918*a*) we make the Γ^i_{jk} independent of the g_{ij}. This can be done as follows. Because the metric $\mathbf{g}$ is a (0, 2)-tensor field, two arrays Γ^i_{jk} and Γ'^i_{jk} defined by (D.1) relative to two different charts x and x' will satisfy, on the intersection of the domains of x and x', equations (C.4.3) of p. 273. Suppose now that on the domain of each chart of an n-manifold M we define an array of n^3 scalar fields Γ^i_{jk} subject to the "transformation law" (C.4.3). If the mutual dependences thus defined are smooth we have bestowed on M a "linear connection" Γ, whose "components" relative to each chart of M are the respective Γ^i_{jk}. If the fields of each array satisfy the condition $\Gamma^i_{jk} = \Gamma^i_{kj}$, we say that Γ is *symmetric*. Γ is said to endow M with an "affine structure" (though it does not make M into an affine space in the standard sense of this expression).

A Riemannian metric $\mathbf{g}$ on the n-manifold M evidently bestows on it the unique linear connection defined by equations (D.1). This is known as the *Levi-Civitả connection*, and is the only *symmetric* connection that is consistent with $\mathbf{g}$ in the following sense: If two vectors v and w at a point $P \in M$ are congruent or orthogonal (i.e. if $\mathbf{g}_P(v, v) = \mathbf{g}_P(w, w)$, or if $\mathbf{g}_P(v, w) = 0$), the vectors parallel to v and w at any $Q \in M$ along any curve joining P and Q are also congruent—in the former case—or orthogonal—in the latter. Since a given symmetric connection can stand in this special relation to several distinct metrics, the affine structure of a relativistic spacetime, defined by its Levi-Civitả connection, is a more general and so to speak a "deeper" structure than its metric. This is important, for all the main notions of Einstein's theory of gravity are linked directly to the connection. Thus, the "gravitational field strengths" (pp. 151 and 317, n. 20) are just the connection components. The operation of covariant differentiation through which, according to Einstein, the influence of the gravitational field on physical processes is made explicit depends on the connection alone and can be defined in terms of it by equations (5.5.1). Also the spacetime curvature, as embodied in the (1, 3) Riemann tensor, and the Ricci tensor, that occurs in Einstein's field equations, can be conceived purely in terms of the connection. (See equations (C.6.9) on p. 277. The Ricci tensor is obtained by substituting h for i on these equations and summing, as usual, over repeated indices. Observe that the Ricci tensor thus derived from an arbitrary symmetric linear connection need not be symmetric, though indeed it will be so if the connection is related to a Riemannian metric by (D.1) i.e. if it is a Levi-Civita connection.) Let us say hereafter that a curve γ in a manifold M endowed with a linear connection Γ is a *geodesic* of Γ if its tangent vectors are mutually parallel along it—or, as one often says, if γ is "self-parallel". γ is then an integral curve of a vector field whose values on γ's range are parallel along it. It turns out that the Riemannian geodesics of a Riemannian manifold are precisely the geodesics of its Levi-Cività connection.[8]

Weyl's definition of a linear connection as so many arrays of scalar fields transforming into one another according to (C.4.3) is unimpeachable if, as the saying goes, a mathematical entity is just what the mathematician chooses it to be—neither more nor less. But non-mathematicians like myself, who do not revel in the metamorphoses of numbers and have been stunned by the suggestion that gravity is just the visible expression of the linear connection of the world, are comforted by the knowledge that other mathematicians, working out some insights of Elie Cartan, have produced a more perspicuous definition. I explain it in Appendix C. Though it is not necessary to study it in order to understand what follows, all that is hereafter said about linear connections applies to the concept developed there—which, by the way, is extensionally equivalent to Weyl's (p. 273). One interesting consequence of the said definition, which may be worth noting here, is that parallelism along curves in an n-manifold M is directly linked to and as it were the reflection of the existence of an absolute parallelism in the bundle of n-ads over M (p. 273).

The idea of path-dependent vector parallelism inspired Hermann Weyl's attempt at a unified, purely geometric theory of gravitation and electricity. In this theory, "all effective reality in the world is a manifestation of the world metric".[9] The latter, however, is not Riemannian, but is conceived after Weyl's idea of a "pure infinitesimal geometry", a genuine "geometry of proximity" (*Nahe-Geometrie*), in which separate geometrical objects can only be compared through intermediaries lying in between. In a multidimensional manifold such mediated comparisons depend on the path along which the intermediaries are placed. Hence, in Weyl's geometry—as opposed to Euclid's—there are no absolute relations between distant objects, and vector congruence, no less than vector parallelism, is path-dependent. Applied to the real world, this means that two units of length or time (e.g. two wavelengths, or two wave periods, instanced by appropriate standard devices) will not generally agree every time they meet unless they have gone along nearby worldlines from one encounter to the next.[10] Specifically, Weyl's "world metric" depends on a symmetric, non-degenerate (0, 2) tensor field **g**—in effect, a Riemannian metric—and a covector field φ, which jointly govern vector congruence as follows. Let x be a chart defined at a point P and let P' be a worldpoint infinitely close to P such that $x^i(P') = x^i(P) + \delta x^i$. Let φ_i be the components, relative to x, of φ at P, and let g_{ij} and $g_{ij} + \delta g_{ij}$ be those of **g** at P and P', respectively. Consider a vector at P and another vector at P', whose components relative to x are respectively V^i and W^i. The two vectors are *congruent* if and only if

$$g_{ij}V^iV^j = (1 + \varphi_k \delta x^k)(g_{ij} + \delta g_{ij})W^iW^j \tag{6.1.2}$$

This criterion provides the means of judging the congruence of vectors at separate points, *relative to a specific path* joining them. Evidently, congruence is not affected if $\varphi - d\theta$ and $e^\theta \mathbf{g}$ are substituted for φ and **g**, respectively, with θ an arbitrary scalar field. As in Einstein's theory, gravitational phenomena are governed by a linear connection coupled to the world metric in the following way: if v and w are congruent vectors at a given point, the vectors parallel to v and w at any other point along any given curve are also congruent *along that particular curve*. This implies that the connection components Γ^i_{jk} relative to the chart x introduced above must satisfy the following conditions:

$$2g_{ij}\Gamma^j_{kh} = \frac{\partial g_{ik}}{\partial x^h} + \frac{\partial g_{hi}}{\partial x^k} - \frac{\partial g_{kh}}{\partial x^i} + g_{ik}\varphi_h + g_{hi}\varphi_k - g_{kh}\varphi_i \tag{6.1.3}$$

(Note that (6.1.3) becomes (D.1) if φ is identically 0.) The linear connection is thus fully determined by **g** and φ. Weyl assumes that the covector field φ is none other than the electromagnetic potential.[11]

The possibility of disregarding metrics inherent in the concept of linear connection was first fully utilized in Eddington's unified theory of gravity and electricity, with which Einstein sympathized for a short time.[12] Spacetime geometry rests here on a symmetric linear connection, whose 40 independent components are to be the unknowns of the field equations. As I noted on p. 189, the Ricci tensor defined by this connection need not be symmetric, but may be regarded, like every (0, 2)-tensor field, as the sum of a symmetric and an antisymmetric tensor field.[13] The theory equates the former with the spacetime metric, the latter with the electromagnetic field tensor.[14]

The linear connection of spacetime is also the key notion of the unified theory of gravity and electricity introduced by Einstein himself in the late twenties.[15] This theory postulates that spacetime has a field of orthonormal tetrads (pp. 99, and 303, note 1). Such a field determines a canonic isomorphism between any two tangent spaces, whence vector parallelism holds absolutely, irrespective of any "path of comparison" (we have *Fernparallelismus*—"parallelism at a distance"—as Einstein calls it), even though the Riemannian metric implicit in the assumption that the tetrad field is orthonormal is not held to be Minkowskian, but only Lorentzian. As Cartan pointed out to Einstein, what all this amounts to is simply that spacetime has a linear connection which is flat (zero curvature) but generally non-symmetric (non-zero torsion).[16] A non-symmetric linear connection is also a characteristic of Einstein's final theory, the Relativistic Theory of the Non-Symmetric Field.[17]

The demetricized concept of linear connection enabled Cartan to produce the geometrical formulation of Newton's theory of gravity mentioned on page 34. The empty Newtonian spacetime described in Section 1.6 can be characterized by means of a flat symmetric linear connection Γ, a scalar field t and a symmetric (2, 0)-tensor field $\mathbf{g}$ on Minkowski's world $\mathscr{M}$. t—which we may regard as determined only up to a linear transformation $t \mapsto \alpha t + \beta$ ($\alpha, \beta \in \mathbf{R}, \alpha \neq 0$)—assigns a time to each worldpoint. The fibre of t over each of its values is a 3-manifold diffeomorphic to $\mathbf{R}^3$. $\mathbf{g}$ satisfies the condition $\mathbf{g}(\mathrm{d}t, \mathrm{d}t) = 0$, and induces a Euclidean metric on each 3-manifold $t = const.$ A curve γ in $\mathscr{M}$ is "timelike" (in the sense of p. 27) if $\dot{\gamma} \neq 0$ on the entire range of γ (cf. p. 259). A free test particle describes a timelike geodesic of the connection Γ. Let x be a global chart of $\mathscr{M}$. We assume that $(\partial t/\partial x^i)_{;j}$ and $g^{ij}{}_{;k}$ vanish identically (where the g^{ij} are the components of $\mathbf{g}$ relative to x and the semicolon indicates covariant differentiation in $(\mathscr{M}, \Gamma)$). This completes the characterization of empty Newtonian spacetime. If matter is present in significant amounts our test particles come under the influence of gravity. This is accounted for by postulating that spacetime is "warped" by matter, i.e. that the connection Γ is no longer flat. Freely falling test particles still obey the geodesic law. A few more assumptions must be added concerning Γ and its linkage to $\mathbf{g}$ and t.[18] But it is time to stop this digression and return to the consideration of General Relativity.

Just as many Riemannian metrics on a given manifold can share the same Levi-Cività connection, so many linear connections will define the same set of geodesics (p. 278). The selection of one such set—or of the class of linear connections that agree on it—endows a manifold with a particular structure, which Weyl called *projective*. (Although it does not make the manifold into a projective space!) It is a trivial, though not insignificant remark, that the possible worldlines of the simplest (non-spinning, monopole) freely falling particles in General Relativity are determined by the projective structure of spacetime, regardless of its affine and metric structures—though, of course, when the latter structures are given, the projective structure is defined uniquely.

Another structure underlying every Riemannian manifold is the so-called conformal structure. Two Riemannian metrics **g** and **g**′ on a manifold M are said to be *conformal* if $\mathbf{g}' = \lambda^2\mathbf{g}$, where λ is an arbitrary scalar field on M. Conformal metrics obviously have the same signature. If they are proper (i.e. positive or negative definite), they all assign the same value to the angle between two given vectors at a point.[19] Thus, angle measure pertains to the conformal structure of a proper Riemannian manifold, rather than to the full metric structure. If, on the other hand, the conformal metrics under consideration are indefinite, they all determine the same partition of the tangent bundle into spacelike, timelike and null vectors.[20] In either case, conformal metrics agree in their definition of orthogonality.[21] Let us say that a worldpoint P in the relativistic spacetime $(\mathscr{W}, \mathbf{g})$ *causally precedes* the worldpoint Q ($P < Q$) if information can flow from P to Q (cf. p. 121). Due to the local validity of Special Relativity, it is clear that $P < Q$ only if $P = Q$ or there is a nowhere spacelike curve joining P to Q. Let us stipulate further that P *chronologically precedes* Q ($P \ll Q$) if $P < Q$ and P and Q are joined by a timelike curve. It is not hard to verify that $(\mathscr{W}, <, \ll)$ satisfies conditions (i)–(vi) in the definition of causal spaces on page 123. If, moreover, $(\mathscr{W}, \mathbf{g})$ does not contain any closed nowhere spacelike curves, $(\mathscr{W}, <, \ll)$ also satisfies condition (vii), so that the spacetime $(\mathscr{W}, \mathbf{g})$ has a causal structure determined by the metric **g**. Evidently, two metrics **g** and **g**′ bestow the same causal structure on the spacetime manifold if and only if they are conformal. (It is worth noting that in any relativistic spacetime each worldpoint P has a neighbourhood U such that (U, $<$, $\ll$)—with $<$ and $\ll$ defined as above but restricted to U—is a causal space.)

Let P be a projective and C a conformal structure bestowed on an n-manifold M. We may equate P and C, respectively, with the class of linear connections and the class of conformal Riemannian metrics that define them. Let us assume that the metrics of C have signature -2. (We may then say that C is a "Lorentzian conformal structure".) A metric **g** of C is compatible with a linear connection Γ of P if vector parallelism as defined by Γ preserves congruence and orthogonality as defined by **g**, in the manner explained on page 188 (whence, if Γ is symmetric, it must be the Levi-Cività connection of $(M, \mathbf{g})$). Let us say that C itself is compatible with $\Gamma \in \mathrm{P}$ if Γ-parallelism preserves C-orthogonality. We can also think of a more basic, direct agreement between P and C. A hypersurface H in (M, C) is said to be *null* if it is everywhere tangent to the local cone of null vectors. (At each $P \in H$, the tangent space H_P meets but does not cut the past null cone.) We shall say that P is compatible with C if every P-geodesic which is C-null lies on a C-null hypersurface. If P is compatible with C, C is compatible with one and only one symmetric linear connection of P. If this connection has a symmetric Ricci tensor there is a Riemannian metric of C, unique up to a constant scale factor, that is compatible with it. In this way, the conformal structure C and the projective structure P of a relativistic spacetime jointly determine its metric.

This theorem was proved 60 years ago by Hermann Weyl, who at once drew the inference that "the world metric can be established (*festgelegt*) solely by the observation of the 'natural' motion of material particles and the propagation of physical effects, especially of light", without resorting to rods and clocks.[22] Curiously enough, this remark and the mathematical result that prompted it were persistently overlooked by philosophers—although it is clearly relevant, say, to Bridgman's complaint that General Relativity is devoid of operational meaning, or to Reichenbach's claim that spacetime geometry is conventional—until it gave rise, in the early seventies, to the fascinating axiomatic formulation of General Relativity put forward by J. Ehlers, F. A. E. Pirani and A. Schild as "The Geometry of Free Fall and Light Propagation". Just as Helmholtz tried to base ordinary space geometry on the (suitably idealized) facts of (practically) rigid motion, Ehlers, Pirani and Schild "attempt to derive the conformal, projective, affine, and metric structures of space-time from some qualitative (incidence and differential-topological) properties of the phenomena of light propagation and free fall that are strongly suggested by experience.[23] Indeed, our authors go even further, for instead of taking the manifold structure of the world for granted, as Weyl did, they construct it from a few assumptions regarding particles and their interactions. While it would take us too long to explain their entire axiom system, we can afford to take a look at this interesting construction.[24]

Let $\mathscr{W}$ stand for the unstructured set of worldpoints. By a "particle" I mean a subset of $\mathscr{W}$ which could be the spacetime trajectory (the life history) of a freely falling material particle. By a "signal" I mean a subset of $\mathscr{W}$ which could be the spacetime trajectory of a light-signal *in vacuo*. Let us postulate:

I. A particle is an oriented 1-manifold, diffeomorphic with $\mathbf{R}$.

The orientation of a particle is meant to reflect the time order of events. If a and b are particles, an *echo* on a from b is a mapping f of a into a, such that for each $P \in a$, P precedes $f(P)$ and there are two signals, σ_P and σ_{fP}, which meet a at P and at $f(P)$, respectively, and intersect at a point of b. An echo f on a from b obviously determines a mapping of a into b by $P \mapsto \sigma_P \cap \sigma_{fP}$. Such a mapping is called a *message* from a to b. We postulate further:

II. Any message from a particle to another is smooth. Any echo on a particle from another is a diffeomorphism.

Let $U \subset V \subset \mathscr{W}$. Let a and b be two particles. Suppose that each $P \in U$ is the meeting point of exactly two signals that meet a inside V and exactly two signals that meet b inside V. Denote the intersections of a and b with the said signals through P by P_a^1 and P_a^2, P_b^1 and P_b^2, respectively, reserving the index 1 for the earlier intersection. Choose a chart α of a and a chart β of b. U is mapped into $\mathbf{R}^4$ by $P \mapsto (\alpha(P_a^1), \alpha(P_a^2), \beta(P_b^1), \beta(P_b^2))$. If this mapping is injective and its range is open in $\mathbf{R}^4$, we call it a *radar chart*, defined on U, relative to V, and based on a and b. We finally postulate:

> III. Each $P \in \mathscr{W}$ lies on the domain of a radar chart, relative to an appropriate subset of $\mathscr{W}$ and based on a suitable pair of particles. The collection of all such radar charts is a smooth atlas.[25]

The atlas of radar charts defines a topology and a differentiable structure in $\mathscr{W}$, in the manner explained on page 348, note 3 to Appendix A.[26]

6.2 Mach's Principle and the Advent of Relativistic Cosmology

The linear connection of Minkowski spacetime accounts for the essential difference between inertial and accelerated motion in Special Relativity. An object that moves inertially sticks to its spacetime direction and describes a geodesic; an accelerated object departs incessantly from the geodesic trajectory of its mirf (p. 96). While the former's worldvelocity remains parallel to itself along the object's worldline, the latter's is being continually deviated by extraneous forces. The same account can be extended to Newtonian dynamics by means of the four-dimensional formulation sketched in Section 1.6. In either theory, Foucault's pendulum holds aloof from the whirling earth and faithfully sweeps a plane fixed in an inertial system because it follows the dictates of the spacetime geometry. Thus, Minkowski's vision, perfected by the work of Levi-Cività and Weyl, at last provides a truly geometric understanding of the strains and stresses that show up in accelerated systems, which Newton had attributed, obscurely yet insightfully, to the peculiar relation of those systems to absolute space.[1] But, although chronogeometry sheds light on Newton's explanation of such phenomena, it does not make it a whit less uncanny. Newtonian dynamics and Special Relativity put us in the grip of a universal structure, capable of shattering to pieces a body that resists its sway, and yet impervious to the body's reaction. In this respect, both theories contrast sharply with General Relativity. In the latter, again, geodesic motion, as in free fall, differs essentially from non-geodesic motion, as illustrated, say, by "rest" in a gravitational field. There can be no question of dumping all motions together into a single equivalence class, as the expression "general relativity" might suggest. But the linear connection responsible for their differences is not itself indifferent to the whereabouts of the bodies subject to its guidance, but is governed, through the Einstein field equations, by the actual distribution of matter. As Hermann Weyl eloquently said:

> If the guidance field [i.e. the linear connection—R.T.] manifests itself in mechanical processes as an effective power of eventually shattering might, at war with the forces of nature, we must regard this field as something *real*. It cannot be—as Newtonian mechanics and the Special Theory of Relativity assumed—a formal, unconditionally given structure of the world, independent of matter and its states, but must in turn *suffer under the action of matter, be determined by matter, and change as it changes.*[2]

The assimilation of gravity and inertia by the Equivalence Principle thus leads not only to a geometrical conception of gravity in the likeness of inertia, but also to a dynamical conception of inertia in the likeness of gravity. Einstein repeatedly described this feature of General Relativity as its chief advantage over its predecessors; most emphatically perhaps in a letter to Max Born of May 12th, 1952, in which he comments on an apparent failure of one of the three classical tests of the theory:

> Even if no deviation of light, no perihelion advance and no shift of spectral lines were known at all, the gravitational field equations would be convincing, because they avoid the inertial system—that ghost which acts on everything but on which things do not react.[3]

The field equations equate the Einstein tensor, characteristic of the spacetime geometry, with the energy tensor that specifies the matter distribution. Since equality is a symmetrical relation, neither term of which can as such claim priority over the other, the field equations bespeak the *mutual* determination of the metric and the matter fields. Their *interaction* is what puts flesh and blood into the otherwise ghostlike geometry. However, Einstein's earlier pronouncements on this question demand that the metric field be *completely* and as it were *unilaterally* determined by matter. This demand, of course, goes beyond what the gravitational field equations can express. Its motivation is philosophical, as Einstein clearly explained to Gustav Mie, in two letters written in February 1918. In the first of them, dated February 8th, he gives the following statement of "the standpoint that I call relativistic":

> The behaviour (continuation of the temporal evolution of states) of every natural body is such, that it is univocally determined by its own state and that of all the other bodies. (The statement must naturally be interpreted without instantaneous action at a distance.) By "natural bodies" we must understand here only the perceivable bodies of our sensible world (*die wahrnehmbaren Körper unserer Sinnenwelt*).[4]

In the second letter, dated February 22nd, he asserts that the "relativistic standpoint" prescribes the following conception of inertia:

> If L is the actual trajectory of a certain freely moving body and L' is a trajectory that deviates from it but has the same initial conditions, the relativistic conception demands that the trajectory L which is in fact followed be preferred to the logically equally possible L' due to a *real cause*. [. . .] Such a real cause can only consist in the relative positions and states of motion of all the other bodies that exist in the world. The inertial behaviour of our mass must be completely and univocally determined by them. Mathematically this means that the g_{ij} must be completely determined by the T_{ij}—naturally, only up to the four arbitrary functions that correspond to the possibility of freely choosing the coordinates.[5]

In a paper on "Questions of Principle regarding the General Theory of Relativity" submitted on March 6th, 1918, the aforesaid demand is distinguished from the "Relativity Principle"—which is here conceived, as on page 153 above, as a requirement of general covariance—and is listed after the latter and the Equivalence Principle as one of the "three main points of view"

(*Hauptgesichtspunkte*) on which General Relativity is based. It is stated as follows:

> **Mach's Principle** (*Machsches Prinzip*): The [metric] field is *exhaustively* determined by the masses of bodies. Since mass and energy are the same, according to the results of the Special Theory of Relativity, and since the energy is formally described by the symmetric energy tensor (T_{ij}), this means that the [metric] field is conditioned and determined by the energy tensor of matter.[6]

The designation for this principle was chosen, says Einstein, because it is "a generalization of Mach's demand that inertia be explained by the interaction of bodies".[7] For a like reason he might as well have dubbed it Newton's Principle, as it also generalizes Newton's similar demand regarding the explanation of gravity. But at that time Einstein was not too keen on the Newtonian heritage.[8] Indeed, Mach's Principle was directly associated in his mind with Mach's own criticism of Newton. In the obituary he wrote for the *Physikalische Zeitschrift* when Mach died in 1916, Einstein quotes extensively from the relevant section of *Die Mechanik* (1883), which shows, in his opinion, that Mach "was not far from postulating a general theory of relativity half a century ago":[9] The leading idea of Mach's criticism was that the state of motion of a body can only be judged with respect to other bodies. As Einstein had pointed out already in 1913,

> in view of this, it seems meaningless to attribute to a body an absolute resistance to acceleration (inertial resistance of bodies in the sense of classical mechanics); the emergence of an inertial resistance should be linked rather to the relative acceleration of the body in question with respect to other bodies. It must be postulated that the inertial resistance of a body can be increased by placing unaccelerated inertial masses around it; and this increase of the inertial resistance must again fall away if those masses share the acceleration of the body.[10]

That the inertia of a body would increase as other massive bodies were piled up in its neighbourhood was a fairly obvious consequence of Einstein's tentative theory of the static gravitational field.[11] According to Einstein, the same effect held in the Einstein–Grossmann theory. In this theory, too, as required by the above analysis, the increment of inertia will not take place if the piled up masses participate in the acceleration of the body in question. This effect can also be described as follows: The accelerations of the neighbouring masses induce on the body an accelerating force in their same direction.[12] Moreover, as Einstein announced to Mach himself on June 25th, 1913, the Einstein–Grossmann theory entails that the rotation ("relative to the fixed stars") of a massive spherical shell about an axis through its centre generates a Coriolis field in its interior, which drags along the plane of Foucault's pendulum.[13] In the Princeton lectures of 1921 Einstein derived all three effects from the Geodesic Law and the linearized field equations of General

Relativity.[14] Though too small to be accessible to experiment, the three Machian effects are formally entailed, according to Einstein, by the general theory and are therefore corroborated by its empirical success. They, in turn, lend strong support to Mach's ideas about the relativity of inertia. Einstein noted that "if we think these ideas consistently through to the end we must expect the *whole* inertia, that is, the *whole* g_{ij}-field, to be determined by the matter of the universe".[15] In this way, Mach's thought does indeed stand behind Mach's Principle and is the key to Einstein's understanding of it.

We saw on page 112 how Einstein's proof that a body at rest loses proper mass as it radiates energy had led him to assert that the mass of a body is a measure of its energy content. Likewise, from the alleged fact that the inertial resistance of a particle is strengthened by piling up masses in its neighbourhood he extrapolated, no less daringly, that all inertial effects must vanish if the surrounding masses are pulled back to infinity. As Einstein said in the "Cosmological Considerations" of 1917:

> In a consistent theory of relativity there can be no inertia *relative to "space"* but only an inertia of masses *relative to each other*. Hence if I take a mass sufficiently far away from all other masses in the world its inertia must fall down to zero.[16]

It is hard to imagine how the inertia of a particle will fade away as it is removed from the neighbourhood of all other bodies. Does its worldvelocity become more and more erratic, its worldline less and less determinate? What was usually assumed—for instance, in Einstein's own approximate and in K. Schwarzchild's rigorous treatment of the gravitational field of a single star[17]—was that as a test particle is withdrawn from the influence of a central body its guidance gradually devolves on the flat Minkowski metric. The assumption implies, however, that inertia is merely *influenced*, not *determined*, by matter.[18] Einstein therefore tried to work out the implications of a different assumption, namely, that the metric degenerates at infinity in a definite way, instead of just flattening out.[19] This implied that no material particle or light ray could ever escape from the bulk of material bodies. Thus the question of its behaviour at infinity would not even arise. And the danger of the world eventually evaporating—which had always loomed large in Newtonian cosmology—would disappear. Einstein, however, was able to show, with the assistance of J. Grommer, that a spherically symmetric static metric degenerating at infinity in the agreed way cannot be reconciled with the fact, certified beyond question by pre-1917 astronomical data, that the stars move quite slowly relative to each other (in comparison with the speed of light).[20] To solve the ensuing difficulty Einstein proposed then one of his most memorable hypotheses, which marks the beginning of the wondrous development of 20th-century cosmology, namely, that physical space is finite, even though it has no bounds.

There is no need to decide about the inertial behaviour of a particle infinitely distant from the rest if there is no room for them to come so far apart. We do not have to prescribe the conditions at infinity if the world is finite in size. Now the idea of a finite yet limitless space was implicit in the very notion of a Riemannian manifold, as Riemann himself had already noted in his inaugural lecture of 1854.[21] The arbitrary Lorentz metrics countenanced by General Relativity might for instance be defined on a compact 4-manifold or at any rate on a 4-manifold decomposable into a family of compact 3-manifolds which the metric makes into spacelike hypersurfaces of finite volume.[22] Einstein (1917) showed that the field equations of General Relativity, in their modified form (D.6) admit an exact solution defined on a 4-manifold diffeomorphic with $\mathbf{R} \times \mathbf{S}^3$,[23] provided that the following additional assumptions are made:

I. The distribution of matter is static and homogeneous.
II. The constant λ in equations (D.6) is different from 0.

This was the first cosmological solution of the field equations and is known as Einstein's spherical or static universe. In accordance with (I) the matter distribution can be thought of as a pressureless fluid ("dust") uniformly smeared over all spacetime. Every tangent to the worldline of each infinitesimal speck of matter can be regarded as the local value of a vector field, the worldvelocity field of the fluid. The spacetime of Einstein's solution can be partitioned into compact spacelike hypersurfaces, orthogonal to the worldvelocity field, and isometric with $\mathbf{S}^3$ (up to a constant scale factor; see note 23). Thus, if H denotes any such spacelike section of the world and $\mathbf{g}$ is the restriction to H of the spacetime metric, $(H, \mathbf{g})$ is a proper Riemannian 3-manifold with constant positive curvature.[24]

The double assumption that the distribution of matter is (*a*) static, and (*b*) homogeneous, is meant of course as a first approximation, true only of what we may call the "coarse structure" of matter—its "fine structure" being quite obviously variegated and changing. To justify assumption (*a*) Einstein again invokes the slow motion of the stars. On the other hand, assumption (*b*) is adopted on purely speculative grounds. Einstein puts it thus: "If we suppose that the world is spatially closed it is plausible to assume that the density of matter is independent of its position."[25] Extragalactic astronomy has subsequently confirmed that the universe is indeed homogeneous in the large, and delivered us from the illusion that it is in any sense static. Ironically, the intragalactic observations that were available to Einstein and persuaded him of the truth of (*a*) also clashed openly with (*b*), for—as Einstein once remarked[26]—the stars of the Galaxy are not evenly distributed in the sky. Apparently Einstein was willing to extrapolate from the observations at hand only when they agreed with his own broad vision of what the world must be like.

The story of assumption (II) is no less instructive from a methodological point of view. The λ-term in equations (D.6) was added to the Einstein tensor for the sole purpose of obtaining a physically viable solution conformable to Mach's Principle. This seemingly *ad hoc* change turned out to be very reasonable from a mathematical and philosophical point of view, for, as we have noted earlier, the modifed Einstein equations (D.6) are virtually the only admissible system of field equations for a geometrical theory of gravity patterned after Einstein's "Outline" (see pp. 160 and 323, note 34). But there was not the slightest physical inducement for setting $\lambda \neq 0$. Indeed, in order to preserve the agreement of the field equations with the observed facts Einstein had to assume that the "cosmological constant" λ was so small as to be empirically indistinguishable from 0. (Observations of distant galaxies by Sandage (1961, 1968) place an upper bound on $|\lambda|$ of the order of 10^{-56} cm^{-2}.) Moreover, as D. J. Raine (1975), p. 515 pointedly observes, a non-zero cosmological constant would represent a physical field which acted on everything but was not in turn acted upon, just like the "ghostly" linear connection of a flat spacetime rejected by the Machian philosophy. Worse still, the hypothesis that $\lambda \neq 0$ is neither necessary nor sufficient to secure the achievement of Einstein's purpose. It is not *necessary*, for—as Einstein himself remarked in his (1917), p. 152—the unmodified field equations (with $\lambda = 0$) are also satisfied by a universe sliced into spacelike 3-spheres provided that the distribution of matter is not required to be almost static. (As we shall see in Section 6.3 this line of thought was successfully pursued by A. Friedmann in 1922.) It is not *sufficient*, for, as W. de Sitter (1917*a*, *b*, *c*) was quick to show, the conditions (I) and (II) on page 198 yield besides the solution proposed by Einstein still another non-flat solution, consisting of a spacetime altogether devoid of matter.[27] To postulate an empty universe may seem the *ne plus ultra* of scientific extravagance, but it is no worse an approximation than Einstein's *plenum* to what is actually seen through the telescope. In a sufficiently grand perspective the visible stars—and indeed the galaxies themselves—could be regarded as test particles exerting a negligible influence on the spacetime metric.[28] Whatever its intrinsic merits, de Sitter's solution demonstrated that, contrary to Einstein's expectations, there is indeed "a singularity-free spacetime continuum with everywhere vanishing energy tensor of matter" that satisfies the *modified* field equations.[29]

There are further indications that General Relativity, as embodied in the original or in the modified Einstein field equations, not only does not entail Mach's Principle but is ultimately inimical to Mach's idea. Let us note some of them.[30]

First several new non-flat matter-free solutions of the field equations have been discovered, which show that an empty universe built in accordance with General Relativity can be a very lively place, the interplay of pure gravitational energy being enough to produce some remarkable phenomena if a handful of test particles is supplied.[31]

Secondly, it was shown by Carl H. Brans (1962) that the first Machian effect derived by Einstein from his field equations, viz. that the inertia of a particle is increased by piling up masses in its neighbourhood, can always be wiped out by a suitable choice of a coordinate system and does not even show up in a coordinate-free description. It is therefore a chart-dependent mirage, not a true physical effect. Brans' result cannot take us by surprise, for, as we observed on page 319, note 6, in a relativistic spacetime it is always possible to define a chart along the worldline of any particle, relative to which all the gravitational field components Γ^{i}_{jk} vanish at every point of that worldline. In memory of its discoverer, such a chart is called a Fermi chart. When referred to it, the particle's behaviour appears to be unaffected by the gravitational influence of the rest of matter, and its inertial tendency to persevere in or relapse into a geodetic path cannot be linked to that influence. In particular, if the particle is falling freely, a Fermi chart along its worldline is what we called earlier a local Lorentz chart (p. 136). Although the particle's geodesic trajectory and the local Lorentz charts definable along it strictly depend on gravity, the Principle of Equivalence implies that physical phenomena referred to such charts obey the laws of Special Relativity and show no effects of the surrounding mass distribution. In this sense, as Steven Weinberg has ably observed, "the equivalence principle and Mach's principle are in direct opposition".[30] On the other hand, the remaining two Machian effects claimed by Einstein (p. 196) do indeed take place in General Relativity under suitable initial and boundary conditions. H. Thirring (1918, 1921) and J. Lense and H. Thirring (1918) showed, respectively, that a massive rotating hollow sphere generates a Coriolis and a centrifugal field in its interior, and that a massive rotating central body induces an acceleration on nearby test particles. Their results are indeed questionable from our point of view, for they used the linearized field equations (p. 325, note 35), thereby presupposing a flat background metric, in flagrant violation of the Machian idea.[33] However, according to A. K. Raychaudhuri (1979, p. 241), A. Lansberg has proved—to first order in the angular velocity of the rotating shell—that Thirring's effect also obtains in Einstein's static universe.

Thirdly, the conceptual system of General Relativity provides the means of giving a definition of rotation in the neighbourhood of a particle which is not tied to a frame of reference and is altogether independent of the distribution of matter. This is easily done if the particle describes a timelike geodesic γ. Let e_0 be the worldvelocity of the particle at a worldpoint P. We can always find three spacelike vectors e_1, e_2, e_3 at P which form with e_0 an orthonormal tetrad (i.e. a basis of the tangent space at P such that $\mathbf{g}_P(e_i, e_j) = \eta_{ij}$, where $\mathbf{g}$ denotes as usual the spacetime metric). Let E_α be a vector field, parallel along γ, whose value at P is e_α. We say that E_1, E_2 and E_3 constitute a set of non-rotating axes or a "compass of inertia" along the geodesic γ. It will be readily seen that at any point of γ the worldvelocity of the particle forms an orthonormal tetrad with the local values of the E_α. A compass of inertia with similar properties can be defined along an arbitrary timelike curve by means of the concept of Fermi transport.[34] Let γ be such a curve, parametrized by proper time, and let $E = (E_1, E_2, E_3)$ be a compass of inertia along it. Consider a field V of unit vectors whose value at each point P of γ is orthogonal to the local tangent γ_P. We shall naturally say that V rotates with respect to E unless the three "direction cosines" $\mathbf{g}(V, E_\alpha)$ are constant along γ. It will be noted that if V rotates with respect to E it also rotates with respect to any other compass of inertia E′ constructed along γ according to our definition. The instantaneous rotation of V at each P in the range of γ can be represented by a vector in the space spanned by E at P, which is of course spanned also by the local value of E′. (One may thus meaningfully and indeed correctly assert, within the framework of General Relativity, that Copernicus and not Ptolemy was right.) Let U denote the worldvelocity field of a fluid. The local rotation of each small volume of fluid with respect to the compass of inertia along the integral curves of U is measured by the fluid's vorticity.[35] The Einstein field equations admit exact solutions in which the cosmic distribution of matter has non-zero vorticity. In such cases it makes good sense to say that matter rotates in space. Nobody claims that we actually live in such a cosmic merry-go-round, but the fact that a rotating universe can be rigorously conceived as just another instance of the same mathematical structure of which our world is very nearly a model, should make every Machian shudder.[36]

Soon after 1918 Einstein became convinced that General Relativity does not subordinate the metric field to the energy tensor of matter but brings them both into a dynamic interplay. In a lecture on "Ether and the Theory of Relativity", delivered on May 5th, 1920, while still proclaiming the Machian inspiration of the theory, Einstein likened the spacetime continuum to an ether, i.e. a physical medium "which is itself devoid of *all* mechanical and kinematical qualities, but helps to determine mechanical (and electromagnetic) events".[37] In turn, the metrical properties characteristic of this medium differ in the neighbourhood of its several points "and are *partly* conditioned by the matter existing outside the region under consideration".[38] Although he was never happy about this duality of metric and matter, his successive attempts at a unified field theory sought to overcome it by deriving the properties of matter from the spacetime geometry, rather than the other way around. In the course of his reflections he grew increasingly wary of Mach's brand of empiricism, whose strongly mentalistic undercurrent had never really got hold of him. In 1937 he wrote for Carlton Berenda Weinberg a sort of balance sheet of Mach's pros and cons. Mach had shattered the dogmatism of 19th century physics by exposing the hypothetical or logically arbitrary nature of its basic concepts and laws. He had rightly criticized Kant's apriorism with regard to the categories and some fundamental principles. But he had erroneously believed that conceptual knowledge grows "out of sense data without *free* creation of concepts".[39] Near the end of his life Einstein sent to Felix Pirani a detailed appraisal of Mach's Principle. Mach "sought to abolish space and replace it by the relative mutual inertia of ponderable bodies". "This certainly did not work, quite apart from the fact that time remained with its absolute character." When people "speak today about Mach's Principle (*Machs Prinzip*) they do not mean to abolish the continuum but to preserve the field. But they think that the field ought to be *completely* determined by matter. This however is a ticklish affair (*eine heikle Sache*) for the T_{ik} which are to represent 'matter' always presuppose the g_{ik}."[40] "In effect, T_{ik} is the part of the total field of which we know that, for the time being, we can only superficially (phenomenologically) take its structure into consideration." Einstein sums up his final position thus:

> In my opinion we ought not to speak about the Machian Principle (*dem Mach'schen Prinzip*) any more. It proceeds from the time in which one thought that the "ponderable bodies" were the only physical reality and that all [concepts that could not be traced back to them] elements that could not be fully determined by them ought to be avoided in the theory. I am well aware that for a long time I too was influenced by this fixed idea.[41]

Nevertheless, during the fifties and sixties Mach's Principle persistently engaged the attention of scientists. Some of them acknowledged that it clashed with General Relativity and proposed new theories of gravity which presumably satisfied it.[42] Others altered its meaning and scope to make it usable in General Relativity.[43,44] Enthusiasm for Mach's Principle has waned in the seventies, together with the ideological motivation that kept it alive. M. Reinhardt (1973) summarizes the conclusion of his excellent account of the subject in the statement that Mach's Principle, though an extremely stimulating thought, has at present little claim to be a basic physical principle. An interesting exception is D. J. Raine (1975) who, having characterized Mach's Principle as the triple assertion that only relative motion is observable, that inertial forces arise from a gravitational interaction of matter alone, and that the spacetime metric is totally dependent on the matter content of the universe, sets up mathematical criteria for selecting the global solutions of the Einstein field equations which satisfy that principle. Only such solutions should be considered physically viable according to Raine.[45]

6.3 The Friedmann Worlds.[1]

Relativistic cosmology arose, as we saw, from philosophical reflections concerning the relationship between spacetime and matter. Initially it only invoked basic, qualitative data, which had been available for a long time. Deftly using the conceptual resources of Riemannian geometry it gave, at its very outset, a highly satisfactory solution to the problem of the world's extension in space. If, as Einstein proposed, the universe is finite yet boundless, there is no point in asking with Archytas what will become of a missile thrown outwards from the farthermost star, nor—it seems—need one fear with Kepler that if stars exist everywhere with the same average density as in our neighbourhood, the firmament will outshine the sun.[2] Einstein's cosmology contributed perhaps more than anything else to his fame among the educated classes right after the first World War. Weltering in novelty as no generation before them, the contemporaries of Pound and Joyce, Schönberg and Bartok, Matisse and Mondrian, greeted the man who had given a shape to the universe as the epitome of creative genius. And yet, mind-boggling as it seemed then, Einstein's spherical world was a staid, almost philistine invention in comparison to what came next. A new spate of astronomical data, following a great technical breakthrough, found in General Relativity the fitting framework for its interpretation. The unexpected agreement of theory and observation put new life and drama into the hackneyed idea of evolution and supported, for the first time in the history of rational inquiry, the ancient belief that the universe has existed only during a limited time, that no piece of matter, however elementary, has been around for more than so many years.

Very sketchily, the story can be told as follows. In the early twenties, the brightest spiral nebulae were resolved into stars, including some of the Cepheid type. These are variable stars whose brightness oscillates with a period related to their intrinsic luminosity. Assuming that the period-luminosity relation established for nearby Cepheids by H. Leavitt also held for the newly discovered ones, E. Hubble estimated that the great Andromeda nebula (M 31 = NGC 224) lay at a distance of 900,000 lightyears and was therefore a star system not substantially smaller than the Galaxy. Assuming that the fainter nebulae were roughly of the same size as the brighter ones, he calculated distances up to eight times larger. This meant a colossal increase in the size of the visible world and the amount of matter contained in it.[3] V. M. Slipher had been studying for some time the spectra of nebulae—or galaxies, as we call them now—in order to determine their radial velocities. It turned out that most of them exhibited significant redshifts, indicative of recession at surprising speeds. A list furnished by Slipher to Eddington in February 1922 and printed in the 1923 edition of the latter's *Mathematical Theory of Relativity* included 41 galaxies of which 36 were receding with speeds of up to 1800 km/s.[4] Half this list, with four additions and five minor corrections, provided the sample on which E. Hubble (1929) based his momentous claim that the radial velocity of the receding galaxies is proportional to their distance from the Earth.[5] A couple of years later, M. L. Humason (1931) published a list of galaxies with receding velocities of almost 20,000 km/s. W. de Sitter (1917*b*) had noted that his solution of the Einstein field equations entailed "a systematic displacement towards the red of the spectral lines of nebulae".[6] At first, therefore, Slipher's data were generally regarded as an indication that de Sitter's spacetime, notwithstanding its emptiness, was a much better model of the real world than Einstein's.[7] However, the Soviet mathematician A. Friedmann had shown as early as 1922 that the universes of Einstein and de Sitter were merely limiting cases of an infinite family of solutions of the field equations for a positive but varying matter density, any one of which would entail, at least for some time, a general recession—or a general oncoming, for the solutions are indifferent to time reversal—of the galaxies. Friedmann's work was carried out in Leningrad during the Russian civil war, presumably in complete ignorance of Slipher's measurements. His simplifying assumptions are much weaker than either Einstein's or de Sitter's and thus define a likelier idealization of the real world. Einstein's theory of gravity predicts the expansion—or contraction—of any universe that roughly agrees with Friedmann's assumptions. If the agreement is perfect and the cosmological constant $\lambda \leq 0$ (cf. pp. 198 f.) the worldlines of matter—i.e. the integral curves of the worldvelocity field of the matter distribution—are all focused on a mathematical point. If the universe is now expanding only a finite proper time can have elapsed along each of them since they parted their ways. (This is true also when $\lambda > 0$, if certain other conditions are fulfilled.)

The Friedmann solutions are hailed nowadays as "the most provocative and important cosmological model which has been devised since Bruno".[8] However, when Friedmann published them in the German *Zeitschrift für Physik* they attracted little attention. Some potential readers may have been deterred by Einstein's laconic remark, in the same journal, that Friedmann's non-stationary universes are incompatible with the field equations.[9] At any rate the Belgian G. Lemaître had apparently never heard of the Friedmann solutions when he rederived them in 1927 with the explicit aim of "accounting for the radial velocity of extra-galactic nebulae".[10] Soon thereafter the American cosmologist H. P. Robertson (1929) set out to discover the most general metric for a relativistic spacetime in which no preferred spatial directions are discernible about any average worldline of matter, and came up with the same family of solutions that had been found by Friedmann and Lemaître.[11] That the world, on a cosmic scale, does indeed look the same to us in every direction is a generally accepted fact. Hence, if we profess the "Copernican" view that there is nothing special about our own vantage point, we must conclude that, as Robertson assumed, the universe is isotropic everywhere. This conclusion is confirmed almost beyond question by the nearly perfect isotropy of the microwave background radiation discovered by A. A. Penzias and R. W. Wilson (1965). What Robertson showed was that a universe shaped according to Einstein's field equations, if it is thus roughly homogeneous and isotropic, must approximately model one of the Friedmann solutions. This cannot be the static one discovered by Einstein in 1917 for, as Eddington (1930) proved, it is essentially unstable and will expand to infinity or contract to a point at the slightest disturbance. Enlightened by Hubble's discovery and probably shaken by Eddington's proof, Einstein (1931) declared his preference for the non-static Friedmann solutions with $\lambda = 0$, which, he noted, enabled one "to do justice to the facts without introducing the theoretically anyway unsatisfactory λ-term".[12] In the Appendix added to the second edition of *The Meaning of Relativity* (1945), Einstein said:

> As Friedman was the first to show one can reconcile an everywhere finite density of matter with the original form of the equations of gravity if one admits the time variability of the metric distance of two mass points. The demand for spatial isotropy of the universe alone leads to Friedman's form. It is therefore undoubtedly the general form, which fits the cosmological problem.[13]

Friedmann (1922) based his work on two kinds of assumptions, which concern (I) the gravitational field and the state and motion of matter, and (II) "the general, so to speak geometrical character of the world." The assumptions of the first class are:

(I.1) The gravitational field satisfies the modified Einstein equations (D.7), where λ can take any value, including 0.

(I.2) Matter can be represented as a pressureless fluid, at rest relatively to the coordinate system (x^0, x^1, x^2, x^3) that will be introduced below.

(I.2) implies that the integral curves of $\partial/\partial x^0$ agree, up to a reparametrization, with the worldlines of matter.

With a minor amendment (see note 14), Friedmann's assumptions of the second class can be stated as follows:

(II.1*a*) Spacetime admits a global time coordinate function x^0.

(II.1*b*) The restriction of the spacetime metric to any spacelike hypersurface $x^0 = const.$ constitutes on the latter a proper Riemannian metric with constant curvature K, which depends on the value of x^0 and is either more than, equal to, or less than 0 for every such hypersurface.[14]

(II.2) The hypersurfaces $x^0 = const.$ are everywhere orthogonal to the integral curves of $\partial/\partial x^0$.

Assumption (II.1*a*) expresses the traditional belief that spacetime can be split into time and space in at least one way. Friedmann apparently found it so obvious that he did not bother to mention it. We now know that it holds good in a relativistic spacetime if and only if the latter is "stably causal", in the sense defined in note 23 to Section 6.4 (see page 337). Assumption (II.1*b*) implies that each three-dimensional space of simultaneous worldpoints is homogeneous and isotropic as far as the geometry is concerned—i.e. that each point on it is equivalent to every other point and that there are no preferred directions about any of them. Friedmann may have adopted (II.1*b*) as a geometrical statement of the observed isotropy of the universe together with the "Copernican" thesis that all contemporary standpoints are interchangeable; or else as a statement of the then common opinion that physical (space) geometry places no restrictions on rigid motion.[15] Friedmann notes that "no physical or philosophical reasons can apparently be given for assumption [II.2]. It serves only to simplify the calculations."[16] This is perhaps the weakest spot in Friedmann's work, for (II.2) in conjunction with the assumptions of class I imposes severe restrictions on the distribution of matter, which are ultimately responsible for the peculiar properties of the Friedmann solutions, while assumptions (II.1) are compatible with a much wider variety of world-models.[17] Therefore (II.2) cannot be accepted merely because it is expedient, but only on solid rational and empirical grounds. Such grounds, however, were supplied by Robertson.

Suppose that we take (II.1*a*) for granted and, as Robertson puts it

> we demand that to any stationary observer ("test body") in this idealized universe all (special) directions about him shall be fully equivalent in the sense that he shall be unable to distinguish between them by any intrinsic property of space-time, and he shall similarly be unable to detect any difference between his observations and those of any contemporary observer.[18]

It follows at once that spacetime must admit a group of motions (i.e. a connected Lie group of isometries) that map each hypersurface $x^0 = const.$ onto itself, either by rotating it about an arbitrary point (a momentary "observer") or by translating any such point to any other one. Since arbitrary rotations and translations in a 3-manifold involve 6 degrees of freedom, the group in question must be a 6-parameter Lie group. Robertson invokes a theorem by G. Fubini (1904) to the effect that a 6-parameter group of motions acting on a Riemannian 4-manifold acts transitively as a group of motions on each member of a family of hypersurfaces with the following properties:

(F1) the hypersurfaces are orthogonal to a congruence of geodesics;
(F2) each pair of hypersurfaces intercepts segments of equal length on every geodesic of the congruence;
(F3) the mapping that sends each point P of the hypersurface H to the intersection of the geodesic through P with the hypersurface H' is a conformal mapping of H onto H'.[19]

Obviously, Robertson's demand entails Friedmann's assumption (II.2), *provided that the integral curves of $\partial/\partial x^0$ are geodesics.* The requirement is a natural one in General Relativity if, as we stipulated, $\partial/\partial x^0$ is everywhere tangent to the worldlines of matter.[20] (Note that, by (F2), x^0 must then measure proper time, up to a constant scale factor.[21]) Robertson's demand also entails (II.1*b*).[22] For in order to admit a 6-parameter group of motions, each 3-dimensional hypersurface $x^0 = const.$ must be a Riemannian manifold of constant curvature;[23] and (F3) implies that the curvatures corresponding to the several values of x^0 have a positive ratio to one another—unless they are all zero.

Later, Robertson (1935) derived assumption (II.1*a*) from the following premises, which may be thought to supplement or merely to clarify his earlier demand regarding the equivalence of all standpoints and perspectives: (*a*) the "observers" which are claimed to be equivalent describe a congruence of worldlines (i.e. each worldpoint lies on the worldline of one and only one of them); (*b*) two worldpoints can be joined at most by one "light-path" or worldline of a direct light signal. (*c*) the path of every light signal that an "observer" A can receive directly from an "observer" B lies on a 2-manifold (AB); (*d*) (AB) is identical with (BA). Robertson shows that if these four conditions are met all observers can synchronize their clocks and thus define a global time coordinate function.[24]

Conditions (II.1*a*), (F1), (F2) and (F3) entail that there exists a spacetime chart x—with the time coordinate function x^0—relative to which the line element takes the form

$$ds^2 = M^2(dx^0)^2 - S^2\bar{g}_{\alpha\beta}dx^\alpha dx^\beta \tag{6.3.1}$$

where the $\bar{g}_{\alpha\beta}$ depend only on the space coordinates x^α, while M and S are functions of x^0 alone.[25] We can set $M = 1$ by equating x^0 with proper time along every worldline of matter (see note 21). It is then customary to write τ for x^0. S is the scale factor relating the conformal metrics of the several spacelike hypersurfaces $\tau = const.$ Since S cannot be equal to 0 for any value of the coordinate function τ, it does not change signs, and we shall assume that it is always positive. (On the implications of the contrary assumption see note 31.) Due to the isotropy of the hypersurface $\tau = const.$ about each point, the space coordinate functions x^α can be so chosen that, if we write (r, θ, φ) for (x^1, x^2, x^3), $\bar{g}_{11} = (1 - kr^2)^{-1}$, $\bar{g}_{22} = r^2$, $\bar{g}_{33} = r^2 \sin^2\theta$ and $\bar{g}_{\alpha\beta} = 0$ whenever $\alpha \neq \beta$; k being equal to 1, -1, or 0 if the curvature of the hypersurfaces $\tau = const.$ is, respectively, positive, negative, or flat.[26] Referred to these coordinates, the line element (6.3.1) becomes

$$ds^2 = d\tau^2 - S^2\left(\frac{dr^2}{1 - kr^2} + r^2(d\theta^2 + \sin^2\theta\, d\varphi^2)\right) \tag{6.3.2}$$

We call (6.3.2) the *Robertson–Walker line element*. It clearly brings out the fact that in a spacetime compatible with Robertson's requirements, the spacelike separation between simultaneous worldpoints along two given worldlines of matter may grow or diminish in the course of time *as a consequence of the geometry alone* (just like the distance between places of equal latitude along two given meridians increases from the poles to the Equator *because such is the shape of the Earth.*) The frequency shift of light sent from one such worldline to another can be directly obtained from (6.3.2) if it is assumed that light describes null geodesics. Suppose that a material particle A radiates light at time $\tau = \tau_1$ during a short time $\Delta\tau_1$, which is received by a particle B at time $\tau = \tau_2$ during a short time $\Delta\tau_2$. Since the number of lightwaves emitted is the same as the number received, the frequencies ν_i $(i = 1, 2)$ at emission and reception satisfy the relation $\nu_1 \Delta\tau_1 = \nu_2 \Delta\tau_2$. We assume, of course, that on its way from A to B the light does not suffer any accidents that can alter its frequency, so that any difference between ν_1 and ν_2 is exclusively due to the shape of the metric field. The worldline of a light signal must satisfy the differential equation:

$$d\tau^2 - S^2 d\sigma^2 = 0 \tag{6.3.3}$$

where I have written $d\sigma^2$ for $\bar{g}_{\alpha\beta} dx^\alpha dx^\beta$. The coordinate distance σ from A to B can be calculated by integrating (6.3.3):

$$\sigma = \int_{\tau_1}^{\tau_2} S^{-1} d\tau \tag{6.3.4}$$

Since σ is constant it is also given by

$$\sigma = \int_{\tau_1 + \Delta\tau_1}^{\tau_2 + \Delta\tau_2} S^{-1} d\tau \tag{6.3.5}$$

If we now subtract (6.3.5) from (6.3.4) and ignore the small change in S during the short periods of time $\Delta\tau_i$ we find that:

$$\frac{\nu_2}{\nu_1} = \frac{\Delta\tau_1}{\Delta\tau_2} = \frac{S_1}{S_2} \tag{6.3.6}$$

where S_i stands for the value of the scale factor S at time τ_i. The *frequency shift* z, conventionally defined as the ratio of the frequency loss to the observed frequency, is therefore given by:

$$z = \frac{\nu_1 - \nu_2}{\nu_2} = \frac{S_2}{S_1} - 1 \tag{6.3.7}$$

z is positive—a "redshift"—if the scale factor S is larger at τ_2 than it was at τ_1—i.e. if the universe has "expanded" in the mean time.[27]

From the metric components g_{ij} implicit in (6.3.2) we obtain, by straightforward application of formulae (D.1), (D.2) and (D.4) on p. 281, the Ricci components

$$R_{00} = 3\ddot{S}/S \qquad R_{0\alpha} = 0 \tag{6.3.8a}$$

$$R_{\alpha\beta} = \bar{R}_{\alpha\beta} - (S\ddot{S} + 2\dot{S}^2)\bar{g}_{\alpha\beta} \tag{6.3.8b}$$

where I have written $\dot{S}$ and $\ddot{S}$ for $dS/d\tau$ and $d^2S/d\tau^2$, respectively, and the $\bar{R}_{\alpha\beta}$ are the Ricci components determined by the "spatial" metric $d\sigma$.[28] It can be shown that $\bar{R}_{\alpha\beta} = -2k\bar{g}_{\alpha\beta}$,[29] so that

$$R_{\alpha\beta} = -(S\ddot{S} + 2\dot{S}^2 + 2k)\bar{g}_{\alpha\beta} \tag{6.3.8b'}$$

By assumption (I.2) the components of the energy tensor relative to the chosen chart are $T_{00} = \rho$, and $T_{ij} = 0$ unless $i = j = 0$. Since the metric is linked to the distribution of matter and the hypersurfaces $\tau = const.$ are homogeneous and isotropic, the energy density ρ must depend on τ only.[30]

Substituting the above values for g_{ij}, R_{ij} and T_{ij} into the Einstein field equations (D.6), we obtain the following—omitting the equations in which both sides turn out to be identically 0—:

$$\frac{3\ddot{S}}{S} - \lambda = -\tfrac{1}{2}\kappa\rho \tag{6.3.9a}$$

$$\frac{\ddot{S}}{S} + \frac{2\dot{S}^2}{S^2} + \frac{2k}{S^2} - \lambda = \tfrac{1}{2}\kappa\rho \tag{6.3.9b}$$

These are the gravitational field equations for the Friedmann worlds. (6.3.9*a*) is equivalent to

$$\ddot{S} = -S\left(\frac{\kappa\rho}{6} - \frac{\lambda}{3}\right) \tag{6.3.10}$$

Consequently if in a given Friedmann world ($\mathscr{W}$, **g**), $\lambda \leq 0$, $\ddot{S}$ must be negative. In that case, S cannot have a minimum and must tend to 0 as τ approaches some finite value τ_B. (The B may stand for "Beginning" or "Big Bang".) Specifically, if S_0 denotes the value of S at time $\tau = 0$,

$$\tau_B = \int_{S_0}^{0} \dot{S}^{-1}\,\mathrm{d}S \tag{6.3.11}$$

and will be negative ("to the past of time 0") if the "expansion rate" $\dot{S}_0 = \mathrm{d}S/\mathrm{d}\tau|_0 > 0$.[31] As I remarked in passing on page 204, a similar situation occurs under certain conditions even if $\lambda > 0$; for instance, if $k \leq 0$.[32]

Since the line element (6.3.2) collapses if $S = 0$, it is clear that the domain of the chosen chart cannot contain a worldpoint P such that $\tau(P) = \tau_B$. This in itself is not surprising, for a spacetime chart need not and normally will not be global. Thus, for example, a quick glance at (6.3.2) will persuade us that, if $k = 1$, the domain of the chosen chart cannot contain a worldpoint P such that $r(P) = 1$. Nevertheless, a point sequence defined along a given parametric line of r by the rule $P_n = r^{-1}(1 - 2^{-n})$ will converge, as n goes to ∞, to a definite worldpoint P_∞ lying in the domain of another chart.[33] However, the situation that arises as τ approaches τ_B is not just a consequence of our choice of a particular chart, for, as we shall now see, the curvature scalar R_i^i—which is, of course, chart-independent—increases beyond all bounds as τ goes to τ_B. Multiplying both sides of the field equations (D.6) by g^{ij} and summing over j we verify that $\mathrm{R}_i^i = \kappa\mathrm{T}_i^i + 4\lambda$. Solving the system (6.3.9) for $\mathrm{T}_i^i = \rho$ we find that

$$\rho = \frac{3}{\kappa}\left(\frac{\dot{S}^2}{S^2} + \frac{k}{S^2} - \frac{\lambda}{3}\right) \tag{6.3.12}$$

Multiplying both sides of (6.3.12) by S^3 and differentiating with respect to τ, we obtain:

$$\frac{\mathrm{d}}{\mathrm{d}\tau}(\rho S^3) = \frac{3\dot{S}S^2}{\kappa}\left(\frac{2\ddot{S}}{S} + \frac{\dot{S}^2}{S^2} + \frac{k}{S^2} - \lambda\right) \tag{6.3.13}$$

Now, the right-hand side of (6.3.13) is equal to 0, as can be readily seen by adding (6.3.9*a*) to (6.3.9*b*). Hence, ρS^3 is a constant and the energy density ρ must increase without bounds as S approaches 0. Since κ and λ are constants, the same can be said of the curvature scalar $\mathbf{R}_i^i = \kappa\rho + 4\lambda$. Thus in our Friedmann world $(\mathscr{W}, \mathbf{g})$ the sequence of worldpoints $P_n = \tau^{-1}(\tau_B + 2^{-n})$, chosen along a worldline of matter, does not converge to a point P_∞ as n goes to ∞, for if such a point existed, the curvature scalar would not be defined at it.[34] Hence, if one has a liking for rhetoric, one may well say that $(\mathscr{W}, \mathbf{g})$ came out of nothing at time $\tau = \tau_B$ and that $|\tau_B|$ is, at $\tau = 0$, the time elapsed "from the Creation of the World".[35] But one ought to bear in mind that in the chosen chronology τ_B is not really a time at all, but only the greatest lower bound of the range of values of the time coordinate, and that at any *assignable* time the Friedmann world $(\mathscr{W}, \mathbf{g})$ already has a past.

6.4 Singularities

The situation discussed at the end of the preceding section is commonly described by saying that a Friedmann world of the type in question *has a singularity* at $\tau = \tau_B$. This manner of speech derives from the classical field theories of gravity and electricity, where the functions that characterize the field often do "have singularities" in a sense akin to that of function theory.[1] Take for example the Newtonian gravitational field of a single massive particle. The field strength at any given point P is inversely proportional to the squared distance r^2 from P to the particle and therefore increases beyond all bounds as r tends to 0 and is undefined at the point where the particle is located. This, of course, does not in the least compromise the existence of that point, which is firmly held in place by its geometrical relations to the rest (cf. the quotation from Newton on p. 285, note 7), and can be aptly described as a singular point of the field. In General Relativity, however, one cannot meaningfully say that the gravitational field is undefined or "becomes singular" at a worldpoint, because the field itself is responsible for the geometry and is therefore constitutive of spacetime. Obviously a point P can only belong to the structure $(\mathscr{W}, \mathbf{g})$ if the metric $\mathbf{g}$ is defined at P. Thus, the Friedmann world on page 209 can only improperly be said to "have a singular point" where the time coordinate τ of (6.3.2) takes the value τ_B defined by (6.3.11). For, as I have already noted, under the stipulated conditions, that world can contain no points at which $\tau = \tau_B$. The impropriety is indeed no worse than other abuses of language endemic in mathematical literature, and I would not have cared to mention it were it not for the confusions it has caused in philosophy, and even in ordinary conversation, with regard to the so-called "beginning" of such Friedmann worlds. However, as things stand, one ought to be meticulously precise on this matter, at least for some time, until the confusions are dispelled.

If the so-called singularities of the Friedmann worlds and other solutions[2] to the Einstein field equations cannot be simply understood on the analogy of the singular points of the classical field theories and the theory of functions, it is all the more necessary to have an appropriate concept with which to grasp them. Several such concepts have been proposed since the sixties, when the subject became the focus of active and successful research, and although none of them is altogether satisfactory they are apt to be very illuminating, as I hope to show. All these concepts rest on the perception that a spacetime in a way *lacks* something there where one would naively say that it *has* a singular point. They aim at defining the nature of this lack and specifying its relation to the points that spacetime actually does have. Now in physics we want to understand nature, as far as possible, purely in terms of itself. We may not therefore go beyond spacetime and its content in order to conceive its privations. It may sound as if this would make our task impossible, but mathematical thought has already proved its mettle by defining and even overcoming the incompleteness of certain structures without resorting to anything outside them. The standard example of this maneuver is the completion of incomplete metric spaces. Let (M, d) be a metric space, in the sense defined on page 303, note 1. A sequence $P_1, P_2, \ldots$ of points of M is said to be a *Cauchy sequence* if, for every positive real number δ, no matter how small, there is an integer N such that δ exceeds the distance between any two points which occur in the sequence after P_N. ($d(P_r, P_s) < \delta$ whenever $N < r, s$). The points of a Cauchy sequence (P_n) are therefore packed closer and closer as n increases, and one would expect them to pile up about a definite point $P \in M$. This, however, is not always the case. (M, d) is said to be *metrically incomplete* unless every Cauchy sequence of points of M converges to a point in M. Cauchy sequences do not only enable us to spot the gaps in (M, d) without having to look beyond it, but also provide the means for filling those gaps, so to speak, out of (M, d) itself. We define the distance between two Cauchy sequences (P_n) and (Q_n) as the limit, in $\mathbf{R}$, of $d(P_n, Q_n)$. Two Cauchy sequences are equivalent if their distance is equal to 0. Let $\bar{M}$ be the collection of all equivalence classes of Cauchy sequences of (M, d). Let $\bar{d}$ assign to any two "points" of $\bar{M}$ the distance between two representative elements. $(\bar{M}, \bar{d})$ is then a metric space. We embed M into $\bar{M}$ by equating each $P \in M$ to the class of Cauchy sequences converging to P. $\bar{d}$ evidently agrees with d on $M \subset \bar{M}$. It is not hard to see that $(\bar{M}, \bar{d})$ is metrically complete. We call it the *Cauchy completion* of (M, d). Clearly it is made of ingredients drawn from the latter.[3]

This example is not only philosophically interesting by itself, but also directly relevant to our problem, as we shall now see. Let M be an n-manifold with some additional structure and let $\gamma: I \to M$ be a curve of some specified kind (e.g. a causal curve, if M is a causal space, etc.). γ is said to be an *inextensible* curve of its kind unless there is a curve $\gamma': I' \to M$ of the same kind such that $I' \neq I, I \subset I'$ and γ is the restriction of γ' to I. γ is an *incomplete* curve of its kind if it is inextensible and I is a finite interval. Let M have a linear connection Γ. We say that Γ is incomplete and that (M, Γ) is *geodesically incomplete* if there are incomplete geodesics in M. If M is a proper Riemannian manifold it is also a metric space, with the distance function defined on page 95. It can be shown that M is then metrically incomplete, in the sense discussed above, if and only if it is geodesically incomplete (with respect to the Levi-Cività connection).[4] The strict equivalence of metric and geodesic incompleteness in the proper Riemannian case suggests that we treat the latter as the natural generalization of the former to improper Riemannian manifolds—such as relativistic spacetimes—to which, for lack of a distance function, the metrical concept does not apply.[5] This suggestion inspired the first fruitful approach to the question of singularities in General Relativity, leading to the remarkable "singularity theorems" of Roger Penrose and Stephen W. Hawking. This approach, which lies at the root of all that have followed, is based on the idea that a singularity-free spacetime is one that is geodesically complete, or, more exactly, one in which all inextensible *timelike and null* geodesics are complete, i.e. defined on the entire field of real numbers.[6] The italicized qualification is due to the fact that not all spacetime geodesics are equally meaningful from a physical point of view. Thus, while the presence of an incomplete timelike geodesic implies that there might be "freely moving observers or particles whose histories did not exist after (or before) a finite interval of proper time",[7] the actual significance of incomplete spacelike geodesics is less clear. The said approach should have sufficed to dispel the air of paradox surrounding the notion of a spacetime singularity since the early days of General Relativity,[8] for evidently there is nothing impossible or even strange about a geodesically incomplete spacetime. After all, barring metaphysical insight or divine revelation, I do not see how we could ever learn that spacetime is geodesically complete—and all the more reputable sources of supernatural information decidedly point to the opposite conclusion. Indeed the venerable vision of man as a microcosm, and hence of the universe as Man Writ Large, can only be upheld within the framework of General Relativity if spacetime contains incomplete timelike geodesics along which, so to speak, time runs out.

In order to see more clearly how the incompleteness of null or timelike geodesics can account at least in part for the situations traditionally described as spacetime singularities, let us again consider the Friedmann world $(\mathscr{W}, \mathbf{g})$ discussed on pages 209 f. Take a finite open interval J such that $\tau_B \in J \subset \mathbf{R}$. The spacetime region $\tau^{-1}(J) \subset \mathscr{W}$ is then an open neighbourhood of each of its points, homeomorphic to $\mathbf{R}^4$.[9] Every worldline of matter contained in $\tau^{-1}(J)$ is a timelike geodesic whose maximal extension within the region—when parametrized by the proper time τ—is defined on the intersection of J with the set $\{u | \tau_B < u\} \subset \mathbf{R}$. Hence $\tau^{-1}(J)$ is by itself a geodesically incomplete 4-manifold. It is contained, of course, in a vaster spacetime region, into which the worldlines of matter are further extended, but only "towards the future", i.e. in the direction of increasing τ. But no inextensible matter worldline of $(\mathscr{W}, \mathbf{g})$ can be defined on all $\mathbf{R}$. The real number τ_B, or whichever takes its place in a reparametrization, is necessarily excluded from the domain of definition of all such worldlines.[10] And there is no way of isometrically embedding $(\mathscr{W}, \mathbf{g})$ into a vaster spacetime in which its incomplete timelike geodesics would be completed. Thus the incompleteness of $\tau^{-1}(J)$ is not remedied by extending this region to the full Friedmann world $(\mathscr{W}, \mathbf{g})$, for the latter is constitutively and incorrigibly geodesically incomplete. As I said before, this is something that a Riemannian manifold can jolly well be.

The region $\tau^{-1}(J)$ in which, so to speak, the incomplete geodesics of our Friedmann world are born is one where, if J is very small, the average energy density will be so great that matter cannot subsist there in any known guise. This suggests that the laws of physics cannot hold in that region in their familiar form. In particular, the Einstein field equations, which have never been tested under such extreme conditions, may stand in need of correction. The same considerations apply to all spacetimes in which geodesics run into regions where the curvature scalar increases indefinitely (see note 2) and entitle us to question the reality of geodesic incompleteness in all such cases.[11] However, if physical theories were to be countenanced only within the range of circumstances in which they have been actually tested, the scope of theoretical physics would become so narrow as to stifle all its chances of growth. Cosmology, like other branches of natural science, is speculative and conjectural, and its conclusions are hypothetical, i.e. true *if* the postulated conditions hold good. However, to disqualify the latter while they still agree with all available facts of observation merely because their consequences run contrary to our preferences or prejudices is a gross violation of the rules of the game. Of course, if a particular solution of the Einstein field equations turns out to be geodesically incomplete this will be due to the initial and boundary conditions on which that solution depends, rather than to the field equations themselves, which also admit geodesically complete solutions. Since the conditions that have to be assumed in order to integrate the equations are uncommonly simple and neat, one might reasonably expect that under the rougher circumstances of real life the predicted incompleteness would not arise.[12] In the case in point, however, such expectations suffered a rude blow when A. K. Raychaudhuri (1955) proved that adding density perturbations and anisotropic expansion (shear) to a Friedmann world cannot prevent the occurrence of incomplete timelike geodesics. They were finally quashed by the theorems of Penrose and Hawking.

The singularity theorems assert that a relativistic spacetime is always geodesically incomplete when some very general and fairly realistic conditions are fulfilled. To state these conditions concisely I must first give some definitions. Let $(\mathscr{W}, \mathbf{g})$ be a relativistic spacetime satisfying Einstein's field equations (D.7). We shall say that $(\mathscr{W}, \mathbf{g})$ is *generic* if each timelike or null geodesic γ goes at least through one worldpoint P at which the spacetime curvature induces expansion or contraction or shear on each local congruence of geodesics that includes γ. (By "local congruence" I mean a congruence in the open submanifold constituted by a small neighbourhood of P.)[13] This means that every geodesic worldline of matter or light includes at least one event at which gravitational curvature is, so to speak, effective. Every plausible, not inordinately symmetric spacetime is generic in this sense, and even one which is not so, becomes generic if the metric is slightly perturbed. We shall say that $(\mathscr{W}, \mathbf{g})$ meets the *weak convergence condition* if the Ricci tensor everywhere satisfies the inequality $\mathrm{R}_{ij}\mathrm{V}^i\mathrm{V}^j \leq 0$ for every null vector V. $(\mathscr{W}, \mathbf{g})$ meets the *strong convergence condition* if $\mathrm{R}_{ij}\mathrm{V}^i\mathrm{V}^j \leq 0$ for every timelike vector V. The convergence conditions imply that matter—in whose presence only does R_{ij} differ from 0—has a focusing (or rather, a non-unfocusing) effect on congruences of null or, respectively, timelike geodesics. By continuity, the strong condition entails the weak one. They may be regarded as a formal statement of the fact that matter attracts nearby matter and radiation.[14] A *trapped surface* in $(\mathscr{W}, \mathbf{g})$ is a compact spacelike smooth 2-surface S such that both the incoming and outgoing future-directed null geodesics that meet S orthogonally converge locally at S. (A trapped surface could be, for instance, the boundary of a gravitational "sink", a spatial region which is the seat of such a strong gravitational pull that all radiation originating inside it is sucked back into it.)

I shall now state without proof the main singularity theorems:[15]

I. The relativistic spacetime $(\mathscr{W}, \mathbf{g})$ contains incomplete null geodesics if it (i) satisfies the weak convergence condition and contains (ii) a noncompact Cauchy surface and (iii) a trapped surface. (Penrose (1965).)

II. $(\mathscr{W}, \mathbf{g})$ contains incomplete timelike geodesics if it (i) satisfies the strong convergence condition and (ii) contains a compact spacelike hypersurface H without edge, such that (iii) the unit vectors orthogonal to H are everywhere converging (or everywhere diverging) on H. (Hawking (1967), Theorem I.)

III. $(\mathscr{W}, \mathbf{g})$ contains incomplete null and timelike geodesics if it (i) satisfies the strong convergence condition, (ii) is strongly causal (in the sense defined on page 125), and (iii) there exists a worldpoint $P \in \mathscr{W}$ such that all future-directed (or all past-directed) timelike geodesics through P are focused by curvature and start reconverging within a compact region in the future (respectively, the past) of P. (Hawking (1967), Theorem II.)

IV. $(\mathscr{W}, \mathbf{g})$ contains incomplete null and timelike geodesics if it (i) satisfies the strong convergence condition, (ii) is generic, (iii) does not have any closed timelike curves and (iv) contains at least one of the following: (a) a compact spacelike hypersurface without edge, (*b*) a trapped surface, or (*c*) a worldpoint *P* such that all future-directed (or all past-directed) null geodesics through *P* are focused by curvature and start reconverging somewhere in the future (respectively, the past) of *P*. (Hawking and Penrose (1970), Corollary.)[16]

These theorems imply that among relativistic spacetimes geodesic incompleteness is at least as likely as geodesic completeness, in the following sense: If the set of Lorentz metrics on any given spacetime manifold is endowed with a reasonable topology, the metrics involving geodesic incompleteness (of the physically significant kind) constitute an open set.[17]

If we examine the hypotheses on which the singularity theorems depend we see that some of them, like the presence of a trapped surface, concern local properties of spacetime, which might be gathered from observations. Others, such as the convergence conditions, though supposed to hold everywhere, can be extrapolated from local circumstances. But the theorems also postulate hypotheses of a strictly global nature, such as the existence of a Cauchy surface, or the absence of closed timelike curves. Although these hypotheses traditionally have been taken for granted in physics, it does not seem possible to ascertain their likelihood by the standard methods of statistical inference. They cannot, however, be dismissed as irrelevant, for their denial runs counter to the same deeply ingrained physical intuitions or beliefs that apparently clash with geodesic incompleteness. Thus the total absence of Cauchy surfaces from the world is no less fatal to (global) scientific prediction than the presence of singularities. And closed timelike curves, along which the past is reached again and again *via* the future, so that any event on them can act on itself as its own remote cause, are probably harder to reconcile with received views than incomplete causal geodesics running as it were into or out of nowhere. (It is worth noting that recent work by Tipler (1976, 1977) suggests that closed timelike curves rather than preventing the occurrence of singularities would contribute to produce them.)

The interpretation of spacetime singularities in terms of geodesic incompleteness, though decisive for establishing their comparative abundance in the set of solutions to the Einstein field equations, is now thought to be both too wide and too narrow. Some find it too wide because, as Misner (1963) has shown, there are compact Riemannian manifolds which are geodesically incomplete and it is felt that a compact manifold, which as such "can be covered by a finite number of well-behaved coordinate patches" (i.e. chart domains), must be regarded as singularity-free.[18] A different interpretation has therefore been proposed, according to which a spacetime is singularity-free if and only if every geodesic defined on a finite interval maps it into a compact subset of the spacetime manifold. This criterion is not exempt from difficulty, however. Since a singularity-free world in this sense can be nevertheless geodesically incomplete it may contain a test particle that emerges from or vanishes into nothingness—a phenomenon we have hitherto linked to the occurrence of a spacetime singularity.[19] Anyway, the problem which the new criterion seeks to avoid is purely academic, for compact spacetimes always contain closed timelike curves and may therefore be dismissed as physically implausible—at any rate until there is some evidence that we happen to live in one. On the other hand, the concept of geodesic incompleteness is certainly too narrow for our purpose, since it fails to cover some spacetimes in which a worldline of matter is liable to run out of time. Thus R. Geroch (1968*a*) proposed an example of a geodesically complete relativistic spacetime containing an inextendible timelike curve of bounded acceleration and finite total length.[20] Such a curve could be the worldline of an accelerated test particle that vanishes without trace in finite proper time. This difficulty has been duly taken care of by the more recent conceptions of the spacetime singularities to which I shall refer below.

The study of singularities cannot aim only at showing that they must occur in this or that type of spacetime. It must also seek to clarify their relationship to each particular spacetime region. We would like to know how those regions are diversely affected by the existence of singularities, when particles are likely to vanish "into thin air", where curvature increases beyond all bounds, etc. In other words, we wish to "localize" the singularities with respect to the ordinary worldpoints. The standard way of achieving this is to represent the former by ideal points, defined in terms of the spacetime structure, which are then gathered together with the "real" worldpoints in a conveniently structured manifold-with-boundary (cf. page 325, note 32). The first proposal of this kind was independently put forward by S. W. Hawking, in his unpublished Adams prize essay, "Singularities and the geometry of space–time", of 1966, and by R. Geroch (1968*b*). The following sketch of its main idea is paraphrased from Geroch. Each point P in the manifold $\mathscr{W}$ and each vector v in the tangent space $\mathscr{W}_P$ determine a unique geodesic with initial condition (P, v) (see p. 278). By slightly varying P and v we obtain a family of geodesics filling a world-tube. We call this tube a *thickening* of the geodesic (P, v). Two geodesics are equivalent if each of them enters into and remains within every thickening of the other. An ideal point is an equivalence class of incomplete geodesics. Though this idea sounds simple and natural, its rigorous definition involves some ambiguities which can only be resolved by more or less arbitrary decisions. Moreover, it rests squarely on the concept of geodesic incompleteness and may fail to provide ideal points corresponding to non-geodesic inextensible curves of finite length, such as we mentioned above. For these reasons, Hawking considered that his representation of spacetime singularities by equivalence classes of incomplete geodesics had been superseded by B. G. Schmidt (1971), whose elegant construction does not suffer from the said shortcomings.[21]

B. G. Schmidt's method can be readily described by means of the concepts explained in Parts B and C of the Appendix. Let $F\mathscr{W}$ denote the principal bundle of tetrads over the relativistic spacetime $(\mathscr{W}, \mathbf{g})$. On page 272 we show how to define 16 real-valued 1-forms ω^i_j on $F\mathscr{W}$ determined by the relevant Levi-Cività connection form ω. Let θ be the canonical form of $F\mathscr{W}$ (p. 272) and set $\theta^i = \pi_i \cdot \theta$, where π_i is the i-th projection function on $\mathbf{R}^4$. We endow $F\mathscr{W}$ with a *proper* Riemannian metric $\mathbf{h}$ such that, for any pair of vector fields X and Y on $F\mathscr{W}$,

$$\mathbf{h}(\mathrm{X}, \mathrm{Y}) = \theta^i(\mathrm{X})\theta^i(\mathrm{Y}) + \omega^i_j(\mathrm{X})\omega^i_j(\mathrm{Y}) \tag{6.4.1}$$

Let $\overline{F\mathscr{W}}$ denote the Cauchy completion of $(F\mathscr{W}, \mathbf{h})$. It is not hard to show that $\overline{F\mathscr{W}} = F\mathscr{W}$ and $(F\mathscr{W}, \mathbf{h})$ is metrically complete only if $(\mathscr{W}, \mathbf{g})$ is geodesically complete. The action of the structure group $GL(4, \mathbf{R})$ on $F\mathscr{W}$ can be readily extended to an action on $\overline{F\mathscr{W}}$. Let $\overline{\mathscr{W}}$ denote the set of orbits of the latter action. $\overline{F\mathscr{W}}$ can be naturally regarded as a principal fibre bundle over $\overline{\mathscr{W}}$. Some of the points of $\overline{\mathscr{W}}$ are the orbits of the action of $GL(4, \mathbf{R})$ on $F\mathscr{W}$ and can therefore be identified with the worldpoints of $\mathscr{W}$. The remaining points of $\overline{\mathscr{W}}$ are the ideal points representing the singularities of $\mathscr{W}$. We call $\overline{\mathscr{W}}$ the *Schmidt completion* of $(\mathscr{W}, \mathbf{g})$. From this point of view, $(\mathscr{W}, \mathbf{g})$ is singularity-free if and only if $\overline{\mathscr{W}} - \mathscr{W} = \phi$. Since $\overline{\mathscr{W}}$ has a natural topology as a quotient space of $\overline{F\mathscr{W}}$ the ideal points are automatically located with respect to the ordinary worldpoints.[22] Although Schmidt's treatment of spacetime singularities is wonderfully simple in its mathematical conception it is extremely difficult to work with it in specific cases. When the Schmidt completion of the expanding and recollapsing Friedmann world described in note 9 was finally computed by B. Bosshardt (1976), it turned out that the initial and the final singularity were represented by the same ideal point. Moreover, R. Johnson (1977) showed that the Schmidt completion of the Schwarzschild and the expanding and recollapsing Friedmann solutions includes an ideal point whose only open neighbourhood is the *whole* completed spacetime (so that $\overline{\mathscr{W}}$, in these cases, is not Hausdorff space). These findings have understandably cooled the enthusiasm for Schmidt's invention.

From a philosophical standpoint the most interesting method of adding ideal points to a relativistic spacetime is that proposed by Geroch, Kronheimer and Penrose (1972), for it rests exclusively on the spacetime's causal structure. Though it can be applied to a wide range of (abstract) causal spaces, we shall concentrate, for brevity's sake, on the case of a typical relativistic spacetime $(\mathscr{W}, \mathbf{g})$. Until the end of this Section, we shall mean by the *past* (*future*) of a point $P \in \mathscr{W}$ the chronological past $I^-(P)$ (respectively, the chronological future $I^+(P)$) defined in Section 4.6. We must assume that $(\mathscr{W}, \mathbf{g})$ is both *past* and *future distinguishing*, i.e. that for any two distinct points P, Q $\in \mathscr{W}$, $I^-(P) \neq I^-(Q)$ and $I^+(P) \neq I^+(Q)$. (Each worldpoint's past and future are exclusively its own—a condition that, as the reader will recall, is not met by Newtonian spacetime.)[23] A subset $A \subset \mathscr{W}$ is said to be a *past set* if it is identical with its own past, i.e. if $A = I^-(A)$. (The past of a set was defined as the union of the pasts of its points.) A non-empty past set is said to be *indecomposable*, or an IP, if it cannot be represented as the union of two past sets distinct from itself. All IPs fall into one of two categories, namely, the *point* IPs or PIPs, which are identical with the past of some point $P \in \mathscr{W}$, and the *terminal* IPs or TIPs, which do not have this property. It is not hard to see that the past of each $P \in \mathscr{W}$ is a PIP. Since $\mathscr{W}$ is past-distinguishing it can therefore be identified with the set of its PIPs. We agree to view the TIPs as ideal points. (The union of $\mathscr{W}$ and its TIPs we denote by $\hat{\mathscr{W}}$.) The rationale of this move will now be made apparent. The past $I^-(\gamma)$ of a curve γ in $\mathscr{W}$ is the union of the pasts of all points in the range of γ. It can be proved that a subset of $\mathscr{W}$ is an IP if and only if it is the past of a timelike curve.[24] Let us say that $P \in \mathscr{W}$ is the future endpoint of a timelike curve γ if it is the supremum of the range of γ in the partial order induced in $\mathscr{W}$ by chronological precedence; i.e. if P chronologically follows every point of γ (other than itself) and precedes every other point with this property. If a timelike curve γ does not have a future endpoint we say that it is future endless. Evidently, an IP is a PIP if and only if it is the past of a timelike curve γ with a future endpoint P, in which case $I^-(\gamma) = I^-(P)$. On the other hand, if an IP is the past of a future endless timelike curve γ whose domain, when parametrized by proper time, extends to $+\infty$, it is, as we shall say, a ∞-TIP. A TIP that is not a ∞-TIP is said to be *singular*. Obviously, any timelike curve γ such that $I^-(\gamma)$ is a singular TIP is future endless and yet attains only a finite length to the future of any given point $\gamma(t)$. The definitions of an indecomposable future set or IF, a PIF, a ∞-TIF and a singular TIF proceed along similar lines and are left to the reader. The argument in note 24 can be easily adapted to prove that the IFs are precisely the futures of the timelike curves of $\mathscr{W}$. Any timelike curve γ such that $I^+(\gamma)$ is a singular TIF is past endless and yet has no more than a finite length to the past of any given point $\gamma(t)$. $\mathscr{W}$ can be identified with the set of its PIFs, while the TIFs are regarded as ideal points. The union of $\mathscr{W}$ and its TIFs is denoted by $\check{\mathscr{W}}$. Clearly $\mathscr{W} \subset \check{\mathscr{W}} \cap \hat{\mathscr{W}}$. As each PIP is identified with a $P \in \mathscr{W}$, it is also identified with a PIF (and each PIF with a PIP). Generally one would also wish to identify some TIPs with some TIFs. After the appropriate identifications are made, the union of $\hat{\mathscr{W}}$ and $\check{\mathscr{W}}$ constitutes the completion $\bar{\mathscr{W}}$ of $\mathscr{W}$. The definition of a topology on $\bar{\mathscr{W}}$ involves some niceties that we cannot discuss here. Penrose has shown that a relativistic spacetime that satisfies the hypotheses of Theorem IV on page 214 contains a singular TIF or a singular TIP.

A successful representation of the spacetime singularities by ideal points enmeshed with spacetime in a neighbourhood system entitles us to situate them in a definite place and time. We should however always bear in mind that such ideal points are not worldpoints, i.e. sites for possible events, but are conceived at an even higher level of abstraction. It ought to be clear also that in any of the above interpretations, the spacetime singularities may but need not be associated to a wild behaviour of the curvature or the energy tensor in any particular region. Indeed a singularity can be compatible with the physical integrity of matter along every worldline leading to it. Recent research has sought to distinguish different types of singularities, wild and tame, and to clarify the circumstances of their occurrence.[25] Among the fiercest are the "crushing singularities" defined by Eardley and Smarr (1979). In a spacetime endowed with a compact Cauchy surface, and in which all inextensible timelike geodesics are past (or respectively future) incomplete, the past (future) singularity is said to be *crushing* if there is a sequence (S_n) of spacelike Cauchy surfaces whose mean curvature k_n satisfies the relation

$$\max_{P \in S_n} k_n(P) \to -\infty$$

as the S_n approach the singularity. The expanding and recontracting Friedmann world described in note 9 subsists between two crushing singularities.

CHAPTER 7

Disputed Questions

7.1 The Concept of Simultaneity

In the Aristotelian philosophy that still shapes much of our common sense, the present time, called "the now" (*to nun*), separates and connects what has been (*to parelthon*) and what is yet to be (*to mellon*).[1] Though the now, as the link of time (*sunekhei khronou*), the bridge and boundary between past and future, is always the same *formally*, *materially* it is ever different (*heteron kai heteron*), for the states and occurrences on either side of it are continually changing. Indeed, the so-called flow or flight of time is nothing but the ceaseless transit of events across the now, from the future to the past. If an event takes some time, then, while it happens, the now so to speak cuts through it, dividing that part of it which is already gone from that which is still to come. Two events which are thus cleaved by (materially) the same now are said to be *simultaneous*. Simultaneity, defined in this way, is evidently reflexive and symmetric, but it is not transitive. For a somewhat lengthy event—e.g. the French Revolution—can be simultaneous with two shorter ones—such as Faraday's birth in 1791 and Lavoisier's death in 1794—although the latter are not simultaneous with each other. However, if we conceive simultaneity as a relation between (idealized) *durationless* events we automatically ensure that it is transitive and hence an equivalence. Thus conceived, simultaneity partitions the universe of such events into equivalence classes, and time's flow can be readily thought of as the march of those classes in well-aligned squadrons past the now. The succession of events on a given object—a clock—linearly orders the classes to which those events belong. The linearly ordered quotient of the universe of events by simultaneity is what we mean by physical time.

We saw in Chapter 1 how Newton's idea of time is rooted in such ultimately Aristotelian notions, and in Chapter 3 we considered Einstein's critical analysis of them in his first paper on Relativity. But we agreed to postpone the study of the philosophical problem raised by Einstein's criticism until the present chapter. The question at issue can be roughly stated as follows: In the light of Einstein's work, is simultaneity a natural, though frame-dependent relation, or is it a man-made relation, generated by our conventional method of labelling events with time coordinates? The first alternative sounds indeed paradoxical, for reference frames are human devices for the description of physical processes, so that whatever depends on them is in a sense artificial. Frame-dependent simultaneity, which in effect amounts to a ternary relation between two events and a frame, must involve the conventional stipulations that define the latter. On the other hand, as we know, some types of frames and of coordinate systems adapted to them are better suited than others for the description of events set in certain spacetime structures. For example, the "comoving" coordinates used in equation (6.3.2) are especially adequate to a Friedmann universe, while Lorentz charts are naturally congenial to Minkowski spacetime. Since the physical world admits a Lorentz chart if and only if spacetime is Minkowskian, the mere possibility of defining such a chart would imply that the frame-dependent simultaneity defined by it is somehow natural, and rests on solid physical grounds. (It may be argued, of course, that a spacetime structure is itself a human fabrication, but that is another matter, that we shall consider in Section 7.2.)

Philosophical discussions of simultaneity have been greatly influenced by Kant. According to the *Critique of Pure Reason* (1781), the meaning of temporal relations is directly grasped through our intuitive awareness of our own mental states, which are necessarily displayed in time. However, the objective time order of physical phenomena cannot be simply read off the subjective time order of sense appearances but must be constructed by subsuming the latter and the phenomenal objects which they manifest under the categories of substance-and-attribute, cause-and-effect, and interaction. In this way, the sudden dilatation of the air that we hear as thunder and the violent electric discharge that we see as lightning are understood to be objectively simultaneous (or very nearly so), although the perception of the former lags behind the perception of the latter. (It is worth noting that, on this view, the understanding, by enmeshing appearances in its categorical framework, also bestows objectivity on their subjective time order. Thus, according to the established laws of nature some time must elapse between the action of distant lightning on my retina and the action of simultaneous and equally distant thunder on my eardrum. This explains the time lag between my perceptions of these phenomena; the same time lag would be registered by electronic sensors placed near me.) According to Kant, the construction of the objective time order of events chiefly depends on two principles. By the *Principle of Temporal Succession in accordance with the Law of Causality* an event P can be said to precede another event Q if and only if, pursuant to a natural law, P is a cause of Q and Q is not a cause of P. By the *Principle of Coexistence in accordance with the Law of Interaction or Community* two things A and B can be said to exist simultaneously in space if and only if they stand in thoroughgoing mutual interaction. If we assume, as Kant certainly did, that the temporal succession of events linearly orders their simultaneity classes, it is clear that two events P and Q are simultaneous if and only if neither precedes the other. The first Principle allows two situations in which this can obtain, namely, (i) when neither P is a cause of Q nor Q is a cause of P, and (ii) when P is a cause of Q and Q is cause of P. Let us call situations (i) and (ii) the *negative* and the *positive causal criterion of simultaneity*, respectively. Situation (ii) may sound strange or even impossible, yet its reality is clearly suggested by Kant's Principle of Coexistence. The mutual interaction between things A and B deserves to be called thoroughgoing (*durchgängig*) only if the action of each responds to the very reaction it presently provokes in the other. This mode of involvement is typical of the mutual attraction of distant masses in a Newtonian world, which Kant very probably had in mind when he devised his conception of a community of substances held together by their "thoroughgoing interaction". If, as in Relativity, no instant action at a distance is allowed, the positive causal criterion of simultaneity is of course inapplicable, and one must either forgo the Kantian approach to temporal relations or make do with the negative causal criterion. The latter criterion, however, can only yield a partition of events into simultaneity classes if, regardless of the distance between cause and effect, there is no positive lower bound to the time lag between them.[2] Consequently, in a relativistic world, where all causal action propagates with a speed equal to or less than that of light, we face the following options. Either (*a*) we give up the notion that simultaneity is an equivalence, and with it the standard view of time as a continuum—the quotient of events by simultaneity—linearly ordered by temporal succession; or (*b*) we introduce a purely conventional criterion of simultaneity; or (*c*) we seek a substitute for the Kantian approach to this matter. Philosophers who for one reason or other have ignored or avoided the third alternative have usually promoted a combination of the first two.

Thus, the well-known explication of the concept of simultaneity proposed by Hans Reichenbach in the 1920s conjoins a natural definition based on the negative causal criterion with a rule for stipulating conventional definitions compatible with the former. Reichenbach gives a purely causal criterion of temporal succession: "If [an event] E_2 is the effect of [an event] E_1, then E_2 is said to be later than E_1"[3] If E_2 is not later than E_1 and E_1 is not later than E_2, E_1 and E_2 are "indeterminate as to time order". Reichenbach defines:

> *Any two arbitrarily chosen events which are indeterminate as to time order are said to be simultaneous.*[5]

According to Reichenbach, this so-called "topological definition"[5] expresses

> the deeper sense of the feeling that we attach to the concept "simultaneous": two simultaneous events are so situated that no causal chain can run from any of them to the other. The events that are taking place at this very moment in a distant country cannot be influenced by me, not even by telegram; nor can they in turn exert any influence on what happens here right now. *Simultaneity means disconnection of the causal link.*[6]

Due to the existence of "a finite limiting velocity for causal propagation"[7] the natural relation of simultaneity characterized by the "topological definition" does not partition the universe into classes of simultaneous events. Such a partition is however indispensable for establishing a global time variable. The physicist must therefore supplement the "topological" definition of simultaneity by the negative causal criterion, with a convention fixing what Adolf Grünbaum has called the concept of "metrical simultaneity".[8] The convention, however, cannot be wholly arbitrary, for the extension of the latter concept must be included in that of the former. (i.e., if P and Q are "metrically" simultaneous they must also be "topologically" simultaneous.)

Reichenbach set up a rule to which every such convention must conform. Reichenbach's rule is formally stated as part of his axiom system for Special Relativity.[9] Although its wording is not specifically linked to this theory, it appears unlikely that the rule can serve its purpose and ensure the definability of a global time in a substantially larger context—e.g. in an arbitrary relativistic spacetime. I shall therefore regard it (like everybody else, by the way) as a rule for the definition of simultaneity of distant events in a universe governed by Special Relativity. (The physical contents of this theory can, as we know, be given a coordinate-free formulation which does not involve the use of any particular definition of simultaneity.) Restriction of the rule to Special Relativity inevitably curtails the philosophical significance of Reichenbach's doctrine; but without this restriction it is far too indefinite to be argued for or against. We assume that each material particle constitutes—or is equipped with—a natural clock. The *arrival time of the first signal* sent from particle A to particle B at time t is the g.l.b. of the times of arrival at B (as registered in B's clock) of all signals that can be sent from A to B at time t (as registered in A's clock). A *bouncing signal* from A to B is a signal sent from A to B and reflected back to A as soon as it arrives in B. The *arrival time of the first bouncing signal* sent from A to B at t is the g.l.b. of the times of arrival at A (according to A's clock) of all bouncing signals that can be sent from A to B at t (according to A's clock).[10] Reichenbach's rule for the definition of simultaneity can now be stated thus:

> Two events are *simultaneous* if they occur at the same time according to synchronous natural clocks at their respective locations. The clock at a particle A is *synchronous* with the clock at a particle B if the arrival time t_B of the first signal sent from A to B at time t_0 and the arrival time t_A of the first bouncing signal sent from A to B at t_0 are related by
>
> $$t_B - t_0 = \varepsilon(t_A - t_0) \qquad (7.1.1)$$
>
> where ε is a real number such that
>
> $$0 < \varepsilon < 1 \qquad (7.1.2)$$

Reichenbach's own formal statement of the rule adds that the coefficient ε must remain constant for each fixed A, as B ranges over all particles. In order to define simultaneity we therefore need only to choose a value of the "Reichenbach parameter" ε in the allowed interval (0, 1). Restriction to this interval is both necessary and sufficient to ensure that the new conventional definition will be compatible with the natural, "topological" definition based on the negative causal criterion.[11]

According to Reichenbach, the definition of simultaneity between distant events in an inertial frame set forth in Einstein (1905*d*), § 1, is merely an instance of the above rule, with $\varepsilon = 1/2$.[12] I do not think that this statement, which has been repeated *ad nauseam* in the literature, can be accepted without qualification. For Reichenbach's rule says nothing about the state of motion of the clocks being synchronized, while Einstein's definition presupposes that they are both at rest in the same inertial frame. In fact, unless this presupposition is made, Reichenbach's rule may not be consistently applicable. (Suppose, for instance, that the clock from which the bouncing signal is sent is affixed to the rim of a rotating disk.[13]) We shall therefore narrow down the scope of Reichenbach's rule to the definition of simultaneity in inertial frames.[14]

Let F be an inertial frame in which simultaneity is to be defined in accordance with Reichenbach's rule. We are free to choose a Lorentz chart x adapted to F. By the "Lorentz image" of a set of events I shall mean its image by x in $\mathbf{R}^4$. Since Lorentz charts are diffeomorphisms and linear isomorphisms, the Lorentz image of a set exhibits the differentiable and affine properties of that set. As usual, we choose our units so that the Lorentz images of null lines intersect the Lorentz image of the worldline of any particle at rest in F at an angle of 45°. Let (A, t) denote an event that occurs on particle A at time t (by A's clock). We now choose an admissible value of the Reichenbach parameter ε and use it for synchronizing all clocks at rest in F by means of bouncing signals emitted from a given "base particle" A. Pick out any time t and consider the set S_t of events simultaneous with (A, t) by virtue of our stipulation. The Lorentz image of S_t is a one-sheet hypercone with its vertex at $x(A, t)$, each of whose generators makes an angle equal to arc tan $(|1-2\varepsilon|)$ with its perpendicular projection on the hyperplane $\pi_0 = x^0(A, t)$. (Recall that π_0 is the first projection function of $\mathbf{R}^4$ onto $\mathbf{R}$, which sends each real number quadruple to its first element. The stated results can be read off Fig. 2, which illustrates an application of Reichenbach's rule with $\varepsilon = 1/4$. The three vertical lines represent the Lorentz images of the worldlines of the base particle A and two other particles, B and B', at rest in the inertial frame F. The dotted lines represent the Lorentz images of the worldlines of first bouncing signals sent from A to B at t_0 and from A to B' at t'_0. The locus of the Lorentz images of events simultaneous with (A, t_B) is therefore the hypercone generated by rotation about the Lorentz image of A's worldline of the ray from $x(A, t_B)$ through $x(B, t_B)$. This ray is represented in the diagram, where it makes with the horizontal the angle $\theta = \text{arc tan}\,(1/2) = \text{arc tan}\,(|1-2\varepsilon|)$.) If $\varepsilon = 1/2$, the

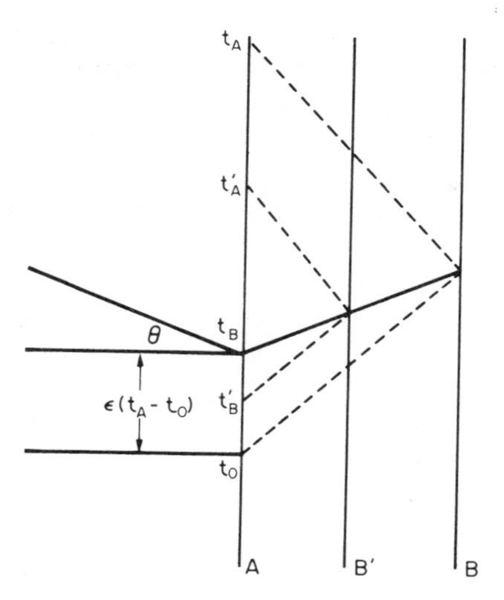

Figure 2.

hypercone degenerates and coincides with the hyperplane $\pi_0 = x^0(A, t)$. If ε is less (resp. greater) than 1/2, the hypercone lies between the said hyperplane and the Lorentz image of the future (resp. past) light cone at (A, t). Thus, unless $\varepsilon = 1/2$, the simultaneity classes defined by our stipulation will not be hyperplanes in Minkowski spacetime and inertial motions referred to our global time will not be uniform or even smooth. We are therefore forced to draw one of the following two conclusions: Either Reichenbach's doctrine concerning the conventionality of simultaneity does not apply to inertial time scales (in Lange's sense, page 17) and thus misses the whole point of Einstein's discussion in §1 of "The Electrodynamics of Moving Bodies" (Section 3.2); or Reichenbach's remark that the arbitrary factor ε must remain constant for each choice of a base particle is not to be taken seriously, but should be modified in accordance with his declared intent of explicating Einstein. I opt for the latter alternative, which I find more charitable. We shall therefore assume that the global time produced by any instance of Reichenbach's rule must constitute an inertial time scale, and we shall examine the conditions that this prescription imposes on the Reichenbach parameter ε.

The simultaneity classes into which an inertial time scale partitions Minkowski spacetime constitute a family of parallel spacelike hyperplanes. Each particular choice of the Reichenbach "parameter" ε is therefore in fact a smooth real-valued function of direction (in the space of the relevant inertial frame), which we shall henceforth call a Reichenbach function. If $\varepsilon(\mathbf{r})$ is the agreed value of the function in the direction of the vector $\mathbf{r}$, $\varepsilon(-\mathbf{r})$ must equal $1 - \varepsilon(\mathbf{r})$. Since the range of each Reichenbach function ε is contained in the interval (0, 1), it has a supremum ε_{max}, equal to its value in a particular direction we shall denote by $\mathbf{r}_{max}$, and an infimum $\varepsilon_{min} = 1 - \varepsilon_{max} = \varepsilon(-\mathbf{r}_{max})$. Every Reichenbach function takes the value 1/2 at the directions perpendicular to $\mathbf{r}_{max}$. (Unless it is constant, these are the only directions at which it takes that value.) Consequently, a Reichenbach function is fully specified by the choice of its supremum ε_{max} and the direction $\mathbf{r}_{max}$ at which it occurs, and may be denoted by $(\varepsilon_{max}, \mathbf{r}_{max})$.

The Reichenbach function $\varepsilon = \text{const.} = 1/2$ determines the so-called standard synchronism of distant clocks at rest in an inertial frame. (Clocks synchronized in this way keep Einstein time.) Clock synchronism in accordance with any other Reichenbach function is usually described as non-standard. There is an interesting relation between standard and non-standard synchronism, as we shall now see. Let F and F' be inertial frames. By the speed of F' in F I shall mean, as usual, the distance travelled in a unit of time by a point of F', as measured by non-stressed rods and clocks in standard synchronism at rest in F. The partition of events into simultaneity classes in accordance with the non-standard synchronism determined in F by the Reichenbach function $(\varepsilon_{max}, \mathbf{r}_{max})$ agrees with their partition according to standard synchronism in F', if F' moves in F in the direction of $\mathbf{r}_{max}$ with speed $v = \tanh(\arctan|1 - 2\varepsilon_{max}|)$. (The reader can satisfy himself of the truth of this statement by comparing Fig. 2 on page 225 with Fig. 1 on page 97.) If each frame is densely packed with clocks in standard synchronism, all that we need to do to put the clocks of F in the non-standard synchronism described above is to set each one of them by the nearest clock of F' when the latter shows a given time. The decision to change standard synchronism in an inertial frame F into some agreed form of non-standard synchronism (compatible with the Law of Inertia) amounts thus in effect to a decision to regulate clock synchronism in F by the standard synchronism proper to some other inertial frame.

Before the discovery of radio waves, distant clocks were often synchronized by carefully and slowly carrying about one good clock by which all others were set. In his renowned treatise on *The Mathematical Theory of Relativity* (1923), A. S. Eddington showed that this procedure is substantially sound and agrees with standard synchronism in the sense that I shall now explain.[15] Let us say that a clock Q_1 affixed to an inertial frame F is *synchronized by clock transport at speed* v with a clock Q_2 at rest in the same frame if a clock travelling in straight line from Q_1 to Q_2 with constant speed v agrees with each as it passes by it. "Speed" can here be understood in the sense indicated in the preceding paragraph, but if we desire to avoid the dependence on a previous definition of synchronism by bouncing first signals, we can also understand it as *self-measured speed*, i.e. as the distance measured by unstressed rods at rest in F travelled in a unit of time measured by the travelling clock itself. If the clock at a given base point A is synchronized by clock transport at a fixed finite speed v

with every clock at rest along each ray drawn from A in the relative space of F, we obtain a global time function T_v. T_v does not agree with any global time defined in accordance with a Reichenbach function and does not constitute an inertial time scale. But the limit T_0 to which T_v converges as v tends to 0 is an Einstein time for F, i.e. a global time defined by standard synchronism in F. T_v is of course practically the same as T_0 if v is very small in comparison with the speed of light. Eddington's result is often expressed by saying that standard synchronism is equivalent to clock synchronization by infinitely slow clock transport.[16]

We obtain a spacetime chart by combining, as on page 54, a time coordinate function defined by standard or non-standard Reichenbach synchronism in an inertial frame F with a Cartesian coordinate system adapted to F. I call such a chart a Winnie chart, after John L. Winnie (1970). The time axis of a Winnie chart is not Minkowski-orthogonal to the hyperplane spanned by its three mutually orthogonal space axes—unless its time coordinate function has been set up by standard synchronism, in which case the Winnie chart is of course a Lorentz chart. If its time coordinate function reflects a non-standard synchronism, the restriction of a Winnie chart to a hypersurface of simultaneity is not an isometry of this hypersurface onto the Euclidean space $\{\text{const.}\} \times \mathbf{R}^3$. The coordinate transformation $y \cdot x^{-1}$ between two Winnie charts x and y is what Winnie calls an ε-Lorentz transformation; but I think it will be fairer and less misleading to call it a Winnie coordinate transformation.[17] By Zeeman's Theorem (p. 127), the Winnie point transformation $y^{-1} \cdot x$ cannot be a causal automorphism of Minkowski spacetime unless the Winnie charts x and y are in fact Lorentz charts. If the latter condition does not obtain, two deterministic processes whose initial conditions, referred respectively to x and y, read the same, can run wildly different courses. Winnie transformations do not constitute a group,[18] so that it is probably pointless to seek for Winnie-invariants or for a Winnie-covariant formulation of the laws of nature.

While Winnie charts thus make up a rather undistinguished subset of the maximal atlas of Minkowski spacetime, Lorentz charts are, one may say, made to order to fit its structure. The discovery of Lorentz charts and Lorentz transformations was prompted by experimental findings in optics and electrodynamics and led in turn to the formulation of new laws of mechanics that have since been brilliantly confirmed. Our study of Special Relativity in Chapters 3 and 4 (especially Sections 3.7 and 4.6) showed that not only inertial frames but also Lorentz charts enjoy a preferred status in that theory. This is not due to some particular bias introduced into the usual formulation of the theory by a convention on synchronism,[19] but to the fact that the Lorentz group is the theory's symmetry group. This, in turn, implies that the standard synchronism used in the definition of Lorentz charts is singled out by the very nature of things in a universe ruled by Special Relativity.[20]

The doctrine of the conventionality of distant simultaneity in Special Relativity owes its resilience in part to a persistent belief in the positive and negative causal criteria as the only grounds on which a natural relation of simultaneity can rest.[21] That this belief is false is most clearly seen in the context of General Relativity. As we noted in Section 6.3, events in a Friedmann world are partitioned into simultaneity classes in a unique, natural way by the global time coordinate of a "comoving" chart (see equation (6.3.2)), which of course has no relation at all to Kant's causal criteria. Insofar as the universe we live in is roughly a Friedmann world, it sports—in the large—a natural cosmic time that bestows a definite meaning on "the age of the universe", "the first three minutes" of its existence, and other such phrases now common in scientific literature. On the other hand, there are relativistic worlds which cannot be partitioned into simultaneity classes in any way, neither naturally nor artificially, though they can certainly contain pairs of events that satisfy the negative causal criterion. In this connection, there is a theorem proved by Hawking (1969): a relativistic spacetime admits a universal time (and hence a partition into simultaneity classes) if and only if it is stably causal, in the sense defined on page 337, note 23. (Roughly: if and only if the spacetime does not contain any closed timelike curves and no such curves will be generated by slightly expanding the null cone at each point.)[22] The theorem shows that simultaneity has after all an essential relationship to the causal structure of relativistic spacetime, although a more subtle one than Reichenbach surmised. This relationship is particularly intimate if the spacetime is flat—i.e. Minkowskian—for in that case, by Zeeman's theorem, the causal structure determines the metric up to a scale factor. Since the Minkowski metric, as we pointed out above, bestows a preferential status on the simultaneity relation determined by standard synchronism in an inertial frame, it cannot surprise us to learn that this relation can be defined in the Minkowski geometry in purely causal terms. As I mentioned on page 123, A. A. Robb (1914) and H. Mehlberg (1937) have produced axiom systems for Minkowski geometry in which the only primitive predicate—apart from "element" and "$\in$"—has roughly the meaning of "chronologically follows" (in Robb's system) and "is causally connectible with" (in Mehlberg's). Simultaneity by standard synchronism is of course definable in both systems.[23] In an updated English version of his system of 1937, Mehlberg (1966) used the word "connected" as his sole undefined term. This is intended to stand for the relation between two events E_1 and E_2, on particles α_1 and α_2, when each event concurs with the collision, at different times, of a third particle α_3 with its respective particle. In practice, one may substitute for α_3 a series of successively colliding particles. Since the particles considered include photons, E_1 is connected in this sense with E_2 if and only if $E_1 \neq E_2$ and E_1 lies either in the causal past or in the causal future of E_2. Mehlberg uses the word "event" to designate the elements of the theory. The following definitions are transcribed almost literally from Mehlberg (1966), p. 475. (I have only suppressed a few words I consider superfluous.)

1. The event E occurs spatially between the unconnected events E' and E'' if every event connected with both E' and E'' is also connected with E.
2. The events E, E', and E'' are *collinear* if one of them occurs spatially between the other two.
3. An *inertial frame of reference* F is a partition of all events which satisfies the following conditions:
   a. If the events E and E' belong to the same class of the frame F then E and E' are unconnected.
   b. If the event E is not a member of a class t of F then E is connected with an event belonging to t.
   c. If the events E and E' both belong to the class t of F then event E'' collinear with the aforementioned two events also belongs to t.

The definition of simultaneity can be made extremely simple: *Two events are simultaneous relative to an inertial frame F if and only if they belong to the same class of F*. That this definition yields in fact the notion of Einstein simultaneity by standard synchronism can only be seen in the light of Mehlberg's seven axioms. It is connectedness *as characterized by these axioms* that occurs in the definiens as a primitive term. The axioms can be briefly stated only with the aid of several additional definitions and cannot be properly understood without further comment. I must therefore refer the reader to Mehlberg's writings.[24]

Although Mehlberg's "Essai" of 1935/37 is duly mentioned in the well-known books by Adolf Grünbaum (1963) and Bas van Fraassen (1970), most philosophers of science did not realize that it was possible to define Einstein simultaneity in terms of causal connectibility until David Malament (1977) drove this point home in a short and trenchant paper. More importantly perhaps, Malament proved that simultaneity by standard synchronism in an inertial frame F is the *only* non-universal equivalence between events at different point of F that is definable ("in any sense of 'definable' no matter how weak") in terms of causal connectibility alone, for a given F. (By a *universal equivalence* between events I mean a relation that holds between any two events in the universe.) Malament also showed that any equivalence between events that is definable in terms of causal connectibility alone, without specifying an inertial frame—in other words, any candidate to the status of a frame-independent or absolute simultaneity—is either universal or trivial. (By a *trivial equivalence* I mean a relation which everything has with itself and with nothing else.) Malament's proofs apply of course only to a universe governed by Special Relativity. Indeed, their ease and brilliance results from what we may call, with mild abuse of language, the "arithmetization" of spacetime geometry by means of a Lorentz chart. Let $\langle p, q \rangle$ denote the Minkowski inner product of two real number quadruples p and q (see equation (4.2.1)). We write $|p|$ for $\langle p, p \rangle$. We say that two real number quadruples p and q are *connectible* if and only if $|q-p| \geq 0$. Let $\mathscr{M}$ denote a universe governed by Special Relativity. Obviously, two events P and Q in $\mathscr{M}$ are causally connectible (in the sense defined on page 121) if and only if, for any given Lorentz chart x: $\mathcal{M} \rightarrow \mathbf{R}^4$, $x(P)$ is connectible with $x(Q)$. A Lorentz chart is thus an isomorphism between $\mathcal{M}$ as structured by causal connectibility, and $\mathbf{R}^4$ as structured by connectibility. Consequently, a relation between events can be defined in terms of their causal connectibility in $\mathcal{M}$, if and only if its Lorentz image by a given Lorentz chart is definable in terms of connectibility in $\mathbf{R}^4$. Malament's work turns on this simple truth. Since the isomorphism between $\mathcal{M}$ and $\mathbf{R}^4$ defined by a Lorentz chart is not canonic it may seem that his argument is tainted by a surreptitious convention. But Malament's argument does not depend on the particular choice of a Lorentz chart but only on the fact that such a choice is possible. This possibility lies at the heart of Special Relativity and has never been disputed by conventionalists.

The upshot of our discussion of simultaneity may be summed up as follows. In the natural philosophy of Relativity, time relations between events are subordinate to and must be abstracted from their spacetime relations. A partition of the universe into simultaneity classes amounts to a decomposition of spacetime into disjoint spacelike hypersurfaces each one of which separates its complement into two disconnected components. (In this respect, each class of simultaneous events behaves in Relativity like the Aristotelian "now".) Such a decomposition is impossible unless the universe is stably causal; but where one is possible, many more are available as well. Not all of them, however, will be equally significant from a physical point of view. Thus, in a universe that admits a Robertson-Walker line element, the hypersurfaces orthogonal to the worldflow of matter provide an unrivalled partition of events into natural simultaneity classes. In the Minkowski spacetime of Special Relativity, the hypersurfaces orthogonal to the congruence of timelike geodesics which makes up any given inertial frame define a partition of events that is naturally adapted to that frame. Simultaneity may thus be regarded as a unique physical relation in a Robertson–Walker world, while in Special Relativity one must put up with a three-parameter family of such physical relations. (Reichenbach's rule, as normally understood, does no more than expand it to a six-parameter family by the cheap expedient of associating *every* inertial frame with the *full* three-parameter family of simultaneity relations adapted to each.) In a more realistic world model things must be much less simple. It seems likely therefore that the concept of simultaneity can only claim physical significance locally, in the small neighbourhoods in which Special Relativity is valid, or very roughly, in the large scale in which the universe agrees with a Robertson–Walker model.[25]

7.2 Geometric Conventionalism.

We have fought the alleged conventionality of simultaneity in Special Relativity invoking the structure of Minkowski spacetime. Our argument is therefore seemingly powerless against those who believe that physical geometry itself is fixed by convention. This, however, as I said above, is a completely different question, that must be resolved on properly philosophical—i.e. universal, not theory-dependent—grounds. On this issue I contend that the conventionalist philosophy that puts the free institution of a system of geometry on a par with the free choice of a unit of measurement or a coordinate system which obtain their meaning from that geometry, confuses two very different senses in which freedom is at play in the enterprise of science. Take, for instance, the basic relativistic conception of the universe as a four-dimensional differentiable manifold with a Lorentz metric linked to the distribution of matter by the Einstein field equations. No one can doubt any longer that this idea was not imposed or even proposed by the observation records available *circa* 1912, but, like other such grand schemes for the investigation and understanding of physical phenomena, had to be freely produced. Einstein's own awareness of this fact was expressed in his statement that the basic concepts and the fundamental laws of physics are not inductively gleaned from sense data but are "free inventions of the human spirit".[1] The act of freedom from which such inventions arise is not, however, the arbitrary decision of a free will opting at no risk between several equally licit—though perhaps unequally convenient—alternatives set forth by a previously accepted conceptual framework. It is the venturesome commitment to a "way of worldmaking", a plan for the intellection of nature, that will guide the scientist in his efforts, outline his routine tasks and options, and ultimately measure out his chances of success and failure. To say that such a plan is "adopted by convention", as one might agree to use the Système International, merely blurs the contrast between two distinct levels of decision and hinders our understanding of the nature and scope of intellectual endeavour.

That scientific experience is not just heaped up as it comes but is built by combining the data of sense according to the plan of reason, was first clearly stated by Kant. Although Kant knew full well that science has a history and even gave the exact dates at which several of its main branches came of age,[2] he apparently never thought that reason's plan may undergo mutations, that the building of experience may have its very foundations redesigned while it is still being worked on. Indeed, he went as far as to say that a different plan, another system of forms and categories and goals, is not only utterly unthinkable, but would, in case we could grasp it, have nothing to do with experience, which is the only kind of knowledge through which objects are given us.[3] Kant's "pure theoretical reason" is therefore rightly viewed as an impassive, changeless, unhistorical power, that came to life fully armed like Athene out of her father's head; and although Kant was wont to describe it as a bottomless abyss, whose elements and structure can in no way be accounted for,[4] few have ventured to think of it as the outcome or the exercise of inscrutable freedom. Kant wondered "why we have precisely this kind of sensibility and an understanding of such nature, that through their combination experience becomes possible"; why moreover these two "otherwise totally heterogeneous sources of knowledge always concur in such good harmony to bring about [. . .] the possibility of an experience of nature [. . .], as if nature had been expressly designed for us to grasp".[5] But he did not ask himself what would happen if in some field of knowledge the incoming data become less and less pliable to the plan of reason, if our efforts to weave them, following the established pattern, into the unfinished tapestry of experience begin to yield diminishing returns. That was, however, the state of affairs in theoretical physics a century after Kant's death, and the question that he forgot to ask was answered *de facto* within the next decades. All the cherished principles that Kant had judged indispensable for the construction of experience—mass conservation, causal determinism, instantaneous distant interaction, the continuity of intensive quantities, the axioms of Euclidean geometry—were discarded and replaced by a tentative, imperfectly clear, not altogether consistent, but dramatically fruitful new set of basic concepts and laws.[6]

Philosophers offered diverse accounts of the breakdown of Kantian reason. The shallowest readily believed that by means of cleverly designed experiments and observations—e.g. the Michelson–Morley interferometer experiment of 1887, the observations of the 1919 solar eclipse by the astronomers of the British expedition—scientists had proved the old principles wrong and their substitutes right. The conventionalist account was subtler. It granted that the mute occurrences around us can only become empirical facts through being conceived in terms of an intellectual system; that sense appearances have to be spelled out by the understanding in order to be read as scientific experience.[7] But reason's "spelling" is no less subject to revision than ordinary orthography. Since, like the latter, it has to be accorded freely, the suggestion is inevitable that it is also a matter of convenience only.

For a less metaphorical statement of these views we turn to Reichenbach's book on *The Theory of Relativity and A Priori Knowledge* (1920).[8] Taking his cue from Hilbert's *Foundations of Geometry*, Reichenbach favours the axiomatic formulation of physical theories. He notes, however, that there is an important difference between axiomatized mathematical theories such as the one put forward in Hilbert's book, and axiomatized physical theories. While the former characterize their undefined terms—the *primitives*—in a series of unproved propositions—the *axioms*—the latter must go one step further in order to put the conceptual structure determined by the axioms in connection with reality. According to Reichenbach this is achieved by means of stipulations that coordinate primitive or defined terms of the theory with this or the other aspect of reality. Thus a two-place predicate of the theory might be coordinated with a binary relation between physical events, etc. Moritz Schlick (1918) had asserted that coordination (*Zuordnung*) is the only kind of relation that can be established by the human mind.[9] Reichenbach remarks that there is a great difference between the coordination of two concepts or conceptual objects, as practiced in mathematics, and the coordination of a concept with reality, as required by physics. In the former case—e.g. when defining a function from real numbers to real number triples—the coordinated objects are defined and known independently of the coordination. It may happen that by coordinating elements of one set with those of another the latter are endowed with new structural properties and relations (which are then said to be "induced" by the former). But according to Reichenbach even in such cases both sets must be given as collections of distinct elements prior to the coordination. The coordination of concepts with reality in empirical science must proceed very differently. For "only the coordination of the concept defines what is an individual thing in the 'continuum' of reality; and only the conceptual network decides, on the basis of perceptions, whether an individual thing that is envisaged in thought 'exists out there in reality'."[10] Empirical knowledge rests therefore on a coordination between two sets, "one of which is not merely ordered through the coordination but *is actually defined—has its elements determined—by the coordination*".[11] "The coordination must procure itself one of the sequences of elements to be coordinated."[12] As a consequence of this some of the axioms of a physical theory must play a special role. They are the axioms that characterize the primitive terms coordinated with otherwise undefined elements of reality. Such "principles of coordination [. . .] define the individual elements of reality and are in this sense *constitutive* of the real object. In Kant's words: 'Because only by means of them can any object of experience be conceived'."[13] Not every axiomatic system, however, provides admissible rules of coordination.

> Although the coordination maps to undefined elements, it can take place only in a perfectly definite way and not arbitrarily. We call this, the determination of knowledge by experience. We notice the curious fact that it is the defined side that determines the individual things of the undefined side, and that vice versa it is the undefined side that prescribes the order of the defined side. *This reciprocity of coordination expresses the existence of reality.*[14]

Admissible or "correct" coordinations can be distinguished from "incorrect" ones because only the former are "consistent". Inconsistencies are discovered by observation. "The theory that continually leads to consistent coordinations is called *true*."[15]

According to Reichenbach the gist of Kant's rationalism lies in the claim that a fixed set of constitutive principles, allegedly prescribed by human reason, are guaranteed to yield consistent coordinations forever. If Kant's claim were tenable all empirical knowledge of actual physical objects could be expected to conform with the said set of principles and these would be a source of *a priori* knowledge of the physical world. But, says Reichenbach, the Theory of Relativity has proved that Kant's claim is false. Einstein has shown that the interpretation of certain experimental results—such as Michelson and Morley's—cannot be consistently carried out in the light of the standard principles of classical physics, which include some that Kant regarded as *a priori* true. Kant was right in stressing the need for constitutive principles, which are reason's contribution to knowledge. But he was wrong in maintaining that such principles are immutable and cannot be affected by the growth of experience. "*The contribution of reason is not expressed by the fact that the system of coordination contains unchanging elements, but by the fact that arbitrary elements occur in it.*"[16] However, such arbitrary elements cannot be chosen at random but must be adjusted to yield consistent descriptions of observed facts. Since inconsistencies turn up as knowledge increases, the constitutive principles of empirical knowledge are subject to historical change. Every change in them "brings about a change in the concepts of thing and event".

> And herein lies the difference between Kant's view and ours: whereas for him only the determination of each *particular concept* is an infinite task, we contend that even our concepts of the object of science as such, of the real and of how it can be determined, can only gradually become more precise.[17]

So far, so good. Some of Reichenbach's notions, like that of "consistent coordination", and indeed the entire process by which the concept singles out its empirical correlate from the vast mass of what Reichenbach blandly calls "reality", would certainly have deserved some further elucidation. But we cannot accuse Reichenbach (1920) of mixing up the free commitment to a constitutive principle with the arbitrary choice of one convention or other. Since the former involves the risk of failure through "inconsistency", it is not, like the latter, indifferent to truth. Though not true in itself in any ordinary sense, a successful system of coordinating principles is the critical factor in transmuting the stubborn, silent presence of "reality" into a manifestation of empirical truth. Nevertheless, when Reichenbach (1924) went on to show in detail how the Theory of Relativity ought to be construed in accordance with his epistemology, he promoted from the outset the very confusion that we here regard as typical of conventionalism. Coordination is now to be achieved by means of "definitions". The difference between mathematical and physical definitions is described as follows:

> Mathematical definition is *conceptual definition* [*Begriffsdefinition*], i.e. it defines concepts by concepts. Physical definition assumes the relevant concept to be already defined and coordinates with it an actual thing [*ein Ding der Wirklichkeit*]; it is a *coordinative definition* [*Zuordnungsdefinition*]. Physical definition consists therefore in the coordination of a mathematical definition with a "piece of reality" [*zu einem "Stück Realität"*].[18]

Reichenbach's own example of a coordinative definition is the 1889 convention by which the unit of length was identified with the Paris metre rod. The coordinative definitions proposed in his book do not have quite the same character as this example, for they do not coordinate a concept of the theory with an actual thing or event, but rather with a *class* of physical phenomena. In order to be recognized as such the latter has of course to be subsumed under an earlier concept, pertaining to another theory, or even perhaps to what Reichenbach calls the "theory of everyday life".[19] Thus, for instance, Definition 9, which says that "light rays are straight lines", coordinates the mathematical concept of a straight line not with one or more individual things or processes, but with the entire class of light rays, and must therefore rely on geometrical optics or on sheer common sense for the criteria of appurtenance to this class.[20] Coordination as actually practised by Reichenbach can thus hardly be said to "define the individual elements of reality". The coordinative definitions of his *Axiomatics* (1924) are more in the nature of dictionary entries that translate some terms of the theory into plain German. This reinforces the view of them as conventions, arbitrarily chosen from a stock of preconceived alternatives. It also suggests erroneously that the conceptual framework of a scientific theory can be readily decomposed into elementary ingredients liable to be revised and replaced one by one.

In the *Philosophy of Space and Time* (1928) Reichenbach further emphasizes the view that physics grows in piecemeal fashion, by successive acts of coordination. Such acts, by which one merely states that "such and such a concept is coordinated to this thing here", cannot be replaced by "explanations of meaning" (*inhaltliche Bestimmungen*).[21] Generally, a coordination is not arbitrary, but rather, due to the fact that the concepts involved have interrelated meanings (*da die Begriffe untereinander inhaltlich verflochten sind*), it can turn out to be true or false, if one requires that "the same concept shall always denote the same thing".

> But certain preliminary coordinations must be established before the coordination procedure can be carried through any further. These first coordinations are therefore definitions, which are appositely called *coordinative definitions*. They are *arbitrary*, like all definitions. On their choice depends the conceptual system that develops with the progress of knowledge.[22]

Arbitrary coordinations constitute therefore the very first steps of a science. Thereafter, admissible coordinations are restricted by those already agreed on, or rather by the articulation the latter bestow on the actual course of phenomena.

Reichenbach's first and best known example of the role of arbitrary coordinative definitions in science concerns the foundations of physical geometry. According to him the distance function of physical space must be empirically ascertained by measurements performed with rigid rods, i.e. with solid bodies of unchanging size and shape. If the span between two places A and B can be covered precisely by n rigid rods of length k, lying end to end, and not by any number less than n, the distance from A to B is nk. The proposition clearly depends on what we mean by "a rigid rod of length k". This, in turn, is the outcome of two arbitrary conventions. Obviously, we must fix the unit of length. In Reichenbach's time it was defined as the distance, at 0°C, between two marks on a particular metal rod kept in Paris. As I mentioned earlier, in

1924 Reichenbach had used precisely this example to illustrate his idea of a coordinative definition.[23] But the truly decisive convention that lies at the foundation of physical geometry according to Reichenbach is the definition of the rigid rod itself. What bodies shall we repute invariable as to shape and size? One is tempted to answer: none at all. For we know that every sizable piece of matter will change its shape under our very eyes if subjected to sufficiently powerful forces. This common sense remark gives the cue to Reichenbach's treatment of the question. By "force" he understands anything that is held responsible for a "geometrical change", i.e., I presume, a change of size or shape.[24] It seems therefore that we might define a rigid rod as a body isolated from the deforming influence of forces. However, according to Reichenbach, there are two kinds of forces. *Differential* forces act differently on different materials and their influence can be warded off by shielding. By comparing their effect on different bodies or under diverse degrees of insulation we can construct with increasing exactitude the ideal norm of a body totally impervious to such forces and correct by this standard the deformations which they inflict on real bodies. On the other hand, *universal* forces "act equally on all materials" and "there are no insulating walls against them".[25] Consequently, the deformations caused by universal forces are undetectable *in principle*, for they produce identical changes on all bodies, so that their effects cannot be quantified by comparison and corrected by calculation. In the light of his classification of forces, Reichenbach opted for the following definition:

> Rigid bodies are solid bodies which are not subjected to any differential forces, or from which the influence of differential forces has been corrected away by calculation. Universal forces are disregarded.[26]

Measurements performed with solid bodies conforming to this criterion will yield, according to Reichenbach, a distance function for physical space. The results of such measurements obviously depend on a convention, namely, our agreement to "set universal forces equal to zero" and fully ignore their effects. However, once this convention is adopted, "the geometry obtained is prescribed by objective reality alone".[27] But it is always possible to obtain a different geometry by postulating a suitable field of universal forces. Reichenbach refers to this possibility as "the relativity of geometry".

Later in the book, when he comes to deal with "Gravitation and Geometry" (§ 40), Reichenbach declares that gravity is "the prototype of a universal force".[28] For, he says, it acts on all bodies in the same way. And, of course, there is no shielding against it. Hence, if we follow the convention proposed above, we ought to put the gravitational force everywhere equal to zero. Reichenbach implies that this was what Einstein did in his theory of gravity. I quote:

> Thus, gravity exerts on measuring instruments [*Messkörper*] such a uniform [*einheitliche*] action, that the latter jointly define a unique geometry. We can abide by it. In this sense, gravity is *geometrized*. We do not speak of an alteration of the measuring instruments by the gravitational field, but regard the measuring instruments as "free from deforming forces" in spite of the influence of gravity.[29]

The obvious suggestion is that one could still avoid Einstein's unorthodox geometry by assigning a suitable non-constant value to the universal force of gravity.

I cannot agree with this reading of General Relativity. It is of course very unsatisfactory to speak, as we have been doing, of the force of gravity as if it were a definite entity, existing out there *in rerum natura*, to which different theories refer invariably, while assigning to it different values. In fact, the truly remarkable relation between Newton's theory of gravity and Einstein's is that both assign practically the same quantitative outcomes to nearly all feasible experiments, though they postulate very different structures to account for the phenomena in their common purview. But no matter what we take to stand for "the force of gravity" in either theory, it is not a "universal force" in Reichenbach's sense.

Thus, we may not conclude that Newtonian attraction acts in the same way on all bodies just because they all fall with the same acceleration at a given potential. For the *force* exerted per unit volume on different bodies is not the same; as we see at once when we try to balance a gallon of mercury with a gallon of oil. If *equal* gravitational forces act on the mercury and on the oil—as would happen, for instance, if they lie at suitable different potentials—the effect on each will obviously be quite different. And a given gravitational force certainly does not cause the same deformations on all bodies, irrespective of the material they are made of. Otherwise, scale manufacturers could achieve substantial savings by using springs made of crepe paper instead of steel.

Einstein's theory of gravity is not easily expounded in terms of force—which is not the same as to say that it sets *the* force of gravity equal to zero. The nearest substitute we can find here for the classical gravitational field strength (equal, at a point, to classical force per unit mass) are the connection components Γ^i_{jk}, which Einstein once called "the components of the gravitational field".[30] As we know, these quantities are chart dependent, and hence ultimately of small physical significance. If referred to a Fermi chart, they all vanish along the chart's time axis. Since the latter can be the worldline of any massive particle, we may, by the choice of a suitable chart, set the "components of the gravitational field" on such a particle equal to zero. But this has precious little to do with Reichenbach's "disregard of universal forces" in his definition of the rigid rod. For the Γ^i_{jk} do not generally vanish on all the domain of the chosen chart, but show up as tidal forces on any extended body that includes the particle in question. And of course such tidal forces do not deform all materials alike, but cause different "geometrical changes" on rubber, say, and on glass.

Another candidate for the title of "gravitational force" in General Relativity is the worldforce on a freely falling particle. (By *worldforce* I mean the generalization to a vector tangent to an arbitrary spacetime of the 4-force defined on Minkowski spacetime in (4.5.2).) Worldforces on particles can be held responsible for the "geometrical change" from a geodesic to a non-geodesic worldline, and one should in principle be able to refer to them the deformations of extended bodies that Reichenbach had in mind when he spoke of universal and differential forces. But it would be wrong to assume that General Relativity puts all gravitational worldforces, all worldforces on free-falling particles, equal to zero. Such world-forces do indeed vanish invariantly in the case of a non-spinning monopole of constant rest mass, for a freely falling particle of this kind has a geodesic worldline along which its worldvelocity is self-parallel. But they do not generally vanish in the case of a spinning particle or, generally, a multipole.[31]

If, in spite of Reichenbach's allegations, gravity cannot be properly described as a "universal force", one is hard put to find an example within the extension of this term. The truth is that a "universal force", a force which cannot be detected by any means because it modifies the shape of the instruments of detection in the exact way required to conceal its presence, belongs to the realm of science fiction and cannot be seriously countenanced in real science. At least in physics, a difference to *be* a difference must *make* a difference. Even the notorious Fitzgerald–Lorentz contraction was supposedly the effect of changes in molecular forces that a clever experiment might test (see p. 292, note 32). While Christian theology has long known of changes of *substance* which no *outward* feature will disclose, it was not until the Rise of Scientific Philosophy that we got to hear of a change of *shape* which nothing can ever make *apparent*.[32]

Reichenbach's misconstrual of gravity as a universal force is not the only reason why his philosophy of geometry is inadequate for understanding General Relativity. Another shortcoming is that it was originally designed for the study of space without time and is ill-suited to cope with spacetime. Reichenbach is obeisant to Hermann von Helmholtz's conception of geometry as the theory of congruence of rigid bodies, and it is not easy to see how this notoriously narrow conception can be extended to encompass spacetime geometry with a Lorentz metric. In fact it will not even fit the needs of proper Riemannian space geometry. Helmholtz's fixation on measurement by rigidly transported rods prevented him from understanding Riemann's idea of metric relations in a manifold in which "every line has a length independent of the way it lies, and therefore any line can measure any other line". But without this idea, and the concomitant shift of attention from bodies and the volumes they fill, to points and the paths that join them, the development of spacetime geometry would not have been possible. It follows from a remark of Riemann's that free mobility of rigid bodies can only exist in a proper Riemannian manifold of constant curvature. Because of this mathematical fact, it was almost unanimously believed before Einstein that physical geometry had to be chosen among the four Riemannian geometries of constant curvature, namely, Euclidean, Bolyai-Lobachevsky, spherical and elliptic geometry. (Note 8 explains why this belief lent strong support to geometric conventionalism.) As Lord Russell once observed, the advent of General Relativity made this opinion look "somewhat foolish".[33] Reichenbach knew of course that there can be no free transport of rigid rods in a typical relativistic world. He therefore interpreted the fundamental coordinative definition of geometry as applying in practice not to rigid rods, but to *infinitesimal* rigid rods. Although the concept of an infinitesimal measuring rod seems at first sight paradoxical—for, indeed, if the length of a rod is negligible, of what use can it be for measuring?—there is a sense in which practically infinitesimal measuring rods have an important role to play in General Relativity. But, as we shall see, infinitesimal rods in this sense are not coordinatively defined standards by which to judge the geometry of physical space. On the contrary, they themselves must be constructed and controlled in accordance with the infinitesimal geometry to which General Relativity is committed. Roughly speaking, an infinitesimal measuring rod is a freely falling solid body of negligible thickness that is short enough not to be significantly stressed and strained by the inhomogeneities of the gravitational field. What is "short enough" will naturally depend on the allowable error and on the local spacetime curvature. The beams of a spacecraft that were infinitesimal as it left the Solar System may cease to be so if it approaches a Black Hole. Though inherently inexact, the concept of an infinitesimal rod can be given an exact foundation in differential geometry. Let P be any point of the arbitrary relativistic spacetime $(\mathscr{W}, \mathbf{g})$. The exponential mapping Exp_P, defined on page 278, maps a neighbourhood of the zero vector in the tangent space $\mathscr{W}_P$ diffeomorphically onto a neighbourhood of P. On a usually smaller neighbourhood of zero the said diffeomorphism differs from an isometry by less than an assigned error. We shall denote the latter neighbourhood by N_P. Choose a timelike vector $v \in N_P$, and let D_P denote the set of vectors in N_P which are orthogonal to v. $\mathrm{Exp}_P(D_P)$ is then a smooth spacelike hypersurface through P, which satisfies, within the assigned error, the relevant propositions of Euclidean geometry. Figures in $\mathrm{Exp}_P(D_P)$ are indeed sets of instantaneous events, but we can protract them to yield bodies by means of a simple device. Let γ be the one and only geodesic in $\mathscr{W}$ such that $\gamma(0) = P$ and $\dot{\gamma}_P = v$ (p. 278). Define N_Q and D_Q as above, for each point Q in the range of γ. As on page 278 we let τ^{γ}_{PQ} denote parallel transport along γ from P to Q. We shall consider only a short segment γ' of γ, containing P, such that for any point Q in γ', N_Q differs but little from $\tau^{\gamma}_{PQ}(N_P)$. Let N be the intersection of the family $\{\tau^{\gamma}_{QP}(N_Q) \mid Q \text{ is a point in } \gamma'\}$. ($N$ is a subset of N_P.) Let D be the intersection of N with D_P. Take any 3-dimensional figure $F \subset \mathrm{Exp}_P(D)$. Put $F' = \mathrm{Exp}_P^{-1}(F)$, $F'_Q = \tau^{\gamma}_{PQ}(F')$ and $F_Q = \mathrm{Exp}_Q(F'_Q)$. As Q ranges over γ', F_Q describes a worldtube in $\mathscr{W}$. Such a worldtube is the history of a small body in the relativistic world $(\mathscr{W}, \mathbf{g})$. The body obeys, in an obvious sense, the laws of Euclidean geometry. If F is a thin parallellepiped, a solid body derived from it in the stated manner would be what I described as an infinitesimal rigid rod. Three such rods meeting orthogonally at a common endpoint whose worldline is γ' provide the frame for the Cartesian coordinates of a local Lorentz chart. The possibility of erecting such a frame along any timelike geodesic bears witness to the local (approximate) validity of Euclidean space geometry near each freely falling particle. It is the local Euclidean geometry that decides which objects do and which do not qualify as infinitesimal measuring rods at each place.[34] Its validity evidently is not determined by measurements performed with rods, but flows as a mathematical theorem from the assumption that spacetime is a Riemannian manifold.

Thus, the foundations of physical geometry do not rest for Relativity on the fact that rigid bodies are freely movable—as Helmholtz argued; nor—as Reichenbach claimed—on a *convention* to ignore universal forces; but—as Riemann anticipated—on *hypotheses* that lay at the heart of the scheme of things by which we seek to judge and understand the course of events. Two are the basic hypotheses:

I. *Spacetime*—i.e. the "arena" or stage of physical events—*is a* 4-*dimensional differentiable manifold.*

II. *The phenomena of gravitation and inertia*—i.e. the effects attributed by Newton to the "impressed force" of gravity and the "innate force" of inertia—*are to be accounted for by a symmetric* (0, 2) *tensor field* **g** *on the spacetime manifold, which is linked to the distribution of matter by the Einstein field equations and defines a Minkowski inner product on each tangent space.* (The condition last mentioned says that **g** is a Lorentz metric, and therefore determines a unique connection on the manifold, with its characteristic curvature and set of geodesics. The field equations entail that, in the absence of non-gravitational forces, the worldlines of the simplest massive particles are timelike geodesics of that set. This accounts for the classical gravitational phenomena of free fall, tidal strain, and planetary motion, and predicts new phenomena, such as the recession of the galaxies and gravitational collapse. The Minkowski inner product on each tangent space induces—through the exponential mapping—a local approximate Minkowski geometry on a small neighbourhood of each worldpoint. This accounts for the Lorentz invariance of the laws of nature referred to local Lorentz charts. By thus deriving the local Minkowski geometries that explain the successes of Special Relativity from the global metric tensor **g**, the hypothesis accounts for the equivalence of inertial motion and free fall in a homogeneous gravitational field—implicit in Newton's handling of the Earth–Moon system—and predicts the optical effects of gravity.)

Hypothesis I is common to all extant theories of mathematical physics. It is inconceivable that a computer programmed for inductive inference could ever derive it from crude observational data. Nevertheless, it is not an arbitrary, harmless convention, but a truly cosmopoietic principle which has hitherto guided the interpretation of all experimental results and shaped the factual system that physics has built from them.

Hypothesis II is characteristic of General Relativity. We studied its development and motivation in Chapter 5. There are two ways of describing its geometrical import. Insofar as the tensor field **g** is in effect a Riemannian metric on the spacetime manifold, we may say that *Hypothesis II explains gravity by geometry*. We should bear in mind, however, that "gravity" stands here for gravitational *effects*, and "geometry" is not being used in its ordinary, pre-Minkowskian meaning. In a still wider sense, every concept of physics is "geometrical"—as Einstein once pointed out to Lincoln Barnett—and the gravitational field **g** is neither more nor less a "geometric object" than, say, the electromagnetic field tensor.[35] On the other hand, insofar as the tensor field **g** assigns to each worldpoint an inner product on the respective tangent space, which in turn induces in a neighbouring region a (roughly) Minkowskian geometry from which a Euclidean geometry on parallel Einstein-simultaneous spacelike sections can be readily derived, we may say that *Hypothesis II explains geometry by gravity*. Here "geometry" may be understood in its old familiar sense, while "gravity" denotes the *cause* of the so-called gravitational phenomena. The behaviour of unstressed rods in a local inertial frame is now to be counted among the latter, insofar as it is also accounted for by the tensor field **g**.

By grounding geometry (and chronometry)—as practised in scientific laboratories and factories of precision instruments—on a physical field governed by testable natural laws, General Relativity escaped the trilemma of apriorism, empiricism and conventionalism in which the epistemology of geometry was caught at the turn of the century.[36] It seems to me, therefore, that Reichenbach puts the cart before the horse when he says that, as "it turns out that the non-Euclidean geometry obtained thanks to our coordinative definition of the rigid rod deviates quantitatively very little from Euclidean geometry when small regions are concerned, [. . .] we may therefore employ the following method of inference: Under the assumption that Euclidean geometry is valid locally, we can prove that a non-Euclidean geometry holds globally, which everywhere converges to the infinitesimal Euclidean geometry".[37] In General Relativity the inference goes the other way around: Under the assumption that spacetime is Riemannian (specifically: Lorentzian) we can prove that a flat (specifically: Minkowskian) geometry must hold locally—to a first approximation. This flat geometry is the standard by which our measuring instruments are judged. The latter can indeed be used to test Hypothesis II, e.g. by measuring the components of the actual spacetime curvature as manifested through gravitational strains.[38] But Hypothesis II is not *inferred* from such measurements; for it must be posited—no less than Hypothesis I—in order to make sense of them.[39]

In our generation geometric conventionalism has been kept alive mainly by Adolf Grünbaum's acumen and perseverance. Grünbaum's conventionalism does not respond, like Reichenbach's, to some alleged necessity of the scientific method, but purportedly follows from the actual nature of the subject matter of physical geometry, namely, real space and time. Grünbaum's prose is renowned for its complexity, and he has often complained that his critics miss his meaning due to inaccurate reading. Rather than risk this reproach, I renounce in advance all claim to give here a faithful picture of the historic Grünbaum (which anyway would probably require another volume like this one). I shall comment only on the conventionalist doctrine that is usually associated with his name. References to Grünbaum's writings are only intended as a guide to the reader who wishes to pursue the matter on his own.[40]

The main tenet of what we shall call, for short, GGC or Grünbaum Geometric Conventionalism, is that physical space and time do not possess a metric by virtue of their own structure, and can therefore only obtain one "extrinsically", by arbitrary stipulations regarding the congruence of segments and intervals. By a metric on a set S Grünbaum means either what he calls a D-metric, i.e. a distance function from S^2 to $\mathbf{R}$, or an M-metric, i.e. a measure function from 2^S to $\mathbf{R}$, or an R-metric (R for Riemann), i.e. a certain kind of cross-section of the bundle of symmetric (0, 2)-tensors over S (which obviously can exist only if S is a differentiable manifold).[41] I take it that all three meanings of metric are intended in the following early statement of GGC:

> The continuity we postulate for physical space and time furnishes a *sufficient* condition for their *intrinsic metrical amorphousness*.[42]

This statement is baffling, for it entails that intrinsic metric amorphousness is a *necessary* condition of continuity, whence it follows, by contraposition, that every structure to which a metric, in any of the above senses, pertains intrinsically, is not continuous. And yet we know that the standard distance function $d: \mathbf{R}^2 \to \mathbf{R}$, by $(a, b) \mapsto |a - b|$, induces the topology of the linear continuum in the field of real numbers. Our counterexample can be dismissed, however, if "the continuity we postulate for physical space and time" differs in some essential respect from the (admittedly paradigmatic) continuity of the real line. Indeed, the "continuity" in question would have to be quite peculiar, and differ from the continuity of all topological manifolds; for otherwise a structure $(M, \mathbf{g})$, where M is an n-manifold and $\mathbf{g}$ a Riemannian metric on M, would be, according to GGC, a contradiction in terms.

A few years later the main tenet of GGC was restated as follows:

> Be it topological or nontopological, there exists no kind of *implicit* (intrinsic) basis for nontrivial metric relations in continuous P-space. [P for *physical*; I assume that a similar statement applies to physical *time* or "T-space"—R.T.][43]

The new formulation does not say that the *continuity* of space—which, no matter how narrowly we conceive it, is always, I presume, a topological property—is *sufficient* to vouchsafe its material amorphousness; but that *all properties* of space, be they topological or not, are *insufficient* to make it metrically definite. (Such insufficiency can be invoked to explain why the metric of space must then be determined by *choice* among the many metrics compatible with space's own properties.) Since the properties of space are not listed, the second statement of GGC is not, like the first, a *proposition* that can be proved or disproved by analyzing its terms, but an *injunction* not to include in the intension of the concept of physical space any properties that entail that space has a metric.

Our task will be made easier if, instead of inquiring into the somewhat elusive reasons for this remarkable injunction,[44] we ask at once about its relevance to General Relativity. For if it turns out that Einstein's theory mindlessly ignores it, we may still indeed look forward to the eventual discovery of a plausible theory of gravity more conformable to the enjoined requirement, but we need not concern ourselves any further with GGC in the present book. Now, the fundamental structure posited by General Relativity is neither space, nor time, but spacetime, which is, of course, by definition, a 4-manifold $\mathscr{W}$ endowed with a Riemannian metric **g**. The following quotation shows that, at least at the time it was written, Grünbaum admitted that **g**, in General Relativity, is not a matter of convention, but an essential ingredient of the theory's concept of spacetime. (Here "ds_4" denotes the line element determined by the spacetime metric **g** as per (4.3.1*b*).)

> The space-time measure ds_4 is an invariant both as between different reference frames and with respect to the use of different space and time coordinates in any given frame. It follows that two *space-time* intervals which are congruent in one reference frame in virtue of having equal measures ds_4 in that frame will likewise have equal measures ds_4 and hence will also be congruent in any other frame. Accordingly, there are no alternative infinitesimal *space-time* congruences in the GTR [i.e. General Relativity—R.T.] and the latter kinds of congruence are thus unique.[45]

It follows that the concept of a relativistic spacetime does not obey the injunction of GGC. But the latter, apparently, was never meant to apply to physical *spacetime*, but only to physical space *and* time. According to Grünbaum,

> the uniqueness of the four-dimensional space-time congruences of the GTR [cannot] gainsay or detract from the significance of the following claim of mine: the GTR actually employs alternative congruences in *each* of the physically important continua of time and space in precisely the sense countenanced philosophically by [GGC]. It is now plain that nothing about my claim that the insight contained in [GGC] has substantial relevance to the GTR requires that there be alternative *space-time* congruences *in addition to* the undeniably present alternative time congruences and alternative space congruences.[46]

There is no denying that, if it is at all possible to carve a space and a time out of a relativistic spacetime, it can be done in countless ways. Any foliation of the spacetime $\mathscr{W}$ into spacelike hypersurfaces defines a time; any congruence of timelike worldlines in $\mathscr{W}$ defines a space.[47] In a few highly streamlined spacetimes some such congruences and foliations are in a sense inbuilt into the worldstructure, as we saw in Section 7.1. But in more realistic models there are no natural grounds for preferring one method of splitting spacetime into space and time over the many other available alternatives. And in the most general case, no congruences of timelike worldlines or foliations into spacelike hypersurfaces exist, and space and time are meaningless *even as abstractions*. The space and the time determined by a given congruence-*cum*-foliation are, respectively, a 3-manifold and a 1-manifold, which can therefore be endowed with proper Riemannian metrics. Following Grünbaum, we shall denote the line element of a typical space metric by ds_3, and that of a typical time metric by ds_1. In some special cases, there is a ds_3 and a ds_1 uniquely determined by the spacetime metric in accordance with the particular structure of the relevant congruence-*cum*-foliation. Thus, if the line element of spacetime can be put into the Robertson–Walker form $(ds_4)^2 = d\tau^2 - (S(\tau))^2 d\sigma^2$, the fibres of the time coordinate function τ furnish the leaves of a suitable foliation, while τ's parametric lines constitute the matching congruence. The proper Riemannian line elements $ds_1 = d\tau$, $ds_3 = d\sigma$, are then clearly prescribed (up to a scale factor) by ds_4.[48] Also, if the spacetime is flat, with line element $(ds_4)^2 = \eta_{ij} dx^i dx^j$, we have a congruence-*cum*-foliation determined, as in the former example, by the time coordinate function x^0, and natural line elements $ds_3 = \sqrt{(-\eta_{\alpha\beta} dx^\alpha dx^\beta)}$ and $ds_1 = dx^0$ for the resulting space and time. The existence of such natural metrics in these cases obviously has to do with the fact that the timelike separation between any two leaves of the foliation is the same at each point of either, while the spacelike separation between any two lines of the congruence is constant or varies uniformly with time. These conditions are not fulfilled in less special cases, in which, therefore, the given spacetime line element ds_4 does not determine (up to trivial rescaling) a ds_3 and a ds_1. The space and time metrics are then entirely left to our choice, but, for this very reason, no matter how we choose them, they will not be physically significant. Now, as Graham Nerlich has remarked, if the metric is an imposed convention, and factually idle, it is quite inscrutable why we should choose one at all.[49] It would seem therefore that GGC has nothing to teach us about General Relativity.

In the latest and, in my opinion, clearest statement of his views on geometry, Grünbaum (1977) contends that the metric of a relativistic spacetime is not given by the mere existence of the latter, but is "constituted", as he says, by "external devices". Although he does not attempt to derive a conventionalist conclusion from this premise (at any rate, he does not imply that there could be a viable substitute for the metric in question if the factual claims of General Relativity are valid), it will be instructive to examine Grünbaum's argument. As he is not easy to paraphrase, I shall quote him abundantly. Grünbaum does not dispute that any given relativistic spacetime is characterized by a particular solution to Einstein's field equations, i.e. a unique symmetric (0, 2)-tensor field, representing the gravitational field. But, he says,

> if a physical field f_{ij} is a symmetric, second-rank, covariant tensor field over space-time, this fact alone surely does *not* guarantee that f_{ij} *must* have the physical significance of being the metric tensor of space-time: after all, qua physical tensor field, the stress-energy tensor field T_{ij}, for example, has the same formal properties as f_{ij} while certainly *not* being the metric tensor of space-time.[50]

Grünbaum forgets to mention that, by hypothesis, a particular solution to Einstein's field equations defines a Minkowski inner product on every tangent space of space-time. Due to this "formal property", which sets it apart from all non-gravitational tensor fields of current mathematical physics—including, of course, the stress-energy tensor field mentioned by Grünbaum—a solution to Einstein's field equations *is* what we call a *Lorentz metric*. This name is not just one more caprice of mathematical terminology, but is meant to hint at the unavoidable chronogeometrical implications of a physical tensor field of this kind. If the gravitational field is adequately represented by a Lorentz metric it cannot fail to have the strictly geometrical effects that General Relativity ascribes to it. Having omitted to say as much, Grünbaum proceeds to ask:

> If g_{ij} is a solution set of Einstein's equations under specified physical conditions, and thus represents some kind of symmetric, covariant, second-rank tensor field over space-time, what warrants the *particular interpretation* that the scalar $(g_{ij}dx^i dx^j)^{1/2}$ *also* has the specific physical significance of being the infinitesimal measure for intervals of the physical space–time manifold? [...] What qualifies the tensor g_{ij}—qua solution of Einstein's field equations—to do *double duty* physically by representing not only a physical field but also the metric tensor of space–time?[51]

The following quotation gives the gist of Grünbaum's reply:

> What is otherwise just another symmetric, covariant, second-rank physical tensor field g_{ij} does control the physical space–time trajectories of photons and freely falling ("free") massive particles, as well as the rates of atomic clocks and perhaps the world-lines of other (as yet unknown) entities as well. And *at least in NE–GQ space–time* [i.e. "*non*-empty GTR space–times that are, however, *devoid* of gravitational radiation"—R.T.], it is the idealized behavior of this particular important class of external entities that is taken to be *canonical* for the metricality of these space–times in the sense that it is fundamentally *constitutive* ontologically for their metric structures. Thus what is otherwise just a physical tensor field g_{ij} thereby *acquires* specifically geometric, additional, physical significance as being the metric tensor of the space–time manifold. And in *NE–GQ* space–times, g_{ij} acquires this geometric significance because—and only because—the scalar $|g_{ij}dx^i dx^j|^{1/2}$ generated from that field is (ideally) *physically realized* by the behavior of that open-ended set of external devices that are taken to be canonical for the metric structure of space–time.[52]

We encounter here a curious combination of philosophical realism (with regard to the existence of the gravitational field and the differentiable manifold on which it is defined) and vestigial operationism (with regard to the metrical significance of that field). Lack of time and space prevents me from examining its rationale. (Howard Stein has done it splendidly in his incisive Letter to Adolf Grünbaum on Space–Time and Ontology (1977*b*).) But there are a few questions I would like to raise before closing this Section. The "constitution" of the spacetime metric is attributed by Grünbaum to "external devices", members of a "family of concordant external metric standards" (FCS) which includes "photons, atomic clocks, infinitesimal rods, and free massive (gravitationally monopole) particles".[53] Why are these physical entities described as *external*? After all, they are not something that supervenes upon the relativistic spacetime and the gravitational field that characterizes it, but are part and parcel of the matter distribution that generates that field. The contribution to the energy tensor of matter of the FCS members actually used in our laboratories is surely negligible; but, unless I badly misread Grünbaum, his FCS includes *every* photon and atomic clock, not only those that we have managed to harness. Now, if the FCS comprises virtually every source of gravity—so that, for example, it is the relative abundance of FCS members in a Friedmann world that decides whether it will end up in conflagration or in thermal death—what sense does it make to say that the gravitational field *acquires* from the FCS an *additional* significance it does not have by itself? The real gravitational field **g** is what it is by virtue of the actual matter distribution T. But, for this very reason, once **g** is *fixed*, there is nothing else that the matching T can *add* to it. Grünbaum distinguishes between the field **g** *qua* metric field and the same **g** *qua* "pregeometric gravitational field".[54] To me it is as if in the formulation of Newton's theory of gravity one carefully demarcated the Galilean force, by which earthly things fall down, from the Keplerian force, that keeps heavenly bodies in orbit. Then, a mere gallon of fuel added to a ballistic missile would suffice, at the margin, to "constitute" as a proper Keplerian force what is otherwise just a pre-Keplerian Galilean force on the missile. Grünbaum's distinction is of less practical consequence, because all feasible tests of the gravitational field **g** must use FCS members as probes. Should we ever succeed in detecting gravitational waves and learn from them something about the metric structure of the world, a philosopher might still argue that the properties thereby disclosed are "geometric" only by virtue of the ontologically constitutive intervention of the FCS members excited by the waves, but "purely gravitational" in the light of the waves alone. In fact, Grünbaum has already served us warning that, if the null geodesic trajectories of gravitational radiation in an empty spacetime ($T \equiv 0$) were declared to be not just "gravitationally null", but "*geometrically* null" as well, "with a view to establishing thereby the geometric physical significance of the g_{ik} tensor, this would be *physically* tantamount to no more than just *calling* this tensor 'metric'."[55] This remark leads to my last question. If spacetime is a 4-manifold it is—whether we perceive it or not—criss-crossed by curves. If the gravitational field **g** is defined on the manifold $\mathscr{W}$, there is a unique symmetric linear connection ∇ in $\mathscr{W}$ such that $\nabla \mathbf{g} = 0$. A curve γ in $\mathscr{W}$ is a geodesic of the connection if and only if, at every point of γ, $\nabla_{\dot{\gamma}}\dot{\gamma} = 0$. If this condition is met, γ is of course energy-critical on every closed subinterval of its domain, i.e. it is a Riemannian geodesic (of **g**). All this follows solely from the "formal mathematical properties" of $\mathscr{W}$, **g** and γ, whatever the make-up of the physical entities on γ's path. Shall we say that γ is "geometrically geodesic" only if it is the worldline of a gravitating particle, but that it is merely "gravitationally geodesic" if it runs across a perfect vacuum? Even those who revel in such fine distinctions will find something very odd about the terms chosen to express this one, for the curve is a real gravitational trajectory only in the former case, while in the other it is precisely what anyone would call a purely geometrical line. But there is no question that you *can* make words mean so many different things, and I, for one, will not grudge a philosopher's claim to be their master.[56]

7.3 Remarks on Time and Causality

Common sense ideas of space and time are tightly knit together with our ordinary notions and experiences of causality. We all know that—other things being equal—it is easier to change what is near at hand than what resides far away; and, until the advent of radiotelecommunications, distance was a serious obstacle to the prompt fulfilment of our wishes. Agents such as light and gravity, that can influence very distant objects very fast, are also extremely weak; while those means of action which keep their full strength as they travel from their source—e.g. a bomb mailed from one country to another to blow up a political exile—are slow to reach their goals. Before Einstein, however, nobody appears to have seriously disputed that any two events might be causally related to each other, regardless of their spatial and temporal distance. The denial of this seemingly modest statement is perhaps the deepest innovation in natural philosophy brought about by Relativity. It has completely upset our traditional views of time, space and causality, as the following remarks will illustrate.

In a relativistic spacetime two events can be causally related only if their respective worldpoints are joined by what we have called a *causal* curve, i.e. a smooth curve that is nowhere tangent to a spacelike vector. In this way, the spacetime metric strictly regulates what events can and cannot influence or be influenced by a given event. In the notation of Section 4.6, if the spacetime manifold is denoted by $\mathscr{W}$, the set of all worldpoints joined to a given $P \in \mathscr{W}$ by causal curves is $J(P)$, that is, the union of P's "causal future" and "causal past". The complement of $J(P)$ in $\mathscr{W}$ is the set $\|(P)$ of worldpoints "separate" from P (p. 121). For some worldpoints P in some spacetimes ($\mathscr{W}$, **g**), $\|(P)$ may indeed be empty. But, as we shall see, in such spacetimes one cannot properly define a time, and causal relations have quite unusual features. Thus, in a relativistic world where some events could be casually connected with all the rest, our familiar notions would be fully inapplicable. And, of course, in the more familiar case of Special Relativity—where $(\mathscr{W}, \mathbf{g}) = (\mathscr{M}, \boldsymbol{\eta})$—we have that, for each worldpoint P, *every* neighbourhood of P has a non-empty intersection with $\|(P)$. This means that, in the nearest proximity of any event, other events take place which cannot affect it or be affected by it. This alone is sufficient to shatter the received views that sustain our common talk and even shape our grammar.

We speak a tensed language, that neatly distinguishes between past and future, and, somewhat less neatly, between them and present. Our use of tenses is presumably the source of Aristotle's conception of the "now" (p. 220). But our tensed speech is hardly appropriate for dealing with the idealized instantaneous events that are essential to the mathematical physicist's handling of Aristotelian time. While we can and do refer with all the required exactness to such punctual events in the past and future tenses, all our statements in the present tense are necessarily about states and processes that last some time. "The bell rang at 9:37:15 A.M." may refer with split-second precision to the time at which the first electric pulse that set the bell ringing stopped a chronometer that was connected to the bell's circuit for this purpose. (The numbers can even make allowance for any known time lag between the current's arrival at each instrument.) "The bell is ringing now" cannot refer to an instantaneous event, but only to a process that has already begun and will still, no matter how briefly, go on. (If the bell stops ringing while I speak, the statement in the present tense is false and must be corrected by reformulating it in the past tense.) If we try to introduce mathematical exactness into ordinary language by restricting the scope of the present tense to the Aristotelian now, we are led to the paradoxical conclusion that all that actually exists is nothing but the vanishing boundary between that which was and is no longer and that which will be and is not yet—a classical instance of philosophical self-entrapment through misuse of language.[1] Obviously, our common usage was not designed to serve the purposes and needs of physics. Quite justifiably, no attempt has been made to adapt it to the advances of physical knowledge. We say in good English, in a clear night, that such-and-such a star "is shining", even though, for all we know, it could have been annihilated four or more years ago. To speak otherwise, to say, for example, pointing at the splendid pattern of Orion, "Look! Rigel was shining while Frederick Barbarossa was on his fourth expedition to Italy!"[2] would not only be preposterously pedantic, but would, from a relativistic standpoint, raise ticklish questions regarding our choice of a particular dating system. For assuming—to avoid even greater complications—that spacetime is Minkowskian, there is no reason to refer all events to the inertial system in which our dear selves happen to be momentarily at rest. We can also choose a Lorentz chart whose origin coincides with our statement, relative to which the light we see left Rigel at a much more recent date.[3]

The muddles in which one may get entangled by naîvely stretching to Special Relativity the standard use of the present tense are very well exemplified by the doctrine of chronogeometrical determinism put forward by C. W. Rietdijk (1966, 1976) and H. Putnam (1967). The doctrine recalls the logical determinism that is often attributed to the Stoics. According to the latter, every proposition has a truth value, *true* or *false*. Consequently, it is argued, the future is fully determined by the true propositions that describe it and I am not free to change it. To counter this facile argument, some libertarian philosophers deny that all propositions have a truth value. Contingent (i.e. non-necessary) propositions about the future acquire whatever truth value they will have as time goes by. This thesis is obviously tailored to fit the Aristotelian idea of time:[4] the "now" is the omnipresent catalyser that activates the transmutation of indeterminate propositions into truth or falsehood. The thesis is therefore hard to reconcile with the relativistic view of time. But there is another, subtler answer to the logical determinist's ploy: that every proposition *p* is necessarily either true or false does not entail that *p* is either necessarily true or necessarily false.[5] The future will be what it will be, but this does not mean that it is already necessitated to be that way, regardless of what might happen between now and then. Truth-functional logic does indeed require every fact to be definite, but not that it be defined by other facts that precede it. An act of freedom is described by a proposition that is always true, but utterly groundless, and thus impossible to predict according to general laws from the true description of earlier facts. This second reply to logical determinism, which was worked out by medieval philosophers bent on reconciling human freedom with divine foreknowledge, is independent of any particular conception of time, and is therefore preferable to the former from a relativistic standpoint. It will also be helpful in dispelling the confusions of chronogeometrical determinism.

Rietdijk's and Putnam's arguments rest entirely on the following fact: If *A* and *B* are two events joined by a timelike worldline in Minkowski spacetime, there is an event *C* which is Einstein simultaneous with *A* in one inertial frame and with *B* in another.[6] In particular, if *A* is an event in my life, here and now, and *B* is an event in my future, there is an event *C* which in my own momentary inertial rest frame (mirf) is happening right now and yet takes place in someone else's mirf at the same time as *B*. What happens now is real (Putnam) and determinate (Rietdijk). Thus, *C* is no less real and determinate than *A*, for both occur now (in my mirf); and *B* is no less real and determinate than *C*, for both occur now (in another frame). Since the reality and determinateness of an event does not depend on the inertial frame to which it is referred, *B* is just as real and determinate as *A*. Therefore I may conclude, much like the logical determinist, that even as I live through *A*, I am no longer free to shape *B*.

The argument shifts without warning from the present tense of ordinary conversation—appropriately used for stating what happens now in my or your frame—to the tenseless present of science—needed for making absolute ontological assertions. (Unless there is an absolute now, as in the Aristotelian world, there can be no frame-independent use of the present tense.) Putnam assumes that "All (and only) things that exist *now* are real". Here "exist", modified by the adverb "now", is present tense; but "are", as in "are real", is tenseless present.[7] But how can that which exists now, relatively to an observer and his chosen frame, be coextensive with the real? Or should we understand Putnam to mean only that "all things which exist now are real now"—real, that is, in (and only in) the relative space of the inertial frame to which the now is referred? But this interpretation ruins the argument, which depends, as we saw, on the absoluteness (frame-independence) of reality. Each event is (tenselessly) real and determinate, in this absolute sense, at its own worldpoint. No tensed, frame-dependent statement can add to or detract from its reality and determinacy. In our particular example, C, in order to be Einstein simultaneous in different frames with A and B, must be joined to each of these events by a spacelike geodesic. C is therefore separate from both A and B, whose reality and determinacy is thus in no way conditioned by that of C. Whether B is caused by A or is totally unaffected by it is a question to which the relation that each has to C cannot have the slightest relevance.

In order to be determined when and where it happens, an event need not be *pre*-determined. The latter predicate is properly asserted of an event E only if E is (in principle) predictable from its own past, i.e. from the set $J^-(E)-\{E\}$. But there is, of course, no way of proving by geometry alone that every event in a Minkowskian universe is thus predetermined. Rietdijk, who claims that the pre-determination of all such events is entailed by the argument explained above, defines "predetermined" in a different way:

> We say that an event P is (pre-)determined if, for any possible observer W_1 (that is, for all possible observers, and even for all other, e.g., physical, instances), who has P in his absolute future (that is, that the future part of W_1's course through the four-dimensional continuum may eventually pass through P), we can think of a possible observer W_2 (or: there may exist an observer W_2) who can prove, at a certain moment T_p, that W_1 could not possibly have influenced event P in an arbitrary way (e.g., have prevented P) at any moment when P still was future, or was present, for W_1, supposed that W_1 did desire to do so.[8]

To see whether Rietdijk's claim follows from this definition, let us apply it to the example we considered above. Let B be the arbitrary event P of which we want to know whether it is predetermined in Rietdijk's sense. Let W_1 denote an inertial observer who has B in his future as he undergoes event A. Let W_2 be another inertial observer, in whose frame B is Einstein simultaneous with C, an event that is Einstein simultaneous with A in the frame of W_1. Is there a time T_P in which W_2 can prove, from the information provided, "that W_1 could not possibly have influenced event P in an arbitrary way at any moment when P still was future, or was present, for W_1, supposed that W_1 did desire to do so"? I assume, though the definition does not say so, that the time T_p is to be some

(Einstein) time in the frame of W_2. (It would be uncharitable to regard T_p as an absolute time, unwittingly smuggled into the argument.) Under the stated circumstances, W_2 can prove something about B if he is acquainted with B. Consequently, T_p cannot be earlier than the time—call it T_B—at which C and B occur in W_2's frame. W_2 will know B at that time only if B lies on W_2's worldline. In that case, W_1's history from A to B lies at T_B in W_2's past, and W_2 cannot prove from purely chronogeometrical considerations that B is not the outcome of an arbitrary influence exerted through that history. If W_2's worldline does not pass through B, W_2 can only learn about B some time later than T_B. At that time, again, W_1's history before B will lie in W_2's past and W_2 will be unable to prove, without additional, more substantive information, that B does not result from a free decision taken by W_1 at or before the time of B. The fact that W_2, as he finally gets to know B, may also learn that in his frame B is simultaneous with an event C, which is simultaneous with A in W_1's frame, cannot modify our conclusion. Thus, even if we adopt Rietdijk's contorted definition of predetermination, we may not maintain, on the sole strength of Minkowski geometry, that every event is predetermined in a flat relativistic spacetime.

Turning now to the consideration of time and causality in an arbitrary relativistic spacetime ($\mathscr{W}$, **g**), we note first of all that every worldpoint $P \in \mathscr{W}$ has a neighbourhood that mimics the causal structure of an open submanifold of Minkowski spacetime. For the exponential mapping Exp_P maps the null, spacelike and timelike vectors in a neighbourhood of zero in the tangent space at P respectively into null, spacelike and timelike geodesics through P. By designating one of the two sheets of the null cone in the tangent space $\mathscr{W}_P$ as "future", we can classify all causal curves through P into future-directed and past-directed. This classification constitutes what we may call a local time orientation at P. The spacetime ($\mathscr{W}$, **g**) is said to be *time-orientable* if such a classification can be smoothly extended to all $\mathscr{W}$, i.e. if a local time orientation can be established at each $P \in \mathscr{W}$ which agrees smoothly with the local time orientations at all neighbouring points.[9] Evidently, ($\mathscr{W}$, **g**) is time-orientable if and only if $\mathscr{W}$ admits a smooth field of timelike vectors. (If X is such a field, we may choose the local time orientation at each worldpoint so that the local value of X is future-directed. Since X is smooth, the time orientation will be smoothly extended throughout $\mathscr{W}$.) But there are other, more handy criteria of time-orientability. Let us say that a curve γ in $\mathscr{W}$ is *time preserving* if every vector parallel along γ to an arbitrarily chosen future-directed vector v at a point of γ is also future directed. (The reader ought to satisfy himself that the property of being time preserving does not depend on the choice of the vector v; in other words, that if any future-directed vector changes to past-directed by parallel transport along γ all its neighbours in the interior of the future null cone at its point of origin suffer the same mutation.) The relativistic spacetime ($\mathscr{W}$, **g**) is time-orientable if and only if every closed (smooth or piecewise smooth) curve through a fixed, arbitrarily chosen worldpoint P is time preserving.[10] It can be shown moreover that if a closed curve γ through a given $P \in \mathscr{W}$ is time preserving, every other curve through P, homotopic to γ, is also time preserving.[11] It follows that the relativistic spacetime $(\mathscr{W}, \mathbf{g})$ is time-orientable if—though not only if—the manifold is simply connected, i.e. if every closed curve in $\mathscr{W}$ is homotopic to a point. (Any worldpoint may be regarded as the image of some real interval by a constant mapping γ; γ is then trivially a time preserving closed curve.) This result is quite significant, for every relativistic spacetime $(\mathscr{W}, \mathbf{g})$ has a universal covering space $(\tilde{\mathscr{W}}, \tilde{\mathbf{g}})$, i.e. a covering space that is simply connected.[12] $(\tilde{\mathscr{W}}, \tilde{\mathbf{g}})$ is constructed as follows. Choose a point $P \in \mathscr{W}$. For each point $Q \in \mathscr{W}$ we consider all curves from P to Q. If two such curves are homotopic we say that they belong to the same homotopy class (see note 11). The set of all pairs $(Q, [\gamma^Q])$, such that $Q \in \mathscr{W}$ and $[\gamma^Q]$ is an homotopy class of curves from P to Q, is a covering space over $\mathscr{W}$, with the covering map $f: (Q, [\gamma^Q]) \mapsto Q$ (see note 12). We denote it by $\tilde{\mathscr{W}}$. $\tilde{\mathscr{W}}$ has a unique differentiable structure relative to which f is smooth. The pull-back of $\mathbf{g}$ by f, $f^*\mathbf{g}$, is a Lorentz metric on $\tilde{\mathscr{W}}$, which we denote by $\tilde{\mathbf{g}}$. The knowledgeable reader will note that, by the construction of $\tilde{\mathscr{W}}$, the fundamental group of $\mathscr{W}$ (the so-called first homotopy group $\pi_1(\mathscr{W}, P)$) acts freely on $\tilde{\mathscr{W}}$, its action being transitive on each fibre of the covering map f. If $\mathscr{W}$ is simply connected all curves from P to a given point Q are homotopic and there is only one homotopy class $[\gamma^Q]$ to each $Q \in \mathscr{W}$; f is therefore bijective and indeed a diffeomorphism, and we may identify $(\mathscr{W}, \mathbf{g})$ with its own universal covering space. But even if $\mathscr{W}$ is not simply connected, every $P \in \mathscr{W}$ has a neighbourhood diffeomorphic to a neighbourhood of each of the multiple elements of the fibre of f over P, and it is impossible to distinguish between the spacetimes $(\mathscr{W}, \mathbf{g})$ and $(\tilde{\mathscr{W}}, \tilde{\mathbf{g}})$ by means of local experiments. Since every experiment is local—including the collection of light on photographic film placed at the receiving end of a telescope—we can safely assume that the spacetime we live in is simply connected and hence time-orientable. (Note however that if our spacetime were the universal covering space of a multiply-connected spacetime, it would contain several identical copies of a neighbourhood of each worldpoint, so that every individual event, process, thing or person in our experience would be exactly repeated at least twice—indeed as many times as there are homotopy classes of closed curves through a given point of $\mathscr{W}$. This vexing iteration could then be suppressed at a stroke by identifying our spacetime with the multiply connected spacetime in question, rather than with its universal covering space.)[13]

Time order and causal influence would make no sense in a relativistic spacetime that is not time-orientable. But even in one which is time-orientable they may have only a local significance. The relations of causal and chronological precedence (Section 4.6) can be formally introduced in a time-orientable relativistic spacetime $(\mathscr{W}, \mathbf{g})$ by choosing a global time orientation and stipulating that, for any pair of worldpoints P and Q, P causally precedes Q ($P < Q$) if and only if there is a future-directed causal curve from P to Q; and that P chronologically precedes Q ($P \ll Q$) if and only if there is a future-directed timelike curve from P to Q. But $(\mathscr{W}, <, \ll)$ might not be a causal space (p. 123), for it may well contain closed causal curves. As a matter of fact, several exact solutions of the Einstein field equations, such as the Gödel, the Taub-NUT and the extended Kerr solutions, are known to contain such curves.[14] An event in the range of a closed causal curve γ would causally precede *and* be preceded by all other events in γ's range, so that it could be both their cause and their effect. As this would be utterly at variance with our received notion of causality, many authors believe that closed causal curves are unphysical, and that a solution to the Einstein field equations is admissible only if it satisfies the so-called causality condition (p. 337, note 23) i.e. if the spacetime defined by it is truly a causal space. Stephen Hawking (1971) equates "ordinary causality" with "the absence of closed timelike curves". He observes that,

> if there were such curves, one could in theory travel round them and arrive in one's past. The logical difficulties that could arise from such time travel are fairly obvious: for example, one might kill one of one's ancestors. These difficulties could be avoided only by an abandonment of the idea of free-will: by saying that one was not free to behave in an arbitrary fashion if one travelled into the past.[15]

However, as Hawking and Ellis (1973) remark, free will "is not something which can be dropped lightly since the whole of our philosophy of science is based on the assumption that one is free to perform any experiment".[16] By the same token, a physically reasonable relativistic spacetime should not contain any *almost closed* causal curve, i.e. a causal curve which, even though it does not intersect itself, enters more than once in the same infinitesimal neighbourhood—in other words, it must be a *strong* causal space (p. 125 and p. 308, note 6). It is not hard to see that in a strong causal space, every point of which has a neighbourhood that is causally isomorphic, by the local exponential mapping, with an open submanifold of Minkowski spacetime, the set $\| (P)$ of points separate from any given point P cannot be empty, and indeed intersects every neighbourhood of P. This ratifies our earlier statement that $\| (P)$ can be empty for some worldpoint P only in a spacetime in which one cannot define a time, and causality has unusual features. As we now see, such spacetimes are not allowed as physically viable by two major authorities in the field. Hawking and Ellis go in fact even further. A strong causal spacetime might be, as they say, "on the verge of violating" the foregoing requirements, if a small variation in the metric suffices to generate closed timelike curves. "Such a situation would not seem to be physically realistic since General Relativity is presumably the classical limit of some, as yet unknown, quantum theory of space-time and in such a theory the Uncertainty Principle would prevent the metric from having an exact value at every point. Thus in order to be physically significant, a property of space-time ought to have some form of stability, that is to say, it should also be a property of 'nearby' spacetimes."[17] To make this demand precise, the authors introduce a topology in the set of Lorentz metrics admissible by a given 4-manifold (I have defined it on p. 337, note 17), and go on to postulate that a relativistic spacetime ($\mathscr{W}$, **g**) will be deemed to be physically reasonable if and only if it is *stably causal*, i.e. if and only if the metric **g** has a neighbourhood U in the said topological space of Lorentz metrics on $\mathscr{W}$, such that for any $\mathbf{g}' \in U$, the spacetime ($\mathscr{W}$, **g**′) does not contain a closed timelike curve. As I mentioned on page 228, Hawking (1969) showed that stable causality is both necessary and sufficient for the existence of a global time coordinate function. Thus, the same requirement that apparently must be met by any causal structure compatible with the practice of experimental science, ensures the possibility of partitioning events—in many different, generally arbitrary ways—into linearly ordered simultaneity classes; a possibility that is essential to our received idea of physical time and in fact equivalent to stable causality. However, not all researchers subscribe to Hawking and Ellis' outright rejection of closed causal curves. Thus, Geroch and Horowitz (1979) recall that "physical theories frequently suggest new and unexpected phenomena, which are later found to be realized physically", and propose that one should keep an open mind on this matter.[18]

Another feature of relativistic spacetimes which would seem to defy our received view of causality—specifically, of the role of causal interaction in the constitution of the universe—are the so-called event and particle horizons. Following Rindler (1956), we shall define these concepts for a relativistic spacetime ($\mathscr{W}$, **g**) in which the worldlines of massive particles (resp. the integral curves of the worldvelocity field of matter) constitute a congruence. Such a congruence can be made in an obvious way into a 3-manifold M. Consider a worldpoint P in the worldline of a particle π. We say that a worldline $\gamma \in M$ lies *beyond P's horizon* (or beyond π's horizon at P) if no point in the range of γ lies in the causal past of P. The *particle horizon* of P (or of π at P) is the boundary (in M) of the set of worldlines that lie beyond P's horizon.[19] Unless the particle horizon of P is empty, there are particles in the world $\mathscr{W}$—namely, those beyond P's horizon—which have never exercised any influence on π at the time of P. We also say that a worldpoint $P \in \mathscr{W}$ (or an event at such a worldpoint) lies *beyond the horizon* of a particle π if no point in the worldline of π lies in the causal future of P. The *event horizon* of a particle π (or of its respective worldline) is the boundary (in $\mathscr{W}$) of the set of events that lie beyond π's horizon. The reader should satisfy himself that the event horizon of π is none other than the boundary of π's causal past (p. 121). Unless the event horizon of π is empty, there are events in the world $\mathscr{W}$—namely, those beyond π's horizon—which will never have any influence on π. Rindler's concepts of particle and event horizons are nicely symmetric: The particle horizon of an event P is the boundary of the set of particles whose worldlines do not intersect the causal past of P. The event horizon of a particle π is the boundary of the set of events whose causal futures do not intersect the worldline of π.[20] Non-empty particle and event horizons are not a rarity in relativistic spacetimes. An event horizon envelops each "black hole", that is, each spacetime region where the gravitational field is so strong that a signal originating within it is forever trapped inside it.[21] Particle horizons occur in all the likeliest relativistic world models.[22] Thus, they exist in every non-empty expanding Friedmann world with initial singularity, and also in the anisotropic generalizations of the Friedmann solutions studied by Taub (1951). Let M be the 3-parameter family of matter worldlines of such a Big Bang world and pick any $\gamma \in M$. The particle horizon of any $P \in \gamma$ divides M into two components, one of which contracts to γ as P approaches the initial singularity. This means that each speck of matter is "born" at the Big Bang in total isolation from all the rest, and only little by little increases its "circle of acquaintances", i.e. the set of particles on which it has acted and which have acted on it. If, as seems likely, we live in a universe roughly representable by a Friedmann or a modified Friedmann Big Bang model, then, at this very instant, new galaxies are entering our particle horizon from opposite parts of the sky, which hitherto have not had the chance of acting on us or upon each other. Since the phenomena that first disclose to us such emergent galaxies must inevitably be read as instances of the same natural laws that govern all things known hitherto, it is clear that we must regard all matter, manifest or forthcoming, as constituting a single world not held together by the bonds of causal interaction. The existence of non-empty particle and event horizons therefore means that, contrary to our inherited views, causality is not the "cement" of a typical relativistic universe. In itself, this is not surprising, for the very possibility of causation obviously presupposes a system of laws under which it can take place. Things do not have to rub themselves against each other in order to become parts of a common nature. They must rather be, of themselves, sufficiently alike in order to sense their mutual presence. In General Relativity, as in any rational system of the world, the formal causality of structure sets the stage for the efficient causality of things. Horizons merely make more evident, in the worlds where they occur, this anyhow unquestionable order of ontological precedence. I must add, however, that although there is nothing really disquieting about horizons as such, the particle horizon of the Big Bang models is a source of paradox under the particular circumstances of our world. The remarkable isotropy of the microwave background radiation discovered by Penzias and Wilson signifies that widely distant parts of the universe were in virtually the same thermal state when that radiation was emitted. If the available data are interpreted in the light of General Relativity it turns out that at the time of emission some of the sources must have lain beyond each other's horizons. The thermal uniformity of causally unconnected world regions is probably too unlikely a coincidence to be simply taken for granted. C. W. Misner, who has long been worried by this problem,[23] now believes that its solution will involve modifications of the Einstein field equations, or at least their quantization.[24] Some such change should anyway be necessary in order to reconcile Einstein's theory of gravity with the accepted theories of the other fundamental forces of nature. An examination of what is being done with this purpose lies, however, beyond my scope.

Appendix

A Differentiable Manifolds

The theory of differentiable manifolds was founded by Bernhard Riemann in 1854 as a generalization of Carl Friedrich Gauss' treatment of smooth surfaces in Euclidean space.[1] Such surfaces can be metrically and even topologically quite different from the Euclidean plane, yet they can always be represented "patchwise" on a region of the latter, as the earth is charted on the pages of an atlas. This method of representation enabled Gauss to define and study the properties of surfaces "intrinsically"; i.e. without regard to the manner how they lie in space. The theory brought to life by Riemann grew richer and clearer in the century following his death, especially after Einstein used it in General Relativity. The following definitions and remarks agree with the now standard approach to the subject.

A topological space S is an n-dimensional manifold if every point of S has a neighbourhood homeomorphic with $\mathbf{R}^n$. A *chart* of S is a homeomorphism of an open set of S onto an open set of $\mathbf{R}^n$.[2] We designate a typical chart by x, its domain by U_x or dom x. x assigns to each point $P \in U_x$ a list of n real numbers $x^1(P), \ldots, x^n(P)$, called the *coordinates* of P by x. The mapping $x^i: U_x \to \mathbf{R}$; $P \mapsto x^i(P)$ is the i-th *coordinate function* of x $(1 \leq i \leq n)$. If $U_x = S$, x is a *global chart*.

Let x and y be two charts of S. The composite mapping $x \cdot y^{-1}$ is a homeomorphism of one region of $\mathbf{R}^n$ onto another, namely, of $y(U_x \cap U_y)$ onto $x(U_x \cap U_y)$. $x \cdot y^{-1}$ and $y \cdot x^{-1}$ are the *coordinate transformations* between x and y. x and y are C^k*-compatible* if both coordinate transformations between them are differentiable mappings of class C^k $(k \geq 1)$. A collection A of C^k-compatible charts of S, such that each point $P \in S$ lies on the domain of some chart in A, is a C^k*-atlas* of S. (Observe that A may consist of a single chart x, provided that x is global.) Given a C^k-atlas A of B, there is a unique *maximal* C^k-atlas $A_{\max}$ of S, comprising every chart of S that is C^k-compatible with the charts in A. (Exercise: Show that any two charts in $A_{\max}$ are C^k-compatible.) A non-empty subset of $A_{\max}$ which is a C^h-atlas of S in its own right is called a C^h*-subatlas* $(h \geq k)$.

Let S be an n-dimensional manifold. Let A be a C^k-atlas of S. The pair $(S, A_{\max})$ is a *real n-dimensional C^k-differentiable manifold*.[3] $(S, A_{\max})$ is said to be *orientable* if $A_{\max}$ contains a C^k-subatlas A^+, such that the Jacobian determinant of any coordinate transformation between two charts of A^+ is positive at every point of its domain. The atlas A^+ is said to be *oriented* and to *bestow an orientation* on the manifold. In this book, spacetime manifolds are generally assumed to be orientable. To avoid annoying exceptions we assume also that every manifold mentioned has the Hausdorff topology, i.e., that any two points of it lie in mutually disjoint open sets. A real n-dimensional C^∞-differentiable manifold with a Hausdorff topology will be called an *n-manifold*. Note, in particular, that $\mathbf{R}^n$ is an n-manifold (with the atlas comprising the identity mapping that sends each list of n real numbers to itself).

Until the end of this Section, S will denote an n-manifold and S' an m-manifold. If U is open in S, U is an n-manifold with the atlas consisting of the restrictions to U of the charts in the maximal atlas of S. U, with this atlas, is said to be an *open submanifold* of S. The Cartesian product $S \times S'$ is made into an $(n+m)$-manifold by the stipulation that, if x and y belong respectively to the maximal atlases of S and S', the mapping that sends each pair $(P, Q) \in S \times S'$ to $(x(P), y(Q)) \in \mathbf{R}^{n+m}$ is a chart of $S \times S'$. $S \times S'$, with the atlas consisting of all such charts, is called the *product manifold* of S and S'.

A mapping f of S into S' is said to be C^r-differentiable at $P \in S$ ($r \leq \infty$) if, for some chart x of S defined at P and some chart y of S' defined at $f(P)$, the composite mapping $y \cdot f \cdot x^{-1}$ is differentiable of class C^r at $x(P)$. (Show that this definition does not depend on the choice of x and y.) f is C^r-differentiable if it is C^r-differentiable at every point of S. If $n = m$ and f is a bijective C^∞-differentiable mapping with C^∞-differentiable inverse f^{-1}, f is a *diffeomorphism* and S and S' are said to be *diffeomorphic*. (A diffeomorphism is an isomorphism of differentiable manifolds.) In this book we need not worry about orders of differentiability. I speak therefore loosely of *smooth* mappings. By smooth I mean C^k-differentiable for some definite integer k, no less than 1 and as large as is required for the purpose at hand.[4]

If x is a chart of S and f is a smooth mapping of S into $\mathbf{R}$, the composite mapping $f \cdot x^{-1}$ is a smooth mapping of $\mathbf{R}^n$ into $\mathbf{R}$. We write $\overline{f}_{,i}$ for the i-th partial derivative of $f \cdot x^{-1}$,[5] and set

$$\partial f / \partial x^i = \overline{f}_{,i} \cdot x \tag{A.1}$$

A smooth mapping of S into $\mathbf{R}$ is called a *scalar field* on S. The collection $\mathscr{F}(S)$ of all scalar fields on S is made into an algebra over $\mathbf{R}$ (i.e. a real vector space that is also a ring, in such way that addition in the ring agrees with vector addition). Let P be any point of S and let f_P denote the value of the scalar field f at P. Vector addition (i), scalar multiplication (ii) and ring multiplication (iii) are defined as follows, for all $f, g \in \mathscr{F}(S)$ and all $\alpha \in \mathbf{R}$:

$$\text{(i)} \quad (f+g)_P = f_P + g_P$$

$$\text{(ii)} \quad (\alpha f)_P = \alpha f_P$$

$$\text{(iii)} \quad (fg)_P = f_P g_P$$

The collection $\mathscr{F}(P)$ of all scalar fields defined on some neighbourhood of $P \in S$ is also an algebra over **R**, with the operations defined as above, for any pair of elements of $\mathscr{F}(P)$, in the intersection of their domains.

A continuous mapping γ of an interval $I \subset \mathbf{R}$ into S is called a curve in S. (The variable argument of γ is usually called the curve's *parameter*; if f is the restriction to I of a homeomorphism of **R** onto itself, the curve $\gamma' = \gamma \cdot f^{-1}$ is a *reparametrization* of γ.) The range of a curve γ is sometimes called a *path*. In this book, unless otherwise stated, we assume that a curve γ in an n-manifold S is always smooth or at least piecewise smooth (i.e. smooth everywhere except at a finite number of points). If its domain I is not open, γ must be the restriction to I of a smooth curve defined on an open interval—this condition ensures the smoothness of γ at the endpoints of I. Let $P = \gamma(u)$ for some $u \in I$. The *tangent* to γ at P, denoted by $\dot{\gamma}_P$, is the linear mapping of $\mathscr{F}(P)$ into **R** defined, for each $f \in \mathscr{F}(P)$ by:

$$\dot{\gamma}_P(f) = \left.\frac{\mathrm{d}f \cdot \gamma}{\mathrm{d}t}\right|_{t=u} \tag{A.2}$$

Instead of $\dot{\gamma}_P(f)$ we shall write $\dot{\gamma}_P f$. (Observe that the symbol '$\dot{\gamma}_P$' is ambiguous if $P = \gamma(u)$ for more than one value of the parameter u; see page 261.) The set of all tangents at P to curves in S is a subset of the vector space $(\mathscr{F}(P))^*$ of linear functions on $\mathscr{F}(P)$. In fact, as we shall see, it is an n-dimensional subspace of $(\mathscr{F}(P))^*$

Let x be a chart of S defined at P. Each coordinate function $x^i \,(1 \le i \le n)$ determines a curve γ^i, which we shall call the *i-th parametric line* of x through P. γ^i is defined as follows: for each suitable real number t, $\gamma^i(t)$ is the point Q in dom x such that $x^i(Q) = t$ and, for $j \neq i$, $x^j(Q) = x^j(P)$. (Observe that γ^i maps a suitable interval in the range of x^i onto the set of points in S that share with P all x-coordinates except the i-th.) Set $x^i(P) = u$. If $f \in \mathscr{F}(P)$,

$$\begin{aligned}
\dot{\gamma}^i_P f = \left.\frac{\mathrm{d}(f \cdot \gamma_i)}{\mathrm{d}t}\right|_{t=u} &= \lim_{h \to 0} \frac{1}{h}\,(f \cdot \gamma_i\,(u+h) - f \cdot \gamma^i\,(u)) \\
&= \lim_{h \to 0} \frac{1}{h}\,(f \cdot x^{-1}\,(x^1\,(P), \ldots, x^i(P) + h, \ldots, x^n(P)) \\
&\quad - f \cdot x^{-1}\,(x^1\,(P), \ldots, x^i\,(P), \ldots x^n\,(P))) \\
&= \left.\frac{\partial f}{\partial x^i}\right|_P
\end{aligned} \tag{A.3}$$

In agreement with this result, we shall hereafter designate the tangent at P to the i-th parametric line through P of a chart x by $\partial/\partial x^i|_P$. The collection $(\partial/\partial x^1|_P, \ldots, \partial/\partial x^n|_P)$ is a free family of vectors in $(\mathscr{F}(\mathrm{P}))^*$.[6] Hence it spans an n-dimensional subspace of $(\mathscr{F}(P))^*$. We shall now prove that this subspace is identical with the set of tangents at P to curves in S. In anticipation of this result we call it the *tangent space* of S at P. We shall denote it by S_P. We show first that, if $\dot{\gamma}_P$ is the tangent at P to a curve γ, $\dot{\gamma}_P$ is equal to a linear combination of the $\partial/\partial x^i|_P$. If $P = \gamma(u)$, $f \in \mathscr{F}(P)$ and $\overline{f}_{,i}$ is the i-th partial derivative of $f \cdot x^{-1}$,

$$\dot{\gamma}_P f = \left.\frac{\mathrm{d}(f\cdot\gamma)}{\mathrm{d}t}\right|_u = \sum_i \overline{f}_{,i}\left|_{x(P)} \frac{\mathrm{d}(x^i\cdot\gamma)}{\mathrm{d}t}\right|_u = \sum_i \dot{\gamma}_P x^i \left.\frac{\partial f}{\partial x^i}\right|_P \tag{A.4}$$

We now show that if $v \in S_P$ there is a curve γ in S such that $\gamma(0) = P$ and $\dot{\gamma}_P = v$. By hypothesis, $v = \Sigma_i v^i \partial/\partial x^i|_P$ for some set of n real numbers v^i. Set $u_i = uv^i + x^i(P)$. The curve $\gamma: u \mapsto x^{-1}(u_1, \ldots, u_n)$ is then defined in some interval containing 0, and it is clear that $\gamma(0) = P$. We have that

$$\dot{\gamma}_P x^i = \left.\frac{\mathrm{d}(x^i\cdot\gamma)}{\mathrm{d}t}\right|_0 = \lim_{h\to 0} \frac{x^i(P) + hv^i - x^i(P)}{h} = v^i \tag{A.5}$$

Hence, by (A.4), $\dot{\gamma}_P = \Sigma_i (\dot{\gamma}_P x^i)\partial/\partial x^i|_P = \Sigma_i v^i \partial/\partial x^i|_P = v$. Q.E.D.

An element of S_P is called a *vector at* P. A rule V that assigns to each point P in S a vector V_P at P is called a *vector field* on S if it fulfils the following requirement of smoothness: for each $f \in \mathscr{F}(S)$, the mapping $\mathrm{V}f: P \mapsto \mathrm{V}_P f$ is a scalar field on S.[7] Evidently, the correspondence $\partial/\partial x^i$: $P \mapsto \partial/\partial x^i|_P$ is a vector field on the open submanifold dom x. In particular, the identity mapping $t \mapsto t$ on **R** determines a vector field $\partial/\partial t$. If V is a vector field on S, a curve γ on S, such that, for each point P in its range, $\dot{\gamma}_P = \mathrm{V}_P$, is an *integral curve* of V. For each P in S there is an integral curve γ of V such that $\gamma(0) = P$ and every other integral curve of V that meets this requirement is the restriction of γ to a subset of its domain; we call this curve, the *maximal integral curve* of V through P. The vector field V is said to be *complete* if all its maximal integral curves are defined on **R**. The collection $\mathscr{V}^1(S)$ of all vector fields on S is made into a module over the ring $\mathscr{F}(S)$ of all scalar fields on S. For every V, $\mathrm{W} \in \mathscr{V}^1(S)$ and every $f \in \mathscr{F}(S)$, the vector sum of V and W is the vector field $(\mathrm{V} + \mathrm{W})$ which assigns to each $P \in S$ the vector $\mathrm{V}_P + \mathrm{W}_P$; and the product of V by f is the vector field $f\mathrm{V}$ which assigns to each $P \in S$ the vector $f_P \mathrm{V}_P$.

The dual space of S_P—i.e. the vector space of real-valued linear functions on S_P—is called the *cotangent space* of S at P and will be denoted by S_P^*. Its elements are called *covectors* at P. A *covector field* on S is defined like a vector field. The collection $\mathscr{V}_1(S)$ of covector fields on S is a module over $\mathscr{F}(S)$, with operations analogous to those on $\mathscr{V}^1(S)$. If U is an open submanifold of S, each smooth real-valued function $f \in \mathscr{F}(U)$ determines a covector field $\mathrm{d}f$ on U, called the *differential* of f; $\mathrm{d}f$ assigns to each $P \in U$ the covector $\mathrm{d}f_P$, defined, for all $v \in S_P$, by the relation

$$\mathrm{d}f_P(v) = vf \tag{A.6}$$

The differentials $\mathrm{d}x^i$ of the coordinate functions x^i of a chart x define, at each $P \in \mathrm{dom}\ x$, a basis $(\mathrm{d}x_P^i)$ of the cotangent space S_P^*. As $\mathrm{d}x_P^i(\partial/\partial x^j|_P)$

$= \partial x^i/\partial x^j|_P = \delta_{ij}$, $(\mathrm{d}x^i_P)$ is none other than the dual basis to $(\partial/\partial x^i|_P)$.[8]

The concept of a differential can be extended to arbitrary differentiable mappings of manifolds. Let $f: S \to S'$ be smooth. The *differential* of f, denoted by f_*, assigns to each $P \in S$ the linear mapping f_{*P} of S_P into $S'_{f(P)}$, which is defined as follows: Each vector $v \in S_P$ is the tangent at P to a curve γ in S; we set $f_{*P}(v)$ equal to the tangent at $f(P)$ of the curve $f \cdot \gamma$ in S'. (The reader ought to satisfy himself that this definition does not depend on the choice of a particular curve γ among the many to which v is tangent.) We have that, for each scalar field g defined on a neighbourhood of $f(P)$,

$$f_{*P}(v)g = v(g \cdot f) \tag{A.7}$$

Note that if $S' = \mathbf{R}$, $f_* = \mathrm{d}f$. If γ is a curve in S such that $\gamma(u) = P$, $\gamma_{*u}\partial/\partial t = \dot{\gamma}_P$. With some impropriety, I often write γ_{*u} for $\dot{\gamma}_P$. (This notation removes the ambiguity inherent in "$\dot{\gamma}_P$" if $P = \gamma(u)$ for more than one value of the parameter u.) By the *rank* of $f: S \to S'$ at a point $P \in S$ we mean the rank of f_{*P}, i.e. the dimension of the vector space $f_{*P}(S_P) \subset S'_{f(P)}$.

A smooth mapping f of the n-manifold S into the m-manifold S' is said to be *regular* at $P \in S$ if its rank at P is n. (This presupposes that $n \leq m$.) If $S \subset S'$ and the canonic inclusion $\mathbf{i}$, that sends each point of S to itself regarded as a point of S', is everywhere regular, S is said to be a *submanifold* of S'. (Open submanifolds, as defined on page 258, are certainly submanifolds in the present sense.) It is not difficult to prove that each point $P \in S$ lies in the domain of some chart of S' which maps a neighbourhood of P in S into the flat $\{(r_1, \ldots, r_m) | r_{n+1} = r_{n+2} = \ldots = r_m = 0\} \subset \mathbf{R}^m$. The collection of all such charts (for every $P \in S$) is readily made into an atlas for S by restricting them to S and ignoring the last $(m-n)$ coordinate functions of each, which by definition are identically zero. With this "induced" structure, S is an n-manifold. Let P be a point of S. A vector $v \in S'_P$ is said to be tangent to the submanifold S at P if v lies in the range of $\mathbf{i}_{*P}$. An $(n-1)$-dimensional submanifold of an n-manifold is called a *hypersurface*. Hypersurfaces are usually called *surfaces* if $n = 3$, *paths* if $n = 2$.

As a useful application of the foregoing notions, let us now define the concept of the Lie derivative of a vector field along another. We consider a vector field V on an n-manifold S. Let γ^P denote the maximal integral curve of V through a point $P \in S$. The reader should satisfy himself that if V is a complete vector field, the mapping F_t that sends each $P \in S$ to $\gamma^P(t)$ (that is, to the point corresponding to the value t of the parameter of the maximal integral curve of V through P) is defined for every $t \in \mathbf{R}$ and is in effect a bijective mapping of S onto itself. If V is not complete there is at any rate for every $P \in S$ an open interval $I_P \subset \mathbf{R}$, containing 0, such that, for each $t \in I_P$, the mapping $F_t: Q \mapsto \gamma^Q(t)$ is defined on a neighbourhood N_P of P, which it maps bijectively onto a neighbourhood of $\gamma^P(t)$. Let W be another vector field on S. If $P \in S$, W_P is an element of the tangent space S_P. Let us write γt for $\gamma^P(t)$. Obviously, if $|t|$

is sufficiently small, γt exists and F_{-t} is defined on a neighbourhood of it and sends γt to $\gamma 0 = \gamma^P(0) = P$. Hence, the differential $F_{-t*\gamma t}$ of F_{-t} at γt maps the tangent space $S_{\gamma t}$ onto S_P. Thus $F_{-t*\gamma t}(\mathrm{W}_{\gamma t})$ is an element of S_P which can be added to W_P or to its inverse $-\mathrm{W}_P$. The *Lie derivative* $L_V\mathrm{W}$ of the vector field W *along* the vector field V at a point P of S is defined by:

$$L_V\mathrm{W}|_P = \lim_{t \to 0} \frac{1}{t}\,(F_{-t*\gamma t}(\mathrm{W}_{\gamma t}) - \mathrm{W}_P) \tag{A.8}$$

The concept of a Lie derivative along a given vector field can be extended in a natural way to scalar fields and to tensor fields of all orders (on tensor fields, see below and page 99). It turns out that the Lie derivative $L_V f$ of a scalar field f along the vector field V is none other than Vf, the "directional derivative" of f along V. It is clear from equation (A.8) that $L_V\mathrm{W}$ is a vector field on S. It can be shown that, if f is a scalar field on S, $L_V\mathrm{W}(f) = \mathrm{V}(\mathrm{W}f) - \mathrm{W}(\mathrm{V}f)$. If $L_V\mathrm{W} = 0$ we say that W is *Lie dragged* by V. (This terminology is also used if W is a scalar or tensor field). Clearly, the vector fields $\partial/\partial x^i (1 \le i \le n)$ determined by any chart x of S are Lie dragged by one another (for $\partial^2/\partial x^i \partial x^j = \partial^2/\partial x^j \partial x^i$). Suppose now that $(\mathrm{V}^1, \ldots, \mathrm{V}^n)$ are n vector fields on an open submanifold $U \subset S$, such that at each $P \in U$, $(\mathrm{V}^1_P, \ldots, \mathrm{V}^n_P)$ is a basis of U_P. $(\mathrm{V}^1, \ldots, \mathrm{V}^n)$ is an n-ad field on U. It is not hard to see that if the V^i are mutually Lie dragged (i.e. if $L_{V^i}\mathrm{V}^j$ vanishes everywhere for all values of the indices i and j) their integral curves through any given $P \in U$ are the parametric lines of a chart of S, defined on U and mapping P on 0. An n-ad field consisting of n mutually Lie dragged vector fields is said to be *holonomic*.

The vector spaces S_P and S^*_P, attached to any point P of an n-manifold S, give rise to an infinite array of vector spaces, namely, the spaces of multilinear functions on suitable Cartesian products constructed from S_P and S^*_P. (E.g., of bilinear functions on $S_P \times S_P$, on $S_P \times S^*_P$, on $S^*_P \times S_P$ and on $S^*_P \times S^*_P$, etc.) The elements of such spaces are called *tensors* at P. I deal with them in Section 4.3, where I define the concept of a *tensor field* on S. The crucial and indeed the hardest step on our way to this concept is the definition of the tangent space S_P. This can be given in several different ways, besides the one proposed above.[9] The intuitive prototype of S_P is of course the tangent plane at a point of an ordinary smooth surface—which, if the latter is flat, may be said to coincide with it, but otherwise extends on its own into the surrounding space. However, neither our definition nor its surrogates rely on the existence of a manifold of more dimensions than S, in which S is embedded and S_P may spread. This is of little consequence in pure geometry, for every n-manifold is diffeomorphic with a submanifold of a $2n$-dimensional Euclidean space (Whitney, 1944*a*). But it is of the utmost importance in natural philosophy, for it enables one to make sense of the idea of a cosmic manifold, that is not coincident and possibly not even diffeomorphic with its tangent spaces, and yet is not embedded in any physical manifold of higher dimension.

B Fibre Bundles

A *bundle* E over B consists of two topological spaces E and B, and a continuous mapping π of E onto B. B is the *base* of the bundle. π is called the projection. A *cross-section* of E is any continuous mapping f of B into E, such that $\pi \cdot f$ is the identity on B. (For each $x \in B$, f picks out an element $f(x)$ of the fibre of π over x.)

In this book we are interested only in differentiable bundles. We assume that E and B are, respectively, an n-manifold and an m-manifold ($n \geq m$), and that π is smooth. We also expect cross-sections to be differentiable to a sufficiently high order, which, however, may be less than the order of differentiability of π.

Let E be a differentiable bundle over B, with projection π, and let G be a Lie group acting on E on the right.[1] E is a *principal fibre bundle* over B with *structure group* G if the following conditions are fulfilled:

(i) The action of G on E is free, i.e., for any $a \in E$ and any $g \in G$, $ag = a$ if and only if g is the neutral element of G.
(ii) The action of G is transitive on each fibre of π and the partition of E into fibres by π agrees with its partition into orbits by the action of G.[2]
(iii) Each $u \in B$ has a neighbourhood U such that $\pi^{-1}(U)$ is mapped diffeomorphically onto $U \times G$ by $a \mapsto (\pi(a), f(a))$, where f is a mapping such that, for every $a \in U$, $g \in G$, $f(ag) = f(a)g$.

Condition (iii) implies that it is always possible to find an open covering $\{U_\alpha\}_{\alpha \in I}$ of B, such that, for each $\alpha \in I$, $\pi^{-1}(U_\alpha)$ is mapped diffeomorphically onto $U_\alpha \times G$ by $a \mapsto (\pi(a), f_\alpha(a))$, where f_α satisfies the relation $f_\alpha(ag) = f_\alpha(a)g$ for every $a \in U_\alpha$ and every $g \in G$. If $a \in \pi^{-1}(U_\alpha \cap U_\beta)$, $f_\beta(a)g(f_\alpha(a)g)^{-1} = f_\beta(a)(f_\alpha(a))^{-1}$, so that the latter product depends only on $\pi(a)$, not on a. We can therefore define a mapping $f_{\beta\alpha}$: $U_\alpha \cap U_\beta \to G$, by $f_{\beta\alpha}(\pi(a)) = f_\beta(a)(f_\alpha(a))^{-1}$. The mappings $f_{\alpha\beta}$, for all $\alpha, \beta \in I$, are the *transition functions* of the principal fibre bundle (E, B, π, G) corresponding to the open covering $\{U_\alpha\}_{\alpha \in I}$ of B. It is clear that if a lies in the intersection of U_α, U_β and U_γ,

$$f_{\gamma\alpha}(a) = f_{\gamma\beta}(a) f_{\beta\alpha}(a) \tag{B.1}$$

On the other hand, given a Lie group G, an n-manifold B with an open covering $\{U_\alpha\}_{\alpha \in I}$ and a collection of mappings $f_{\beta\alpha}$: $U_\alpha \cap U_\beta \to G$ (for each non-empty intersection $U_\alpha \cap U_\beta$) that satisfy (B.1), there exists a principal fibre bundle E over B with structure group G and transition functions $f_{\alpha\beta}$.[3]

Let E be a principal fibre bundle over B with projection π and structure group G. Let G act effectively on the left on a differentiable manifold F. The mapping of $G \times (E \times F)$ into $(E \times F)$ by $(g, (a, s)) \mapsto (ag, g^{-1}s)$ is plainly an action of G on $E \times F$, on the right.[4] Two pairs (a, s), $(a', s') \in E \times F$ belong to the same orbit of this action if there is a $g \in G$ such that $(a', s') = (ag, g^{-1}s)$. Such orbits are obviously disjoint by pairs. Let E' be the set of all orbits of the action of G on $E \times F$. We denote by $[a, s]$ the element of E' that contains the pair (a, s). By condition (ii) on p. 263, $\pi(a) = \pi(ag)$ for every $a \in E$ and $g \in G$. We can therefore define a mapping π' of E' onto B by setting $\pi'([a, s]) = \pi(a)$. By condition (iii), for each $u \in B$ we can find a neighbourhood U of u and a diffeomorphism of $\pi^{-1}(U)$ onto $U \times G$. We shall now define a bijection f of $\pi'^{-1}(U)$ onto $U \times F$ such that the fibre of π' over each $v \in U$ is mapped onto $\{v\} \times F$. In the fibre of π over each $v \in U$ there is a unique element a which the chosen diffeomorphism of $\pi^{-1}(U)$ onto $U \times G$ maps on $(\pi(a), e)$, where e denotes the neutral element of G. Let us set $f([a, s]) = (\pi(a), s)$. I claim that we have thereby defined the sought-for bijection. To prove it, I observe that for each $[a, s] \in \pi'^{-1}(U)$ there is a pair $(a', s') \in [a, s]$, such that a' is mapped by the chosen diffeomorphism on $(\pi(a'), e)$. Our stipulation therefore defines a mapping of $\pi'^{-1}(U)$ into $U \times F$. The mapping is plainly surjective and sends $\pi'^{-1}(v)$ to $\{v\} \times F$ (for each $v \in U$). To show that it is also injective assume that $[a', s'] \neq [a, s]$. If $\pi(a') \neq \pi(a)$, obviously $f([a', s']) \neq f([a, s])$. On the other hand, if $\pi(a') = \pi(a)$ there is a $g \in G$ such that $a' = ag$. Hence $[a', s'] = [a, gs']$. But gs' is surely different from s, or else $s' = g^{-1}s$ and $[a', s'] = [ag, g^{-1}s] = [a, s]$, contrary to our assumption. Consequently, $f([a', s']) = (\pi(a), gs') \neq (\pi(a), s) = f([a, s])$. Since $U \times F$ is an open submanifold of the product manifold $B \times F$, the charts of the latter can be combined with the bijections we have just defined (for each $u \in B$) to bestow a differentiable structure on E', whereby π' is a smooth mapping. E' is then a differentiable bundle over B with projection π'. We call it the *fibre bundle* with *typical fibre F associated with* the principal fibre bundle (E, B, π, G).

Let M be an n-manifold. We shall define the *principal bundle of n-ads over M*. An n-ad at $P \in M$ is an ordered basis of the tangent space M_P. Let FM denote the aggregate of all n-ads at every point of M. The mapping π: $FM \to M$ assigns to each element of FM the point of M at which it is an n-ad. We shall make FM into an $(n + n^2)$-manifold. For each chart x in the maximal atlas of M we define a chart $\bar{x}$ on $\pi^{-1}(\text{dom } x)$ by the following stipulation: If $F = \{F_1, \ldots, F_n\}$ is an n-ad at $P \in \text{dom } x$, the coordinate functions $\bar{x}^1, \ldots, \bar{x}^{n^2+n}$ of $\bar{x}$ satisfy at F the equations:

$$\bar{x}^i(F) = x^i(P) \qquad \bar{x}^{nj+i}(F) = F_j(x^i) \qquad (1 \leq i, j \leq n) \tag{B.2}$$

FM is a manifold with the maximal atlas determined by all such charts $\bar{x}$. On this manifold, π is certainly smooth. We shall now define an action of the general linear group $GL(n, \mathbf{R})$, on the right, on FM. Let $A = (A^i_j)$ be a non-singular $n \times n$ real-valued matrix and let $F = (F_1, \ldots, F_n)$ be an n-ad at $P \in M$. We set $FA = (A^j_1 F_j, \ldots, A^j_n F_j)$—where summation is understood, as usual, over repeated indices. FA is obviously an n-ad at P. The action thus defined evidently meets conditions (i) and (ii) on p. 263. The diffeomorphisms required by condition (iii) can be readily found. For each $P \in M$ there is a chart x defined on a neighbourhood U of P. For each $Q \in U$, $(\partial/\partial x^i|_Q)$ is an n-ad at Q. We define a mapping f of $\pi^{-1}(U)$ onto $GL(n, \mathbf{R})$ by stipulating (i) that f sends $(\partial/\partial x^i|_Q)$ to the identity matrix, and that (ii) for any $F \in \pi^{-1}(U)$ and any $A \in GL(n, \mathbf{R})$, $f(FA) = f(F)A$. The mapping $F \mapsto (\pi(F), f(F))$ is then a diffeomorphism of $\pi^{-1}(U)$ onto $U \times G$ that meets the prescribed requirements. $(FM, M, \pi, GL(n, \mathbf{R}))$ is the principal bundle of n-ads over M. We denote it hereafter by FM. A cross-section of FM is called an n-ad field on M. If such an n-ad field exists, it defines canonic isomorphisms between every pair of tangent spaces of M. Two vectors at different points of M are the canonic images of each other if they are the same linear combination of the local value of the n-ad field at their respective points. Two vectors are said to be *parallel* if they are thus the canonic image of each other. Parallelism is obviously an equivalence. M is said to be *parallelizable* if FM admits a cross-section. Parallelizable n-manifolds are the exception, not the rule. However, the domain of any chart of an n-manifold M is always itself a parallelizable submanifold (p. 93).

As is well-known, $GL(n, \mathbf{R})$ acts transitively on the left on $\mathbf{R}^n$ as a group of linear transformations. We can therefore define, according to the procedure sketched above, the fibre bundle with typical fibre $\mathbf{R}^n$ associated with the principal bundle FM. We denote it by TM. Since usually there is little danger of confusion, we shall denote the projection on TM onto M by π. A typical element of TM is an equivalence class. $[F, \mathbf{r}]$, where $F = (F_1, \ldots, F_n)$ is a n-ad at $\pi(F) \in M$ and $\mathbf{r} = (r^1, \ldots, r^n) \in \mathbf{R}^n$. Since for each $A \in GL(n, \mathbf{R})$, $[F, \mathbf{r}] = [FA, A^{-1}\mathbf{r}]$, $[F, \mathbf{r}]$ is uniquely associated and may indeed be identified with the vector $r^i F_i$ at $\pi(F)$. Through this identification the fibre of π over each $P \in M$ turns out to be the tangent space M_P. The underlying set of TM is the union of all such tangent spaces. TM is called the *tangent bundle* of M. A cross-section of TM is a *vector field* on M (cf. p. 260).

Let $\mathbf{t}^q_r$ be the linear space of (q, r)-tensors on $\mathbf{R}^n$. $GL(n, \mathbf{R})$ acts transitively on the left on $\mathbf{t}^q_r$ in a natural way.[5] We may therefore define the fibre bundle $T^q_r M$ with typical fibre $\mathbf{t}^q_r$ associated with the principal bundle FM. $T^q_r M$ can be identified with the union of the spaces of (q, r)-tensors at all points of M (p. 000) and is called the (q, r)-tensor bundle of M. A cross-section of this bundle is a (q, r)-tensor field on M.

C Linear Connections

(Unless otherwise noted, summation is implied over repeated indices.)

We have seen above that vectors at different points of a parallelizable manifold can be compared with one another and partitioned in a natural way into equivalence classes of parallel vectors. In a non-parallelizable manifold—

e.g. the surface of a sphere—parallelism between distant vectors cannot be defined thus absolutely, but one can usually introduce a concept of relative, path-dependent parallelism which enables one to compare vectors at different points with respect to any particular path joining those points. As we shall see below (p. 273), this notion of relative parallelism can be defined in an n-manifold M if and only if the principal bundle of n-ads over M is parallelizable, i.e., if and only if a concept of absolute parallelism is applicable to FM. Such is indeed the case if M is topologically paracompact, i.e. if every open covering of M has a locally finite open refinement.

Let S be a smooth surface in Euclidean 3-space. Consider two vectors v and w at different points P and Q in S. If S is a plane we naturally say that w is parallel to v if $w = \tau_{*P}(v)$, where τ is the translation that sends P to Q. If S is a developable surface, such as a cylinder or a cone, we say that v and w are parallel if $f_{*P}(v)$ is parallel to $f_{*Q}(w)$, where f is an isometry of a suitable region of S into a plane. If S is not developable we can still make sense of the idea of parallelism between v and w, relatively to a path joining P and Q, as the following example will show. Let S be a sphere and γ a curve in S from P to Q.

Let a plane touch S at P and allow S to roll, without slipping, in such a way that the plane always touches a point on the range of γ until it reaches Q. The vectors v and w can be naturally identified with tangent arrows emerging from the sphere at P and Q. At the beginning of the motion, v coincides with some directed segment v' on the plane; at the end, w coincides with another such segment w'. We say that v and w are parallel *along the curve* γ if and only if v' is parallel to w'. Note that this concept of parallelism depends essentially on the *range* of γ, not on a particular parametrization. Note further that, if $P = Q$ and γ is a closed curve, v need not be parallel to itself along it, indeed v is parallel, along diverse loops, to *different* vectors at P. In particular, if v is tangent to γ, v is self-parallel along γ if and only if γ is a great circle. Indeed, if γ is a great circle, parametrized by arc length, all its tangent vectors are parallel to one another along it. In this sense, great circles have constant directions and are the straightest lines on a circle.

The notion of path-dependent parallelism is extended to a vast class of manifolds by means of the concept of a linear connection. This concept has been defined in several equivalent ways. The definition we shall propose below was given by Ch. Ehresmann (1957), but is ultimately based on the method and ideas of Elie Cartan. Other definitions are easier to spell out, but this one is, in my opinion, the most intuitive and illuminating.[1] We develop first some notions that will be required for it.

1. Vector-valued Differential Forms

Throughout this paragraph, V denotes an m-dimensional real vector space, endowed with the natural differentiable structure defined by its isomorphisms with $\mathbf{R}^m$. We agree to call a list "repetitious" if it contains the same object more than once. Let W be an n-dimensional real vector space. A p-linear mapping $f: W^p \to V$ is said to be *alternating* if every repetitious list in W^p lies in the kernel of f.[2] (Note that according to this definition all 1-linear mappings of W into V are alternating, for there are no repetitious lists in their domain.) Let S be an n-manifold. We agree to call any smooth mapping of S into V a *V-valued 0-form on S.* A V-valued p-form ω on S $(p \geq 1)$ is a correspondence that assigns to each $P \in S$ an alternating p-linear mapping ω_P of S_P into V, and varies smoothly with P. We observe at once that ω determines a mapping—also called ω—of the p-th Cartesian power of the module $\mathscr{V}^1(S)$ of vector fields on S into the set of V-valued 0-forms on S; namely, if $X^1, \ldots, X^p \in \mathscr{V}^1(S)$, $\omega(X^1, \ldots, X^p)$ is the 0-form $P \mapsto \omega_P(X^1_P, \ldots, X^p_P)$. We also note that if $p > n$, ω is identically 0.[3] Let $\varepsilon_{i_1 i_2 \ldots i_p}$ be equal to 1 or to -1 if $(i_1, i_2, \ldots, i_p)$ is, respectively, an even or an odd permutation of $(1, 2, \ldots, p)$, and let it otherwise be 0. If ω_1 is a V-valued p-form and ω_2 is a V-valued q-form on S, their *exterior product* $\omega_1 \wedge \omega_2$ is the $(p+q)$-form whose effect on a list of $p+q$ vector fields $X^i \in \mathscr{V}^1(S)$ is given by:

$$(\omega_1 \wedge \omega_2)(X^1, \ldots, X^{p+q}) = \frac{1}{(p+q)!} \varepsilon_{i_1 i_2 \ldots i_{p+q}} \omega_1(X^{i_1}, \ldots, X^{i_p}) \times \omega_2(X^{i_{p+1}}, \ldots, X^{i_{p+q}}) \quad \text{(C.1.1)}$$

(where the indices i_k range over $\{1, \ldots, p+q\}$). The operation thus defined is evidently associative. It therefore makes sense to speak of the exterior product $\omega_1 \wedge \omega_2 \wedge \ldots \wedge \omega_r$, of any number of V-valued forms on S.

The p-forms on an n-manifold S defined on page 99 are of course $\mathbf{R}$-valued p-forms. On the domain of a chart x any such $\mathbf{R}$-valued p-form ω is equal to a linear combination of all the different p-forms that can be constructed by taking the exterior product of p differentials of the coordinate functions. Symbolically:

$$\omega = \sum_{i_1 < i_2 < \ldots < i_p} f_{i_1 i_2 \ldots i_p} \, \mathrm{d}x^{i_1} \wedge \mathrm{d}x^{i_2} \wedge \ldots \wedge \mathrm{d}x^{i_p} \quad \text{(C.1.2)}$$

There is a unique linear operator $\mathbf{d}$ on the $\mathbf{R}$-valued forms on S, which satisfies the following conditions:

(i) If ω is an $\mathbf{R}$-valued p-form on S, $\mathbf{d}\omega$ is an $\mathbf{R}$-valued $(p+1)$-form on S $(p \geq 0)$;
(ii) $\mathbf{d}(\mathbf{d}\omega) = 0$;
(iii) if f is an $\mathbf{R}$-valued 0-form (i.e. a scalar field) on S, $\mathbf{d}f = \mathrm{d}f$, the ordinary differential of f;
(iv) if ω_1 is an $\mathbf{R}$-valued p-form and ω_2 an $\mathbf{R}$-valued q-form on S, $\mathbf{d}(\omega_1 \wedge \omega_2) = \mathbf{d}\omega_1 \wedge \omega_2 + (-1)^p \omega_1 \wedge \mathbf{d}\omega_2$.

$\mathbf{d}\omega$ is called the *exterior derivative* of ω. On the domain of a chart x, where ω is given by (C.1.2)

$$\mathbf{d}\omega = \sum_{i_1 < i_2 < \ldots < i_p} \mathrm{d}f_{i_1 i_2 \ldots i_p} \wedge \mathrm{d}x^{i_1} \wedge \mathrm{d}x^{i_2} \wedge \ldots \wedge \mathrm{d}x^{i_p} \tag{C.1.3}$$

If $X_1, \ldots, X_{r+1}$ are any vector fields on S,

$$\begin{aligned}(r+1)\,\mathbf{d}\omega(X_1, \ldots, X_{r+1}) &= \sum_{i=1}^{r+1} (-1)^{i+1} X_i \omega(X_1, \ldots, \overline{X}_i, \ldots, X_{r+1}) \\ &+ \sum_{i<j} (-1)^{i+j} \omega(L_{X_i}X_j, X_1, \ldots, \overline{X}_i, \ldots, \overline{X}_j, \ldots, X_{r+1})\end{aligned} \tag{C.1.4}$$

where $L_{X_i}X_j$ stands for the Lie derivative of X_j along X_i,[4] and the symbol $\overline{X}_i$ indicates that X_i is omitted.[5] In particular, if ω is a 1-form and X and Y are two vector fields on S,

$$2\mathbf{d}\omega(X, Y) = X\omega(Y) - Y\omega(X) - \omega(L_X Y) \tag{C.1.5}$$

Let ω be a V-valued p-form on S and let $(e_1, \ldots, e_m)$ be a basis of V. There are then m $\mathbf{R}$-valued p-forms on S, $\omega^1, \ldots, \omega^m$, such that $\omega = \Sigma_i e_i \omega^i$. (In other words, if $X^1, \ldots, X^p$ are vector fields on S, $\omega_P(X^1_P, \ldots, X^p_P) = \Sigma_i e_i \omega^i_P(X^1_P, \ldots, X^p_P)$, at each $P \in S$.) We define the exterior derivative $\mathbf{d}\omega$ of the V-valued p-form ω as

$$\mathbf{d}\omega = \Sigma_i e_i \,\mathbf{d}\omega^i \tag{C.1.6}$$

2. *The Lie Algebra of a Lie Group*

Let V be a vector space. A *Lie bracket* on V is a bilinear mapping $V \times V \to V$, by $(u, v) \mapsto [u, v]$, such that $[u, v] = -[v, u]$, and $[[u, v], w] + [[w, u], v] + [[v, w], u] = 0$. The vector space V, endowed with a Lie bracket, is said to be a *Lie algebra*. Thus, for example, the Lie derivative of a vector field along another (p. 262) is a Lie bracket on the real vector space of vector fields over an n-manifold, and one often writes $[X, Y]$ for $L_X Y$.

Let G be an n-parameter Lie group. G acts on itself on the left transitively and effectively by $(h, g) \mapsto hg$. Thus, each $h \in G$ defines a diffeomorphism of G onto itself, the left translation by h, L_h, which sends each $g \in G$ to hg. A vector field X on G is said to be *left invariant* if $L_{h*}X = X$ for every $h \in G$. The set of left invariant vector fields on G is a real vector space which we shall denote by $\mathfrak{g}$. (The linearity of the differential entails that, if X and Y are left invariant fields on G and $q \in \mathbf{R}$, qX and $X + Y$ are left invariant fields as well.) $\mathfrak{g}$ is in fact a Lie subalgebra of $\mathscr{V}^1(G)$. (For the Lie derivative $[X, Y]$ is left invariant if X and Y are left invariant.) $\mathfrak{g}$ is called the *Lie algebra* of the Lie group G.

We shall now show that $\mathfrak{g}$ is linearly isomorphic to the tangent space G_e (at the neutral element $e \in G$), and is therefore n-dimensional. The mapping of $\mathfrak{g}$

into G_e by $X \mapsto X_e$ is a linear injection. (If $X_e - Y_e = 0$ for $X, Y \in \mathfrak{g}$, then, for any $h \in G$, $X_h - Y_h = L_{h*e}(X_e) - L_{h*e}(Y_e) = L_{h*e}(X_e - Y_e) = 0$—since L_{h*e} is linear; consequently, $X = Y$.) If v is any vector at e, the vector field V, defined on all G by $V_h = L_{h*e}(v)(h \in G)$ is evidently left invariant and such that $V_e = v$. The mapping $X \mapsto X_e$ is therefore surjective, and hence a linear isomorphism. It is indeed an isomorphism of Lie algebras, if the Lie bracket on G_e is defined by $[X_e, Y_e] = [X, Y]_e$. It is often convenient to identify $\mathfrak{g}$ and G_e through this canonic isomorphism.

Let $(X^1, \ldots, X^n)$ be a basis of $\mathfrak{g}$. The Lie bracket $[X^j, X^k]$ of any two elements of this basis is equal to a linear combination $\Sigma_i C^i_{jk} X^i$. The numbers C^i_{jk} are known as the *structure constants* of the Lie group G (relative to the basis (X^i) of $\mathfrak{g}$).

If f is an automorphism of G (i.e. a diffeomorphism of G onto itself that is at the same time a group homomorphism), $f(e) = e$, and f_{*e} is an automorphism of G_e (i.e. a linear isomorphism that preserves the Lie bracket). *Via* the canonic isomorphism described above, f determines a matching automorphism of the Lie algebra $\mathfrak{g}$. In particular, if f is the inner automorphism $g \mapsto hgh^{-1}$, induced by $h \in G$, the matching automorphism of $\mathfrak{g}$ is denoted by $ad(h)$. The mapping $h \mapsto ad(h)$ is a representation of the Lie group G in the linear space $\mathfrak{g}$, known as the *adjoint representation.* If R_h denotes the "right translation of G by h"—i.e. the permutation of G by $g \mapsto gh$—we have that $h^{-1}gh = R_h L_{h^{-1}} g$. Hence, for every $X \in \mathfrak{g}$, $ad(h^{-1})X = R_{h*}X$.

3. *Connections in a Principal Fibre Bundle*

Let E be a differentiable principal fibre bundle over an n-manifold B, with projection π and structure group G. We shall denote the points of E by small italics and write πq for the value of π at $q \in E$. As we know, G acts on E freely on the right (p. 263). Let $\bar{R}$ denote this action. For each $g \in G$ there is a diffeomorphism $\bar{R}_g$ of E onto itself, by $q \mapsto qg$. We often write $\bar{R}_g q$ for qg. If G is m-dimensional, E must be $(n+m)$-dimensional. The fibre of π over any point of B is diffeomorphic to G and is therefore an m-dimensional submanifold of E.

Consider an arbitrary point $q \in E$. The vectors at q tangent to the fibre of π over πq constitute an m-dimensional subspace V_q of the $(n+m)$-dimensional tangent space E_q. We call V_q the space of *vertical vectors*, or the *vertical subspace* at q. Note that V_q is the kernel of the differential π_{*q} of π at q.[6] There is a natural isomorphism of V_q onto the Lie algebra $\mathfrak{g}$ of G. As G acts on E by $\bar{R}$, each left invariant vector field $X \in \mathfrak{g}$ defines a vector field $\bar{X}$ on E. If γ denotes the maximal integral curve of X through the neutral element $e \in G$, the maximal integral curve of $\bar{X}$ through $q \in E$ is given by $t \mapsto q\gamma(t)$. Since the orbits of G's right action are the fibres of π, the value $\bar{X}_q$ of $\bar{X}$ at q is tangent to the fibre over πq and therefore lies in V_q. The mapping $\mathfrak{g} \to V_q$ by $X \mapsto \bar{X}_q$ is a linear injection of an m-dimensional vector space into another and hence an isomorphism. The inverse of this mapping is the natural isomorphism of V_q into $\mathfrak{g}$.

There are many n-dimensional subspaces of E_q which, together with V_q, span E_q. Any such subspace of E_q—i.e., any complement in E_q of the vertical subspace at q—is called a space of *horizontal vectors* or a *horizontal subspace* at q. A *connection* in the principal fibre bundle E is a rule that picks out a horizontal subspace H_q at each $q \in E$, subject to the following two conditions:

(*a*) *smoothness*—H_q depends differentiably on q;
(*b*) *consistency with the right action of G on E*—if $g \in G$, $H_{qg} = \bar{R}_{g*q}(H_q)$.

It is not hard to see that if U is an open submanifold of B, and if we are given a connection in E, the restriction of this connection to $\pi^{-1}(U)$ is a connection in the principal fibre bundle $\pi^{-1}(U)$ over U with structure group G, *induced*—as we say—by the connection in E.

Let $q \mapsto H_q$ be a connection in E. Relative to this connection, a vector at q is said to be *horizontal* if and only if it lies in H_q. Any vector $u \in E_q$ can be analyzed into a vertical component ${}^v u \in V_q$ and a horizontal component ${}^h u \in H_q$, such that $u = {}^v u + {}^h u$. This terminology and notation naturally extends to vector fields on E. If X is such a vector field, we denote by ${}^v\mathrm{X}$ the vector field $q \mapsto {}^v(\mathrm{X}_q)$. ${}^h\mathrm{X}$ is defined analogously. X is said to be *vertical* if $\mathrm{X} = {}^v\mathrm{X}$, *horizontal* if $\mathrm{X} = {}^h\mathrm{X}$. Likewise a curve in E is *vertical* (*horizontal*) if all its tangent vectors are vertical (horizontal). Given a connection in the principal fibre bundle E we can define the *exterior covariant derivative* $\mathrm{D}\alpha$ of a vector-valued r-form α on E. For any vector fields $\mathrm{X}_1, \ldots, X_{r+1}$ on E

$$\mathrm{D}\alpha(\mathrm{X}_1, \ldots, \mathrm{X}_{r+1}) = \mathbf{d}\alpha({}^h\mathrm{X}_1, \ldots, {}^h\mathrm{X}_{r+1}) \tag{C.3.1}$$

If X is a vector field on the base manifold B there is a unique horizontal vector field X^* on E, such that $\pi_*\mathrm{X}^* = \mathrm{X}$. (The differential of π sends the value of X^* at each $q \in E$ to the value of X at πq.) We call X^* the *lift* of X.[7] If γ is a curve in B defined on an interval I, such that $P = \gamma(t)$ for some $t \in I$, there exists for each $p \in \pi^{-1}(P)$ a unique horizontal curve $\hat{\gamma}: I \to E$, such that $\hat{\gamma}(t) = p$.[8] We call $\hat{\gamma}$ the *lift* of γ through p. Let γ join P and Q in B. The lift of γ through $p \in \pi^{-1}(P)$ meets $\pi^{-1}(Q)$ at a definite point q. We say that q is *parallel to p along* γ. By *parallel transport along* γ *from P to Q* we mean the mapping that assigns to each element of $\pi^{-1}(P)$ its parallel along γ in $\pi^{-1}(Q)$. We denote this mapping by $\overset{\gamma}{\tau}_{PQ}$. It is not hard to see that, for any $g \in G$, $\overset{\gamma}{\tau}_{PQ} \cdot \bar{R}_g = \bar{R}_g \cdot \overset{\gamma}{\tau}_{PQ}$. Hence, $\overset{\gamma}{\tau}_{PQ}$ is a diffeomorphism of the fibre of π over P onto its fibre over Q.

The notions of lift and parallel transport are readily extended to any fibre bundle E' with typical fibre F, associated with the principal bundle (E, B, π, C), if the latter is endowed with a connection. Since there is little danger of confusion I denote the projections of both E and E' by π. We recall that each point of E' can be equated with an equivalence class $[q, s]$, with $q \in E$ and $s \in F$ (p. 264). We denote by f_s the mapping of E onto E' by $q \mapsto [q, s]$. Let $[p, t] \in E'$.

The vertical subspace $V_{[p,t]} \subset E'_{[p,t]}$ is simply the space of vectors at $[p,t]$ tangent to the fibre of π over $\pi[p,t]$. We define the horizontal subspace $H_{[p,t]} \subset E'_{[p,t]}$ as the image of $H_p \subset E_p$ by the differential of f_t:

$$H_{[p,t]} = f_{t*}(H_p) \tag{C.3.2}$$

It can be easily proved that $H_{[p,t]}$ is well defined—i.e. that it is independent of the particular choice of $(p,t) \in [p,t]$— and that $H_{[p,t]}$ and $V_{[p,t]}$ together span $E'_{[p,t]}$ and have only the zero vector in common. A curve $\hat{\gamma}$ in E' is the lift of a curve γ in B if $\gamma = \pi \cdot \hat{\gamma}$. If γ goes through $P \in B$ there is a unique lift of γ through each point in the fibre of π over P. Parallelism and parallel transport along γ can therefore be defined in E' just as in E. A cross-section σ of E' is said to be *parallel* (relative to the given connection) if the differential σ_*, at each $P \in B$, maps the tangent space B_P onto the horizontal subspace at $\sigma(P) \in E'$. If σ is a parallel cross-section of E' and γ is any curve in B through, say, $Q \in B$, $\sigma \cdot \gamma$ is clearly the lift of γ through $\sigma(Q)$. Hence, all points of $\sigma \cdot \gamma$ are parallel along γ by pairs.

Let $\bar{\omega}_q$ denote the natural isomorphism of the vertical subspace $V_q \subset E_q$ onto the Lie algebra $\mathfrak{g}$ of G (p. 269). $\bar{\omega}_q$ is extended to a definite linear mapping ω_q of E_q into $\mathfrak{g}$ by stipulating that $\ker \omega_q = H_q$. Due to the smoothness of the connection $q \mapsto H_q$, the correspondence $\omega: q \mapsto \omega_q$ depends smoothly on q and is therefore a $\mathfrak{g}$-valued 1-form on E. ω is called the *connection form.* Because the connection is consistent with $\bar{R}$, we have that, for each $g \in G$, $q \in E$ and $v \in E_q$,

$$\omega_{qg}(\bar{R}_{g*q} v) = ad\,(g^{-1})\,\omega_q(v) \tag{C.3.3}$$

On the other hand, if ω is any $\mathfrak{g}$-valued 1-form on E that satisfies (C.3.3) and is such that, for each $q \in E$, ω_q agrees on V_q with the natural isomorphism $\bar{\omega}_q$, the correspondence $q \mapsto \ker \omega_q$ is a connection in E.

The *curvature form* of a given connection in E with connection form ω is the $\mathfrak{g}$-valued 2-form Ω on E that is the exterior covariant derivative of ω:

$$\Omega = \mathrm{D}\omega \tag{C.3.4}$$

The connection and curvature forms satisfy the following *equation of structure*: for any vector fields X, Y on E

$$\mathbf{d}\omega(\mathrm{X}, \mathrm{Y}) = -\tfrac{1}{2}[\omega(\mathrm{X}), \omega(\mathrm{Y})] + \Omega(\mathrm{X}, \mathrm{Y}) \tag{C.3.5}$$

Recalling the formula for the exterior derivative of a 1-form given on p. 268, we see that if X and Y are horizontal and $L_X\mathrm{Y}$ stands for the Lie derivative of Y along X, $\omega(L_X\mathrm{Y}) = -2\Omega(\mathrm{X}, \mathrm{Y})$. The curvature form Ω thus measures how the Lie derivative of an horizontal vector field along another departs from horizontality. If the curvature form is identically 0 the connection is said to be *flat*.

The curvature form Ω satisfies the *Bianchi identity*:

$$\mathrm{D}\Omega = 0 \tag{C.3.6}$$

In order to prove it, we choose a basis (e_i) of the Lie algebra $\mathfrak{g}$ and set $\omega = \omega^i e_i$ and $\Omega = \Omega^i e_i$. Let c^i_{jk} be the structure constants of the structure group G relative to the basis (e_i) (p. 269). The equation of structure (C.3.5) is then expressed by the system

$$\mathbf{d}\omega^i = -\tfrac{1}{2} c^{\,i}_{jk}\, \omega^j \wedge \omega^k + \Omega^i \tag{C.3.7}$$

Consequently, by (ii) and (iv) on pp. 267.

$$0 = \mathbf{dd}\omega^i = -\tfrac{1}{2} c^{\,i}_{jk}\, (\mathbf{d}\omega^j \wedge \omega^k - \omega^j \wedge \mathbf{d}\omega^k) + \mathbf{d}\Omega^i \tag{C.3.8}$$

Since all horizontal vector fields on E are sent to zero by the connection form ω and hence also by the forms ω^i, $\mathbf{d}\Omega^i(\mathrm{X}, \mathrm{Y}, \mathrm{Z}) = 0$ whenever X, Y and Z are horizontal. It follows at once from this and (C.3.1) that $\mathrm{D}\Omega$ vanishes identically.

4. *Linear Connections*

Let S be an n-manifold and let FS denote the principal bundle of n-ads over S (p. 264). As usual we denote the projection by π. A connection in FS is called a *linear connection in S*.

Each $q \in FS$ is an ordered basis $(q_1, \ldots, q_n)$ of the tangent space $S_{\pi q}$. Consequently, if v is any vector at πq, $v = v^i q_i$ for some list of real numbers $(v^1, \ldots, v^n)$, called the *components* of v relative to q. The mapping of $S_{\pi q}$ into $\mathbf{R}^n$ that assigns to each vector at πq the list of its components relative to q is a linear isomorphism, which we shall also designate by q.

The *canonical form* θ of FS is the $\mathbf{R}^n$-valued 1-form on FS defined by

$$\theta_q(b) = q \cdot \pi_* b \tag{C.4.1}$$

where b is any vector at $q \in FS$. Note that $\theta_q(b) = 0$ if and only if b is vertical. (Recall that, as π maps FS onto S, its differential π_* maps the tangent bundle of FS onto that of S; all vectors tangent to the fibre of π over some $P \in S$ are sent by π_* to the zero vector in S_P.)

Let us henceforth assume that $\Gamma: q \mapsto H_q (q \in FS)$ is a linear connection in S. We denote respectively by ω and Ω the connection form and the curvature form of Γ. ω and Ω take their values in the Lie algebra $\mathfrak{gl}(n, \mathbf{R})$ of the structure group $GL(n, \mathbf{R})$. Let E^i_j denote the $n \times n$ matrix that has a 1 at the intersection of the i-th column and the j-th row, and a 0 everywhere else. The array of all such matrices is obviously a basis of $\mathfrak{gl}(n, \mathbf{R})$. We may therefore set $\omega = \omega^j_i E^i_j$, where the ω^j_i are n^2 real-valued 1-forms on FS. Let x be a chart defined on $U \subset S$. There is an n-ad field on U which assigns to each $P \in U$ the n-ad $(\partial/\partial x^i|_P)_{1 \leq i \leq n}$. We denote this n-ad field by $(\partial/\partial x)$. Set $\hat{\omega}^i_j = (\partial/\partial x)^* \omega^i_j$

(where $(\partial/\partial x)^*$ stands for the pull-back of $(\partial/\partial x)$, as defined on p. 316, n. 12). The $\hat{\omega}^i_j$ are real-valued 1-forms on U and can therefore be expressed as linear combinations of the 1-forms dx^i:

$$\hat{\omega}^i_j = \Gamma^i_{kj} dx^k \tag{C.4.2}$$

The n^3 scalar fields Γ^i_{kj} are called, with some impropriety, the *components* of the linear connection Γ relative to the chart x. If y is another chart, defined on $U' \subset S$, and the components of Γ relative to y are denoted by Γ'^i_{jk}, we have that, on $U \cap U'$,

$$\Gamma'^a_{bc} = \Gamma^i_{jk} \frac{\partial y^a}{\partial x^i} \frac{\partial x^j}{\partial y^b} \frac{\partial x^k}{\partial y^c} + \frac{\partial y^a}{\partial x^i} \frac{\partial^2 x^i}{\partial y^b \partial y^c} \tag{C.4.3}$$

If a set of n^3 scalar fields (Γ^i_{jk}) is defined on the domain of each chart x of S, so that the diverse sets corresponding to the different charts are mutually related by (C.4.3), there is demonstrably a definite linear connection Γ in S whose components relative to any particular chart of S are the respective Γ^i_{jk}.[9]

The *torsion form* Θ of the linear connection is the $\mathbf{R}^n$-valued 2-form on FS which is the exterior covariant derivative of the canonical form θ:

$$\Theta = \mathrm{D}\theta \tag{C.4.4}$$

The torsion form Θ is related to the canonical form θ and the connection form ω by the *equation of structure*:

$$\mathbf{d}\theta(\mathrm{X}, \mathrm{Y}) = -\tfrac{1}{2}(\omega(\mathrm{X})\theta(\mathrm{Y}) - \omega(\mathrm{Y})\theta(\mathrm{X})) + \Theta(\mathrm{X}, \mathrm{Y}) \tag{C.4.5}$$

where X and Y are any two vector fields on FS. It can be shown that

$$\mathrm{D}\Theta = \Omega \wedge \theta \tag{C.4.6}$$

(Compare equations (C.4.4–6) with (C.3.4–6).)

For each $\mathbf{r} \in \mathbf{R}^n$ there is a unique horizontal vector $\mathrm{B}^{\mathbf{r}}_q$ at $q \in FS$, such that $\theta_q(\mathrm{B}^{\mathbf{r}}_q) = \mathbf{r}$. The mapping $q \mapsto \mathrm{B}^{\mathbf{r}}_q$ is a horizontal vector field on FS, called the *standard horizontal field* corresponding to $\mathbf{r}$, which we denote by $\mathrm{B}^{\mathbf{r}}$. It can be easily shown that if $(\mathbf{e}_1, \ldots, \mathbf{e}_n)$ is a basis of $\mathbf{R}^n$, the standard horizontal fields $\mathrm{B}^{\mathbf{e}_1}, \ldots, \mathrm{B}^{\mathbf{e}_n}$ are linearly independent. Through the natural isomorphism of the vertical subspaces of FS onto the n^2-dimensional Lie algebra $\mathfrak{gl}(n, \mathbf{R})$ (p. 269), we can define n^2 linearly independent vertical vector fields on FS. Together with the n linearly independent standard horizontal fields corresponding to any given basis of $\mathbf{R}^n$ the said vertical fields form a basis of the module of vector fields on FS, which, if ordered, constitutes an (n^2+n)-ad field on FS. Consequently, the existence of a linear connection in a manifold S entails that the principal bundle of n-ads FS is parallelizable. On the other hand, if FS is parallelizable we can always find an (n^2+n)-ad field on FS with exactly n^2 vertical elements, and use the remaining n non-vertical ones for defining a connection. (Let the n-space spanned by the local value of these n non-vertical vector fields at each $q \in FS$ be the horizontal space H_q assigned to q by the connection.)

5. *Covariant Differentiation*

Let Γ be a linear connection in the n-manifold S. If V is a vector field and W a (q,r)-tensor field defined on some neighbourhood of $P \in S$, the *covariant derivative* $\nabla_{\mathrm{V}}\mathrm{W}$ of W along V at P is given by:

$$\nabla_{\mathrm{V}}\mathrm{W}|_P = \lim_{t \to 0} \frac{1}{t}(\tau^{\gamma}_{(\gamma t)(\gamma 0)}(\mathrm{W}_{\gamma t}) - \mathrm{W}_{\gamma 0}) \qquad \text{(C.5.1)}$$

where γ denotes the maximal integral curve of V through P (p. 000), γt abbreviates $\gamma(t)$, and τ signifies parallel transport in the (q, r)-tensor bundle of S (pp. 265, 270). $\nabla_{\mathrm{V}}\mathrm{W}$ is plainly a (q, r)-tensor field defined on the intersection of the domains of V and W. It can be shown that, if W is also a vector field, like V,

$$\nabla_{\mathrm{V}}\mathrm{W}|_P = q^{-1}(\mathrm{V}^*_q(\theta(\mathrm{W}^*))) \qquad \text{(C.5.2)}$$

where V* and W* denote the lifts of V and W, respectively; q is any element of $\pi^{-1}(P) \subset FS$; q^{-1} is the inverse of the linear isomorphism that sends each vector at $P = \pi q$ to its components relative to q, and θ is the canonical form of FS. We stipulate, moreover, that for any scalar field f,

$$\nabla_{\mathrm{V}} f = \mathrm{V} f \qquad \text{(C.5.3)}$$

It can be shown that, for any vector fields X and Y, any (q,r)-tensor fields W and W′, and any scalar field f, defined on S or on an open submanifold of S, the operator ∇ defined by (C.5.1) satisfies the following four conditions:

$$\begin{aligned} &\text{(i)}\ \nabla_{\mathrm{X+Y}}\mathrm{W} &&= \nabla_{\mathrm{X}}\mathrm{W} + \nabla_{\mathrm{Y}}\mathrm{W} \\ &\text{(ii)}\ \nabla_{\mathrm{X}}(\mathrm{W}+\mathrm{W}') &&= \nabla_{\mathrm{X}}\mathrm{W} + \nabla_{\mathrm{X}}\mathrm{W}' \\ &\text{(iii)}\ \nabla_{f\mathrm{X}}\mathrm{W} &&= f\nabla_{\mathrm{X}}\mathrm{W} \\ &\text{(iv)}\ \nabla_{\mathrm{X}} f\mathrm{W} &&= f\nabla_{\mathrm{X}}\mathrm{W} + (\mathrm{X}f)\mathrm{W}. \end{aligned} \qquad \text{(C.5.4)}$$

On the other hand, the existence of an operator ∇ that meets the above conditions determines a unique linear connection in S, relative to which ∇ satisfies (C.5.1). It can be proved, moreover, that such an operator ∇ is fully defined for all types of tensor fields on S, if it is defined for arbitrary *vector fields* on S, X, Y, W and W′, and for arbitrary scalar fields in agreement with (C.5.3). One may thus endow a manifold S with a definite linear connection by introducing such an operator in $\mathscr{F}(S)$ and $\mathscr{V}^1(S)$. The one–one correspondence between linear connections and covariant differential operators evidently authorizes us to call the former by the name of the latter, as it is often done.

The (absolute) *covariant derivative* of the (q, r)-tensor field W—often called the *covariant differential* of W—is the $(q, r+1)$-tensor field ∇W defined as follows: if X, $X_1, \ldots, X_r$ are any vector fields and $\omega_1, \ldots, \omega_q$ are any covector fields or 1-forms on the domain of W,

$$\nabla W(\omega_1, \ldots, \omega_q, X_1, \ldots, X_r, X) = (\nabla_X W)(\omega_1, \ldots, \omega_q, X_1, \ldots, X_r) \tag{C.5.5}$$

In view of (C.5.1) and our remark about parallel cross-sections on p. 271, it is clear that the (q, r)-tensor field W is a parallel cross-section of the (q, r)-tensor bundle over its domain in S (relative to the linear connection given by ∇) only if $\nabla_X W = 0$ for every vector field X on the domain of W. (C.5.5) entails then that W is parallel, relative to ∇, if and only if $\nabla W = 0$.

Let x be a chart defined on $U \subset S$. As we know, any vector field on U is equal to a linear combination of the vector fields $\partial/\partial x^i$. In particular, it can be shown that, if we allow ∇_k to signify covariant differentiation along $\partial/\partial x^k$,

$$\nabla_k \partial/\partial x^j = \Gamma^i_{kj} \partial/\partial x^i \qquad \text{(C.5.6)[10]}$$

where the Γ^i_{kj} are the components, relative to x, of the linear connection uniquely associated with ∇. From (C.5.4–6) one derives an expression for the components, relative to x, of the covariant derivative ∇W of a (q, r)-tensor field W. For example, if W is a vector field equal (on U) to $W^i \partial/\partial x^i$, and we agree to write $W^i_{;j}$ for $\nabla W(dx^i, \partial/\partial x^j)$, it will be readily seen that

$$W^i_{;j} = \frac{\partial W^i}{\partial x^j} + W^k \Gamma^i_{jk} \tag{C.5.7}$$

(Note, by the way, that on U, $\nabla_j W = W^i_{;j} \partial/\partial x^i$; so that the (i, j)-th component, relative to a chart x, of the absolute covariant derivative of a vector field is equal to the i-th component, relative to that chart, of its covariant derivative along $\partial/\partial x^j$.) The general expression, for a (q, r)-tensor field W, is given by equation (5.5.1) on p. 155.

6. *The Torsion and Curvature of a Linear Connection*

Let Γ be a linear connection in the n-manifold S. Its torsion form Θ and its curvature form Ω determine respectively, a (1, 2)-tensor field on S, called the *torsion* and denoted by T, and a (1, 3)-tensor field on S, called the *curvature* and denoted by R. The definition of T and R can be made simpler and more perspicuous if we first observe that any $(q, 1)$-tensor field $Q (q \geq 1)$ on a manifold S can be regarded as a q-linear mapping of $(\mathscr{V}^1(S))^q$ into $\mathscr{V}^1(S)$, insofar as Q assigns to each list of q vector fields on S, $X_1, \ldots, X_q$, the linear mapping of $\mathscr{V}_1(S)$ into $\mathbf{R}$ by $\omega \mapsto Q(\omega, X_1, \ldots, X_q)$. This linear mapping, which belongs to $\mathscr{V}^1(S)$—the dual of $\mathscr{V}_1(S)$—is usually denoted by $Q(X_1, \ldots, X_q)$. By the same token, if $q > 1$, Q can also be regarded as a

$(q-1)$-linear mapping of $(\mathscr{V}^1(S))^{q-1}$ into the linear space of linear mappings of $\mathscr{V}^1(S)$ into itself, inasmuch as Q assigns to each list of $q-1$ vector fields on S, $X_1, \ldots, X_{q-1}$, the mapping $X \mapsto Q(X, X_1, \ldots, X_{q-1})$. This mapping is usually denoted by $Q(X_1, \ldots, X_{q-1})$. We write $Q(X_1, \ldots, X_{q-1})X$ for the value of this mapping at $X \in \mathscr{V}^1(S)$, i.e. for the vector field $Q(X, X_1, \ldots, X_{q-1})$. Though the symbol Q is used hereby with deliberate and seemingly perverse ambiguity, no confusion will arise if due attention is paid to the nature and length of the list of objects to its right.

The (1, 2)-tensor field T and the (1, 3)-tensor field R will be fully determined if we give the value at any $P \in S$ of $T(X, Y)$ and $R(X, Y)Z$, for arbitrary vector fields X, Y and Z, defined on a neighbourhood of P. We set

$$T(X, Y)|_P = q^{-1}(2\Theta_q(X^*, Y^*)) \tag{C.6.1}$$

$$R(X, Y)Z|_P = q^{-1}(2\Omega_q(X^*, Y^*)(qZ_P)) \tag{C.6.2}$$

We use here the same nomenclature as in equation (C.5.2), with q an arbitrary n-ad at P; we write qZ_P for the value of q at Z_P—i.e. for the list of Z_P's components relative to q—; $2\Omega_q(X^*, Y^*)(qZ_P)$ is then the image of qZ_P by the linear mapping $2\Omega_q(X^*, Y^*)$ of $\mathbf{R}^n$ into itself (recall that the curvature form Ω is a $\mathfrak{gl}(n, \mathbf{R})$-valued 2-form). Our stipulations only make sense if the right-hand sides of (C.6.1) and (C.6.2) remain constant as q ranges over $\pi^{-1}(P)$. The proof is not difficult, but it would take up more space than we can afford here.[11]

Due to the skew-symmetry of the curvature form Ω we have that

$$T(X, Y) = -T(Y, X) \tag{C.6.3}$$

and

$$R(X, Y)Z = -R(Y, X)Z \tag{C.6.4}$$

It can also be proved that

$$\sigma(R(X, Y)Z) = \sigma(T(T(X, Y), Z) + (\nabla_X T)(Y, Z)) \qquad \text{(C.6.5)}^{12}$$

where σ signifies the cyclic sum with respect to the three variables X, Y, Z. (Thus, $\sigma(R(X, Y)Z)$ stands for $R(X, Y)Z + R(Y, Z)X + R(Z, X)Y$.)

From (C.6.1–2) and (C.5.4) one infers that

$$T(X, Y) = \nabla_X Y - \nabla_Y X - [X, Y] \tag{C.6.6}$$

and

$$R(X, Y)Z = \nabla_X \nabla_Y Z - \nabla_Y \nabla_X Z - \nabla_{[X, Y]} Z \tag{C.6.7}$$

where $[X, Y]$ denotes the Lie derivative of Y along X. The components of T and R relative to a chart x can be obtained by straightforward computation from the above formulae and (C.5.6). If the components of the linear connection relative to x are denoted, as usual, by Γ^i_{jk}, the components of the torsion and the curvature are, respectively, given by:

$$T^i_{jk} = \Gamma^i_{jk} - \Gamma^i_{kj} \tag{C.6.8}$$

$$\mathrm{R}^{i}_{jkh} = \partial\Gamma^{i}_{hj}/\partial x^{k} - \partial\Gamma^{i}_{kj}/\partial x^{h} + \Gamma^{s}_{hj}\Gamma^{i}_{ks} - \Gamma^{s}_{kj}\Gamma^{i}_{hs} \tag{C.6.9}$$

Direct calculation also yields the following two interesting results: if f is a scalar field and V is a vector field in the domain of x,

$$f_{;kj} - f_{;jk} = \mathrm{T}^{i}_{kj} f_{;i} \tag{C.6.10}$$

$$\mathrm{V}^{i}_{\ ;kj} - \mathrm{V}^{i}_{\ ;jk} = \mathrm{R}^{i}_{\ hjk}\mathrm{V}^{h} + \mathrm{T}^{h}_{kj}\mathrm{V}^{i}_{\ ;h} \tag{C.6.11}$$

where $f_{;kj}$ stands for $(\nabla\nabla f)(\partial/\partial x^{k}, \partial/\partial x^{j})$, etc.

If the torsion $\mathrm{T} = 0$ everywhere, the linear connection is said to be *torsion-free* or *symmetric*. (The latter name stems from the fact that, by (C.6.8), T vanishes on the domain of a chart if and only if the respective connection components are symmetric in the two lower indices.) The linear connection is demonstrably *flat*, in the sense defined on p. 271, if and only if the curvature R is identically 0.

7. *Geodesics*

Throughout this section, S is an n-manifold with linear connection Γ. The concept of a geodesic extends to the vastly arbitrary structure (S, Γ) the intuitive idea of a straight or straightest line, a line that keeps its direction through its entire course. A curve γ in S is a *geodesic* (of the linear connection Γ) if all vectors tangent to γ are parallel to each other along γ. More formally: Let $\gamma : I \to S$ be a curve in S. Let X be a field of tangent vectors on γ (i.e. let X be a vector field defined on a neighbourhood of each point in γ's range $\gamma(I)$, such that $\mathrm{X}_{\gamma t} = \gamma_{*t}$ for each $t \in I$). γ is a *geodesic of the linear connection* if and only if $\nabla_{\mathrm{X}}\mathrm{X}$ vanishes everywhere on the range of γ, that is, if and only if

$$\nabla_{\mathrm{X}}\mathrm{X}|_{\gamma t} = 0 \tag{C.7.1}$$

for each value of the parameter t. If z is a chart defined on a neighbourhood of $\gamma(I)$, it is clear that, on $\gamma(I)$, $\mathrm{X} = (\mathrm{d}z^{i}\cdot\gamma/\mathrm{d}t)\partial/\partial z^{i}$. From (C.7.1) and (C.5.4) we infer then that γ is a geodesic if and only if it satisfies the *geodesic equations*:

$$\frac{\mathrm{d}^{2}z^{i}\cdot\gamma}{\mathrm{d}t^{2}} + (\Gamma^{i}_{jk}\cdot\gamma)\frac{\mathrm{d}z^{j}\cdot\gamma}{\mathrm{d}t}\frac{\mathrm{d}z^{k}\cdot\gamma}{\mathrm{d}t} = 0 \tag{C.7.2}$$

where the Γ^{i}_{jk} are the components of the linear connection relative to z.

Let γ' be a reparametrization of the geodesic γ, i.e., let $\gamma = \gamma'\cdot f$, for some homeomorphism f of **R** onto itself. If Y is a field of tangent vectors on γ' and s denotes the new parameter, we generally have that

$$\nabla_{\mathrm{Y}}\mathrm{Y}|_{\gamma' s} = \lambda(s)\mathrm{Y}|_{\gamma' s} \tag{C.7.3}$$

for some real-valued function λ. This implies that, if s_1 and s_2 are two values of the parameter s, γ'_{*s_1} is linearly dependent on the image of γ'_{*s_2} at $\gamma(s_1)$ by parallel transport along γ, though it need not be strictly equal to it. Thus, every reparametrization of the geodesic γ may be said to "point constantly in the same direction", even if it is not a geodesic according to our definition. Of course, if γ' is a geodesic, i.e. if the function λ is identically 0, γ_{*s_1} is *the* parallel along γ to γ_{*s_2} at $\gamma(s_1)$—in our sense of this expression. Such will be the case if and only if f is an affine transformation—at least on the domain of γ—i.e. if and only if $s = \alpha t + \beta (\alpha, \beta \in \mathbf{R}; \alpha \neq 0)$. s is then said to be an *affine parameter*.

A curve γ in S is a geodesic if and only if any lift of γ is an integral curve of one of the standard horizontal fields of FS (p. 273). This fundamental theorem entails that for each point P in S and each vector v in S_P there is a unique geodesic γ such that $\gamma(0) = P$ and $\gamma_{*0} = v$. We call it the geodesic with *initial condition* (P, v).

The linear connection in S is said to be *complete* if all its geodesics can be extended to geodesics defined on $\mathbf{R}$. The fundamental theorem stated at the beginning of the foregoing paragraph entails that the linear connection is complete if and only if every standard horizontal field of FS is complete in the sense defined on p. 260.

If the connection Γ is complete there is for each $P \in S$ a mapping Exp_P of S_P into S, called the *exponential map* at P. This is defined as follows: for each $v \in S_P$, let $\hat{v}$ denote the geodesic with initial condition (P, v); the

$$\text{Exp}_P(v) = \hat{v}(1) \tag{C.7.4}$$

If Γ is not complete, the exponential map Exp_P is defined by (C.7.4) only on the set of vectors v at P, such that the geodesics $\hat{v}$ with initial condition (P, v) are defined at least on $[0, 1]$. It is not hard to see that this set constitutes a neighbourhood of 0 in S_P. In any case, there is a neighbourhood of 0 in S_P, included in the domain of Exp_P, which Exp_P maps *diffeomorphically* onto a neighbourhood of P in S. On the latter neighbourhood, Exp_P has an inverse, which we denote by Exp_P^{-1}. If q denotes an n-ad at P, and hence also the respective isomorphism of S_P into $\mathbf{R}^n$ (p. 272), $q \cdot \text{Exp}_P^{-1}$ is a chart defined on the said neighbourhood of P. Such a chart is called a *normal chart* or a *system of normal coordinates* at P. Let the components of the linear connection relative to a normal chart at P be denoted by Γ^i_{jk}. It can then be shown that $\Gamma^i_{jk}(P) = -\Gamma^i_{kj}(P)$. Hence, if Γ is symmetric, the Γ^i_{jk} vanish at P.[13]

Though each linear connection in S determines its set of geodesics, the same set can pertain to diverse connections. Indeed, we have the following important result. Let Γ be as hitherto an arbitrary linear connection in S. Let t be a fixed real number such that $0 < t \leq 1$. For every component Γ^i_{jk} of Γ relative to each chart x of S define, on dom x, a scalar field $\Gamma'^i_{jk} = t\Gamma^i_{jk} + (1-t)\Gamma^i_{kj}$. The arrays (Γ'^i_{jk}) corresponding to each chart x determine a linear connection Γ' on S (p. 273). The geodesic equations (C.7.2) imply that Γ and Γ' share the same set of geodesics. Observe, in particular, that by choosing $t = 1/2$ we can obtain a symmetric connection that has the same geodescics as a given linear connection.

8. Metric Connections in Riemannian Manifolds

Let S be an n-manifold with a proper (i.e. positive definite) Riemannian metric $\mathbf{g}$ (pp. 93, 100). An *orthonormal n-ad* at $P \in S$ is an ordered basis $(v_1, \ldots, v_n)$ of S_P such that

$$\mathbf{g}_P(v_i, v_j) = \delta_{ij} \tag{C.8.1}$$

(that is, 1 if $i = j$ and 0 if $i \neq j$). The set of all orthonormal n-ads at each $P \in S$ is made in an obvious way into a principal fibre bundle OS over S with structure group $O(n)$—the orthogonal group. (Cf. the construction of FS on p. 264.) For each connection Γ' in $(OS, S, O(n))$ there is one and only one connection Γ in $(FS, S, GL(n, \mathbf{R}))$ such that the canonical injection of OS into FS sends the horizontal subspaces of Γ' into the horizontal subspaces of Γ.[14] A linear connection in the Riemannian manifold S is said to be *metric* if it is the unique connection in FS which is thus determined by some connection in OS. It can be shown that a linear connection Γ in S is metric if and only if the Riemannian metric $\mathbf{g}$ is parallel relative to Γ, in the sense defined on p. 271 (Recall that $\mathbf{g}$ is a cross-section of the (0, 2)-tensor bundle over S.) Hence, by a remark on p. 275, Γ is metric if and only if

$$\nabla \mathbf{g} = 0 \tag{C.8.2}$$

(where ∇ signifies covariant differentiation in (S, Γ)).

Among the metric connections that can be defined in an n-manifold S with proper Riemannian metric $\mathbf{g}$ there is one and only one that is symmetric or torsion-free. It is called the *Levi–Cività connection* of $(S, \mathbf{g})$. The geodesics of this connection, in the sense defined on p. 277, are precisely the Riemannian geodesics of $(S, \mathbf{g})$, in the sense defined on p. 94. Let x be a chart of S. Let Γ denote the Levi–Cività connection of $(S, \mathbf{g})$. Its components relative to x are given by:

$$\Gamma^i_{jk} = \tfrac{1}{2} g^{ih}(\partial g_{hj}/\partial x^k + \partial g_{kh}/\partial x^j - \partial g_{jk}/\partial x^h) \tag{C.8.3}$$

Here the $g_{ij} = \mathbf{g}(\partial/\partial x^i, \partial/\partial x^j)$ are the components of $\mathbf{g}$ relative to x, while the g^{ij} are defined by the relations

$$g^{ij} g_{jk} = \delta_{ik} \tag{C.8.4}$$

By definition, the torsion of Γ is identically 0. Its curvature R is demonstrably equal to the (1, 3)-Riemann tensor of $(S, \mathbf{g})$. More precisely, if $\bar{\mathrm{R}}$ denotes the covariant tensor field on S constructed from the metric $\mathbf{g}$ in the manner proposed by Riemann in 1861 (p. 145), and X, Y, Z and W are any four vector fields on S,

$$\bar{\mathrm{R}}(\mathrm{W}, \mathrm{Z}, \mathrm{X}, \mathrm{Y}) = \mathbf{g}(\mathrm{W}, \mathrm{R}(\mathrm{X}, \mathrm{Y})\mathrm{Z}) \tag{C.8.5}$$

Thus, the (0, 4) Riemann tensor $\bar{\mathrm{R}}$ is none other than the covariant form of the curvature R (p. 100). If, as is customary, we look on two tensor fields related to one another as $\bar{R}$ is to R by (C.8.5), as two manifestations of the same geometrical object, we must obviously identify the Riemann tensor of a Riemannian manifold with the curvature of its Levi–Cività connection. We therefore use the same letter R to denote them both. (C.6.4) implies that

$$R(W, Z, X, Y) = -R(W, Z, Y, X) \tag{C.8.6}$$

Using (C.6.2) and (C.8.5) one proves that

$$R(W, Z, X, Y) = -R(Z, W, X, Y) \tag{C.8.7}$$

As the Levi–Cività connection is torsion-free, (C.6.5) leads at once to the *first Bianchi identity*:

$$R(W, X, Y, Z) + R(W, Y, Z, X) + R(W, Z, X, Y) = 0 \tag{C.8.8}$$

Equations (C.8.6–8) yield, by straightforward calculation, that

$$R(W, Z, X, Y) = R(X, Y, W, Z) \tag{C.8.9}$$

The *second Bianchi identity*,

$$\nabla_X R(Y, Z)W + \nabla_Y R(Z, X)W + \nabla_Z R(X, Y)W = 0 \tag{C.8.10}$$

can be derived from the Bianchi identity (C.3.6), which is automatically satisfied by the curvature form of the Levi–Cività connection.

The covariant *Ricci tensor* is the contraction of the curvature with respect to its only contravariant and its third covariant indices. The *curvature scalar* is the contraction of the mixed Ricci tensor.[15]

The above results can be extended to Riemannian manifolds with indefinite metrics. In particular, all statements in this Section 3.8 hold good for an n-manifold S with a Lorentz metric **g** (i.e. with a metric of signature $2-n$; see page 93), provided that we replace δ_{ij} in equation (C.8.1) by 1 if $i = j = 0$, and by $-\delta_{ij}$ otherwise, and that we substitute $L(n)$ for $O(n)$ and LS for OS. Here $L(n)$ denotes the general homogeneous Lorentz group, which we can define as follows: Let H be the $n \times n$ matrix with its first diagonal element equal to 1, its other diagonal elements equal to -1, and all other elements equal to 0; $L(n)$ is the subgroup of $GL(n, \mathbf{R})$ represented by all non-singular matrices A which satisfy the equation

$$A^T H A = H \tag{C.8.11}$$

(where A^T is the transpose of A).

In the light of our remark at the end of Section C.6 (p. 277) it is clear that a Riemannian manifold $(S, \mathbf{g})$ if *flat* in the sense of page 145 if and only if the Levi–Cività connection Γ of $(S, \mathbf{g})$ is *flat* in the sense of page 271. In other words: all sectional curvatures vanish at each point of S if and only if the curvature form of Γ is identically zero.

D Useful Formulae

I collect here some formulae for easy reference. Definitions have been given elsewhere. Indices range from 0 to 3. Repeated indices imply summation over their range.

Let g_{ij} be the metric components relative to a particular chart x of a relativistic spacetime. Let g be the determinant of the matrix (g_{ij}). If G_{ij} is the cofactor of g_{ij} in g, we write g^{ij} for G_{ij}/g.

A comma followed by one or more indices will signify partial differentiation with respect to the corresponding coordinate functions. For example, $g_{ij,kh} = \partial^2 g_{ij}/\partial x^k \partial x^h$.

The components of the Levi–Cività connection relative to x are:

$$\Gamma^i_{jk} = \tfrac{1}{2} g^{ih}(g_{hj,k} + g_{kh,j} - g_{jk,h}) \tag{D.1}$$

The components of the $(1, 3)$ Riemann tensor are:

$$R^h_{ijk} = \Gamma^h_{ki,j} - \Gamma^h_{ji,k} + \Gamma^s_{ki}\Gamma^h_{js} - \Gamma^s_{ji}\Gamma^h_{ks} \tag{D.2}$$

The components of the covariant Riemann tensor are:

$$\begin{aligned} R_{hijk} &= g_{hs} R^s_{ijk} \\ &= \tfrac{1}{2}(g_{hk,ij} + g_{ij,hk} - g_{hj,ik} - g_{ik,hj}) + g_{sr}(\Gamma^s_{ij}\Gamma^r_{hk} - \Gamma^s_{hj}\Gamma^r_{ik}) \end{aligned} \tag{D.3}$$

The components of the Ricci tensor are:

$$\begin{aligned} R_{ij} &= R^s_{ijs} \\ &= \tfrac{1}{2}(\log g)_{,ij} - \Gamma^s_{ij,s} + \Gamma^r_{is}\Gamma^s_{rj} - \tfrac{1}{2}\Gamma^s_{ij}(\log g)_{,s} \end{aligned} \tag{D.4}$$

The curvature scalar is:

$$R^i_{\ i} = g^{ij} R_{ij} \tag{D.5}$$

The modified field equations of General Relativity, proposed by Einstein in 1917, are:

$$R_{ij} - \lambda g_{ij} = -\kappa(T_{ij} - \tfrac{1}{2} g_{ij} T^h_h) \tag{D.6}$$

Or, equivalently:

$$R_{ij} - \tfrac{1}{2} g_{ij} R^h_h + \lambda g_{ij} = -\kappa T_{ij} \tag{D.7}$$

The original or unmodified field equations, proposed by Einstein in 1915, are obtained by setting $\lambda = 0$ in (D.6).

On the range of a geodesic γ, within the domain of x, the coordinate functions x^i satisfy the following differential equations, for each value of the affine parameter t:

$$\left.\frac{d^2 x^i}{dt^2}\right|_{\gamma t} + \Gamma^i_{jk}(\gamma t) \left.\frac{dx^j}{dt}\right|_{\gamma t} \left.\frac{dx^k}{dt}\right|_{\gamma t} = 0 \tag{D.8}$$

The symmetries of the Riemann tensor (C.8.6–9) entail that it can have no more than 20 independent components relative to any given chart. 10 of them are combined into the Ricci tensor and are algebraically tied by the field equations (D.6) to the energy tensor of matter. The remaining 10 components determine the 10 independent components of the *Weyl tensor*—introduced by Hermann Weyl (1918*a*, p. 404)—which are given by:

$$C_{ijkh} = 2R_{ijkh} + g_{ik}R_{jh} + g_{jh}R_{ik} - g_{jk}R_{ih} - g_{ih}R_{jk} + \tfrac{1}{3}R^m_m(g_{jk}g_{ih} - g_{ik}g_{jh}) \tag{D.9}$$

It can be seen at a glance that the Weyl tensor has the same symmetries as the Riemann tensor. A quick calculation shows that

$$C_{jk} = g^{ih}C_{ijkh} = 0 \tag{D.10}$$

Notes

Motto

* The quotation from Riemann is taken from a posthumous philosophical fragment (Riemann (1902), p. 521). It may be translated as follows: "Physical science is an attempt to grasp Nature by means of exact concepts." The Einstein quotation is part of the Herbert Spencer lecture, delivered at Oxford on June 10th, 1933 (Einstein (1934), p. 183). It can be approximately rendered thus: "Hitherto our experience entitles us to trust that Nature is the realization of the mathematically simplest. I am convinced that through a purely mathematical construction we can find the concepts and the nomic connections between them which furnish the key to the understanding of natural phenomena." [All references to the literature are decoded on pp. 351 ff.]

Introduction

1. G. F. B. Riemann (1867); delivered as a public lecture in 1854. F. Klein (1893); first published in 1872. I examine the ideas of Riemann and Klein and their impact on late 19th-century thought in my book *Philosophy of Geometry from Riemann to Poincaré*, hereafter referred to as Torretti (1978*a*). See also below, pp. 26 f. (on Klein), 93 ff., 143 ff. (on Riemann). It is worth observing that the Erlangen Programme does not encompass Riemann's theory of metric manifolds, for the group of isometries of a Riemannian manifold generally consists of the identity alone, and is therefore powerless to characterize its structure. Nevertheless, the group-theoretical approach has been fruitfully applied to Riemannian geometry, at a higher level (so to speak), through the theory of fibre bundles. See for example, the remarks on metric connections on pp. 279 f.

2. The action of a group on a set is defined below, on p. 26.

3. Compare Klein (1893), p. 67: "Es ist eine Mannigfaltigkeit und in derselben eine Transformationsgruppe gegeben. Man entwickele die auf die Gruppe bezugliche Invariantentheorie." I should add perhaps that in Klein's investigations the set underlying the group transformations is not quite abstract; it is a "number manifold" (*Zahlenmannigfaltigkeit*) with an inherent topology, inherited from the real line $\mathbf{R}$ or the projective line $\mathbf{R} \cup \{\infty\}$.

4. Compare Poincaré (1887), p. 216: "Geometry is nothing else but the study of a group, and in this sense one might say that the truth of Euclid's geometry is not incompatible with that of Lobachevsky's geometry, for the existence of a group is not incompatible with the existence of another group. Among all possible groups we have chosen one, to which we refer physical phenomena, just as we choose three coordinate axes, to which we refer a geometrical figure."

5. Klein (1911), p. 21: "Was die modernen Physiker *Relativitätstheorie* nennen, ist die Invariantentheorie des vierdimensionalen Raum-Zeit-Gebietes x, y, z, t (der Minkowskischen 'Welt') gegenüber einer bestimmten Gruppe von Kollineationen, eben der 'Lorentzgruppe'." Cf. Minkowski (1915), delivered as a public lecture in 1907, and the Introduction to the second volume of Klein (1967).

6. One such study is the learned book by Arthur I. Miller (1981) on the emergence and early interpretation of Special Relativity. Unfortunately it appeared too late to be of more than cosmetic use in the preparation of this one. However, it caused me to take a new look at my sources wherever I noted a factual discrepancy between Miller's story and mine.

7. A more detailed yet sufficiently concise treatment of the main topics of Chapter 2, with schematic descriptions of the principal experiments, will now be found in A. I. Miller (1981), Chapter 1.

8. Namely E. Guth (1970); M. A. Tonnelat (1971), Chapters IX and X; C. Lanczos (1972), J. Mehra (1974*a*,*b*), V. P. Vizgin and Ya. Smorodinskiĭ (1979). I have indeed learnt much from

them, but my exposition is definitely intended as an alternative to theirs. A more complete coverage of some special points will be found in the articles of J. Earman and C. Glymour (1978*a*, *b*; 1980*a*, *b*), J. Stachel (1980*a*, *b*), and E. Zahar (1980).

9. B. F. Schutz explains most of these concepts at a more leisurely pace in his lucid and very readable book, *Geometrical Methods of Mathematical Physics* (1980). (Schutz collects, moreover, in Chapter 1 and in Section 2.22 of his book many of the definitions from analysis and linear algebra that I take for granted.) The reader who seriously wishes to study modern differential geometry would in my opinion do well to master the first two volumes of Spivak's *Comprehensive Introduction to Differential Geometry* (hereafter referred to as *CIDG*), and then to turn directly to Kobayashi and Nomizu's *Foundations of Differential Geometry* (*FDG*). Though undeniably difficult, Kobayashi and Nomizu's magnificent treatise is much more perspicuous and thus more rewarding than other books, allegedly designed for student use.

10. As presented, say, in Spivak's *Calculus on Manifolds*.

11. Chapters VI–IX, XI and XVI of MacLane and Birkhoff's *Algebra* (1967) contain all that is needed—and much more. Nomizu's *Fundamentals of Linear Algebra* (1979) is probably more suitable for self-study.

12. As developed, say, in A. Lichnerowicz, *Elements of Tensor Calculus* (1962), or in J. L. Synge and A. Schild, *Tensor Calculus* (*TC*).

13. A. Einstein (1954), H. Bondi (1964), E. F. Taylor and J. A. Wheeler (1963). J. G. Taylor's *Special Relativity* (1975) is a short and very clear introduction to the theory. See also in this series, R. B. Angel, *Relativity: The Theory and its Philosophy* (1980).

14. Two quite attractive recent introductory expositions of General Relativity are C. J. S. Clarke's *Elementary General Relativity* (1979) and J. V. Narlikar's *General Relativity and Cosmology* (1979). The serious student must eventually turn to Weinberg's *Gravitation and Cosmology* (1972), Misner, Thorne and Wheeler's *Gravitation* (1973) and Hawking and Ellis's *The Large Scale Structure of Space–Time* (1973). These are as of now the great Relativity treatises of our generation, in the grand tradition of Weyl's *Raum-Zeit-Materie* (1919*a*), Eddington's *The Mathematical Theory of Relativity* (1923), Tolman's *Relativity, Thermodynamics and Cosmology* (1934), Fock's *Theory of Space, Time and Gravitation* (1959), and Synge's *Relativity: The General Theory* (1960).

Chapter 1

Section 1.1

1. Newton, *PNPM*, p. 16.

2. Newton, "De Gravitatione et Aequipondio Fluidorum", Def. 4, in Hall and Hall (1978), p. 91.

3. Newton, "De Gravitatione . . .", Def. 5, in Hall and Hall (1978), p. 114.

4. Hall and Hall (1978), p. 114.

5. *Definition III*: "The innate force of matter is the power of resisting by which every body, as much as in it lies, perseveres in its state, either of rest, or of moving uniformly in a straight line." *Definition IV*: "An impressed force is an action exerted upon a body, in order to change its state, either of rest, or of moving uniformly in a straight line" (Newton, *PNPM*, pp. 40, 41). Newton's comments on Definition III raise some problems that we must ignore here; see McMullin (1978), pp. 33–43 and notes.

6. Newton, *PNPM*, p. 41.

7. Newton, *PNPM*, p. 54. As a matter of fact, Newton says that "the change of the motion (*mutatio motus*)"—not its *rate* of change—is proportional to the motive force impressed. This means that Newton's "impressed force" is what we now call "impulse" and is not synonymous with the current understanding of "Newtonian force". Newton's statement of the Second Law should therefore be transcribed, in modern notation, as $\mathbf{F}\Delta t = \Delta(m\mathbf{v})$. However, if one goes to the limit $\Delta t \to 0$—as Newton does, e.g., in his proof of Prop. I, Th. I (*PNPM*, pp. 88 ff.)—the foregoing equation entails that $\mathbf{F} = \mathrm{d}(m\mathbf{v})/\mathrm{d}t$, which is the familiar version of the Second Law I gave above.

To avoid distracting the readers with niceties irrelevant to our present purpose, I shall hereafter understand Newton's "impressed force" as an instantaneous, accelerative force.

8. Newton *PNPM*, pp. 55 f.
9. Hall and Hall (1962), p. 243.

Section 1.2

1. Aristotle, *Physica*, IV, 4; 212ᵃ20.
2. Descartes, *Principia*, II, 15 (*AT*, VIII, 48 f). See also *Regulae*, XII (*AT*, X, 426).
3. Descartes, *Principia*, II, 25 (*AT*, VIII, 53).
4. Hall and Hall (1978), p. 104.
5. The reader is referred to the excellent books by Max Jammer (1954) and Alexander Koyré (1957).
6. For greater precision, let us recall that the geometry of Euclid (the set of consequences of all the propositions of the *Elements*) can be derived from the axioms proposed in the first edition of Hilbert's *Grundlagen der Geometrie* and can therefore be given a denumerable model (Hilbert, *GG*, §9, pp. 34 ff.). This, however, will not satisfy the needs of a Newtonian physicist, who describes motions by means of *differentiable* mappings of time intervals into space. He requires a model of the stronger theory axiomatized in the second edition of the *Grundlagen*, in which Hilbert added the "Axiom of Completeness". The transition from the Greek "geometry of ruler and compass" to what is in effect the geometry of a flat Riemannian manifold homeomorphic to $\mathbf{R}^3$ cannot be said to have occurred at any particular date. If, as I have argued elsewhere (Torretti (1978*a*), pp. 34 ff.), Descartes' algebra of segments is that of a complete ordered field, the transition was completed before the publication of Descartes' *Géométrie* (1637). Hereafter, when we speak of Euclidean geometry, we shall understand it in its Cartesian sense, that is, as the theory axiomatized in the second and subsequent editions of Hilbert's *Grundlagen*. A model of this theory is a *Euclidean space*.
7. Herein lies, in my opinion, the essential difference between the philosophies of space of Newton and Leibniz; not in Leibniz' thesis that space is a relational structure, to which Newton, as the full-blooded mathematician that he was, certainly subscribed. Newton writes, in the important early manuscript from which I have already quoted: "Just as the parts of time obtain their individuality from their order, so that, e.g., if yesterday could change places with today [. . .] it would no longer be yesterday but today, thus the parts of space are individualized by their position, so that if any two could exchange their positions they would at once exchange their identities and each would become the other. It is only by their mutual order and positions that the parts of time and space are understood to be the very same which in truth they are, and they do not possess any principle of individuation apart from this order and these positions" (Hall and Hall (1978), p. 103; cf. Leibniz, Fifth Paper to Clarke, §47, *GP*, VII, 400 ff).
8. For Newton, a point is the dimensionless limit of a limit (line) of a limit (surface) of a part of space; see Hall and Hall (1978), p. 100.
9. Hall and Hall (1978), p. 99.
10. To avoid misunderstandings I wish to emphasize that the statement "The rigid body *B* exists all by itself in Newtonian space and is presently moving" is meaningless not because it is empirically unverifiable but because it is unintelligible: there is no *conceivable* difference between the situation described by this statement and the one that would be described by saying that the same body *B* is at rest. On the other hand, it makes perfect good sense to say that an isolated mass-point moves in a Riemannian space of non-constant curvature, even if by its very isolation it is unobservable *ex hypothesi*. The varying curvature of the space furnishes a criterion for identifying the successive positions of our mass-point.
11. Newton, *PNPM*, p. 50. Readers unacquainted with the water-bucket and the two-globe experiments ought to take a look at Newton's text. It will be found in English translation in Newton (*Cajori*), pp. 10–12; also in Smart (1964), pp. 85 ff.; in Čapek (1976), pp. 100 ff., etc. It is worthwhile noting that the experiments can tell us, with the aid of dynamical principles, *that* the system in question moves, but not *how* it moves. Thus, in the water-bucket experiment, the relevant phenomena are the same if the bucket's axis is at rest and each particle of water moves about it in a

circle or if the bucket's axis travels with a constant speed in a fixed direction perpendicular to itself and the water particles meander along its path.

Section 1.3

1. Aristotle, *Physica*, IV, 11; 219^{b}2. On the meaning of *arithmos*, which I have rendered as "measure" (though it usually means an integer greater than 1), see *ibid.*, IV, 12; 220^{a}27–32; also, 220^{b}32. Note that "motion" (*kinesis*) includes not only locomotion (*phora*), but growth and qualitative alteration as well.
2. Aristotle, *Physica*, IV, 12; 220^{b}14–16.
3. Aristotle, *Physica*, IV, 14; 223^{b}22–23.
4. Aristotle, *Physica*, IV, 10; 218^{b}3–5; IV, 12; 220^{b}5; cf. 218^{b}13, 223^{b}11.
5. Newton, *PNPM*, p. 46; Hall and Hall (1978), p. 355.
6. Newton, *PNPM*, p. 46.
7. Newton, *PNPM*, p. 48.
8. Credit for the first attempt at "demetaphysicizing" Newtonian space and time is usually given to Euler, but I am not impressed by his efforts in this direction. See the relevant texts, in English translation, in Koslow (1967), pp. 115–25, and Čapek (1976), pp. 113–19; cf. also the critical discussion in L. Lange (1886), pp. 87–97.
9. Aristotle, *Physica*, IV, 10; 218^{b}13.
10. Hall and Hall (1978), p. 104. Not every critic of mechanics failed to mention this important assumption, however. James Thomson (1884*a*) posits it explicitly; see the quotation on p. 294, note 5.
11. See, for instance, A. Sommerfeld (1964), pp. 71 f.
12. The centrifugal force on the dynamometer is proportional to the angular velocity squared. A purist might object that by attaching an elastic contraption to our rotating system we violate the requirement that it be rigid. The usual answer to this objection would be that by taking a sufficiently small dynamometer the violation can be made negligible. Such an answer is sensible enough when one is doing physics, but I am not sure that it is admissible in a discussion of the conceptual framework upon which the very possibility of doing physics rests. At any rate, the objection can be easily disposed of: According to Newton's principles, if our system rotates beyond the range of all external forces and the dynamometer is a permanent fixture of it, the shape and the angular velocity of the system will be constant.

Section 1.4

1. Newton, *PNPM*, p. 46.
2. That is, if $\mathbf{R}^3$ is endowed with the standard Euclidean metric, by which the distance from (a_1, a_2, a_3) to (b_1, b_2, b_3) equals $|(\Sigma_n (a_n - b_n)^2)^{1/2}|$.
3. The meaning of this can be readily explained if x and y are based on tetrahedra with pairwise congruent faces. Then, the tetrahedra themselves are congruent if x and y are similarly oriented. Otherwise, the tetrahedra are mirror images of each other.

Section 1.5

1. C. G. Neumann (1870), p. 14.
2. C. G. Neumann (1870), p. 15.
3. C. G. Neumann (1870), p. 26. Neumann suggests that the Alpha body might be represented by the principal axes of inertia of the universe, but adds that this representation lacks substance (*Inhalt*) because it cannot be supported or shaken by empirical evidence.
4. C. G. Neumann (1870), pp. 16, 18.
5. C. G. Neumann (1870), p. 18. See above, p. 12.
6. See Corollary V to the Laws of Motion, quoted on p. 19.

7. L. Lange (1885*b*), pp. 337 ff. Lange speaks of "inertial systems" (*Inertial-systeme*). I take from J. Thomson (1884*a*) both the word "frame" and the concept I express with it.

8. Readers unfamiliar with the epithets in parenthesis should not be put off by them. The last two make it possible to define a differential calculus on the space. See, for instance, H. Cartan (1967).

9. $\partial f(0,t)/\partial t$ is the derivative of a mapping of $\mathbf{R}$ into $S_{F'}$ and is therefore a mapping of $\mathbf{R}$ into $\mathscr{L}(\mathbf{R}, S_{F'})$, the space of linear mappings of $\mathbf{R}$ into $(S_{F'}, \mathbf{0}')$. $\mathscr{L}(\mathbf{R}, S_{F'})$ is canonically equated with $(S_{F'}, \mathbf{0}')$ by setting each $u \in S_{F'}$ in correspondence with the linear mapping $\lambda \mapsto \lambda u$ $(\lambda \in \mathbf{R})$. This correspondence is an isomorphism of the *normed* vector spaces $(S_{F'}, 0')$ and $\mathscr{L}(\mathbf{R}, S_{F'})$ (H. Cartan (1967), p. 20). The following result is used in the derivation of equations (1.5.2) and (1.5.4): If g is a differentiable mapping of $\mathbf{R}$ into S_F and f is a differentiable mapping of S_F into $S_{F'}$ and h denotes the composite mapping $f \cdot g$, the derivative $h'(a)$ of h at $a \in \mathbf{R}$ is the composite linear mapping $f'(g(a)) \cdot g'(a)$, where $f'(g(a))$ is the derivative of f at $g(a) \in S_F$ and $g'(a)$ is the derivative of g at a (see H. Cartan 1967), p. 31).

10. In particular, if $\mathbf{k}$ is the resultant of several component forces, $f_0(\mathbf{k})$ is the resultant of the images of those components by f_0.

11. Newton, *PNPM*, p. 63.

12. See M. Strauss (1972), p. 21.

13. The velocity with which the frame $F_{\mathbf{v}}$ moves in the frame $F_{\mathbf{u}}$ is properly represented by a vector in the relative space of the latter frame. This vector is mapped on $\mathbf{v} - \mathbf{u}$ by the isomorphism that takes each point of the space of $F_{\mathbf{u}}$ to the point of S_F with which it coincides at the time when the zero of that space coincides with the zero of S_F. Such irritating niceties are unavoidable if we desire to speak with precision and yet to abide by the present artificial approach that refers the unity of becoming to a plurality of spaces moving inside one another. Fortunately, in Section 1.6 we shall be able to take a more natural approach.

14. Newton, *PNPM*, p. 64. Curiously enough, Corollary VI is usually ignored in critical discussions of Newtonian relativity. See however H. Dingler (1921), pp. 152 ff.; E. Cartan (1923), p. 335 n. 3. H. Stein (1977*a*), p. 19 f. suggests that Mach's silence on this matter is a proof of his tact, and explains how the apparently relativistic implications of Corollary VI are cancelled in Newton's theory by the Third Law of Motion. (I follow him in the text above.) Stein correctly points out that, in the light of its use in Newton's *Principia*, Corollary VI is tantamount to Einstein's Equivalence Principle (see the proof of Prop. III, Th. III of Bk. I in Newton, *PNPM*, p. 94).

15. McMullin (1978), Chapter II, shows that Newton probably shared the view, common in his time, that matter is essentially passive and that all action in the world is due to immaterial agents. But the Third Law clearly entails that, even if every action on matter is really due to goblins, phenomenally it must be traceable to a material object, with which the acting goblin is somehow associated, and on which the goblin associated with the matter acted upon will immediately react.

16. The criterion furnished by the Third Law does not, of course, amount to an "operational definition" of a *freely moving particle* and an *inertial frame*. In the above example, the acceleration of β by α's reaction will generally be only a component of β's total acceleration and it might not be easy to discern it. But the criterion surely bestows a definite, intelligible meaning on the italicized expressions.

17. The (strictly) inertial force and the centrifugal force also show up as stresses in bodies fixed in the frame in question. The Coriolis and d'Alembert forces appear on particles in motion in a rotating frame and depend, respectively, on the velocity of the particle and the angular velocity of the frame, and on the rate of change of the latter with time. See Landau and Lifshitz (1960), p. 128, eqn. 39.7.

Section 1.6

1. See also K. Friedrichs (1927), P. Havas (1964), A. Trautman (1965), pp. 104 ff., (1966). H. Stein (1967) drew the attention of philosophers to the concept of Newtonian spacetime, which has subsequently found its way into the better philosophical books in the field. See, for instance, Sklar (1974), Nerlich (1976).

2. Einstein (1956), p. 31.

3. Galilei, *Discorsi*, *EN*, VIII, 272.

4. D'Alembert, article "Dimension" of the *Encylopédie* (vol. IV, 1754); Lagrange (1797), p. 223. The original texts were transcribed by R. Archibald (1914) in the *Am. Math. Soc. Bulletin*. See also Kant (1770), p. 17 n. (in Kant, *WW*, II, 401 n.).

5. A. M. Bork (1964) collects and discusses several passages of late 19th-century physical literature in which reference is made to a fourth dimension. Practically all of them refer to a fourth spatial dimension, not to time as a fourth dimension. Reading Bork's quotations one has the impression that the authors were inhibited from conceiving a spacetime by their assumption that a four-dimensional manifold—as Maxwell put it—"has four dimensions all coequal" (J. C. Maxwell to C. J. Monroe, 15th March 1871; quoted by Bork (1964), p. 329.)

6. According to H. Freudenthal (1968), p. 227, non-metric differential geometry was created by Sophus Lie (1842–99).

7. See, for instance, H. Poincaré (1893), and my comments in Torretti (1978*a*), pp. 345 f.

8. Hereafter, I shall call the set of arguments at which a mapping f takes a particular value a, *the fibre of f over a.*

9. The said mapping sends the value of $\bar{x}$ at p to the value of $\bar{z}$ at the point of S_G with which p coincides at time 0. It must be a rotation or a rotation-*cum*-parity of $\mathbf{R}^3$, because the value of $\bar{x}^{-1}$ at (0, 0, 0) coincides at time 0 with the value of $\bar{z}^{-1}$ at (0, 0, 0). (*Parity* means reflection in (0, 0, 0)).

10. The nine numbers $a_{\alpha\beta}$ must satisfy the six conditions of orthogonality $\Sigma_\gamma a_{\alpha\gamma} a_{\beta\gamma} = \delta_{\alpha\beta}$.

11. Let σ be such a hyperplane and let $\theta = y \cdot x^{-1}$ be a Galilei coordinate transformation. That $\theta(\sigma) \| \sigma$ follows from the fact that $y^0 = x^0 + k^0$. A glance at equations (1.6.3) should persuade us that, if $ABCD$ is any tetrahedron in σ, $\theta(ABCD)$ differs from a perpendicular projection of $ABCD$ on $\theta(\sigma)$ only by a translation followed by a rotation. The restriction of θ to σ is therefore an isometry of σ onto $\theta(\sigma)$. Note also that both σ and $\theta(\sigma)$, as hyperplanes of $\mathbf{R}^4$, are isometric to $\mathbf{R}^3$, with the usual metric, and are therefore copies of Euclidean space.

12. An n-manifold S is said to be analytic if the coordinate transformations between the charts of its maximal atlas are analytic, i.e. if they can be expanded about each point in their domain into a convergent power series. Analytic mappings of one analytic manifold into another are defined like differentiable mappings between ordinary n-manifolds (p. 258). Let G be a group which is also an analytic n-manifold. G is an *n-parameter Lie group* if the group product is an analytic mapping of $G \times G$ into G. The proper orthochronous Galilei group is a connected 10-parameter Lie group. (Connected in the topological sense, i.e. not the union of two disjoint non-empty open sets.) The *full* Galilei group is obtained in the same way, if we lift the restrictions (ii) and (iii) that we imposed on the charts of $A_{\mathscr{G}}$ on p. 25. The full Galilei group will therefore include the transformations $(t, x, y, z) \mapsto (-t, x, y, z)$ (time reversal), and $(t, x, y, z) \mapsto (t, -x, -y, -z)$ (parity) that were excluded by those restrictions. The full Galilei group is a disconnected 10-parameter Lie group.

13. In the case described, $\mathscr{H}$ is more accurately said to act on S "on the left". F is a "left action" $\mathscr{H}$ acts on S "on the right" and F is a "right action" if $F(gh, s) = F(h, F(g, s))$, for every $g, h \in \mathscr{H}$ and every $s \in S$. If F is a right action it is evidently convenient to write sg for $F(g, s)$. The reader will not fail to notice that, for any group G, the *group product* constitutes a transitive and effective left action of G on itself.

14. See, for instance, Sudarshan and Mukunda (1974), pp. 255 ff.

15. See, for instance, Lie (1893), pp. 130 f., 216 f. Note that the arbitrary element of $\mathscr{G}_{\mathbf{R}}$ defined by equations (1.6.3) is an isometry if and only if the parameters v^α are all equal to 0, that is, if no change of frame is involved.

16. See p. 191. A pleasantly simple intrinsic characterization of the geometry of Newtonian spacetime will be found in H. Field (1980), Chapter 6.

17. Herivel (1965), p. 149.

18. Galilei, *Dialogo*, *EN*, VII, 212 f.; quoted after Stillman Drake's translation (Galilei (1967), pp. 186 f.). A passage very similar to the above occurs in Galilei's letter to Francesco Ingoli of September 1624 (*EN*, VI, 547 f.).

Section 1.7

1. Newton, *PNPM*, p. 433 (Bk. II, Prop. XXIX, Th. XIX, Cor. VI); p. 572 (Bk. III, Prop. VI, Th. VI). Newton's result has been confirmed of late to one part in 10^{12}, a degree of precision rarely achieved in physics. See p. 310, note 5.

2. Note, however, that, if $\mathbf{F}(y)$ is the vector representing the force of 2 on 1 in terms of y, $\mathbf{F}(x) = \mathrm{A}^{-1}\mathbf{F}(y)$, and therefore differs from $\mathbf{F}(y)$ unless A is the identity matrix, that is, unless $x \cdot y^{-1}$ does not involve a spatial rotation. The magnitudes $|\mathbf{F}(x)|$ and $|\mathbf{F}(y)|$ are equal, of course, because A is orthogonal.

3. Newton to Bentley, 25th February 1693; in Newton (1756), pp. 25 f.

4. Newton, *PNPM*, p. 764 (General Scholium to Bk. III).

5. Cf. Boscovich (1759), Kant (1756, 1786).

6. See, for instance, Robertson and Noonan (1968), pp. 16–18. H. Stein (1970*b*) forcefully argues that the field-theoretical approach to gravity can be traced back to Newton's own thought and words.

Chapter 2

Section 2.1

1. Whittaker (1951/53), vol. I, pp. 53, 56 f.

2. Ampère (1827). English translation and commentary in Tricker (1965). Philosophical commentary—in French—in Merleau-Ponty (1974).

3. Ampère, in Tricker (1965), p. 155.

4. The electrostatic unit (esu) of charge is defined as the charge which, placed at unit distance *in vacuo* from an exactly equal charge, repels it with unit force. The electromagnetic unit (emu) of charge is the charge carried in unit time by one emu of electric current, the latter being defined as follows: (i) the emu of magnetic pole strength is the strength of a magnetic pole which, placed at unit distance *in vacuo* from an exactly equal pole, repels it with unit force; (ii) the emu of current is the current carried by either of two exactly equal closed circuits whose mutual potential is the same as that of two magnetic shells of unit pole strength bounded by the circuits. (The mutual potential of two circuits or shells is the work done by the field on one of them when it is transported parallel to itself from infinity to its present position, while the other remains where it is.) That the ratio c of the emu to the esu must have the dimension [length/time] can be seen by the following consideration due to Maxwell (*TEM*, §768): Let two parallel conducting wires be so placed that they attract each other with a force equal to the product of the currents I, I', conveyed by them. The amount of charge transmitted by the current I in time t is It [emu] or Itc [esu]. Let two small conductors be charged with the electricity transmitted by the currents I and I', respectively, in time t, and place them at a distance r, such that the repulsive force between the conductors exactly balances the attractive force between the conducting wires. Then $II' = II't^2c^2/r^2$. Hence $c = r/t$.

5. See Whittaker (1951/53), vol. I, pp. 201 ff.; Woodruff (1962). Helmholtz objected to the fact that the electric forces were made to depend on the velocities and not only on the positions of their sources. He claimed—wrongly—that Weber's theory violated the Conservation of Energy. He also pointed out—correctly—that in particular circumstances it predicted some physically implausible situations (involving, be it noted, faster-than-light motions—cf. Maxwell, *TEM*, §854).

6. Maxwell, *TEM*, §§846 ff.; cf. §529 and Preface, pp. VIII, X.

7. Hertz, *EW*. See S. D'Agostino (1975).

8. Maxwell, *TEM*, p. IX.

9. Cf. Faraday, *ERE*, §3071; Maxwell, *TEM*, §47.

10. Maxwell, *TEM*, pt. IV, ch. IX. Maxwell's original equations are transcribed in vector notation in Gillespie, *DSB*, vol. IX, p. 212. The slightly simpler system published in Maxwell (1865) is transcribed in vector notation in Berkson (1974). p. 176.

11. Weber and Kohlrausch's value was 3.1074×10^{10} [cm/s] Fizeau's value for the speed of light was 3.14×10^{10} [cm/s]. In 1890 J. J. Thomson and Searle improved the value of the ratio of the emu to the esu to 2.9955×10^{10} [cm/s]. Michelson, in 1882, had obtained 2.9976×10^{10} [cm/s] for the speed of light. Let me add that in his original formulation of equation (2.1.1) Weber used electrodynamic instead of electromagnetic units; the constant that turned up was therefore equal to $c\sqrt{2}$, not a particularly suggestive number. (Note that in the formula of Weber's force law in A. I. Miller (1981), p. 92, c stands for this peculiar constant; *not* for the speed of light.)

12. Maxwell (1865), §§15, 16; my italics. Cf. §74. It is worth noting that Faraday, who after discovering the magnetic rotation of the plane of polarization of light in 1845 became convinced that light was electromagnetic by nature, expected his lines of force to "supply for the vibrating theory the place of an ether". "The view which I am so bold as to put forth considers, therefore, radiation as a high species of vibration in the lines of force which are known to connect particles and also masses of matter together. *It endeavours to dismiss the ether, but not the vibrations.*" (My italics.) With unfailing instinct he remarks that "a uniform medium like the ether does not appear apt, or more apt than air or water" for the propagation of purely transverse waves (Faraday, "Thoughts on Ray Vibrations", *ERE*, vol. III, p. 450). On Faraday's final steps to field theory, see D. Gooding (1981).

13. Maxwell (1861/62).

14. Maxwell (1865), §73.

15. Maxwell, *TEM*, §574.

16. Maxwell, *TEM*, §570; cf. §552. See Poincaré's very clear explanation (*SH*, ch. XII, pp. 219–25): By choosing a potential energy function U, depending on some observable parametres q_i, and a kinetic energy function T, depending on the q_i and their time derivatives $\dot{q}_i$, such that the Lagrange equations $(d/dt)(\partial(T-U)/\partial\dot{q}_i) = \partial(T-U)/\partial q_i$ agree with the data of experience, Maxwell proved that it was possible to give a mechanical explanation of electricity—indeed, infinitely many such explanations. See also Larmor (1900), §§49 ff., pp. 82 ff. Green (1838) and MacCullagh (1848) had already used Lagrangian methods in dealing with the optical aether.

17. See Schaffner (1972), pp. 68 ff. Schaffner is probably right in thinking that Kelvin meant his models merely as illustrative mechanical analogies.

18. Quoted by Schaffner (1972), p. 90, from the original letter in the Deutsches Museum, Munich. Heaviside's suggestion is the core of the so-called electromagnetic conception of matter, proclaimed by W. Wien (1900) and much favoured in the first decade of the 20th century, especially after W. Kaufmann's alleged experimental proof of the purely electromagnetic origin of inertial mass. See R. McCormmach (1970). On Kaufmann's experiments, see A. I. Miller (1981), pp. 47–54, 61–7, 228–32; see also E. Zahar (1978) for a subtle logical analysis of the interpretation of Kaufmann's 1905 results.

Section 2.2

1. See, for instance, Hertz, *EW*, p. 246. Quoted on p. 293, note 17.

2. Maxwell, *TEM*, §601. "Electromotive intensity", or the force on a particle with unit charge, is given according to Maxwell by the equation $\mathbf{E} = \mathbf{v} \times \mathbf{B} - \dot{\mathbf{A}} - \nabla\psi$, where $\mathbf{v}$ is the particle's velocity, $\mathbf{B}$ the magnetic induction, $\mathbf{A}$ the vector potential and ψ the electrostatic potential. The change made necessary by the transformation described above consists in adding to the potential ψ the scalar field $\psi' = -\mathbf{A}\cdot\mathbf{w}$, where $\mathbf{w}$ is the linear velocity of the moving frame with respect to the fixed frame. Now $\oint \nabla\psi'\cdot d\mathbf{s} = 0$, whence Maxwell's conclusion.

3. Towards the end of his life Maxwell proposed a method for calculating the velocity of the solar system through the aether from the velocity of light measured by Roemer's method at periods when Jupiter lay in opposite directions from the earth. Though, as it turned out, astronomy was then unable to meet the exacting demands of Maxwell's method, he saw no other way because, in his opinion, the motion of the solar system through the aether could not be ascertained in the optical laboratory. "In the terrestrial methods of determining the velocity of light—he wrote to D. P. Todd, of the U.S. Nautical Almanac Office—the light comes back along the same path again, so that the velocity of the earth with respect to the ether would alter the time of the double passage by a quantity depending on the square of the ratio of the earth's velocity to that of light, and this is quite too small to be observed" (Maxwell to Todd, 19th March 1879; Maxwell (1879/80), p. 109). Soon thereafter, though unfortunately not before Maxwell's death, A. A. Michelson, also of the U.S. Nautical Almanac Office, devised an interferometer that was capable of detecting just such tiny effects.

4. Hertz (1908*a*); English translation in Hertz, *EW*, pp. 195–240. Arnold Sommerfeld remarked years later: "It was as though scales fell from my eyes when I read Hertz's great paper" (Sommerfeld (1964), p. 2).

6. Hertz, *EW*, pp. 241 f.

7. Hertz, *EW*, p. 242.

8. Hertz's version of the Maxwell equations is transcribed in vector notation in A. I. Miller (1981), pp. 11 f. The change they undergo when referred back from a moving system to the laboratory rest frame is stated and briefly explained on p. 12. For a less laconic explanation, see Silberstein (1924), pp. 30 f., 56 ff.

9. W. C. Roentgen (1885, 1888, 1890); A. Eichenwald (1903); H. A. Wilson (1905). For a short discussion of these experiments, see Laue (1955), pp. 11 f.

10. Let h be the height of the objective above the eyepiece. Then the time the corpuscle takes to travel from the former to the latter is hc^{-1}. In the same time the eyepiece travels the distance vhc^{-1}. Clearly $\tan\alpha = vhc^{-1}/h = v/c$. Though Bradley analyzes the same special case proposed here, he derives a formula valid in the general case, in which light might not fall vertically. Let θ be the real and θ' the apparent inclination of the star. Then $v/c = \sin(\theta - \theta')/\sin\theta'$.

11. Stokes (1845); reprinted in Schaffner (1972), p. 131.

12. Schaffner (1972), p. 137.

13. Predicting this null result was no mean achievement, "for if the tangent of the angle of aberration is the ratio of the velocity of the earth to the velocity of light, then since the latter velocity in water is three-fourths its velocity in a vacuum, the aberration observed with a water telescope should be four-thirds of its true value" (Michelson and Morley (1887), p. 449).

14. Michelson (1881); reprinted in Kilmister (1970), p. 103.

15. Michelson and Morley (1887), p. 459. The Supplement at the end of this paper suggests that the aether drag may diminish on hill-tops, an effect subsequently studied by Michelson with null results. This suggestion, however, does not stem from a lingering faith in Stokes' hypothesis, but from the consideration that —as Michelson wrote to Lord Rayleigh—the aether may be dragged at sea level by the nearby mountain chains, for "Fizeau's experiment holds good for transparent bodies only, and I hardly think we have a right to extend the conclusions to opaque bodies" (Letter of 7th March 1887, reproduced by Shankland (1964), p. 29). The same consideration motivates the proposal made by the authors in the said Supplement that the experiment be repeated using a glass cover instead of a wooden one on the instrument.

16. I have found the writings of T. Hirosige (1969, 1976) and K. Schaffner (1969, 1976) particularly useful as a guide to Lorentz's thought. A. I. Miller (1981), Chapter 1, is also very illuminating, but was not available when I wrote this Section.

17. Lorentz (1892*a*), in Lorentz, *CP*, vol. II, p. 206. Where Lorentz writes "ce nom [scil. 'matière'—R. T.] sera donc appliqué a l'ether tout aussi bien qu'à la matière pondérable", S. Goldberg (1969), p. 984, translates: "The name ether will be applied to all else besides ponderable matter." Goldberg's false rendering suggests that Lorentz, like some of his British contemporaries, believed the aether to be immaterial.

18. Lorentz (1892*a*), in *CP*, vol. II, p. 228.

19. Lorentz (1892*a*), *CP*, vol. II, p. 246. The vectorial notation used above was adopted by Lorentz later.

20. Lorentz (1898), in *CP*, vol. VII, p. 114. Cf. Poincaré (1900), and Lorentz's reaction in his letter to Poincaré of 20th January 1901 (published by A. I. Miller (1980), pp. 70 f.).

21. McCormmach (1970), p. 467, gives the following intuitive explanation of how, according to Lorentz, a refringent body impresses on light a fraction of its speed, without actually dragging the aether: "When a ray strikes the first charged particle of a dielectric, Lorentz's electrical force is called into play at once, displacing the particle from its equilibrium position. Because of its motion, the particle becomes the center of an electromagnetic disturbance which propagates outward at the speed of light, and this new pulse interferes with the already existing state of the ether. This sequence is repeated at the second charged particle and so on; and the resulting superposition of primary and induced vibrations accounts for the slowing and bending of light inside dielectrics."

22. Lorentz (1895), in *CP*, vol. V, p. 4. Lorentz adds: "If I say for brevity's sake that the ether is at rest I only mean to say that one part of this medium does not move with respect to the other and that all perceivable motions of the celestial bodies are relative motions with respect to the aether."

23. Lorentz (1892*a*), in *CP*, vol. II, pp. 292, 297. See also Schaffner (1969), pp. 501 f.; Zahar (1973), p. 112.

24. As the reader will surely notice, equations (2.2.5) differ but slightly from the standard Lorentz transformation applicable in a situation like the one described above. Cf. equations (3.4.18) on p. 61.

25. The above is a paraphrase of the conciser statement in Lorentz (1895), *CP*, vol. V, p. 84. For a short proof of the Theorem, see Whittaker (1951/53), vol. I, pp. 406 f.

26. Lorentz (1895), *CP*, vol. V, p. 90. Lorentz (1899) restates the Theorem of Corresponding States in almost the same terms, but understands it in a somewhat different sense. The primed coordinates are now related to a Galilei chart adapted to the rest frame of the moving material system by the transformation (2.2.9) on p. 47, which is similar to, but not the same as (2.2.4). See Lorentz, *CP*, vol. V, pp. 149 f.

27. Lorentz (1916), p. 321. See, however, Lorentz (1899), in *CP*, vol. V, p. 150; Lorentz (1899*), quoted on p. 47 and n. 35. Lorentz (1916), p. 224, clearly implies that local time, as defined in that book, is the time that natural clocks in fact keep. Joseph Larmor arrived independently to a similar conception—see, for instance, Larmor (1900), p. 179.

28. Lorentz's letter to Lord Rayleigh is quoted by Schaffner (1972), p. 103.

29. Lorentz (1892*b*), in *CP*, vol. IV, p. 221.

30. These paragraphs are reproduced in the well-known collection, Lorentz *et al.*, *RP*.

31. For a short discussion of the Michelson–Morley experiment see, for instance, A. P. French (1968), pp. 49–57, 63–4.

32. More significant than the stationary oscillations of the molecules of a solid body at rest or moving inertially in the aether is the change in the oscillations that should result from a change in velocity if indeed the molecular forces holding the body together are altered thereby. Wood, Tomlison and Essex (1936) tested this consequence of the Fitzgerald–Lorentz contraction hypothesis in a quartz rod, obtaining a null result.

33. Larmor (1897) had anticipated him (see especially *loc. cit.*, pp. 221, 230), but there is no evidence that Lorentz was aware of his work.

34. Lorentz (1899), in *CP*, vol. V, p. 152.

35. Lorentz (1899*), in Schaffner (1972), p. 290. Lorentz adds: "The transformation of which I have now spoken is precisely such a one as is required in my explication of Michelson's experiment. In this explication the factor ε may be left indeterminate. We need hardly remark that for the real transformation produced by a translatory motion, the factor should have a definite value. I see however no means to determine it." These passages of the English version of Lorentz (1899)—which in all likelihood was written or at least revised by Lorentz himself—do not occur in this form in the French version in *CP*, vol. V.

36. Lorentz (1899), in *CP*, vol. V, p. 154.

Chapter 3

Section 3.1

1. Einstein's letter to Habicht is reproduced in C. Seelig (1954), p. 75.

2. Einstein (1905*b*), p. 153; English translation in D. ter Haar (1967), p. 92.

3. "Die vierte Arbeit liegt erst im Konzept vor und ist eine Elektrodynamik bewegter Körper unter Benutzung von Modificationen der Lehre von Raum und Zeit" (Einstein to Habicht, *loc. cit.*) The remaining two papers dealt with molecular dimensions and Brownian motion, and contributed decisively to the final acceptance of the atomic view of matter (Einstein (1905*a*, *b*). The best available English translation of Einstein (1905*d*) is to be found in A. I. Miller (1981), pp. 392–415.

4. Shankland (1963), p. 56.

5. Einstein (1949), p. 44.

6. Einstein (1949), p. 46.

7. Einstein's readiness to accept this conclusion may have been due, in part, to the influence of Ernst Mach. See Hirosige (1976), pp. 57 ff.

8. That Einstein was well aware of this implication of his (1905*b*) can be gathered from the introductory remarks to his paper of 1907 on the inertia of energy (Einstein (1907*a*), pp. 372 ff.). See also Einstein's letter to Max von Laue of 17th January 1952 (quoted by Holton (1973), pp. 201 ff). In 1955 Einstein wrote to Carl Seelig that the novelty of his (1905*d*) with respect to Lorentz (1904) and Poincaré (1905, 1906) consisted in "the realisation of the fact that the bearing of the Lorentz

transformation transcended its connection with Maxwell's equations and had to do with the nature of space and time in general. [. . .] This was for me of particular importance because I had already found that Maxwell's theory did not account for the micro-structure of radiation and could not therefore have general validity" (quoted by Born (1956), p. 194).

9. On the contrast between "phenomenological" and "deep" theories, see Bunge (1964).

10. Einstein (1949), p. 52.

11. Einstein (1949), p. 56.

12. De Sitter (1913*a*,*b*,*c*) published observations of binary stars that were long regarded as proof that the speed of light is not affected by the velocity of its source. This interpretation of de Sitter's results is questionable, however, if, as suggested by J. G. Fox (1962), the double stars are surrounded by a cloud of gas which does not partake of their motion and which absorbs and retransmits all the radiation we receive from the said stars. Terrestrial experiments by G. C. Babcock and T. G. Bergman (1964) and T. Alväger *et al.* (1964) have confirmed Einstein's LP most convincingly. More recently, K. Brecher (1977) has verified the LP to 1 part in 10^9, by the observation of X-ray sources in binary star systems. (Such sources are less liable than ordinary light sources to Fox's strictures.)

13. Let the inertial frame F move with velocity v in the inertial frame F. According to the Galilei Rule, a light pulse travelling in F with velocity ck (where k is a unit vector), travels in F' with velocity $c\mathbf{k}+\mathbf{v}$. If $\mathbf{v}$ and c are given, $|c\mathbf{k}+\mathbf{v}|$ depends on $\mathbf{k}$. Consequently, if the speed of light is constant in every direction in F, it cannot be so in F'.

14. Einstein (1949), p 52. See also M. Wertheimer (1959), p. 214. W. Berkson (1974), p. 296, recalls that Oliver Heaviside had worked out a theory concerning the field generated by electrons moving with velocities equal to or greater than c.

15. Einstein (1905*d*), p. 891; English translation in Kilmister (1970), p. 187. A. I. Miller (1981), Chapter 3 analyses this passage in depth and comments on its historical background.

16. Einstein (1905*d*), p. 891. This is the only text in Einstein (1905*d*) that might refer to Michelson and Morley's experiment. Of course, it can also refer to other, more recent second-order experiments (for example those by Rayleigh (1902), Trouton and Noble (1904), Brace (1904)), or even to the first-order experiments of Hoek (1868, 1869), Airy (1871), Mascart (1872/74), etc., all of which yielded null results. On the role of Michelson and Morley's experiment in the genesis of Special Relativity, see Holton (1969), Shankland (1973), Einstein (1982).

17. In Hertz's theory, "the absolute motion of a rigid system of bodies has no effect upon any internal electromagnetic processes whatever in it, provided that *all* the bodies under consideration, including the ether as well, actually share the motion" (Hertz, *EW*, p. 246).

18. Einstein (1905*d*), p. 892; English translation in Kilmister (1970), p. 188.

19. Shankland (1963), p. 49. Ritz's theory is discussed by J. G. Fox (1965).

Section 3.2

1. The thesis that Special Relativity forgoes the universality of natural time is maintained not only by such well-known conventionalists as Hans Reichenbach and Adolf Grünbaum, but also, with varying degrees of emphasis, by A. A. Robb (1914), p. 6; H. Weyl (1923), pp. 144 f., 163 ff.; A. S. Eddington (1923), pp. 15 f., 27; L. Silberstein (1924), p. 92; R. B. Lindsay and H. Margenau (1936), pp. 75, 334; P. G. Bergmann (1942), p. 32 n.; H. P. Robertson (1949), p. 380; C. Møller (1952), p. 32; H. Arzeliès (1966), p. 19; W. Rindler (1977), pp. 28 f., etc. The thesis that Einstein merely questioned the uniqueness of natural time, and that in Special Relativity each inertial frame is furnished by nature with a universal time of its own, is implied by M. von Laue (1955), pp. 6f., 30 f.; and probably also by L. D. Landau and E. M. Lifschitz (1962), § 1, whose treatment of the question is not altogether clear to me. It also underlies the recurrent proposals for measuring the one-way speed of light without making any stipulations about distant synchrony (see, for instance, E. Feenberg (1974)), or for basing such stipulations on "our intuitive understanding of simultaneity" (F. Jackson and R. Pargetter (1979), p. 311). It is worth noting that near the end of his life Einstein still regarded the knowledge that "there is no simultaneity of distant events" as one of the "definitive insights" that we owe to Special Relativity (Einstein (1949), p. 60).

2. My proposal is based on Einstein's own practice, in his more mature writings; thus Einstein (1916), p. 772, characterizes "a Galilean system of reference" as one relative to which "a mass,

sufficiently distant from other masses, moves with uniform motion in a straight line" (English translation in Kilmister (1973), p. 144); see also Einstein (1954), p. 11. The footnote added to Einstein (1905*d*), § 1, in Lorentz *et al.*, *RP*, p. 27, to the effect that the stationary system is one in which the equations of Newtonian mechanics "hold good in a first approximation", avoids the inconsistency at a rather high price, for the concept of a stationary frame thereby becomes irretrievably ambiguous.

3. Einstein (1911), § 4.

4. Einstein (1905*d*), p. 892; English translation in Kilmister (1970), p. 189.

5. "Men have very good means of knowing in some cases, and of imagining in other cases, the distance between the points of space simultaneously occupied by the centres of two balls; if, at least, we be content to waive the difficulty as to imperfection of our means of ascertaining or specifying, or clearly idealising, simultaneity at distant places. For this we do commonly use signals by sound, by light, by electricity, by connecting wires or bars, or by various other means. The time required in the transmission of the signal involves an imperfection in human powers of ascertaining simultaneity of occurrences in distant places. It seems, however, probably not to involve any difficulty of idealising or imagining the existence of simultaneity" (James Thomson (1884*a*), p. 569).

6. Einstein (1905*d*), p. 894; Kilmister (1970), p. 19.

7. He probably also assumed tacitly that synchronizations performed by his method do not depend on the time at which they are carried out. Let $S(A, B, C, t)$ signify that two clocks at rest at points A and B of a given inertial frame run synchronously according to Einstein's criterion, when the latter is applied by means of light-signals sent from a point C of the same frame at time t (measured by a clock at C). Then we must expect that: (1) $S(A, B, A, t_0)$ at some definite time t_0 only if, for all t, $S(B, A, B, t)$; (2) $S(A, B, A, t_0)$ and $S(A, C, A, t_1)$—at definite times t_0 and t_1—jointly imply $S(B, C, B, t)$, for all t. I presume that this is the meaning of the symmetry and transitivity of synchronism postulated by Einstein near the end of his (1905*d*), § 1; for if it only meant that the four-place relation S is symmetric and transitive in its first two terms—when the last two are fixed—it would follow trivially from the very meaning of synchronism and would not constitute an additional assumption, as Einstein says.

Section 3.3

1. Let f be the Lorentz chart described above. Then f is defined by the following set of relations: $t = \pi_0 \cdot f$, $x = \pi_1 \cdot f$, $y = \pi_2 \cdot f$, $z = \pi_3 \cdot f$; where π_i is the i-th projection function on $\mathbf{R}^4$, which maps each real number quadruple on its $(i+1)$-th element. Note that the Lorentz charts mentioned in the above statement of the RP can of course be adapted to different inertial frames, but must all employ the same units of length and time, embodied in physically equivalent standards at rest in the respective frames.

2. Einstein (1905*d*), p. 895; Kilmister (1970), p. 191. The above statement of the RP is an almost literal translation of Einstein's. I have only substituted "Lorentz charts" for "coordinate systems in uniform translatory motion relative to each other". The two expressions are not synonymous, but the one used by Einstein covers more systems than he actually intended, e.g., systems whose time coordinate functions are related by equation (3.2.1). My version of the LP is less literal but is entirely faithful to the meaning of the original. It is understood that the LP refers to light transmitted through empty space.

Section 3.4

1. Einstein himself derives the Lorentz transformation equations from the RP and the LP by a seemingly simpler method in his (1954), Appendix I; by a more elegant one in his (1956), pp. 29 ff. The original derivation in his (1905*d*), § 3 has been analysed by J. Merleau-Ponty (1974), R. B. Williamson (1977), M. Jammer (1979) and A. I. Miller (1981).

2. The latter assumption is not necessary for Einstein's argument and is not made by him. We

introduce it here only because it will save us some tedious explanations. I recall in passing that today's metrology defines the unit of time as a given multiple of the period, and the unit of length as a given multiple of the wavelength of some standard monochromatic light signals. If, contrary to current fact, the standards adopted in both definitions were the same the value of c would be fixed by metrological convention.

3. Conditions (ii) and (iii) can be stated in a less simple but probably more familiar fashion, in terms of the Cartesian coordinate systems x and $\bar{y}$. (ii) says that, for $\alpha = 1, 2, 3$, the α-th axes of both Cartesian systems lie on each other and are similarly oriented at time $x^0 = 0$. (iii) says that the origin of $\bar{y}$, i.e. the point P of G such that $\bar{y}(P) = (0,0,0)$, moves in frame F with velocity components $\mathrm{d}\bar{x}^1(P)/\mathrm{d}x^0 = v$, $\mathrm{d}\bar{x}^2(P)/\mathrm{d}x^0 = \mathrm{d}\bar{x}^3(P)/\mathrm{d}x^0 = 0$. As to condition (iv), it is designed to exclude time reversal. As M. Jammer (1979), p. 212, aptly remarks, the rejection of time reversal is implicit in Einstein's argument for eqn. (3.4.16); see p. 61 and note 9, below.

4. Einstein (1905*d*), p. 898; Kilmister (1970), p. 194.

5. If we express Einstein's formulae in our notation, we shall see that he sets $x^\alpha(A) = y^\alpha(A) = 0$, but $x^0(A) = t$. However, by setting $x^0(A) = y^0(A) = 0$ we do not risk any loss of generality.

6. Einstein designates the auxiliary coordinate function u^1 and the parameter $\hat{u} = u^1(B)$ by the same variable x'. When he asks us to choose x' "infinitesimally small" he means the parameter.

7. Note Einstein's careful reasoning. The LP was *assumed* to be valid in F, whence F is to be regarded as the "stationary system". On the strength of the RP, one can then infer that the LP is also valid in G. Observe also that equation (3.4.3) is introduced as a consequence of the definition of the Lorentz chart y, adapted to G.

8. An arbitrary point $(4t, a, b, c)$ is equal to $(t, vt, 0, 0) + (t, a - 3vt, 0, 0) + (t, vt, b, 0) + (t, vt, 0, c)$, which is a linear combination of points of the four kinds considered above.

9. By initial conditions (i) and (ii), the α-th axes of x and z—for all values of α—lie on each other and are similarly oriented at time $x^0 = 0$, and hence at all times. By condition (iv), time reversal is excluded. Since x and z are determined by measurements performed with clocks and unstressed rods at rest on the same frame, they fully agree with each other.

10. Einstein (1905*d*), p. 902; Kilmister (1970), p. 198. Einstein's "reasons of symmetry" can be spelled out as follows: by means of equations $w^0 = x^0$, $w^1 = -x^1$, $w^2 = x^2$, $w^3 = x^3$, we introduce a new Lorentz chart adapted to F, relative to which G moves with velocity components $(-v, 0, 0)$; at time $w^0 = t$, the distance between the positions of the rod's endpoints in F is $\lambda/\varphi(-v) = \bar{w}^2(Q,t) - \bar{w}^2(P,t) = \bar{x}^2(Q,t) - \bar{x}^2(P,t) = \lambda/\varphi(v)$.

11. See, for instance, Arzeliès (1966), pp. 72 ff; Robertson and Noonan (1968), p. 45; Møller (1952), pp. 41 ff. The condition $\Sigma_\alpha v_\alpha^2 \leq c^2$ is discussed on pp. 71, 296 n. 9; equality is of course excluded because no inertial frame can move in another one with the speed of light (p. 57).

12. It is evident that this collection of matrices represents a group, for (i) it includes the identity matrix I; (ii) if A and B belong to it, $(AB)^T H(AB) = B^T(A^T HA)B = B^T HB = H$, so that AB also belongs to it; (iii) if $A^T HA = H$ and $AB = I$, $B^T HB = B^T A^T HAB = (AB)^T H(AB) = I^T HI = H$.

13. The construction of the homogeneous Lorentz group as a Lie group is explained, for instance, in Kobayashi and Nomizu, *FDG*, vol. II, pp. 268 ff., Example 10.2. A proof that it is indeed a Lie subgroup of $GL(4, \mathbf{R})$ can be given on the lines of Chevalley's proof of this property for the rotation group $SO(n)$. (Chevalley (1946), pp. 119 f., Proposition 2.)

14. If a group G acts on the left on a set S (p. 26 and p. 288, n. 13), the *orbit* of an object $s \in S$ under this action is the set $\{gs | g \in G\}$.

15. Consequently, a succession of two or more pk transformations between Lorentz charts adapted to different, non-collinearly moving inertial frames necessarily involves a rotation of the space coordinate axes. This structural feature of the Lorentz group explains the physical phenomenon known as the Thomas precession. See, for instance, Arzeliès (1966), pp. 178 f.; Møller (1952), pp. 54 ff.

16. Robertson and Noonan (1968), p. 43; Møller (1980), p. 476.

17. Einstein evidently regarded both formulations as equivalent. See his (1911*b*), p. 1.

18. Møller (1952), p. 4, nicely combines both versions of the RP: "All physical phenomena should have the same course of development in all systems of inertia, and observers installed in different systems of inertia should thus as a result of their experiments arrive at the establishment of the same laws of nature." The frame that matters is, of course, not the one in which the observer is installed, but the one to which he actually refers his observations. They need not be the same. Thus, terrestrial astronomers use the fixed stars as their frame of reference.

19. See, for instance, the articles "Parity" and "*CPT* Theorem" in Lerner and Trigg's *Encyclopedia of Physics* (1981).

Section 3.5

1. C. Giannoni (1971), p. 580, observes that "the very concept of measuring a moving rod while not moving with it requires a conceptual innovation". Apparently nobody thought of such an operation before Einstein who with characteristic sureness lighted on the only plausible way of performing it. Since Q—in our example—is moving in F_v, it would be quite senseless to consider the distances between non-simultaneous positions of its vertices.

2. The relation between C and C_y as they meet for the second time is not governed by the transformation $y \cdot x^{-1}$, but by the transformation $y \cdot x'^{-1}$, between y and a suitable chart x' adapted to F'_0.

3. Prokhovnik (1967), p. 41, attributes this opinion to Herbert Dingle.

4. The problem we have been considering was first raised by Einstein himself, in his (1905*d*), p. 904, and has since been discussed incessantly, under the name of "the Clock Paradox" or "the Paradox of Twins". Marder (1971) gives a bibliography of some 250 items, to which more could be added now. The effect predicted by Einstein has been verified by observations of the decay rate of fast particles (B. Rossi, and D. B. Hall (1941), J. Bailey *et al.* (1977)) and of the behaviour of atomic clocks (J. C. Hafele and R. E. Keating (1972), C. O. Alley *et al.* (1977)). Good textbook analyses of the problem will be found in A. P. French (1968), pp. 154–9, and J. L. Anderson (1967), pp. 174–8. R. P. Perrin (1979) gives detailed calculations based on the realistic assumption that C undergoes acceleration during a brief but finite period as it changes frames.

5. Eqn (3.5.5) was first derived from Einstein's Rule by Laue (1907).

6. Einstein (1905*d*), §6. See also Larmor (1900), pp. 173 ff; Lorentz (1904); Poincaré (1905, 1906); Minkowski (1908); Einstein and Laub (1908*a*, *b*). Cf. pp. 109, 115 ff.

7. Einstein (1905*e*), (1906), (1907*a*, *b*), Planck (1906, 1907), Lewis and Tolman (1909), Frank (1909), Born (1909*a*, *b*, *c*; 1910), Laue (1911, 1911*b*), Epstein (1911), Tolman (1912), etc.

8. For example, if in (3.5.4) we take $v = 0.99c$, $w_1 = 0.99c$, and $w_2 = w_3 = 0$, then $\Sigma_\alpha u_\alpha^2 = 0.999899\, c^2$.

9. By definition, the four coordinate functions of a Lorentz chart must be real-valued, for they consist of three Cartesian space coordinate functions that measure distances along their parametric lines and a time coordinate function whose values must be linearly ordered. But if $\Sigma_\alpha v_\alpha^2 > c^2$ in equations (3.4.19) some at least of the coordinates related by them would be complex valued.

10. On tachyons and related matters see H. Schmidt (1958); S. Tanaka (1960); O. M. P. Bilaniuk *et al.* (1962); G. Feinberg (1967); Y. P. Terletskii (1968), Chapters V and VI; F. A. E. Pirani (1970); R. Fox *et al.* (1970); and E. Recami and R. Mignani (1974). P. Fitzgerald (1971) and J. Earman (1972*b*) address themselves to philosophical readers. Cf. also the interesting remarks in G. N. Whitrow (1980), pp. 356–60.

Section 3.6

1. If this is granted, the linearity of *all* Lorentz coordinate transformations follows easily, for each inhomogeneous Lorentz coordinate transformation is the product of an homogeneous one by a translation of $\mathbf{R}^4$.

2. See, for instance, Iyanaga and Kawada, *EDM*, 358.B; cf. also 201, 248 f. The transitive action of a group on a set was defined on p. 26. Note that if S is homogeneous with respect to G, the orbit of each $x \in S$, i.e. the set $\{gx \mid g \in G\}$, is identical with S. The action of G does not therefore partition S into several distinct classes.

3. They are also homogeneous relatively to their respective isometry group, but this fact is irrelevant to the question of linearity.

4. By the choice of an arbitrary point as zero, an affine space is made into a vector space, isomorphic with the one whose additive group acts on it. Cf. p. 18.

5. Remember that if V and W are vector spaces over the same field, their direct sum $V \oplus W$ is the vector space formed by all ordered pairs (v, w)—for each $v \in V$, $w \in W$—with the linear operations defined by:

$$\alpha(v_1, w_1) + \beta(v_2, w_2) = (\alpha v_1 + \beta v_2, \alpha w_1 + \beta w_2)$$

for any vectors $v_1, v_2 \in V$, $w_1, w_2 \in W$, and any pair of scalars α, β.

6. To prove this statement we may assume, without significant loss of generality, that the time coordinate functions y_1^0 and y_2^0 are identical, that the space coordinate functions $\bar{y}_1^\alpha$ and $\bar{y}_2^\alpha$ ($\alpha = 1, 2, 3$) agree at time $y^0 = 0$, and that the angular velocity of F_1 in F_2 is the unit vector in the direction of increasing $\bar{y}_2^1$. The transformation is then given by the *non-linear* system:

$$\begin{aligned} y_1^0 &= y_2^0 \qquad & y_1^2 &= y_2^2 \cos y_2^0 - y_2^3 \sin y_2^0 \\ y_1^1 &= y_2^1 \qquad & y_1^3 &= y_2^2 \sin y_2^0 + y_2^3 \cos y_2^0. \end{aligned}$$

7. The following proof is given by Berzi and Gorini (1969), p. 1524. If λ is a positive integer, (3.6.2) follows from (3.6.1) by induction, if we set $k = v$. As $f(0) = 0$, we have that $f(-v) = -f(v)$. Thus, (3.6.2) holds for any integer λ. If $\lambda = \mu/\nu$, where μ and ν are integers, set $\mu v = \nu w$. Then $\mu f(v) = f(\mu v) = f(\nu w) = \nu f(w)$. Thus, $(\mu/\nu) f(v) = f((\mu/\nu) v)$, and (3.6.2) holds for any rational number λ. Assume now that f is continuous at 0. (3.6.1) entails then that f is continuous everywhere. Let λ be any real number and (λ_n) a sequence of rationals converging to λ. Then, as n goes to ∞, $\lambda_n v$ converges to λv and $\lambda_n f(v)$ converges to $\lambda f(v)$. But $\lambda_n f(v) = f(\lambda_n v)$, which, by continuity, converges to $f(\lambda v)$. Thus, (3.6.2) holds for any $\lambda \in \mathbf{R}$.

8. The argument on p. 75 can be traced back to Lalan (1937). In the above form it is due to Berzi and Gorini (1969), p. 1519. Like Lalan, they base it on the "homogeneity of space-time" (not, as in Einstein's text, "of space *and* time"). This homogeneity, i.e. the fact that the common domain of all Lorentz charts is an affine space acted on by the additive group of $\mathbf{R}^4$, is simply taken for granted by these authors, as indeed it may be by someone who already knows that the Lorentz charts which the said domain by definition admits transform into one another by Lorentz coordinate transformations (in the mathematical sense). However, to *assume* the homogeneity of space-time in order to *prove* that Lorentz coordinate transformations (in the physical sense) are linear, does seem to beg the question. After all, as J. M. Lévy-Leblond (1976), p. 273, aptly remarks, "the homogeneity of space-time [. . .] does *not* hold in every conceivable physical theory". Terletskii (1968), pp. 18 f., gives a similar proof of the linearity of the transformations, appealing to "the uniformity of space and time, i.e. the independence of the properties of space and time of the choice of the initial points of measurement (the origin of the coordinates and the origin of t for time measurements)". Mittelstaedt (1976), p. 154, also invokes "the homogeneity of space and time", which—as he explains—"means that no place in space and no instant of time is especially distinguished". The facts adduced by Terletskii and Mittelstaedt would doubtless be highly relevant to the discussion of a coordinate transformation between two Lorentz charts adapted to one and the same inertial frame, for then indeed there would be no more than one homogeneous space and one homogeneous time under consideration. But where two such frames are involved we have to do with two Euclidean spaces and two Einstein times, and the homogeneity of each is not by itself sufficient to determine how each space/time pair is to be set in correspondence with the other.

9. This line of argument is developed in detail by V. I. Fock (1959), pp. 12–16, 377–84. The invariance of equations (3.6.3) and more generally of the Maxwell equations under the 15-parameter Lie group generated by the transformations (3.6.5/6) was first pointed out by H. Bateman (1910) and E. Cunningham (1910). See however S. Lie and F. Engel, *TTG*, vol. III (published in 1893), p. 334, Theorem 24. Other methods for proving that the Lorentz coordinate transformations must be linear or for deriving them "without linearity assumptions" are explained by P. Suppes (1959), R. Weinstock (1964) and W. Benz (1967). In my opinion they are less significant than the two we have considered here. The proof of linearity by M. Dutta, T. K. Mukherjee and M. K. Sen (1970) baffles me. For, on the face of it, the authors only assume that the transformations under study are analytic permutations of $\mathbf{R}^4$. Yet surely the non-linear transformation $t \mapsto t$, $x \mapsto x$, $y \mapsto y$, $z \mapsto z + \sin t$ satisfies this hypothesis. The proof must therefore rely on some additional, tacit assumption.

Section 3.7

1. See the bibliographies in Arzeliès (1966) pp. 80–2, and Berzi and Gorini (1969), note 2. To the works listed there let me only add—besides Berzi and Gorini's own rigorous and lucid paper—G. Süssmann (1969), A. R. Lee and T. M. Kalotas (1975, 1976), and J. M. Lévy-Leblond (1976). My own presentation is mainly based on Ignatowsky (1910, 1911*a*, *b*), Frank and Rothe (1911, 1912), Pars (1921), Lalan (1937), Berzi and Gorini (1969) and Levy-Léblond (1976).

2. See pp. 74 f. I do not mean to say that we already possess the means of ascertaining how v determines the transformation. For although we have assumed that T_F and $T_{F'}$ are unique, it might turn out that our assumptions, up to this point, are still insufficient to determine them.

3. As a matter of fact, Ignatowsky's argument does not explicitly or implicitly depend on the premise that the transformations are linear. (I thank John Stachel for reminding me of this.) But it does presuppose that *all* points of an inertial frame move in another such frame at *all* times with the same velocity, i.e. the very assumption from which we inferred the linearity of the Lorentz transformations in Section 3.6 (see note 9 to the present Section). Frank and Rothe (1912) derive linearity from the following two hypotheses: (I) The Principle of Inertia holds good in all inertial frames. (This leads to (A), on p. 75, and hence to the conclusion that the transformations must be projectivities of the form (3.6.4).) (II) If any two particles move with the same velocity $\mathbf{v}$ in an inertial frame F, they move in any other inertial frame F' with the same velocity $\mathbf{v}'$. (This implies that the projectivity which transforms a coordinate system adapted to F to one adapted to F' must be affine—i.e. that it must send points at infinity to points at infinity—so that its restriction to the finite domain is in effect a permutation of $\mathbf{R}^4$ and therefore a linear mapping.)

4. The speed $|v|$ with which either frame moves in the other one must be understood to mean the distance traversed in the stationary frame by a point fixed in the moving frame, in unit time *measured by clocks at rest in the stationary frame*. The Reciprocity Principle yields therefore a prescription for the synchronization of clocks in an inertial frame, which does not rely on optical means. If speed were taken to mean "self-measured" speed, as defined by Bridgman (1963), p. 26, i.e. the ratio of the distance traversed to the time measured by a clock carried by the moving object, the Reciprocity Principle would not say a word about the time coordinate functions attached to distinct inertial frames and could not therefore be used for inferring anything with regard to the manner how those time coordinate functions transform into each other.

5. Thus, for example, the Relativity Principle is satisfied if the inverse of $v \in I$ is $-v/(1-v)$. Ignatowsky (1911*a*), p. 5, argues for the Reciprocity Principle as follows: Let AB and $A'B'$ be unit measuring rods at rest respectively in the inertial frames F and F'. Let $A'B'$ slide along AB, so that B' passes A at time t_0, and passes B at time t_1 (of the unprimed coordinate system). Let t'_0 and t'_1 be the times (of the primed system) at which A passes B' and A', respectively. Then, according to Ignatowsky, the Relativity Principle demands that $\Delta t = t_1 - t_0 = t'_1 - t'_0 = \Delta t'$. Consequently, the speed $1/\Delta t$ of F' in F equals the speed $1/\Delta t'$ of F in F', and their respective velocities can only differ in sign. But Ignatowsky has obviously overestimated the strength of the Relativity Principle. From it you can infer that Δt is *the same function* of $\Delta t'$ as $\Delta t'$ is of Δt, but not that that function is the identity.

6. The reader may wonder why this should depend also on the speed $|v|$ of F' in F, and not—by analogy with (*b*)—on the speed—call it $|v'|$—of F in F'. But v', being the inverse of v in the velocity group, is uniquely correlated with it. And it is not being claimed that the dependence (*c*) has the same form as (*b*).

7. The continuity of the mapping is a truism if, as one would tend to assume, the velocity group is a one-parameter Lie group (or at least a one-dimensional topological group). Berzi and Gorini's proof makes use of the theorem that every continuous bijective mapping of a connected set of real numbers onto itself is either strictly increasing or strictly decreasing. They argue thus. If the mapping $v \mapsto v'$ is strictly increasing, and $v < v'$, $v' < (v')' = v$, which is absurd. Likewise, if $v' < v$, $v = (v')' < v'$. Therefore $v = v'$. Hence, unless the mapping in question is the identity—an alternative tacitly excluded by the authors—it must be strictly decreasing. Set $v^* = -v'$. By (3.7.3) $-(-v')' = v$, whence the preceding argument can be applied to the strictly increasing function $v \mapsto v^*$, to show that $v^* = v$. Therefore, $v' = -v$.

8. To verify this statement, pre- and postmultiply the matrix $[a_{ij}]$ by the matrix of a rotation of

the (y, z)-plane by an arbitrary angle θ, viz.

$$\begin{bmatrix} 1 & 0 & 0 & 0 \\ 0 & 1 & 0 & 0 \\ 0 & 0 & \cos\theta & \sin\theta \\ 0 & 0 & -\sin\theta & \cos\theta \end{bmatrix}$$

and equate the results.

9. From this point on, to eqn. (3.7.14), we closely follow Pars (1921). As I remarked in note 3, Ignatowsky (1911*a*) did not initially claim that the transformation from the unprimed to the primed system was linear. He did, on the other hand, argue that due to the isotropy of the planes perpendicular to the direction of motion, the coordinates y, y', z and z' can only occur in the transformation equations implicitly, through the distances r and r' from any given spatial point to the x-axis and, respectively, the x'-axis. The transformation equations read therefore:

$$t' = f(t, x, r, v) \qquad x' = g(t, x, r, v) \qquad r' = h(t, x, r, v) \qquad (*)$$

where f, g and h are smooth but otherwise arbitrary functions of t, x, r and the velocity v. Ignatowsky invokes the Relativity Principle to conclude that

$$t = f(t', x', r', v') \qquad x = g(t', x', r', v') \qquad r = h(t', x', r', v')$$

where v'—the velocity of F in F'—is, as we know, equal to $-v$ (see note 5; this value for v' is used by Ignatowsky (1911*a*) on p. 11, in the transition from eqn. (2) to eqn. (3)). Ignatowsky's next step is to take differentials of both sides of eqns. (*):

$$\begin{aligned} dt' &= f_{00}\,dt + f_{01}\,dx + f_{02}\,dr \\ dx' &= f_{10}\,dt + f_{11}\,dx + f_{12}\,dr \\ dr' &= f_{20}\,dt + f_{21}\,dx + f_{22}\,dr \end{aligned} \qquad (**)$$

where the f_{ij} are functions of t, x, r and v. He then argues that $f_{20} = f_{21} = 0$ and that $f_{22} = 1$, so that $dr' = dr$, whence $f_{02} = f_{12} = 0$. The rest of the argument is less straightforward than Pars' proof, essentially because Ignatowsky does not rely on the linearity of the system (**). But the reader cannot fail to observe that eqns (**) follow from eqns (*) only if $dv = 0$, i.e. if the velocity v is the same for every point of the relative space of the moving frame, at every instant of its relative time. As I argued in Section 3.6, this entails that the transformation is linear.

10. If times and lengths are measured in units of the kind we have agreed to use (p. 56), k is of course a pure number.

11. The reader may wish to complete the computation of the matrix of the transformation (3.7.5). We know already that, in (3.7.8), $\beta = \mu$, so we have only to determine β and ν. Putting $w = -v$ in (3.7.13*b*) we obtain

$$\beta(0) = (1 - kv^2)(\beta(v))^2$$

Setting $v = 0$ in the foregoing equation we see that

$$\beta(0) = (\beta(0))^2 = 1.$$

Thus,

$$\beta(v) = \frac{1}{\sqrt{1 - kv^2}} \qquad \nu(v) = \frac{kv}{\sqrt{1 - kv^2}}$$

Hence, equations (3.7.5) finally become:

$$t' = \frac{t - kvx}{\sqrt{1 - kv^2}} \qquad y' = y$$

$$x' = \frac{x - vt}{\sqrt{1 - kv^2}} \qquad z' = z \qquad (T)$$

If $0 < k = 1/c^2$, they are indeed the same as the pk Lorentz transformation equations (3.4.18). If $k = 0$ they reduce to a homogeneous pk Galilei transformation without rotation (cf. eqns (1.6.2)).

12. If $k < 0$, we can choose our units so that $k = -1$. Then $u = v*w = (v+w)(1-vw)^{-1}$. If we set $v = w = 1/2, u = 4/3 > 1$. Consequently, $1 \in I$. If we set $v = w = 1$, it turns out that $u = \infty$. If v and w are finite but greater than 1, $u < 0$. We have indeed assumed that $1/2 \in I$. But such must be the case if I includes any positive number, no matter how small, as will be readily gathered from our sketch of the structure of the group $(I, *)$ in note 13.

13. Lalan (1937), pp. 90–3. To see why a negative value of k is irreconcilable with the Chronology Principle, we can make use of the above remark that, if $k < 0$, the group $(I, *)$ is homeomorphic with the projective line and hence with the quotient space formed from the circle S^1 by identifying diametrically opposite points. Let $\bar{v}$ stand for arc tan v. Set $k = -1$. (3.7.14) can then be written as:

$$(\tan \bar{v})*(\tan \bar{w}) = \frac{\tan \bar{v} + \tan \bar{w}}{1 - \tan \bar{v} \tan \bar{w}} = \tan(\bar{v} + \bar{w})$$

The mapping $\bar{v} \mapsto \tan \bar{v}$ is thus a continuous homomorphism of the 1-parameter group of rotations $O(1)$ onto the group $(I, *)$. Substituting -1 for k and $\tan \bar{v}$ in eqns (T) in note 11, we obtain:

$$\begin{aligned} t' &= t \cos \bar{v} + x \sin \bar{v} \qquad & y' &= y \\ x' &= -t \sin \bar{v} + x \cos \bar{v} \qquad & z' &= z \end{aligned} \tag{T*}$$

Thus, if $k < 0$, the group G of transformations between matching coordinate systems adapted to collinearly moving inertial frames is none other than the realization of $O(1)$ as the group of rotations of the (t, x)-plane. No wonder therefore that a pair of events that share the same space coordinates in one system may turn up in another in reversed time order. Note that if we restrict the rotation angle $\bar{v}$ to $(0, \pi)$, arguing perhaps that the image of this interval by tan contains every number we can think of as a velocity, the set of coordinate transformations governed by equations (T*)—that is, by eqns (3.7.5)—is not a group.

Section 3.8

1. Whittaker (1951/53), vol. II, p. 40. The "amplifications" mentioned by Whittaker are Einstein's formulae for stellar aberration and the Doppler effect.

2. See, for instance, M. Born (1956), pp. 100 ff., G. Holton (1960), (1964), (1968), C. Scribner (1964), S. Goldberg (1967), (1969), E. Zahar (1973), T. Hirosige (1976), K. Schaffner (1976), A. I. Miller (1981). G. H. Keswani (1964/5, 1965/6) sides with Whittaker, though he somewhat tempers the latter's views. While Keswani is generally accurate in his handling of the sources—in contrast with Whittaker, who sometimes uses inverted commas rather superciliously—some of his mathematical arguments are untenable. For instance, in (1965/6), p. 27, Keswani criticizes Einstein's derivation of (3.4.5) and (3.4.10) because he "uses velocities of light propagation equal to $(c-v)$, $(c+v)$ and $\sqrt{(c^2-v^2)}$ in utter disregard of his own second postulate". Keswani apparently overlooks that the second postulate—the LP—applies to velocities referred to an inertial coordinate system whose time coordinate function is defined in agreement with Einstein (1905*d*), § 1, whereas the velocities $(c-v)$, etc., objected to by Keswani are referred to the auxiliary system u, defined by (3.4.2).

3. Poincaré (1906). The paper was submitted on July 23rd, 1905. A five-page abstract appeared in the *Comptes Rendus* of the Paris Academy of June 5th, 1905. Einstein (1905*d*) was submitted to *Annalen der Physik* on June 30th, 1905. Consequently, if their theories were indeed the same, Poincaré would take precedence over Einstein. Second inventors have no rights.

4. Poincaré (1954), p. 536. The quotation is taken from a course given at the Sorbonne in 1899.

5. In his painstaking study of Poincaré (1906), A. I. Miller writes that Poincaré "considered the retardation of moving clocks as experimentally unobservable. Thus, it is not a real effect; rather,

the hypothesis of a local time coordinate is purely mathematical" (Miller (1973), p. 243). But, as noted on p. 45, unless natural clocks keep Lorentz local time, the latter cannot be used for calculating the actual outcome of experiments.

6. Poincaré, *SH*, p. 101. Cf. the following statement of the "law of relativity", published in 1899: "The readings we shall be able to make on our instruments at any instant will depend only on the readings we could have made on these same instruments at the initial instant" (Poincaré, *SH*, p. 99).

7. Poincaré, *Oeuvres*, vol. IX, p. 412.

8. Poincaré, *SH*, p. 182.

9. Poincaré, *VS*, p. 127. A few pages later Poincaré dismisses the suggestion that the aether should be viewed as part of the physical systems to which the principle of relativity applies. If motion relative to the aether could be detected the principle ought to be pronounced false. However, "every attempt to measure the velocity of the Earth relative to the aether has led to negative results. This time experimental physics has been more faithful to the principles than mathematical physics; the theoreticians would have readily discarded [the Relativity Principle] in order to achieve consistency in their other general views; but experience has obstinately confirmed it" (Poincaré, *VS*, p. 132).

10. Poincaré, *SH*, pp. 111 f.

11. Poincaré, *VS*, pp. 52 f.

12. Poincaré, *SH*, pp. 182 f.

13. Lorentz (1904), in *CP*, vol. V, p. 174; Kilmister (1970), p. 121.

14. Lorentz (1904), in *CP*, vol. V, p. 176; Kilmister (1970), p. 123. It will be seen that, if we set $\varepsilon^{-1} = l$ in (2.2.10)—which is indeed compatible with the conditions prescribed for both quantities—(3.8.1) turns out to be the product of the transformations (2.2.10) and (2.2.9) of Lorentz (1899).

15. Lorentz (1904), in *CP*, vol. V, p. 188; Kilmister (1970), p. 138.

16. Lorentz (1904), in *CP*, vol. V, pp. 189 f.; Kilmister (1970), p. 139.

17. See the note added by Lorentz in 1912 to the German translation of his (1904), in Lorentz *et al.*, *RP*, p. 10.

18. Lorentz's statement of his two main hypotheses may give us an inkling of his intellectual style: (*a*) Charged particles or "electrons" are spherical when at rest in the aether, and "have their dimensions changed by the effect of a translation, the dimensions in the direction of motion becoming βl times and those in perpendicular directions l times smaller." (*b*) "The forces between uncharged particles, as well as those between such particles and electrons, are influenced by a translation in quite the same way as the electric forces in an electrostatic system" (Lorentz (1904), in *CP*, vol. V, pp. 182 f.; Kilmister (1970), p. 131). The effects of translation on the electric forces can be calculated as follows: Let Σ' be an isolated material system at rest in the aether, which differs from the moving system Σ only in that the dimensions parallel to the x-axis are multiplied by βl and those perpendicular to it are multiplied by l; "then we shall obtain the forces acting on the electrons of the moving system Σ, if we first determine the corresponding forces in Σ', and next multiply their components in the direction of the axis of x by l^2, and their components perpendicular to that axis by l^2/β" (Lorentz (1904), in *CP*, vol. V, p. 179; Kilmister (1970), p. 127).

19. Lorentz (1916), pp. v, 229.

20. Lorentz (1916), p. 230; my italics.

21. Lorentz (1916), p. 230.

22. Lorentz (1916), p. 321. See also Lorentz, *CP*, VII, p. 262.

23. Poincaré (1906), p. 129; English translation in Kilmister (1970), p. 145.

24. Poincaré (1906), p. 129; Kilmister (1970), p. 146.

25. Poincaré (1906), p. 130; Kilmister (1970), p. 146.
The first such modification concerns the transformation of electric charge density; Poincaré proves that Lorentz's formula cannot be correct, for his transformed density does not satisfy the equation of continuity (Poincaré (1906), pp. 133 f.; Kilmister (1970), pp. 152 f.). Poincaré pointed this out to Lorentz in a letter written in 1904, published by A. I. Miller (1980), p. 76.

26. See, for instance, Poincaré (1906), pp. 163, 165, and especially p. 151: "The Lorentz transformation substitutes an ideal electron at rest for the real electron in motion." Cf. A. I. Miller's remarks on this passage in his (1973), p. 266.

27. Poincaré (1906), p. 132; Kilmister (1970), p. 150.

28. In a letter to Lorentz written late in 1904 or early in 1905 (reproduced by A. I. Miller (1980), pp. 79 f.), Poincaré *proved* that the l factor in (3.8.1) must be equal to 1 if it is *assumed* that the transformations between the several Lorentz primed coordinate systems defined by (3.8.1) as v ranges over the interval $(-c, c)$ form a one-parameter Lie group.

29. Poincaré (1906), p. 168; Kilmister (1970), pp. 175 f. Note however then in §4 of his paper Poincaré designates as "the *Lorentz group*" the broader group generated by Lorentz transformations and dilatations. The group characterized in the text above is obtained in §4 as a subgroup of this extended "Lorentz group" by imposing an additional requirement to the Lorentz equations (3.7.1), namely, that the constant l be a function of v. Poincaré shows that l must then be equal to 1, if it is assumed that the transformations satisfying (3.7.1) subject to this requirement must still form a group. See Poincaré (1906), p. 146; Kilmister (1970), pp. 171 f.

30. Poincaré (1906), p. 168; Kilmister (1970), p. 176.

31. Poincaré (1906), §9; cf. A. I. Miller (1973), p. 308.

32. Poincaré, *SH*, p. 215.

33. A. I. Miller (1973), p. 320.

34. I was happy to learn, after writing the present Section, that S. G. Suvorov (1979), pp. 542 ff., had reached a similar conclusion with regard to the source of Poincaré's shortcomings. In my view, however, Poincaré's conventionalism was not so radical as Suvorov depicts it; see Torretti (1978*a*), pp. 320 ff. See below, p. 350, Addenda to this note.

35. In a lecture given in London on May 4, 1912, Poincaré spoke of "the new conception of space and time", resulting from "Lorentz's Principle of Relativity", according to which "space and time are no longer two separate entities [. . .] but two parts of the same whole, which are so intimately bound together, that they cannot be easily separated". He describes it as "a new convention" that some physicists wish to adopt. "Not that they are constrained to do so; they feel that this new convention is more comfortable, that's all; and those who do not share their opinion may legitimately retain the old one, to avoid disturbing their ancient habits. Between ourselves, let me say that I feel they will continue to do so for a long time still" (Poincaré, *DP*, p. 109).

36. C. Seelig (1954), p. 134.

37. For instance, in two lectures of 1912; Poincaré, *DP*, pp. 82, 108. On Poincaré's silence, see S. Goldberg (1970).

Chapter 4

Section 4.1

1. See D. Hilbert (1910).

2. L. Pyenson (1977) and P. L. Galison (1979) study Minkowski's contribution to Relativity. Both articles contain much useful historical information, but tend on the whole to underrate the physical significance of Minkowski's mathematical and philosophical insight. For a fairer judgment of the importance of Minkowski's work and valuable references to its early reception, see T. Hirosige (1968), pp. 46 ff. I thank Kenneth Schaffner for calling my attention to this paper by the great Japanese scholar.

3. This seminar is the subject of an interesting article by L. Pyenson (1979).

4. Minkowski (1915).

5. Minkowski apparently expected that some sort of "electromagnetic world-view" could be built on the Relativity Principle. Cf. Minkowski (1909), p. 111, last paragraph.

6. Minkowski (1915), p. 372.

7. Minkowski (1910), submitted to the Kgl. Gesellschaft der Wissenschaften zu Gottingen on 21 December 1907; first printed in its *Nachrichten* in 1908. Our references are to the reprint in *Mathematische Annalen*. An appendix to this paper sketches a four-dimensional formulation of relativistic mechanics.

8. Minkowski (1909); English translation in Lorentz *et al.*, *PR*. The reader is advised to peruse this lecture now, if he has never read it.

9. Minkowski (1915), p. 373.

10. Minkowski, who was still under the influence of the Lie–Klein conception of geometrical

space as a "manifold of numbers" (*Zahlenmannigfaltigkeit*), equates "a worldpoint" with "a point of space at a point of time, i.e., a system of values x, y, z, t" (Minkowski (1909), p. 104; cf. his (1910), p. 475, last two lines). It might seem that "worldpoint" is only a fancy name for a number quadruple and that Minkowski's "world" is none other than $\mathbf{R}^4$. But a few lines later he adds: "In order that there is nowhere a gaping void, we shall assume that something perceivable exists at each place at all times. To avoid saying *matter* or *electricity*, we designate this something by the word *substance*. Let us turn our attention to the substantial point that is present at the worldpoint x, y, z, t." Since no physical, perceivable, substance can be present at a number quadruple, we must conclude that Minkowski speaks ambiguously. I resolve the ambiguity by distinguishing between a physical manifold of "worldpoints", or "world", and the "number manifold" $\mathbf{R}^4$, used for charting the former. I have discussed the concept of a *Zahlenmannigfaltigkeit* in Torretti (1978*a*), pp. 137 f., 173, 393 n. 42, 399 n. 81.

11. The concept of a quadratic form is defined below, in note 1 (*b*) to Section 4.2.
12. See above, p. 64.
13. Poincaré was the first to note this; see above, p. 86. It can be proved by deriving the Lorentz transformation from the assumed invariance of (4.1.1); see, for instance, Einstein (1956), pp. 32 ff.
14. One may always choose $\mathbf{r}$ so that its Euclidean length $|\mathbf{r}| = \theta$, and its sense is such that $\mathbf{r}$ represents, according to the accepted conventions, the rotation induced by ρ on each of the said hyperplanes. The three components r^α suffice then to characterize ρ. The matrix $(a_{\alpha\beta})$ is expressed as a function of $\mathbf{r}$ by Sudarshan and Mukunda (1974), p. 238, eqn. (11).
15. Let x and y be two Lorentz charts and denote $y^{-1}\cdot x(P)$ by P'. Since f_2 is invariant under the coordinate transformation $y\cdot x^{-1}$, $F(P', Q') = f_2(y(P'), y(Q')) = f_2(x(P), x(Q)) = F(P, Q)$.

Section 4.2

1. The following mathematical concepts will be essential in the sequel.

(*a*) Let V be a real vector space. An *inner product* on V is a symmetric bilinear function $f: V\times V \to \mathbf{R}$, such that $f(v, w) = 0$ for every $w \in V$ only if $v = 0$. f is *positive definite* if $f(v, v) > 0$ for every $v \in V$ unless $v = 0$. It is *negative definite* if $f(v, v) < 0$ for every $v \in v$ unless $v = 0$. If f is neither positive nor negative definite it is said to be *indefinite*. The structure $\langle V, f\rangle$ is called an *inner product space*. Let V be n-dimensional. A basis $(e_1, \ldots, e_n)$ of V is said to be *orthonormal* if $f(e_i, e_j) = \pm\delta_{ij}$, i.e. ± 1 if $i = j$ and 0 if $i \neq j$ ($1 \leqslant i, j \leqslant n$). Every finite-dimensional inner product space has an orthonormal basis. Let $(e_1, \ldots, e_n)$ be an orthonormal basis of $\langle V, f\rangle$. The trace of the matrix $[f(e_i, e_j)]$—i.e. the sum $\Sigma_i f(e_i, e_i)$ of its diagonal elements—is called the *signature* of the inner product f. (Note that it does not depend on the choice of a particular orthonormal basis.) Evidently the signature of f is equal to n if f is positive definite, to $-n$ if f is negative definite, and to some integer less than n but greater than $-n$ if f is indefinite.

(*b*) Let V be a finite dimensional vector space over a field F of characteristic $\neq 2$. A *quadratic form* on V is a function $q: V \to F$ such that (i) $q(v) = q(-v)$ for all $v \in V$, and (ii) the function $h: V\times V \to F$ given by $h(v, w) = (1/2)(q(v+w) - q(v) - q(w))$ is bilinear. h is the bilinear function obtained by *polarizing* the quadratic form q.

(*c*) If S is a set and d is a real valued function on $S\times S$, $\langle S, d\rangle$ is a *metric space* if for any elements $a, b, c \in S$ (i) $d(a, a) = 0$, (ii) $d(a, b) > 0$ unless $a = b$, and (iii) $d(a, b) + d(b, c) \geqslant d(a, c)$. A function d meeting these requirements is called a *distance function* on S. The reader should satisfy himself that if $\langle V, f\rangle$ is an inner product space with positive definite inner product f and we set $d(v, w) = f(v-w, v-w)$ for every $v, w \in V$, d is a distance function on V. A metric space $\langle S, d\rangle$ is made into a topological space by the following stipulations: For each $a \in S$, $r \in \mathbf{R}$, we define the *open ball* $B(a, r)$ to be the set $\{x \in S \mid d(a, x) < r\}$; a subset of S is *open* if and only if it is empty or it is a union of open balls. The topology thus defined is said to be *induced by the metric*.

2. The null cone in $\mathbf{R}^4$ is the set of real number quadruples $\{(k, kr_1, kr_2, kr_3) \mid k \in \mathbf{R}; r^\alpha r^\alpha = 1\}$.
3. To ensure that the space coordinate functions of the said chart form a right-handed system, as prescribed on page 66, we must impose an additional physical condition on H or replace (*c*) by

the following: (c') If $\langle v, v \rangle = 1$ and $R_\alpha = u_\alpha P = w_\alpha Q$, where $\langle w_\alpha, w_\beta \rangle = \eta_{\alpha\beta}$, and $\langle u_\alpha, u_\alpha \rangle = \langle v, w_\alpha \rangle = 0$ ($\alpha, \beta = 1, 2, 3$), there could exist an unstressed right-handed inertial tetrahedron $OX_1X_2X_3$ of the kind described on page 14, such that P and Q occur at O and R_α at X_α.

4. By (4.2.3), if $\varphi, \psi \in \mathscr{L}_0$, $H_\varphi(\psi(v), P) = H_\varphi(\psi(v), P) = H(\varphi \cdot \psi(v), P)$. Consequently, $L(\psi, L(\varphi, H)) = L(\psi, H_\varphi) = L(\varphi \cdot \psi, H)$. Note that L is a "right action", in the sense defined on p. 288, note 13.

5. This unfortunate name stems from a particular application of differential geometry to mechanics. To avoid confusion with the physical concept of energy I write "energy" in quotes when it stands for the geometrical concept.

6. The concept of a critical point of E can be summarily explained as follows: Let K be the set of all parametrized curves $\sigma: [p, q] \to S$ whose endpoints are $\gamma(p)$ and $\gamma(q)$. The mapping $\sigma \mapsto E(\sigma)$ is a real valued function on K. Denote $[p, q]$ by I_2, and let I_1 be a closed interval containing 0. A family $(\sigma_u) \subset K$, indexed by I_1 is said to be a *variation* of γ if $\gamma = \sigma_0$ and there is a smooth mapping f of $I_1 \times I_2$ into S such that $\sigma_u(t) = f(u, t)$ for each $u \in I_1$ and each $t \in I_2$. γ is a *critical point* of the function E if, for every variation (σ_u) of γ, the derivative $dE(\sigma_u)/du$ is equal to 0 at $u = 0$.

7. Customarily a curve γ in a proper Riemannian manifold $\langle S, \boldsymbol{\mu} \rangle$ is said to be a "geodesic" if it is length-critical on every closed subinterval of its domain. If γ is geodesic in this sense there exists always a reparametrization of γ which is a Riemannian geodesic in the sense defined above (and hence also a geodesic of the Levi-Cività connection determined by $\boldsymbol{\mu}$—see Appendix, pp. 277, 279).

8. See the definition of a metric space in note 1.

9. To show this one need only solve the geodesic equations (D.8) on p. 281 in the particular case in which x is a Lorentz chart, so that the $g_{ij} = \eta_{ij}$ and the Γ^i_{jk} are all 0. In this case, the right hand side reduces to 0. Hence, if γ is a Riemannian geodesic of $\langle \mathscr{M}, \boldsymbol{\eta} \rangle$, $x \cdot \gamma$ is a straight line in $\mathbf{R}^4$, whose image by x^{-1} must also be straight in the affine space $\langle \mathscr{M}, \mathbf{R}^4, H \rangle$.

10. Minkowski (1909), p. 108.

11. W. Rindler, (1977), p. 43.

12. The length of OB can be easily calculated from our data. Let θ be the angle between μ and the x^0-axis (or between the x^1-axis and ν). Obviously, $\tan\theta = v$ (the speed of Y in X). Set $\measuredangle OAB = \omega$. Clearly $\theta + \omega/2 = \pi/4$. Hence $\sin\omega = \cos 2\theta$. Since $\measuredangle OBA = \theta + \pi/2$, $OB/OA = \sin\omega/\cos\theta = \cos 2\theta/\cos\theta$. Noting that the perpendicular projections of OA on the x^0-axis and on the x^1-axis are, respectively, $OA\sin\theta$ and $OA\cos\theta$, we infer from the hyperbola equation $(x^1)^2 - (x^0)^2 = 1$ that $(OA)^2 = (\cos^2\theta - \sin^2\theta)^{-1} = 1/\cos 2\theta$. Hence $OB = (\cos 2\theta/\cos^2\theta)^{1/2} = \sqrt{(1 - \tan^2\theta)} = \sqrt{(1 - v^2)}$, in agreement with the length contraction formula (3.5.1). The length of OS can be calculated similarly. $OS = OB\cos\theta$. But $(OB)^2 = (OA)^2 = (\cos^2\theta - \sin^2\theta)^{-1}$. Hence, $OS = (\cos 2\theta/\cos^2\theta)^{-1/2} = 1/\sqrt{(1 - v^2)}$, in agreement with (3.5.2).

Section 4.3

1. Fibre bundles are defined in the Appendix, pp. 263 f. However, it is not necessary to master their definition in order to understand the present section. See note 3.

2. A (q, r)-tensor at P is said to be q times contravariant and r times covariant. It is not, however, the only kind of mixed tensor at P that can be so described. It differs, for instance, from a $(q+r)$-linear function on $(S_P)^r \times (S_P^*)^q$, or one on $(S_P^*) \times (S_P) \times (S_P^*)^{q-1} \times (S_P)^{r-1}$, etc. To avoid cumbersome explanations, we shall consider only (q, r)-tensors in the sense defined above. The reader can easily adapt all definitions to the other kinds of mixed tensors one might conceive.

3. The tangent bundle of S is defined on pp. 265. Higher order tensor bundles can be defined analogously. But the crucial notion of a (q, r)-tensor field on S can also be defined without resorting to the concept of a fibre bundle. Let x be a chart defined at $P \in S$. Let e_i and e_i^* stand, respectively, for the vector $\partial/\partial x^i|_P$ and its dual $dx^i(P)$. $(e_1, \ldots, e_n)$ is a basis of S_P and $(e_1^*, \ldots, e_n^*)$ is the dual basis of S_P^* (pp. 259–60). Hence, a (q, r)-tensor t at P is fully determined by the n^{q+r} real numbers $\mathrm{t}^{i_1 \ldots i_q}_{j_1 \ldots j_r} = \mathrm{t}(e^*_{i_1}, \ldots, e^*_{i_q}, e_{j_1}, \ldots, e_{j_r})$ $(1 \leq i_h, j_k \leq n; 1 \leq h \leq q;\ 1 \leq k \leq r)$. These real numbers are called the *components* of t relative to the chart x. Let T assign to each $P \in S$ a (q, r)-tensor T_P at P. Then, if $U \subset S$ is the domain of a chart x, T assigns to each $P \in U$ the array of T_P's

components relative to x. Thus, T defines on U an array of n^{q+r} real-valued functions, which we may call the components of T relative to x. T is a (q, r)-*tensor field* if every component of T relative to each chart of S is smooth.

4. The module structure of $\mathscr{V}^1(S)$ is defined in the Appendix, p. 260. The definition is readily extended to $\mathscr{V}^q_r(S)$: bear in mind that the values of all elements of $\mathscr{V}^q_r(S)$ at a given point $P \in S$ belong to the same vector space attached to that point, formed by all pairs (P, T_P), where T_P is a (q, r)-tensor at P.

5. Principal fibre bundles are defined on p. 263.

6. The following remarks on components will be useful later. In this note, small Latin indices range from 1 to n. If the chart x is defined on $U_x \subset S$, the n-ad field $(\partial/\partial x^i)$ is a basis of $\mathscr{V}^1(U_x)$. If $\mathrm{d}x^j$ stands, as on p. 260 for the differential of the coordinate function x^j, $\mathrm{d}x^j(\partial/\partial x^i) = \partial x^j/\partial x^i = \delta_{ij}$. Hence, the co-$n$-ad field $(\mathrm{d}x^i)$ is the basis of $\mathscr{V}_1(U_x)$ dual to $(\partial/\partial x^i)$. If V is a vector field on S, its components V^i relative to the chart x are given by: $\mathrm{V}^i = \mathrm{V}(\mathrm{d}x^i) = \mathrm{d}x^i(\mathrm{V}) = \mathrm{V}x^i$. On U_x, V is equal to a linear combination $L^i\partial/\partial x^i$. We see at once that $L^i = L^j\delta_{ij} = L^j\mathrm{d}x^i(\partial/\partial x^j) = \mathrm{d}x^i(L^j\partial/\partial x^j) = \mathrm{d}x^i(\mathrm{V}) = \mathrm{V}^i$. For further details on the components of the covariant, contravariant and mixed forms of a given tensor field, relative to a particular chart, and on the transformations rules that link the arrays of such components, relative to different charts, the reader will do well to consult a good textbook on the tensor calculus, such as Lichnerowicz (1962), Synge and Schild, *TC*, or Lovelock and Rund (1975).

7. It will be noticed that I am using the same letter T to designate three different things, namely, a mapping of the manifold S on its (q, r)-tensor bundle, a $(q+r)$-linear mapping of a Cartesian product constructed from the modules of vector and covector fields on $U \subset S$ into the ring of scalar fields on U, and a mapping of the set of n-ad fields on U into a Cartesian product constructed from the said ring of scalar fields. But the first two of these three things are uniquely associated to one another by an isomorphism of the linear spaces to which each belongs (see, for instance, Abraham and Marsden (1978), p. 82, Theorem 2.2.8), while the third is unambiguously determined by the second. It is therefore more sensible to denote all three by the same letter than to burden our memories with three different symbols. It is also worth noting that, if two (q, r)-tensor fields T and T′ have the same array of components relative to a particular n-ad field X on S, T and T′ are equal. For the difference between the values of T and T′ at any suitable list of elements of X^* and X is the value of the tensor field $(\mathrm{T} - \mathrm{T}')$ at that list. Now, if $(\mathrm{T} - \mathrm{T}')$ equals 0 at each relevant list, $(\mathrm{T} - \mathrm{T}')$ is the zero vector in the module of (q, r)-tensor fields on S, and $\mathrm{T} = \mathrm{T}'$. Evidently, in this case T and T′ have the same array of components relative to each n-ad field on S (or part of S). A similar argument can be invoked to establish the symmetry (skew-symmetry) of a tensor field of order 2 if the matrix of its components relative to a particular n-ad field on its domain is symmetric (skew-symmetric).

8. See Appendix, pp. 261 f.

9. See Appendix, pp. 267 f.

10. These concepts are explained in the Appendix, pp. 274 f.

11. See Appendix, Section C. 8, pp. 279 f.

12. Paraphrased from Minkowski (1910), p. 484. Minkowski actually uses here the spacetime chart with imaginary time coordinates introduced by Poincaré (see p. 86). Hence, for him r_0 (which he calls r_4) is not a real but an imaginary number.

13. Compare this simple definition with the formidable construction in Minkowski (1910), pp. 484 f.

14. Let y be another Lorentz chart adapted to F. $\mathrm{V} = \mathrm{V}y^i\partial/\partial x^i$. As $y^0 = x^0$, $\mathrm{V}y^0 = \mathrm{V}x^0 = \mathrm{V}^0$, and $\mathrm{V}y^\alpha\partial/\partial y^\alpha = \mathrm{V} - \mathrm{V}^0\partial/\partial x^0 = \mathbf{V}$.

15. Let $\mathbf{V}|F_P$ be the restriction of $\mathbf{V}$ to F_P, and let $\bar{x}$ be, as on p. 56, the Cartesian chart on S_F associated to x. The Cartesian vector field in question, which we may denote by $\overline{\mathbf{V}}$, is defined by the equations: $\mathrm{d}\bar{x}^\alpha(\overline{\mathbf{V}}) = \mathrm{d}x^\alpha(\mathbf{V}|F_P)$.

16. $\gamma_{*\tau}$ abbreviates $\gamma_{*\tau}(\partial/\partial\tau|_\tau)$; see p. 261. Hence, at $\gamma(\tau)$, by (A.7),

$$\mathrm{U}^i = \mathrm{U}y^i = \gamma_{*\tau}\left(\frac{\partial}{\partial\tau}\bigg|_\tau\right)y^i = \frac{\mathrm{d}y^i}{\mathrm{d}\tau}\bigg|_{\gamma(\tau)}$$

17. Note that $E(\mathrm{U}) = 1$ only if $c = 1$; otherwise, $E(\mathrm{U}) = c^2$. Our choice of units normalizes the 4-velocities of all material particles to unit length. The argument leading to (4.3.5) has been

simplified to avoid distracting the reader with too many symbols. Strictly speaking, the components of **u**, for each time t, are, according to (4.3.2), $u^\alpha \cdot \bar{\gamma} = \mathrm{d}(\bar{y}^\alpha \cdot \bar{\gamma})/\mathrm{d}t$, while those of U, for each proper time τ, according to (4.3.3), are $U^i \cdot \gamma = \mathrm{d}(y^i \cdot \gamma)/\mathrm{d}\tau$. If t is regarded as a function of τ (i.e., if t stands, as, in $\mathrm{d}t/\mathrm{d}\tau$, for the mapping $\tau \mapsto t$), we can readily verify that $\bar{y}^\alpha \cdot \bar{\gamma} \cdot t = y^\alpha \cdot \gamma$. Hence $\mathrm{U}^\alpha \cdot \dot{\gamma} = \mathrm{d}(y^\alpha \cdot \gamma)/\mathrm{d}\tau = (\mathrm{d}(\bar{y}^\alpha \cdot \bar{\gamma})/\mathrm{d}t)(\mathrm{d}t/\mathrm{d}\tau) = \beta_{u \cdot \gamma}(u^\alpha \cdot \gamma)$.

18. This definition, of course, is admissible only if A is in fact a 4-vector. The easiest way of proving it is to verify that its components obey the appropriate transformation rule. If we denote the components of U relative to y by $\mathrm{U}^i_{(y)}$, and x is another Lorentz chart, it is clear that $\mathrm{U}^i_{(x)} = (\partial x^i/\partial y^j)\mathrm{U}^j_{(y)}$, for U is, by definition, a 4-vector. Hence $\mathrm{A}^i_{(x)} = \mathrm{dU}^i_{(x)}/\mathrm{d}\tau = \mathrm{d}/\mathrm{d}\tau(\mathrm{U}^j_{(y)}\partial x^i/\partial y^j) = (\partial x^i/\partial y^j)\mathrm{A}^i_{(y)}$. Observe that the last equation results from the fact that if $x \cdot y^{-1}$ is a Lorentz transformation, the $\partial x^i/\partial y^j$ are constant functions.

19. See, for instance, F. Kottler (1912); W. Pauli (1958), pp. 156 f.; V. Fock (1959), Chapter IV, etc.

20. See, for instance, H. Heintzmann and P. Mittelstaedt (1968). Reading this article, especially pp. 216 ff, 223 ff., it seemed to me that the authors assume that the components of a geometrical object relative to the frame on which they are being observed ought to be measured always by the same experimental devices, regardless of the type of frame involved. Yet nobody would wish to measure time on a spaceship with pendulum clocks, no matter how accurate they have proved to be in terrestrial labs. When extending 4-tensors to ordinary tensors, one should bear in mind that inertial and non-inertial frames are not physically equivalent, even if the tensor calculus provides the means for dealing with them mathematically on the same footing.

Section 4.4

1. The early development of Special Relativity is studied in detail by A. I. Miller (1981).

2. In our units ($c = 1$). Except for this and for the change in notation, eqns (4.4.4) are the same as the "Maxwell–Hertz equations" given in Einstein (1905*d*), p. 907. In the standard vector notation they read thus:

$$\partial\mathbf{E}/\partial x^0 = \nabla \times \mathbf{H} \qquad -\partial\mathbf{H}/\partial x^0 = \nabla \times \mathbf{E}.$$

3. Equations (4.4.5) can be derived from (4.4.4) by subjecting the differential operators in the latter to the Lorentz transformation (3.4.18). See, for instance, L. Silberstein (1924), pp. 225 f. Of course, **E**′ and **H**′ are divergence-free, like **E** and **H**.

4. Cf. for instance, Lorentz (1892*a*), §42 (*CP*, II, p. 198); (1895), §§9, 12 (*CP*, V, pp. 18, 21); Einstein (1905*d*), p. 909.

5. Einstein (1905*d*), pp. 909 f.

6. Cf. the discussion of time-dependent vector fields on p. 103.

7. More precisely: to their associated Cartesian charts.

8. "If a unit electric point charge is in motion in an electromagnetic field, the force acting upon it is equal to the electric force which is present at the locality of the charge, and which we ascertain by transformation of the field to a system of coordinates at rest relatively to the electrical charge" (Einstein (1905*d*), p. 910; Kilmister (1970), p. 206).

9. Einstein (1905*d*), p. 909, says that this is a consequence of the Relativity Principle, but evidently the RP *alone* yields no conclusions as to the invariance of charge.

10. "These results as to the mass are also valid for all ponderable material points, because a ponderable material point can be made into an electron (in our sense of the word) by the addition of an *arbitrarily small* electric charge" (Einstein, (1905*d*), p. 919; Kilmister (1970), p. 216).

11. J. Stachel and R. Torretti (1982) discuss the derivation of the mass-energy equation in Einstein (1905*e*). We there dissect and dismiss the allegation by H. I. Ives (1952) that Einstein's reasoning is fallacious. Other derivations of the mass-energy relation were given by Einstein (1906), (1907*a*), (1907*b*), pp. 440 ff. and by Planck (1907); cf. also Einstein (1935).

12. Einstein (1905*e*), p. 641.

13. M. Bunge (1967) argues that the "numerical equivalence" of the (proper or relativistic) mass and the energy of a body does not imply that they are the same thing, for "numerical equalities do not entail the identity of the predicates involved" (p. 202). On the other hand, Einstein believed

that "science is fully justified in assigning . . . a numerical equality only after this numerical equality is reduced to an equality of the real nature of the two concepts" under consideration. (Einstein (1956), pp. 56 f.) Be that as it may, if a kitchen refrigerator can extract mass from a jug of water and transfer it by heat radiation or convection to the kitchen wall behind it, a trenchant metaphysical distinction between the mass and the energy of matter does seem far-fetched. Of course, if lengths and times are measured with different, unrelated units, the "mass" $m_0 \beta$, measured, say, in grams, differs conceptually from the "energy" $m_0 \beta c^2$, measured in ergs. But this difference can be understood as a consequence of the convenient but deceitful act of the mind by which we abstract time and space from nature. Equation (4.4.13) and the equivalence of mass and energy can be vindicated by purely mechanical arguments, without appealing to the laws of electrodynamics; see, in particular, the beautiful paper by R. Penrose and W. Rindler (1965).

14. Even E. G. Cullwick (1981), who sees in the 1905 definition of force the root of some alleged "inconsistencies" in Einstein's electrodynamics, describes the definition itself as *consistent but incorrect* (*loc. cit.*, p. 173).

15. See, for instance, R. C. Tolman (1934), pp. 43 ff.; W. Rindler (1960), pp. 77 ff.; A. P. French (1968), pp. 167 ff.

16. Page 110. Lorentz (1904) had given the now accepted value of the transverse mass (Lorentz, *CP*, V, p. 185, eqn. (30)). The discrepancy with Einstein's value suggests indeed that in 1905 Einstein had not yet read Lorentz's paper, but does not prove it conclusively. Einstein's value follows from his carefully contrived definition of force, and he expressly warns us that a different definition would yield another value.

Section 4.5

1. The following sketch cannot claim to be much more than an aid to memory, though I hope it will give the reader who first meets the subject a feeling of its drift and will move him to seek better instruction elsewhere. Besides the introductory monograph by W. Rindler (1960), the encyclopedia article by P. G. Bergmann (1962*a*) and the classical treatise by J. L. Synge (1955), I recommend the lucid and profound book by W. G. Dixon (1978), which clarifies and refines, particularly in Chapter 3, several key points that are shoddily dealt with here or in the foregoing Section.

2. Minkowski (1910), p. 521, eqn. (22). Minkowski places the proper mass outside the scope of the differential operator, because in the present context he explicitly *assumes* that it is constant. In that case, as he correctly points out, the 4-force K is orthogonal to the 4-velocity U. But he was well aware that it was a special case, as one may gather from a remark he makes in another context, to the effect that "the quantity that deserves to be called the body's mass . . . changes . . . with each heat transfer" (Minkowski (1915), p. 381).

3. The components of the (1, 1) mixed form of F are $\mathrm{F}^i_j = \eta_{jk}\mathrm{F}^{ik}$; those of its covariant form are $\mathrm{F}_{ij} = \eta_{ik}\mathrm{F}^k_j$. It follows that the F^i_j are identical with the F^{ij}, except that $\mathrm{F}^\alpha_0 = -\mathrm{F}^{\alpha 0} = E_\alpha$, while the F_{ij} are identical with the F^i_j, except that $\mathrm{F}_{0\alpha} = -\mathrm{F}^0_\alpha = -E_\alpha$.

4. Use the transformation rule $\mathrm{F}'^{ij} = (\partial y^i/\partial x^h)(\partial y^j/\partial x^k)\mathrm{F}^{hk}$, and (3.4.18′).

5. The system on the left epitomizes the Maxwell equations usually given as:

$$\nabla \times \mathbf{E} = -\partial \mathbf{H}/\partial x^0 \qquad \nabla \cdot \mathbf{H} = 0.$$

The system on the right epitomizes the Maxwell equations:

$$\nabla \times \mathbf{H} = 4\pi \mathbf{J} + \partial \mathbf{E}/\partial x^0 \qquad \nabla \cdot \mathbf{E} = 4\pi\rho.$$

A more elegant formulation is achieved by introducing the electrostatic potential φ and the vector potential **A**, such that (i) $\mathbf{H} = \nabla \times \mathbf{A}$, (ii) $\mathbf{E} = -\partial \mathrm{A}/\partial t - \nabla\varphi$, and (iii) $\nabla \cdot \mathrm{A} + \partial\varphi/\partial t = 0$. It can be shown that φ and A are the timelike and spacelike parts, relative to their frame of reference, of a 4-vector, the *electromagnetic 4-potential* Φ. Let x be the Lorentz chart such that $t = x^0$ and $\nabla = (\partial/\partial \bar{x}^\alpha)$ in (i)–(iii). Relative to x, $\Phi^0 = \varphi$ and $\Phi^\alpha = A_\alpha$. We write $\Phi^i_{,j}$ for $\partial\Phi^i/\partial x^j$. The gauge condition (iii) reads then: $\Phi^i_{,i} = 0$. We define a skew-symmetric covariant tensor field, componentwise, by the equations (iv) $\mathrm{F}_{ij} = \Phi_{j,i} - \Phi_{i,j}$ (where the Φ_i are the components of the covariant form of Φ). It follows from (i), (ii) and note 3, that the F_{ij} are the components of the covariant form of the electromagnetic field tensor F, whose contravariant form was given by (4.5.4). Equations (iv) guarantee that F is indeed a 4-tensor field. If □ symbolizes the D'Alembertian operator $\eta_{ij}\partial^2/\partial x^i \partial x^j$, classical electrodynamics is summarized by the following system of equations, of which the Maxwell equations (4.5.5) are mere corollaries:

$$\Box \Phi^i = -4\pi \mathrm{J}^i.$$

6. Maxwell, *TEM*, §§641 ff.
7. See, for instance, P. G. Bergmann (1942), p. 131. Our (4.5.11) is Bergmann's (8.27*a*) with different units and notation. Bergmann's general solution (8.30) differs from the particular solution (4.5.12) above, by an arbitrary constant 4-tensor field.
8. That the S^{ij} are indeed the components of a 4-tensor field follows from well-known rules of tensor algebra. See, for instance, Lichnerowicz (1962), § 39.
9. H. Minkowski (1908), § 13; M. Abraham (1909*a*, 1910). Max von Laue (1911), p. 528 refers to the second article by Abraham and to Sommerfeld (1910).
10. M. v. Laue (1911), p. 529.
11. The existence of such a tensor had been anticipated by M. Abraham (1909*b*), p. 739.
12. For a simple proof, see W. Rindler (1960), pp. 153 f.
13. For a proof, see, for instance, R. C. Tolman (1934), pp. 60–2.
14. The energy tensor can also be defined for a discrete material system, using Dirac's delta function. See, for instance, S. Weinberg (1972), p. 43.
15. Noting that $\mu\mathbf{u}$ stands for momentum density in the equations of motion (4.5.13*a*), and for density of energy flux in the equation of energy (4.5.13*b*), we see that the symmetry of M implies the equality of these two, and thus provides an additional argument for the equivalence of mass and energy, as Laue (1911), p. 530, points out. Our notation, indeed, takes the said equivalence for granted.
16. Through Lovelock's theorem. See p. 160, and Lovelock (1972).

Section 4.6

1. Robb (1921) is an abridged exposition of the theory, which lists and thoroughly motivates its 21 axioms and states the main theorems without proof. Robb (1936) is the revised second edition of Robb (1914).
2. Mehlberg (1966) includes an improved and highly compressed formulation of the system (in English). R. W. Latzer (1972) and J. W. Schutz (1973) axiomatize Minkowski spacetime geometry using *worldpoint* ("point", "instant") and *horismotical connectibility* ("λ-relation", "signal relation") as their sole primitives.
3. Reichenbach (1958), p. 268.
4. Conditions (i)–(vii) are equivalent to Kronheimer and Penrose's axioms for causal spaces. The irreflexivity of $\ll$, which they postulate instead of our (iv), follows immediately from (iii), (iv) and (vii).
5. Strictly speaking, the *Alexandrov topology* of a causal space S is the topology generated by the chronological pasts and futures of the points of S (i.e. the weakest topology in which both $I^-(P)$ and $I^+(P)$ are open, for every $P \in S$). The chronological neighbourhoods of the points of S constitute a base of this topology if and only if (i) for every $P \in S$ there are points $Q, R \in S$ such that $Q \ll P \ll R$, and (ii) for every $P, Q_1, Q_2, R_1, R_2 \in S$ there are points $X, Y \in S$ such that $Q_i \ll X \ll P \ll Y \ll R_i$ $(i = 1, 2)$. These conditions are met in all cases that interest us, but a causal space consisting only of the three points P, Q, and R, such that $P \ll Q \ll R$, does not meet them. Here the Alexandrov topology is the discrete topology, and there is but one chronological neighbourhood, namely $I^+(P) \cap I^-(R) = \{Q\}$, which is not a base for a topology. Another topology that can be naturally introduced in a causal space is the weakest topology in which all causal pasts and futures are closed sets.
6. We say that a curve $\gamma: I \to S$ enters n times into an open subset $U \subset S$ if there are n real numbers $t_i \in I$ $(1 \leq i \leq n)$ such that $\gamma(t_i)$ lies on the boundary of U and t_i is the g.l.b. of an open interval which γ maps into U.
7. In Robb's system, such an assumption is only warranted by the entire set of axioms. The last

of them, Postulate XXI, which bestows on null straights the linear order of **R** and plays a role comparable to that of Hilbert's Axiom of Completeness (page 285, note 6), is apparently motivated less by the observation of optical phenomena than by a desire to secure the "right" geometrical structure.

8. *Proof*: Let y be a Lorentz chart of Minkowski spacetime. If x is a global chart of $\mathscr{M}$ that satisfies the stated requirement, $y^{-1} \cdot x$ is an isomorphism of causal spaces. On the other hand, if there is a causal isomorphism f of the given structure onto Minkowski spacetime, $y \cdot f$ is a chart of $\mathscr{M}$ which satisfies the stated requirement.

9. Winnie's definition of congruence omits condition (ii). Hence, two non-causal vectors which together span a subspace of $\mathscr{M}_P$ containing a chronological vector cannot be congruent. Not only is this consequence undesirable; it also invalidates the proof of Winnie's Proposition 6.6. See Winnie (1977), pp. 175 f.

10. See, for instance, Robb (1914), pp. 36, 47. Acceleration planes are called "inertia planes" in Robb (1936). Latzer offers another, very attractive, characterization of spacelike straights: Three distinct, pairwise separate points lie on the same straight generated by a spacelike vector if and only if no point of $\mathscr{M}$ is horismotically connectible to all three points (Latzer (1972), Def. 2.1 and Prop. 2.11, pp. 245s., 250s). Winnie characterizes timelike straights, somewhat more intricately, as follows: Let $E(A)$ the set of all points horismotically connectible with $A \in \mathscr{M}$: if $A \ll B$, the intersection $E(A) \cap E(B)$ will be homeomorphic to a sphere; let $S(A, B)$ denote the union of all spacelike straights that join any two points in $E(A) \cap E(B)$; three points $A, B, C \in \mathscr{M}$ such that $A \ll B \ll C$ lie on a straight generated by a timelike vector if and only if $S(A, B) \cap S(B, C)$ is empty. (Winnie (1977), pp. 160 ff., in particular, Prop. 5.4.)

11. A. D. Alexandrov and V. V. Ovchinnikova, "Notes on the Foundations of Relativity", *Vestnik Leningrad. Univ.*, 11 (1953) 95 (in Russian). I have not seen this paper. I owe the reference to A. D. Alexandrov (1967).

12. Zeeman (1964) uses $<$ for our $\ll$ and "causal" for our "chronological". In view of Lemma I, these differences are innocuous.

13. It follows that any $g \in G$ must be an orthochronous similarity. For, as a causal automorphism, g maps null cones onto null cones; g is therefore a conformal linear mapping, i.e. a similarity. Since g preserves causal precedence, it must be orthochronous. See Zeeman (1964), p. 493.

14. Zeeman (1964), p. 491.

15. If T is a topology on a set S, the topology *induced* by T on $U \subset S$ is the weakest topology relative to which the canonical inclusion of U into S is a continuous mapping.

16. Zeeman (1967), p. 162, counters this criticism with the reminder "that we are dealing with special relativity"—a baffling remark, for some of the best examples of the applicability of this theory are furnished by particle accelerators.

17. Both the Zeeman and the path topology are Hausdorff, connected and locally connected, and neither normal nor locally compact. The path topology is, and the Zeeman topology is not, paracompact. Hawking, King and McCarthy (1976), p. 178, use this property for proving that the continuous curves of the path topology are precisely the "Feynman paths", i.e. curves continuous in the manifold topology, whose ranges are constrained to lie within the local null cones but may zigzag with respect to time orientation like the worldline of an accelerated electron in Feynman's quantum electrodynamics (Theorem 2).

Chapter 5

Section 5.1

1. A short non-mathematical discussion of the unsolved problems of Newtonian gravitational theory *circa* 1900 will be found in Whittaker (1951/53), II, pp. 144–9.
2. Einstein (1913*a*), p. 1249.
3. Einstein (1913*a*), pp. 1249 f. Cf. M. Abraham (1912*f*), p. 793.
4. Lorentz (1900), in Lorentz, *CP*, vol. 5, pp. 206, 214.
5. Lorentz (1900); *CP*, vol. 5, p. 211.

6. Laplace had shown that if gravitation propagates with finite speed V and, as he implicitly assumed, terms of the first order in v/V do not cancel out, experience requires that V be of the order of 10^6 times the vacuum speed of light.

7. Other, deeper difficulties inherent in Lorentz's theory of gravity are noted by P. Havas (1979), pp. 83 f.

8. Poincaré (1906), pp. 166–75. A generalization of Poincaré's Law of Gravity was proposed by W. de Sitter (1911). Whitrow and Morduch (1965), pp. 20 f., 24, 29 f., 45 ff., 50, 61, have worked out the predictions of the Poincaré–de Sitter theory for the classical tests of General Relativity. It does not fare so well as the latter when compared with experience (see the comparative tables, *ibid.*, pp. 63 f.). On Poincaré's theory of gravitation see also A. L. Harvey (1965), pp. 452–4.

9. Poincaré (1906), p. 166.

10. Poincaré (1906), p. 167.

11. Poincaré (1906), p. 167.

12. Referring to the calculations of Laplace mentioned in note 6, Poincaré observes that his illustrious countryman "had examined the hypothesis of the finite propagation speed *ceteris non mutatis*; but here, on the contrary, this hypothesis is complicated with many others, and there could well be a more or less perfect compensation between them" (Poincaré (1906), p. 170).

13. Poincaré, of course, writes this and all other equations componentwise, but the coordinate-free language wonderfully fits his thinking.

14. Hermann Minkowski also tried his hand at a relativistic theory of gravity. See Minkowski (1908), Anhang (reprinted in his (1910), pp.. 522 ff.), and the comments by L. Pyenson (1977), pp. 80–4.

Section 5.2

1. Planck (1907), p. 542.

2. In the German original the last sentence reads: "Wenn diese Frage zu verneinen ist, was wohl was Nächstliegende sein dürfte, so ist damit offenbar die durch alle bisherige Erfahrungen bestätigte und allgemein angenommene Identität von träger und ponderabler Masse aufgehoben" (Planck (1907), p. 544). In his eagerness to detract from Einstein's originality, Whittaker quotes the above passage as saying the opposite of what we have just read: "Now, said Planck, all energy has inertial properties, and therefore *all energy must gravitate*" (Whittaker (1951/53), II, p. 152; Whittaker makes reference to "*Berl. Sitz.*, 13 June 1907, p. 542, specially at p. 544").

3. Einstein (1907*b*), p. 442. See above, p. 112.

4. Einstein (1907*b*), p. 443.

5. It had been demonstrated to one part in a thousand by Newton (p. 32) and to a much higher precision by Eötvös (1889), and has subsequently been confirmed to 5 parts in 10^9 by Eötvös, Pekár and Fekete (1922—originally submitted in 1909), to $1:10^{11}$ by Roll, Krotkov and Dicke (1964), and to an astounding $1:10^{12}$ by Braginsky and Panov (1972). Ohanian (1976), pp. 18 ff. gives a schematic description of these experiments. See also the important methodological remarks of J. Ehlers (1973), p. 7.

6. Einstein (1907*b*), p. 454. Einstein does not explain any further what, exactly, is a uniformly accelerated frame in the context of Relativity Theory, but in later presentations of the Equivalence Principle he makes his meaning more precise. In his (1911), he specifies that the frame he has in mind must lie "in a space free of gravitational fields" (Einstein (1911), p. 899). Einstein (1912*a*), p. 356, demands that the frame be uniformly accelerated "in Born's sense", i.e. that the acceleration of each one of its points relative to its respective momentary inertial rest frame be constant.

7. R. H. Dicke (1959), p. 623; (1964*a*), p. 13.

8. Einstein's remark ((1913*a*), p. 1255) that "Eötvös's experiment plays in this connection a role similar to that of Michelson's experiment in relation to the problem of the physical detectability of *uniform* motion", is purely rhetorical.

9. Einstein (1907*b*), pp. 461, 459. The matter is discussed again with great clarity and forcefulness in Einstein (1911*a*), §§3, 4, which is readily available in English translation in Kilmister (1973).

10. "[Ich ging] von der hächstliegenden Auffassung aus, dass die Aequivalenz von träger und schwerer Masse dadurch auf einer Wesensgleichheit dieser beiden elementaren Qualitäten der

Materie bzw. der Energie zurückzuführen sei, dass das statische Gravitationsfeld als physikalisch wesensgleich mit einer Beschleunigung des Bezugsystems aufgefasst wird" (Einstein (1912*c*), p. 1063; cf. Einstein (1911*a*), pp. 899, 903; Kilmister (1973), pp. 130, 134).

11. Einstein (1912*c*), p. 1063.

12. Let h be the distance between the observer and the source. The redshift can be calculated in Σ_1 using the relevant relativistic formula for the Doppler effect: $\nu_{obs}/\nu_{em} = \sqrt{(1-\gamma h)/(1+\gamma h)}$ (Einstein (1905*d*), p. 912), for γh is the speed with which the source's mirf (at the time of emission) recedes from the observer's mirf (at the time of reception). The same redshift must be observed in Σ_2, where h is of course the height of the observer above the source, and γh the potential difference from the former to the latter. To a first approximation, the redshift $\Delta\nu = (\nu_{em} - \nu_{obs})/\nu_{em} \approx \gamma h$. From bottom to top of a tower 22.6 metres high ($h \approx 7.538 \times 10^{-8}$ light-seconds) such as the one used by R. V. Pound and his associates (1960, 1965), $\Delta\nu$ should amount to a frequency loss of some 2.46×10^{-15} Hz—given that $\gamma \approx 9.8\ \text{m/s}^2 \approx 3.269 \times 10^{-8}\ \text{s}^{-1}$. The value measured by Pound and Snider agreed with this prediction to one part in 100. For a detailed historico-critical study of the redshift as a test of Einstein's theory of gravity, see Earman and Glymour (1980*b*).

13. The Equivalence Principle also admits another interpretation, according to which it is weaker than Einstein's EP, yet stronger than the Weak Principle. This "semistrong" principle asserts the equivalence of all non-rotating free-falling local frames in the vicinity of any given worldpoint, but not the equivalence of frames which do not happen to be near the same worldpoint. A local version of Special Relativity is valid then in each sufficiently small spacetime region, but the numerical contents (speed of light, gravitational "constant", fine structure "constant", etc.) of the several versions can vary from one region to another. Imitating Einstein we may say that there is as yet no inducement to espouse the Semistrong Principle. Cf. the apt comments of W. Rindler (1977), pp. 19 f.

14. Einstein (1912*c*), p. 1063.

Section 5.3

1. We should not forget, however, that Newton himself was aware of its usefulness: in the form of Corollary VI to the Laws of Motion it enabled him to deal with the Earth-Moon system, that falls freely in the field of the Sun, as if it were inertial. See page 287, note 14, and H. Stein (1977*a*), p. 19 f.

2. Einstein (1911*a*), p. 900; Kilmister (1973), p. 131.

3. The formulae derived from the EP for the energy loss and the frequency shift experienced by radiation as it overcomes the pull of gravity are beautifully consistent with the law of proportionality between frequency and radiated energy postulated in Einstein (1905*a*) in order to explain the photoelectric effect. Einstein however does not mention it here, probably because his 1905 hypothesis of light quanta was still far from obtaining general acceptance in 1911.

4. Einstein (1911*a*), p. 906; Kilmister (1973), p. 137. Einstein's statement is somewhat misleading, for the Light Principle of Special Relativity was never supposed to hold in the accelerated systems to which systems at rest in homogeneous gravitational fields are assimilated here. There is, however, a genuine and deep difference between "the ordinary Theory of Relativity" and Einstein's evolving theory of gravity, insofar as, according to the latter, the actual world cannot admit a global Lorentz chart (cf. p. 137).

5. Einstein (1911*a*), §3, before eqn. (2*a*); cf. Einstein (1907*b*), p. 459 n. 1.

6. Though the effect is significant it is easily swamped by thermal effects on the surface of the Sun: (Einstein (1913*a*), p. 1266.) As I indicated above in note 12, the gravitational redshift was confirmed, with an estimated uncertainty of 1% by Pound and Snider (1965) in the homogeneous field near the ground, in Cambridge, Mass. In 1977, Vessot and Levine used for the same purpose a hydrogen maser frequency standard placed on a spacecraft that travelled on an orbital arc with a maximum altitude of 10,000 km. Preliminary results reported by I. I. Shapiro (1980*a*), p. 473, indicate that the ratio of the measured to the predicted redshift differs from unity by 5 parts in 10^5, with an uncertainty of 1 part in 10^4.

7. This is half the deviation of 1.75″ predicted by Einstein's mature theory of gravity. See pp. 173; 322, note 23.

8. M. Abraham (1912*a*, *b*, *c*, *f*). Abraham (1912*d*, *e*) are specifically directed against Einstein.

9. G. Nordström (1912*a*, *b*, 1913, 1914). On Nordström's theory, see Einstein (1913*a*), Einstein and Fokker (1914), Whitrow and Morduch (1965), Misner *et al.* (1973), p. 429.

10. G. Mie (1912*a*, *b*; 1913).

11. Einstein (1912*a*, *b*). In a third paper, Einstein (1912*c*) discussed the Machian implications of his theory of the static field—see below, Section 6.2. Einstein (1912*d*) is the very instructive reply to Abraham (1912*d*); Einstein (1912*e*) is a five-line note putting an end to the controversy with Abraham.

12. Einstein and Grossmann (1913); see also Einstein (1913*a*, *b*). As we shall see in Section 5.4, the main steps leading to the new approach can be traced in Einstein's papers of 1912.

13. The three postulates regarding (i) natural clocks, (ii) freely falling particles and (iii) the spacetime direction of light signals give a definite physical meaning to the gravitational metric **g**. (ii) is introduced in Einstein and Grossmann (1913), § 2, (i) and (iii) in § 3. (iii) implies that the null cone determined in each tangent space of the spacetime manifold by the metric **g** meets the requirements of Special Relativity. It is clearly spelled out as a separate postulate in Einstein (1913*a*), p. 1256: "The law of light-propagation will be determined by the equation $\mathrm{d}s = 0$."

14. M. v. Laue (1920), E. T. Whittaker (1928).

15. In Abraham's correction to his (1912*a*), in *Physikalische Zeitschrift*, 13 (1912), p. 176. The magnitude of the 4-velocity, $(\mathrm{d}x^i/\mathrm{d}\tau)(\mathrm{d}x^i/\mathrm{d}\tau)$ is $-c^2$ (Section 4.3, note 17). Hence $\mathrm{d}\Phi/\mathrm{d}\tau = -F^i\mathrm{d}x^i/\mathrm{d}\tau = -(\mathrm{d}^2x^i/\mathrm{d}\tau^2)(\mathrm{d}x^i/\mathrm{d}\tau) = -(1/2)(\mathrm{d}/\mathrm{d}\tau)(\mathrm{d}x^i/\mathrm{d}\tau)(\mathrm{d}x^i/\mathrm{d}\tau) = c\mathrm{d}c/\mathrm{d}\tau$. (I have adjusted all signs to the signature $(-+++)$ used by Abraham.)

16. A Lorentz metric on a 4-manifold is the Minkowski metric $\boldsymbol{\eta}$ only if the Riemann tensor R is identically zero (p. 145). A tedious but straightforward calculation shows that not all the components of R relative to the coordinates x, y, z, t will vanish in Abraham's spacetime. Writing x^0 for t, x^1 for x, etc., we have that

$$\mathrm{R}^{\alpha}_{0\beta 0} = c\frac{\partial^2 c}{\partial x^\alpha \partial x^\beta}$$

which usually will not be zero. Note in particular that the time–time component of the Ricci tensor is $\mathrm{R}_{00} = -c\nabla^2 c$.

17. M. Abraham (1912*c*), p. 312.

18. Let β stand for $|(1-v^2c^{-2})^{-1/2}|$. The equations $\mathrm{d}t' = \beta(\mathrm{d}t - vc^{-2}\mathrm{d}x)$, $\mathrm{d}x' = \beta(-v\mathrm{d}t + \mathrm{d}x)$, $\mathrm{d}y' = \mathrm{d}y$, $\mathrm{d}z' = \mathrm{d}z$ (cf. eqns (3.4.18)) are not integrable, for example, in a static gravitational field in which c depends on x but not on t. The conditions $\partial\beta/\partial t = -\partial\beta v/\partial x$ and $-\partial\beta vc^{-2}/\partial t = \partial\beta/\partial x$ cannot be met in this case, for the left-hand sides vanish identically, while the right-hand sides generally do not vanish (Einstein (1912*a*), p. 368).

19. Einstein (1912*a*), p. 355.

20. M. Abraham (1912*d*), pp. 1057 f.

21. M. Abraham (1912*f*), p. 794.

22. "Eine Missgeburt der Verlegenheit". Einstein to Sommerfeld, 29 October 1912; in Einstein/Sommerfeld (1968), p. 26.

23. G. Nordström (1912*a*), p. 2; cf. (1912*b*), pp. 872 ff.

24. G. Nordström (1913), p. 533; (1912*b*), pp. 873, 878.

25. Nordström (1913), p. 534. He gives credit for this idea to Laue and Einstein.

26. Namely, whenever the gravitating body can be treated as a particle, according to the theory; hence, in particular, when it is falling in a static, homogeneous field (Nordström (1913), p. 554; cf. p. 540). The dependence of the gravitational factor on the potential in Nordström's theory is briefly discussed—in English—by J. Mehra (1974*b*), p. 7.

27. Nordström (1913), p. 547.

28. Nordström (1913), p. 550. Nordström adds: "Hence, it is not necessary to transport measuring rods and clocks from one place to another in order to compare lengths and times at different places."

29. It will be instructive to prove this in detail, in order to see where the different assumptions come in. We shall evaluate the g_{ij} at an arbitrary worldpoint P. Let v be a vector at P, tangent to the worldline of a light pulse that goes through P in the direction of increasing or decreasing x^1. Let $v^i = v(x^i)$ be the i-th component of v relative to x. Evidently, $v^2 = v^3 = 0$, and $v^1 = \pm v^0$, with each solution holding for some admissible choice of v. By the last postulate in proposition C (p. 139), v

must be null. Hence the energy $E(v) = g_{ij}\mathrm{d}x^i(v^h\partial/\partial x^h)\mathrm{d}x^j(v^k\partial/\partial x^k) = g_{ij}v^iv^j = 0$. Substituting the above values for the spatial components v^α, we obtain: $(v^0)^2(g_{00}+g_{11}+2g_{01}) = (v^0)^2(g_{00}+g_{11}-2g_{01}) = 0$. Since $v^0 \neq 0$, it follows that $g_{00} = -g_{11}$ and $g_{01} = 0$. Substituting first x^2 and then x^3 for x^1 in the characterization of the light pulse tangent to v we prove, by the same argument that $g_{00} = -g_{22} = -g_{33}$, and that $g_{02} = g_{03} = 0$. Since the space coordinates x^α are Cartesian, their parametric lines are orthogonal at every intersection and $g_{\alpha\beta} = 0$ whenever $\alpha \neq \beta$.

30. **g** is what is now usually called a *conformally flat metric*, characterized by the vanishing of the Weyl tensor. See, for instance, Eisenhart (1950), pp. 89–92.

31. Einstein and Fokker (1914), p. 327.

32. Mie (1912*a*), p. 527; cf. pp. 513, 524. Mie (1913), p. 30.

33. Mie (1913), p. 45.

34. Einstein (1913*a*), p. 1263.

35. Mie (1913), p. 50. Cf. note 5 to Section 5.2. The above remarks do no justice to Mie's complex and subtle theory. As far as I know, the best summary in English is still that given in Weyl (1950), pp. 206–17.

Section 5.4

1. In his inaugural lecture "On the Hypotheses that lie at the Foundation of Geometry" (1854), Riemann showed that physical geometry cannot be derived a priori from the mere concept of space, for the latter is compatible with many diverse geometrical structures. If space is continuous its metrical properties cannot be figured out by counting points. Hence, he concluded, the foundation of the metrical relations in physical space must be sought in the nature of the physical forces that keep it together (Riemann (1867), p. 23; see below, p. 344, note 44).

2. See Torretti (1978*a*), p. 73, where I quote and discuss the relevant passage of Gauss (1827), §6.

3. The curvature of a plane curve at a point measures the rate at which the curve is changing direction—i.e. "departing from straightness"—as it passes through that point. The curvature's sign—positive or negative—indicates the sense in which it bends. See for instance, J. J. Stoker (1969), pp. 21 ff.; or my own short discussion in Torretti (1978), pp. 68 ff.

4. Gauss (1827), § 11. Let **g** denote the metric on the surface S—i.e. the restriction to S of the Euclidean metric of space—and let u be a chart of S with coordinate functions u^1 and u^2. Then, $E = \mathbf{g}(\partial/\partial u^1, \partial/\partial u^1)$, $F = \mathbf{g}(\partial/\partial u^1, \partial/\partial u^2)$ and $G = \mathbf{g}(\partial/\partial u^2, \partial/\partial u^2)$. The line element on S is therefore given by

$$\mathrm{d}s^2 = E(\mathrm{d}u^1)^2 + 2F\mathrm{d}u^1\mathrm{d}u^2 + G(\mathrm{d}u^2)^2$$

Gauss showed also that the surface element on S is $\mathrm{d}S = \sqrt{(EF - G^2)}\,\mathrm{d}u^1\mathrm{d}u^2$. If we now let H stand for $\sqrt{(EF - G^2)}$, and represent differentiation with respect to the i-th coordinate function by a comma followed by i (e.g.: $E_{,1} = \partial E/\partial u^1$), we have that the Gaussian curvature K satisfies the equation

$$KH = \left(\frac{E_{,1}F - G_{,2}E}{2EH}\right)_{,1} + \left(\frac{2F_{,1}E - E_{,1}F - E_{,2}E}{2EH}\right)_{,2}$$

Thus K depends only on E, F and G and their first and second derivatives with respect to u^1 and u^2.

5. Gauss (1827), § 12.

6. See, for instance, Choquet-Bruhat *et al.*, *AMP*, p. 315. By a geodetic path I mean the range of a geodesic. Cf. p. 94.

7. Schur (1886) proved, moreover, that if the sectional curvatures all have the same value at each $P \in M$, they are everywhere the same, and M is a manifold of constant curvature. For a proof, see Kobayashi and Nomizu, *FDG*, vol. I, p. 202.

8. It is known also as the Riemann curvature tensor, or the Riemann-Christoffel tensor. Its components relative to an arbitrary chart are given in formula (D.3) on p. 281. Unless there is danger of confusion, R will stand for the Riemann tensor in any of its forms, covariant, contravariant or mixed.

9. $1/\alpha = \mu_P(v, v)\mu_P(w, w) - (\mu_P(v, w))^2$. $1/\sqrt{\alpha}$ is the area of the parallellogram spanned by v and w. For a proof of equation (5.4.2), see Spivak, *CIDG*, vol. II, pp. 196 f.

10. Two (0, 4)-tensor fields R and S on M, which satisfy equations (5.4.1), are identical if R(X, Y, X, Y) = S(X, Y, X, Y), for every pair of vector fields X and Y on M. The proof is simple and

constitutes a good exercise for the reader. Consider the tensor field T = R − S, which also satisfies (5.4.1). Show first that if T(X, Y, X, Y) = 0, for arbitrary X and Y, then T(X, Y, Z, Y) = 0, for arbitrary X, Y and Z. From this infer that T(X, Y, Z, W) = 0 for arbitrary X, Y, Z and W. Use the fact that T is linear on each argument.

11. As explained in Appendix (Sections C.6, C.8), the (1,3) Riemann tensor on a manifold endowed with an arbitrary Riemannian metric μ is precisely the curvature tensor defined by the Levi-Cività connection determined by μ. Cf. p. 101.

12. See Appendix, eqns. D.1, D.2, D.3 (p. 281).

13. Some authors use the expression "locally Minkowskian" (or "locally Euclidean") to indicate that a neighbourhood of each point of a manifold is *approximately* isometric to a region of Minkowski spacetime (resp. Euclidean space). This condition is met by *every* Riemannian 4-manifold with a Lorentz metric (resp. by *every* proper Riemannian 3-manifold), but the approximate local isometries are only as good as the local curvature will allow. The mere fact that the tangent space at a point has a Minkowskian (resp. a Euclidean) inner product—as it obviously does everywhere, by definition, on the manifolds under consideration—says nothing whatsoever about the value of the Riemann tensor and the manifold's departure from flatness at that point.

14. A manifold is *path-connected* only if it is connected in the topological sense, i.e. only if it cannot be decomposed into two mutually disjoint non-empty open sets. Two results are pertinent here: (A) If M is a connected n-manifold, the following statements are equivalent: (i) M admits a proper Riemannian metric; (ii) M is metrizable, i.e. M admits a distance function whose open balls constitute a base of M's topology; (iii) the topology of M has a countable base; and (iv) M is paracompact, i.e. every open covering of M has a locally finite open refinement. (B) An n-manifold M which admits a proper Riemannian metric also admits a Lorentz metric if and only if it admits a non-vanishing field of line elements, i.e. a smooth assignment of a pair of equal and opposite vectors at each $P \in M$. Since the latter condition is always met unless M is compact and its Euler number differs from 0, we may conclude that a relativistic spacetime can have almost any topology, provided that it is paracompact and that, if it happens to be compact, its Euler number is 0. (A) is proved in Kobayashi and Nomizu, *FDG*, vol. I, p. 271; for a proof of (B) see Markus (1955) or Hawking and Ellis (1973), p. 39 f.

15. Cf. Einstein's letter to Sommerfeld of October 29th, 1912; Einstein and Sommerfeld (1968), pp. 26 f. English translations of the relevant passage are found in R. McCormmach (1976), p. xxviii; J. Stachel (1980*a*), p. 13, etc.

16. Quoted by J. Stachel (1980*a*), p. 12n.

17. See his comments on "the most important point of contact between Gauss's theory of surfaces and the general theory of relativity" in the Princeton Lectures of 1921 (Einstein (1956), pp. 61–4).

18. Einstein and Fokker (1914), p. 328, concluding paragraph. See however Grossmann's remarks in Einstein and Grossman (1913), p. 36.

19. Einstein (1949), p. 66.

20. The evidence has been gathered and brilliantly interpreted by J. Stachel (1980*a*).

21. Born (1909*c*); cf. Born (1910).

22. Quoted by Stachel (1980*a*), p. 2; not in Einstein/Sommerfeld (1968). On the same day, Paul Ehrenfest submitted to the *Physikalische Zeitschrift* a note on "Uniform Rotation of Rigid Bodies and the Theory of Relativity", in which he presented what soon became known as the "Ehrenfest paradox". Ehrenfest proposes the following criterion of "relative rigidity", which, he claims, is equivalent to Born's: a body B is relatively rigid (*relativ-starr*) if the deformations it suffers in the course of an arbitrary motion are such that each infinitesimal element of B exhibits at each moment, to an observer at rest, the Lorentz contraction prescribed by the momentary velocity of the element's midpoint. He then derives the following "contradiction": A relatively rigid disk of radius R is slowly set in rotation about its axis. Let R' be the disk's radius as measured by an observer at rest when the rotation attains the angular velocity $\boldsymbol{\omega}$. Obviously, $2\pi R' < 2\pi R$, for each line element on the circumference of the disc is Lorentz-contracted as it moves in its own direction with momentary speed $R'|\boldsymbol{\omega}|$. And yet $R' = R$, for each line element along a radius of the disk has zero momentary speed in the longitudinal direction and consequently is not Lorentz-contracted. The only sensible solution to Ehrenfest's paradox was given by Einstein in a letter to J. Petzoldt in 1919: "A rigid circular disk at rest must break up if it is set into rotation, on account of the Lorentz contraction of the tangential fibres and the noncontraction of the radial ones. Similarly, a rigid

disk in rotation (produced by casting) must explode as a consequence of the inverse changes in length, if one attempts to put it at rest" (Quoted by Stachel (1980*a*) pp. 6 f.).

23. Einstein (1912*a*), p. 356.

24. Einstein (1916*a*), p. 774; English translation in Kilmister (1973), p. 146.

25. This assumption is explicitly introduced by Einstein in a footnote to the discussion of the rotating frame in the Princeton Lectures (Einstein (1956), p. 60). We may call it the rod hypothesis, by analogy with the clock hypothesis mentioned on p. 96. I must emphasize, however, that, in contrast with the latter, the rod hypothesis cannot be formulated exactly, for there is no such thing as *the* momentary inertial rest frame of an accelerated body: at an agreed moment its several points generally will be comoving with different inertial frames. That is why rods ultimately cannot hold their own as instruments of fundamental measurement in General Relativity.

26. Cf. Einstein (1916*a*), pp. 774 f. (Kilmister (1973), pp. 146 f.); Einstein (1956), pp. 59 ff., and the "popular exposition", Einstein (1954), pp. 79–82. The above discussion depends essentially on the possibility of *counting* the rods that can be placed end to end along C_2 and made to leave their imprints on C_1. This possibility is evidently implied by Einstein's demand, in the final step of his argument in the Princeton Lectures, "that at a definite time t of $[F_1]$ we determine *the ends* of *all* the rods" (Einstein (1956), p. 60; my italics). If the said possibility is granted, our circumferences can only be measured approximately, to some agreed degree of accuracy. But one can hardly expect exact results from Standard Rod Geometry. The important fact is that the length of C_2 in G_2 will exceed $2\pi R$ by any agreed difference as $R|\boldsymbol{\omega}|$ approaches c. Some authors give a semblance of rigour to their treatment of the problem by expressing the length of the standard rod as a differential which they integrate over C_2. The semblance is deceptive, however, and their arguments not a whit stricter than Einstein's, because, as I pointed out in note 25, the concept of the momentary inertial rest frame of an accelerated body, involved in the evaluation of the Lorentz contraction of the rods placed along C_2, cannot be made exact. (Cf. the otherwise highly commendable discussion in Møller (1952), pp. 222–6, 237–44, where the inevitable inexactness creeps in on p. 237, line 17 from below.) I have deliberately refrained from discussing the question of time on the rotating frame, because it is quite complicated and would distract us from our main concern in this Section; cf. Arzeliès (1966), pp. 217–35.

The rotating frame in flat spacetime is the subject of a vast and growing literature. See the bibliography (to 1961) in Arzeliès (1966), pp. 241–3, partially completed, to their respective dates, by Grøn (1975) and Grünbaum and Janis (1977). Unfortunately, much of this literature is marred by the persistent confusion of Einstein's analysis of the standard rod *geometry* of the *relative space* of a *uniformly* rotating *frame*, with Ehrenfest's problem concerning the *mechanical* behaviour of a *material disk* set in rotation *from rest* (see note 22). Our circle C_2 can of course be represented by a uniformly rotating disk—if that helps the imagination—but we must not forget that, by hypothesis, the rim of this disk must always lie exactly on the circumference of C_1.

27. Einstein (1912*c*), p. 1063 f.

28. Thus, for instance, ordinary two-dimensional polar coordinates map a plane from which a ray has been excised, non-isometrically onto the region $(0, 2\pi) \times (0, \infty) \subset \mathbf{R}^2$.

29. Cf. Einstein (1912*c*), p. 1063, quoted on p. 137.

30. "Ich vermochte lange nicht einzusehen, was dann die Koordinaten in der Physik überhaupt bedeuten sollten" (Einstein (1934), p. 253).

31. Einstein and Grossmann (1913), pp. 4–8; Einstein (1913*a*), p. 1256; (1913*b*), pp. 285 f.; (1914*a*), p. 177; (1914*b*), pp. 1032–4, 1044–6; (1916*a*), § 13.

32. The functions Γ^i_{jk}—given by (5.4.6*b*) in terms of the components g_{ij} of an *arbitrary* Riemannian metric—were introduced by Christoffel (1869). For a modern approach to their definition see Appendix, Part C.

33. See p. 312, note 13.

34. In "Notes on the Origin of the General Theory of Relativity", (Einstein, (1934), p. 254).

35. Einstein (1914*a*), p. 177. The context deserves attention. After writing out eqns (5.4.3) and (5.4.6*a*)—labelled, respectively, (2) and (1*a*)—Einstein comments: "This law of motion ((2), (1*a*)) is derived initially for the case of a particle that moves quite freely, on which therefore no gravitational field is acting (when it is judged from an appropriate reference frame). But, as experience shows that the law of motion of a particle in the gravitational field does not depend on the body's material and as it should always be possible to state that law in the Hamiltonian form $[\delta \int ds = 0]$, it is plausible to consider (2, 1*a*) generally as the law of motion of a particle on which

no forces other than gravitational forces are acting. This is the core of the 'Equivalence Hypothesis'".

36. To suggest as some have done that the worldline of a freely falling particle must be a geodesic because, according to the EP, each little segment of it is mapped onto a straight in $\mathbf{R}^4$ by a suitable local Lorentz chart, is like arguing that the parallels of latitude on the surface of the Earth are no less geodesic than the meridians, because they agree, for instance, with the avenues that cross Chicago from East to West, which are no less straight than those leading from North to South. In a Riemannian manifold *every* curve is "straight in the infinitesimal".

37. Einstein (1934), p. 219.

38. Einstein (1912*b*), p. 458. The paper was submitted on March 23, 1912, but the passage I am quoting was added in proof. It must have been printed before June 5, 1912, when Max Abraham submitted his (1912*d*), where Einstein's (1912*b*) is mentioned on p. 1056 n. 2.

Section 5.5

1. The current conception of a geometrical object as a unique entity which manifests itself through its several chart-dependent arrays of components did not take root until the thirties—as far as I know the expression "geometrical object" was first used in this sense by Schouten and van Kampen (1930) and was popularized by Veblen and Whitehead (1932). But Einstein had something like it in mind when he wrote to Michele Besso on October 31st, 1916: "Definition of tensors: not 'things that transform thus and thus'. But: things that can be described, with respect to an (arbitrary) coordinate system by an array of quantities ($A_{\mu\nu}$) obeying a definite transformation law." Einstein/Besso (1972), p. 85. Cf. Einstein (1916*a*). p. 782.

2. Einstein (1916*a*), p. 776; Kilmister (1973), pp. 147 f. Dominic Edelen (1962), p. 36, maintains that the requirement of general covariance has to be met only by equations governing *observable* physical quantities. However, it is questionable that a statement regarding *unobservable* quantities referred to a particular chart or family of charts of unknown physical meaning can have a definite sense.

3. One can also have the following surprise: the point transformation $x^{-1} \cdot y$ maps timelike onto spacelike curves!

4. See, for instance, R. H. Dicke (1962), pp. 15 f. The great Soviet physicist V. Fock (1957, 1959) was, to my knowledge, the first one to emphasize that Einstein's General Relativity is not really relativistic in the more obvious sense of the word. But Einstein himself was well aware of it too; cf. his criticism of Emden's understanding of the theory in a draft letter to Sommerfeld of January 1926 (Einstein/Sommerfeld (1968), pp. 102 f.).

5. Einstein (1913*a*), p. 1257.

6. The coordinate transformations associated with the maximal atlas of a differentiable manifold constitute what Veblen and Whitehead (1932), pp. 37 f., called a *pseudo-group*.

7. For a lucid, penetrating discussion of "Invariance, Covariance and the Equivalence of frames", see J. Earman (1974).

8. F. Kottler (1912), p. 1689, n. 1. Kottler's work, which is mentioned by Einstein and Grossmann (1913), p. 23, was an attempt to formulate spacetime physics in the language of "a generalized vector analysis [. . .] for arbitrarily many dimensions, arbitrary metric and arbitrary coordinates" (Kottler (1912), p. 1669).

9. See pp. 101 and 274 f.

10. See p. 275.

11. Einstein and Fokker (1914), p. 322. Cf. Einstein (1913*b*), p. 287: "The method of the absolute differential calculus enables us to generalize the equation systems of any physical process, as they are found in the original Theory of Relativity, so that they fit into the scheme of the new theory. The components g_{ik} of the gravitational field always occur in such [generalized] equations. Physically this means that the equations inform us about the influence of the gravitational field on the processes of the region under study. The simplest example of this kind is the law of motion of a particle, given above." (Einstein refers to our equations (5.4.3) and (5.4.3*).) See also Einstein (1914*a*), p. 177, § 5.

12. If f is a smooth mapping of an n-manifold M into an m-manifold M', the "pull-back" f^* sends $(0, r)$-tensor fields on M' to $(0, r)$-tensor fields on M. If ω is a $(0, r)$-tensor field on M'

and $X^1, \ldots, X^r$ are vector fields on M, then, at each $P \in M$, $(f^*\omega)_P(X)^1_P, \ldots, X^r_P)$ $= \omega_{f(P)}(f_{*P}X^1_P, \ldots, f_{*P}X^r_P)$—where f_* is the differential of f, defined on p. 260. If ω is the Euclidean metric on the plane and f is a Mercator projection of the Earth, $f^*\omega$ is the metric defined by assigning to each pair of points on the region of the Earth covered by the projection the distance between their images on the Mercator map. *With this metric*, Canada is about twice as large as Brazil, and the shortest route from Moscow to Washington goes through Berlin.

13. Compare two maps of North America, made by stereographic projection on the same scale, but centred respectively in Denver and Detroit. The distance between the images of Toronto and Montréal in each map will be found to be quite different.

14. Just how small a neighbourhood must be chosen for a preset agreement—or how good an agreement can be guaranteed on a pregiven neighbourhood—depends, as we know, on the behaviour of the Riemann tensor near P. By using the diffeomorphism f defined on p. 144 instead of any arbitrary diffemorphism of U_P into $\mathscr{W}_P$, these relations are optimized and made definite.

15. Consider a vector field Z on the spacetime $(\mathscr{W}, \mathbf{g})$. Let R be the covariant Riemann tensor. If X and Y are vector fields, we write $[X, Y]$ for the Lie derivative of Y along X (p. 262). We shall denote by R(X, Y)Z the vector field defined by the equation: $R(W, Z, X, Y) = \mathbf{g}(W, R(X, Y)Z)$, which must hold for every vector field W, if X, Y and Z are fixed. It can be shown that

$$R(X, Y)Z = \nabla_X \nabla_Y Z - \nabla_Y \nabla_X Z - \nabla_{[X, Y]} Z \qquad (*)$$

(See, for instance, Kobayashi and Nomizu, *FDG*, vol. I, p. 133, Th. 5.1; J. V. Narlikar (1979), p. 46, Lemma 2.) Since $[\partial/\partial x^j, \partial/\partial x^k] = 0$ for any pair of elements $\partial/\partial x^j, \partial/\partial x^k$ of the holonomic tetrad field determined by a chart x, $\nabla_j \nabla_k Z = \nabla_k \nabla_j Z$ if and only if $R(\partial/\partial x^j, \partial/\partial x^k)Z = 0$. Such will be the case if and only if the Riemann tensor is identically 0, i.e. if and only if the metric $\mathbf{g}$ is flat.

16. "Das allgemeinste Gesetz, welches die theoretische Physik kennt". Einstein (1913*b*), p. 287.

17. To obtain (5.5.4*b*) develop $T^{ij}{}_{;j}$ according to (5.5.1). Recall that $\Gamma^j_{hj} = (1/\sqrt{-g})$ $(\partial\sqrt{-g}/\partial x^h)$, and substitute. (Cf. Synge and Schild, *TC*, eqn. 2.542.) Multiply both sides of the resulting equation by $g_{ik}\sqrt{-g}$ and sum over repeated indices.

18. Einstein (1914*b*), p. 1053, remarks that the frequent occurrence in his formulae of tensor components multiplied by $\sqrt{|g|}$ justifies the introduction of a special notation. The product of a tensor component $A^{\mathfrak{a}}_{\mathfrak{b}}$ by the square root of the absolute value of the determinant of the metric components g_{ij} is to be denoted by the corresponding Gothic letter, followed by the same indices, thus: $\mathfrak{A}^{\mathfrak{a}}_{\mathfrak{b}}$. The assignment of an array of such components $(\mathfrak{A}^{\mathfrak{a}}_{\mathfrak{b}})$ to each chart he proposes to call volume tensors or *V-tensors*. He observes that, since $\sqrt{|g|}\, d^4x = \sqrt{|g|}\, dx^0 dx^1 dx^2 dx^3$ "is a scalar", an array of *V*-tensor components multiplied by d^4x yields the array of components of a tensor field in the usual sense. Einstein's *V*-tensors, known in the literature as "tensor densities" or "relative tensors of weight 1", have subsequently been defined as genuine, though somewhat far-fetched, geometrical objects. (See, for instance, M. Spivak, *CIDG*, vol. I, pp. 183 ff.) We cannot go further into this matter. Let me recall however that if x and y are two charts with the same domain, if J stands for the determinant of the Jacobian matrix $(\partial x^i/\partial y^j)$ and if unprimed and primed components are referred, respectively, to x and y, then, given that $g'_{ij} = (\partial x^h/\partial y^i)(\partial x^k/\partial y^j)g_{hk}$, the determinant g' equals $J^2 g$. Consequently, if $\mathfrak{A}$ is a (q, r)-tensor density formed, as indicated above, from an ordinary tensor field A, and $\mathfrak{A}^{i_1 \ldots i_q}_{j_1 \ldots j_r} = \sqrt{|g|}\, A^{i_1 \ldots i_q}_{j_1 \ldots j_r}$ is one of its components relative to x, the homonymous component of $\mathfrak{A}$ relative to y is

$$\mathfrak{A}'^{i_1 \ldots i_q}_{j_1 \ldots j_r} = \sqrt{|g'|}\, A'^{i_1 \ldots i_q}_{j_1 \ldots j_r} = J\mathfrak{A}^{h_1 \ldots h_q}_{k_1 \ldots k_r} \frac{\partial y^{i_1}}{\partial x^{h_1}} \cdots \frac{\partial y^{i_q}}{\partial x^{h_q}} \frac{\partial x^{k_1}}{\partial y^{j_1}} \cdots \frac{\partial x^{k_r}}{\partial y^{j_r}}.$$

On tensor densities, see B. F. Schutz (1980), pp. 128 f.; D. Lovelock and H. Rund (1975), pp. 102–17. C. Misner *et al.* (1973), p. 501, aptly comment on the rationale for resorting to such objects.

19. Einstein (1913*b*), p. 288; (1914*a*), p. 178.

20. Einstein (1915*a*), p. 782. This is especially clear in Einstein (1914*b*), where, by the way, he denotes the H^i_{jk} by Γ^i_{jk}. In his (1915*a*), p. 783, Einstein observes that, in view of the geodesic law (5.4.7*)—above, p. 151—the gravitational field components are more adequately represented by the functions defined by (D.1), p. 281, which he thereafter calls Γ^i_{jk}, as we do to this day. Indeed,

as he points out, this view is confirmed by (5.5.5) for, as a short calculation will show, the right-hand side is equal to $\Gamma^i_{jk}\mathfrak{T}^j_i$. (Readers of the original papers should bear in mind that in 1915 and 1916 Einstein defines Γ^i_{jk} as *minus* the right-hand side of eqn. (D.1). The now standard sign convention is followed in the 1921 Princeton lectures.)

21. Einstein and Grossmann (1913), p. 11, eqn. (11). Einstein uses Greek letters for the contravariant forms of G and T and for the indices ranging over the first four natural numbers.

22. Einstein (*loc. cit.*, pp. 11 f.) only states conditions (i) and (iv). He probably thought that (ii) and (iii) were too obvious to mention. Note that (iii) *does not* entail that the $G^{ij}{}_{;j}$ must vanish *identically*, for *every* choice of the g_{ij}.

23. Einstein and Grossmann (1913), p. 35.

24. The symmetries of the Riemann tensor, (5.4.1*a*), entail that $R^k{}_{kij} = g^{hk}R_{hkij} = -g^{hk}R_{khij} = -R^h{}_{hij} = 0$ and $R^k{}_{ikj} = -R^k{}_{ijk} = -R_{ij}$. Cf. Einstein (1915*a*), p. 781.

25. See, for instance, J. L. Anderson (1967), pp. 360–2, or C. Misner *et al.* (1973), pp. 412–16.

26. The components of the Riemann tensor satisfy the Second Bianchi Identity: $R_{abij;k} + R_{abjk;i} + R_{abki;j} = 0$ (p. 280). Multiplying both sides by $g^{aj}g^{bi}$ and summing over repeated indices, we obtain, using (5.4.1*a*):

$$g^{bi}R_{bi;k} - g^{bi}R_{bk;i} - g^{aj}R_{ak;j} = 0 \qquad (a)$$

If we suitably relabel the dummy indices and recall that $g^{ij}{}_{;k} = 0$ we see that (*a*) implies that:

$$(-2R^j{}_k + \delta^k{}_j R^i{}_i)_{;j} = 0 \qquad (b)$$

Multiplying both sides by $-(1/2)g^{hk}$ and summing over k we finally verify that

$$(R^{jh} - \tfrac{1}{2}g^{jh}R^i{}_i)_{;j} = 0 \qquad (c)$$

27. Einstein (1917), (1931). See below, pp. 6.18 f. Observe that the factor of g^{ij} in (5.5.10) is the sum of the diagonal components or *trace* of the Ricci tensor (in its mixed form): $R^{hk}g_{hk} = R^k{}_k$. This is plainly invariant under arbitrary coordinate transformations and hence belongs to $\mathscr{F}(\mathscr{W})$. It is called the *curvature scalar* of ($\mathscr{W}$, **g**).

28. In the 1921 Princeton lectures Einstein motivated the field equations (5.5.11) by invoking a similar mathematical result, which however, from a physical point of view, is far less compelling than Lovelock's theorem. The modified Einstein tensor is the only tensor field on a Riemannian n-manifold which satisfies the following four conditions: (*a*) it is symmetric; (*b*) its covariant divergence vanishes identically; (*c*) its components involve only the g_{ij} and their first and second derivatives; (*d*) its components are linear in the second derivatives of the g_{ij}. This theorem was proved rigorously by Cartan (1922). In the Appendix to the fourth edition of *Raum-Zeit-Materie*, H. Weyl had stated it without proof, while noting that it could be easily demonstrated by the same method employed there for proving a related theorem of H. Vermeil (1917), concerning the curvature scalar (See Weyl (1921*a*), p. 288). In the physically relevant case, in which $n = 4$, the Cartan-Weyl theorem is substantially weaker than Lovelock's, which does not presuppose conditions (*a*) and (*d*).

29. Einstein and Grossman (1913), pp. 36 and 11. John Stachel (1980*b*) explains Grossmann's opinion on the Ricci tensor as follows: According to Einstein's theory of the static gravitational field—published in his (1912*a*, *b*) and reasserted in Einstein and Grossmann (1913), § 1—the line element on such a field is given in terms of a suitable chart by $ds^2 = c^2(dx^0)^2 - dx^\alpha dx^\alpha$, where c^2 is a function of the space coordinates x^α. To evaluate the components of the Ricci tensor we may, if the field is very weak, set $c^2 \approx 1 - 2\Phi$, for a Newtonian potential Φ, and neglect all but the linear terms in Φ and its derivatives. It turns out that $R_{00} \approx \nabla^2\Phi$, and $R_{\alpha\beta} \approx \partial^2\Phi/\partial x^\alpha \partial x^\beta$, where ∇^2 stands for the Laplacian. "While at the first sight this looks promising, since R_{00} does reduce to the Laplacian of g_{00} (without any coordinate conditions, as it would for any static metric), a moment's thought shows that the result is really a disaster for any attempt to base gravitational field equations on the Ricci tensor. For if we set the Ricci tensor equal to zero (as we must outside the sources of the gravitational field), it follows from $R_{\alpha\beta} = 0$ that g_{00} can depend at most *linearly* on the coordinates. Thus, it cannot represent the exterior gravitational potential of any (finite) distribution of matter, static or otherwise" (Stachel (1980*b*), p. 8). Einstein and Fokker (1914), p. 328n., emphatically dismiss Grossmann's rejection of the Riemann tensor and its derivatives as viable ingredients of the left-hand side of (5.5.7). Unfortunately they do not explain why they thought that Grossman's grounds—whichever they were—were worthless.

30. The basic calculations will be found in Einstein and Grossmann (1913), pp. 37 f. and 15 ff. Equations (5.5.12*a*) were first given in this form in Einstein (1913*a*), p. 1258, and in a note appended to Einstein and Grossmann (1913*), p. 261. Noting that $g_{ik}g^{ij}{}_{,s} = -g^{ij}g_{ik,s}$, we obtain the streamlined form of (5.5.12*a*) introduced in Einstein (1914*b*), p. 1077:

$$(\sqrt{-g}\, g^{rs}\, \mathrm{H}^{j}_{ks})_{,r} = -(\kappa/2)\,(\mathfrak{T}^{j}_{k} + \mathrm{t}^{j}_{k})$$

On the same page, the equations for the t^{j}_{k} are put in the form (5.5.12*b*) for the first time. (Observe, however, that the gravitational constant denoted by κ in the said passage of E.'s (1914*b*) is only half as large as in the earlier papers and in our formulae.)

Section 5.6

1. Section 5.6 was rewritten after two long conversations with John Stachel. In its present form it owes much to Stachel's thought-provoking views and illuminating remarks.

2. Einstein (1913*a*), p. 1257. Here and elsewhere I substitute Latin indices ($i, j, \ldots$) for the Greek indices ($\mu, \nu, \ldots$) used by Einstein, whenever the latter range over the first four natural numbers.

3. Einstein and Grossmann (1913), p. 18. I propose that we read "erstere nur" instead of "nur erstere" in line 5 from below.

4. Einstein to Lorentz, 14th August [1913]; Princeton Einstein Archive, Microfilm Reel 16. Italicized words are underlined in the original. The last sentence reads in German as follows: "Wenn also nicht alle Gleichungssysteme der Theorie, also auch Gleichungen (18) ausser den linearen noch andere Transformationen zulassen, so widerlegt die Theorie ihren eigenen Ausgangspunkt; sie steht dann in der Luft." Equations (18) of Einstein and Grossmann (1913) are equivalent to the field equations (5.5.12*a*) given above.

5. Equations (19) of Einstein and Grossmann (1913) are equivalent to our (5.5.6).

6. Einstein to Lorentz, 16th August [1913], p. 2; Princeton Einstein Archive, Microfilm Reel 16. Thanks to Enrico Fermi we can now give a definitive reason why the exchange of energy and momentum between matter and the gravitational field cannot be expressed by a generally covariant system of differential equations. If γ is an arbitrary worldline of the spacetime $(\mathscr{W}, \mathbf{g})$ there is always a chart x defined on a neighbourhood of γ such that the "gravitational field strengths" Γ^{i}_{jk} derived according to (D.1) from the components of $\mathbf{g}$ relative to x will all vanish at every point of γ (Fermi (1922); cf. Levi-Cività (1927), §4; O'Raifertaigh (1958)). Therefore, if γ is the worldline of a particle, one can always find a chart relative to which the field exerts no force and does no work on the particle. The status of gravitational energy in General Relativity was criticized for this reason by E. Schrödinger (1918*a*) and H. Bauer (1918). Einstein (1918*d*) boldly replied that the energy of the gravitational field need not be localized pointwise (see especially *loc. cit.*, pp. 451 f.) This raises of course a difficult problem with regard to the nature of gravitational radiation, which cannot be represented simply by a sort of analogue of the Poynting vector. See F. A. E. Pirani (1957). On the subject of energy and energy conservation in General Relativity see also A. Trautman (1962).

7. Einstein to Lorentz, 16th August [1913], p. 3.

8. Einstein (1913*a*), pp. 1257 f., immediately after the passage quoted at the beginning of this Section. The same idea turns up in an undated letter to Ernst Mach with New Year's greetings which, according to Vizgin and Smorodinskiĭ (1979), p. 499, was "evidently written on the eve" of January 1, 1913 and hence "contains the first known outline of the geometrical theory of gravitation". The exact dating of Einstein's discovery of the said idea in the letter to Lorentz of August 16, 1913, proves in my opinion that the letter to Mach must have been written one year later, *circa* January 1, 1914. The letter to Mach was first published by F. Herneck (1966). It is given in English translation by Vizgin and Smoridinskii, *loc. cit.*

9. Einstein (1913*a*), p. 1257, column 2, note 2. A similar announcement is made in the main text of Einstein (1913*b*), p. 289. This reproduces a lecture delivered to the Swiss Society of Natural Scientists on September 9, 1913, two weeks before the Vienna lecture. However, we do not know whether the remark in question was not added later.

10. Princeton Einstein Archive, Microfilm Reel 13.

11. Einstein and Grossmann (1913*), p. 260. The "Outline" circulated first as a booklet

published by Teubner in Leipzig, without these additional notes. A copy of it was mailed to Ernst Mach before June 25th, 1913 (see Einstein's letter to Mach of that date in Hönl (1966), p. 27).

12. As I noted on p. 100, this view only makes sense with regard to global charts (on spacetimes that admit them). However, the charts considered in our paraphrase of Einstein's argument are global on the submanifold which is their common domain.

13. Consider the following elementary example. Let $f_1 : \mathbf{R}^2 \to \mathbf{R}$ be given by $f_1(a, b) = a^2 + b$, while $f_2 : \mathbf{R}^2 \to \mathbf{R}$ is given by $f_2(a, b) = f_1(a, b) + f_1(b, a)$. f_2 plainly depends on f_1. However, $2 = f_2(1, 0) \neq f_2(-1, 0) = 0$, although $f_1(1, 0) = f_1(-1, 0) = 1$.

14. I have found the objection in B. Hoffmann (1972), p. 161.

15. Einstein (1914*b*), p. 1067. In Einstein (1914*a*), p. 178, the change in the argument is introduced in a footnote purporting to clarify the meaning of the main text, which essentially repeats the original version in Einstein and Grossmann (1913*), p. 260. This, and the fact that the original version reappears almost unaltered in Einstein and Grossmann (1914), p. 217, which was finished *after* Einstein (1914*a*) had been submitted on January 24th, 1914, may warrant J. Stachel's suggestion that Einstein *always* understood his argument in the same sense, as dealing with the geometric objects themselves, not just with their chart-dependent components.

16. Einstein (1914*b*), p. 1067.

17. The second interpretation seems to me more likely because Einstein says that $G'(x)$ is referred to the coordinate system K.

18. Since the charts y and z agree everywhere except on the hole H whose boundary lies in the interior of their common domain, they must have the same range. Hence the point transformation $f = y^{-1} \cdot z$ is defined on the entire domain of z, which it maps onto itself.

19. See above, p. 316, note 12.

20. Write fP for $f(P)$. The typical element of $G'(x)$ is $g_{fP}(\partial/\partial z^i|_{fP}, \partial/\partial z^j|_{fP}) = (\partial y^h \cdot f/\partial z^i|_P) \quad (\partial y^k \cdot f/\partial z^j|_P) g_{fP}(\partial/\partial y^h|_{fP}, \partial/\partial y^k|_{fP}) = g_{fP}(f_{*P}(\partial/\partial z^i|_P), \quad f_{*P}(\partial/\partial z^j|_P)) = (f^*g)_P(\partial/\partial z^i|_P, \partial/\partial z^j|_P)$.

21. It follows that, for any set of vector fields $X_1, \ldots, X_r$ on M, $\omega(X_1, \ldots, X_r) = \omega((h^{-1} \cdot h)_* X_1, \ldots, \quad (h^{-1} \cdot h)_* X_r) = \omega(h_*^{-1}(h_* X_1), \ldots, h_*^{-1}(h_* X_r)) = (h^{-1})^* \omega(h_* X_1, \ldots, h_* X_r) = h_* \omega(h_* X_1, \ldots, h_* X_r)$. This shows in what sense ω is 'dragged along' by h_*.

22. We prove it directly by a simple computation: $y_*(f_* \mathbf{g}) = (y^{-1})^* (f^{-1})^* \mathbf{g} = (f^{-1} \cdot y^{-1})^* \mathbf{g} = ((y \cdot f)^{-1})^* \mathbf{g} = (z^{-1})^* \mathbf{g} = z_* \mathbf{g}$.

23. Einstein (1915*c*), p. 832.

24. Princeton Einstein Archive, Microfilm Reel 9.

25. Einstein/Besso (1972), pp. 63 f.

26. According to the equation of geodesic deviation. See p. 343, note 38.

27. Note that f is not an isometry of $(U, \mathbf{g})$ onto itself; for $\mathbf{g} \neq f^*\mathbf{g}$. It is, however, an isometric mapping of $(U, \mathbf{g})$ onto the distinct Riemannian manifold $(U, f_*\mathbf{g})$—same underlying manifold, but different metric!—because, by our definition, if X and Y are any two vector fields on U, $\mathbf{g}(X, Y) = f_*\mathbf{g}(f_*X, f_*Y)$.

28. The relevant passage of the letter to Ehrenfest had been reproduced in English translation by Earman and Glymour (1978*a*), p. 273, but with a mistranslation that mars its sense.

29. Newton, "De gravitatione et aequipondio fluidorum", in Hall and Hall (1978), p. 103. See p. 285, note 7.

30. The viability of such a conception of spacetime is cogently argued for by Mortensen and Nerlich (1978).

31. See, for instance, the anthology by S. P. Schwartz (1977).

32. This is not to deny that two things can be really different even if they are empirically indistinguishable and they cannot, on any grounds, be regarded as instances of distinct conceptual structures. Differences may well exist which our concepts and observations cannot grasp. But such differences—which anyway would lie outside the purview of physical science—are irrelevant to the present discussion. Here we are concerned with a hypothetical case proposed by ourselves, and the physical situations considered are in no sense more different from one another than, in the light of our assumptions, we can make them out to be.

33. Einstein's "hole" argument, as presented in his (1914*b*), p. 1067, was criticized also by David Hilbert, in his second paper on "The Foundations of Physics", submitted to the Göttingen Gesellschaft der Wissenschaften on December 23, 1916. According to Hilbert, Einstein's argument involved a misunderstanding of the physical Principle of Determinism (*Kausalitätsprinzip*) as it

applies to generally covariant theories. A generally covariant differential system with n unknowns cannot comprise more than $n-4$ independent equations and must therefore be underdetermined, because otherwise the arbitrary choice of a chart, subject to 4 coordinate conditions, would generate an overdetermined and hence generally inconsistent system (Hilbert (1917), pp. 59–61 and note 1 on p. 61). Thus, for example, the 10 field equations of General Relativity (5.5.10) are constrained by the 4 identities $G^{ij}{}_{;j} = 0$, where G is the proper Einstein tensor. Hence no more than 6 equations are independent and the system does not suffice to determine the 10 unknown functions g_{ij}.

Section 5.7

1. Einstein/Besso (1972), p. 53. Einstein adds: "Hence it turns out that there are transformations representing accelerations of a very varied nature (e.g. rotations) which map the equations on themselves (e.g. rotations also)."
2. Einstein and Grossmann (1914), p. 225: "Die Kovarianz der Gleichungen [ist] eine denkbar weitgehende". Einstein was aware, of course, that any physical theory could be given a generally covariant formulation—he did not have to wait for E. Kretschmann (1917) to enlighten him on this point. However, the hole argument (Section 5.6) had persuaded him that a generally covariant substitute for the Einstein–Grossmann equations (5.5.12) would "not be particularly interesting from a physical or a logical point of view" (Einstein (1914*a*), § 9, p. 179). See also Einstein (1913*a*), p. 1258, column 1, lines 37–9.
3. Einstein and Grossmann (1914), p. 219, n. 2.
4. Einstein to Sommerfeld, 15th July 1915 (Einstein/Sommerfeld (1968), p. 30). Einstein's visit to Göttingen probably stimulated Hilbert to work on a theory of gravity based on Einstein's programme. See p. 175.
5. Einstein, (1915*a*, *b*, *c*, *d*), submitted on November 4th, 11th, 18th and 25th, respectively. The alternative systems of field equations are proposed in the first, second and fourth paper. The third paper derives the precession of Mercury's perihelion from the second system.
6. Einstein/Sommerfeld (1968), pp. 32 f. As I noted in the Introduction, this section contains a few passages which presuppose some acquaintance with the calculus of variations and its application to physical field theories. Readers lacking it may skip over them or turn to Lovelock and Rund (1975), Chapters 6 and 8, which furnish a very clear introduction to the subject, especially attuned to the needs of the student of Relativity.
7. Einstein (1915*a*), p. 778.
8. Einstein (1915*a*), pp. 778 f.
9. Observe, by the way, that if spacetime admits *one* J-atlas it will admit infinitely many *distinct* J-atlases, obtained from the first by changing the scale of the charts. Such atlases belong to *different* maximal J-atlases, for the Jacobians of the rescalings are not equal to 1.
10. Einstein (1915*a*), p. 781. See above, p. 318, note 24.
11. I modify Einstein's notation. In the 1915 papers he denotes the components of the Ricci tensor by G_{ij} and uses R_{ij} for our U_{ij}. The last equality in (5.7.3*c*) follows from a remark on p. 317, note 17, which entails that $\Gamma^{j}_{hj} = (\log\sqrt{-g})_{,h}$.
12. A proof is given in Einstein (1916*a*), § 15 (where L is denoted by H).
13. Remember that in Einstein's printed version of (5.5.12*b*) our H^{i}_{jk} were denoted by Γ^{i}_{jk}. The new formula for the t^{j}_{k} is therefore typographically almost identical with the old. Note however the changed position of j and k in the last term. The factor $\sqrt{-g}$ can be divided out because, as a J-scalar, it is not affected by the allowed transformations. It is therefore otiose to distinguish between J-tensor fields and J-tensor densities.
14. Bear in mind that $g^{ij}_{,k} = -(g^{ir}\Gamma^{j}_{rk} + g^{jr}\Gamma^{i}_{rk})$, as a simple calculation can show.
15. According to our remark in note 14, $g^{ih}\Gamma^{k}_{ij} = (1/2)\,(g^{ih}\Gamma^{k}_{ij} - g^{ik}\Gamma^{h}_{ij} - g^{hk}{}_{,j})$. Substitute this value in the first term of the left-hand side of (5.7.10) and differentiate both sides with respect to x^h. The right-hand side vanishes by (5.7.7).
16. Use the remark in note 14. It entails that $g^{ij}{}_{,k}\, g_{ij} = -2\Gamma^{j}_{jk} = -2(\log\sqrt{-g})_{,k}$.
17. Einstein (1915*a*), p. 785. Einstein adds that the condition $(g^{ij}{}_{,ij} - \kappa t^{k}{}_{k}) = 0$, on which (5.7.13) depends, "says how the coordinate system must be adapted to the manifold". I take this to mean that the said condition must be regarded as a convention restricting the choice of charts—

like eqns (5.7.2) in the Einstein–Grossmann theory. It is worth observing, however, that if $(\kappa t^k_k - g^{ij}{}_{,ij})$ is equated with any *arbitrary* constant C, the same calculation that led to (5.7.13) will show that, if g is constant, $\kappa T^j{}_j = C$, a rather unpalatable result. Cf. note 21.

18. Cf. eqns (4.5.12) on p. 117. Substitute g^{ij} for η^{ij}, multiply both sides by g_{ij} and sum over i and j.

19. Einstein (1915*b*), pp. 799 f. On the "electromagnetic view of matter" see p. 290, note 18. Gustav Mie's Theory of Matter was initially predicated on the tentative assumption that "the electric and magnetic fields, the electric charge and the charge current are entirely sufficient for describing all phenomena in the material world" (Mie (1912*a*), p. 513). These objects were supplemented, in the Theory's third instalment, with quantities representing the gravitational field. Cf. Mie (1913), pp. 27 f.

20. Einstein (1915*b*), p. 800.

21. Einstein (1915*b*), p. 800. Why is the adoption of (5.7.15) "made possible" by the hypothesis that $T^h_h = 0$? A straightforward answer to this question can be found using the Second Bianchi Identities, as on p. 318, note 26. Equation (*a*) in that note entails that

$$R^h_{h;k} = 2R^j_{k;j} \qquad (d)$$

substituting $-\kappa T^j_k$ for R^j_k, in agreement with (5.7.15), and dividing through by $-\kappa$, we find that

$$T^h_{h;k} = 2T^j_{k;j} \qquad (e)$$

According to the fundamental law (5.7.18), the right-hand side of (*e*) must vanish. Hence, $0 = T^h_{h;k} = T^h_{h,k}$ (for T^h_h is a scalar field), and $T^h_h = const.$ By making a virtue out of this necessity, the hypothesis in Einstein (1915*b*) paves the way for eqns (5.7.15). It is unlikely however that an argument along these lines deterred Einstein from adopting the generally covariant field equations (5.7.15) in the first paper of November 1915. The Second Bianchi Identities, though published by Bianchi in an Italian journal in 1897, are notoriously absent in the German literature of the 1910s. See J. Mehra (1974), p. 49 and note 242.

22. An exact solution of the same problem was soon found, independently, by K. Schwarzschild (1916—submitted January 13th) and J. Droste (1916—submitted May 27th).

23. The results of the British Expedition, solemnly announced by the Astronomer Royal on November 6th, 1919, were hailed by the world press as sufficient proof of Einstein's theory of gravity. However, their level of accuracy was well below the usual standards of astronomy and optics. (Earman and Glymour (1980*a*) have moreover accused Eddington of doctoring the data.) The bending of light-rays in a gravitational field predicted by General Relativity has been verified with reasonable accuracy only in the last decade, by the method of very-long-base-line radio interferometry. Let $1.75'' (1+\gamma)/2$ be the observed value of the deviation. Fomalont and Sramek (1976) obtained $\gamma = 1.01 \pm 0.02$. See the brief discussion in I. I. Shapiro (1980*a*), pp. 474–7.

24. Einstein (1915*c*). The observed value of Mercury's secular perihelion advance amounts to some 5600″. Almost 90% is accounted for by the Earth's own motion. Over 9% is attributable, according to Newton's Law, to the perturbing action of the other planets. But there remains a balance, estimated by G. M. Clemence (1947) as $43.11'' \pm 0.45''$ per century, which classical astronomy was unable to explain. The beautiful agreement of this figure with Einstein's is unfortunately somewhat messed up by uncertainties regarding the shape of the Sun. See R. H. Dicke and H. M. Goldenberg (1974), A. R. Hill and R. T. Stebbins (1975), C. M. Will (1979), pp. 55–57. I. I. Shapiro (1980*a*), pp. 481–3.

25. Multiply both sides of (5.7.17) by g^{ij}, sum over i and relabel dummy indices to obtain $R^h_h = \kappa T^h_h$. Substituting this value for κT^h_h in (5.7.17) yields eqns (5.5.10) in covariant form.

26. Einstein (1915*d*) p. 846. The sneer quotes are Einstein's.

27. Compare eqns (5.7.3), (5.7.6*).

28. In the following calculations use the hints given in notes 14, 15, 16.

29. Both (5.7.10) and (5.7.21) formally correspond to the Einstein–Grossmann field equations. Compare them with the formulation of the latter given on p. 319, note 30.

30. Einstein (1951*d*), pp. 846 f. I take it that in this passage Einstein states a reason for preferring the field equations (5.7.17) to eqns. (5.7.15). Thus, I cannot agree with J. Earman and C. Glymour (1978*b*), p. 305, for whom "the only explicit reason" that Einstein gives here for changing the field equations is that T^k_k and t^k_k occur symmetrically in (5.7.22) and (5.7.23).

31. Einstein (1916*a*), p. 803 f.; Kilmister (1973), p. 156.
32. Einstein (1916*a*), p. 807; Kilmister (1973), p. 160.
33. Einstein (1916*a*), §§ 17, 18.
34. For a detailed proof see Lovelock and Rund (1975), pp. 305–323. $\sqrt{-g}\,R^k_k$ is not the only scalar density that can be constructed from the components of the metric tensor and their first and second derivatives, but it is certainly the simplest. The following facts contribute further to single out the Einstein field equations among all conceivable alternatives: (i) There does not exist a scalar density which depends solely on the g_{ij} and their first derivatives. (ii) If L is a scalar density depending only on the g_{ij} and their first and second derivatives and if the corresponding Euler-Lagrange equations $E^{ij}(L) = 0$ are of the second order—and not of the fourth order, as one would normally expect—then $E^{ij}(L)$ equals $R^{ij} - (1/2)g^{ij}\,R^k_k + \lambda g^{ij}$, i.e. the modified Einstein tensor (Lovelock (1969); Rund and Lovelock (1972), pp. 43 f.).
35. See, for instance, A. Papapetrou (1974), pp. 122 f.
36. Hilbert (1915). The so-called Einstein tensor, with components $R_{ij} - (1/2)g_{ij}R^k_k$, first came out in print on p. 404, line 1 from below. Hilbert's paper is analysed by J. Mehra (1974), pp. 24–31. On Hilbert's relations with Einstein in 1915–16, see J. Earman and C. Glymour (1978*b*).

Section 5.8

1. For a detailed up-to-date discussion of the subject with abundant references to the enormous literature, see J. Ehlers, ed. (1979), especially the contributions by P. Havas (pp. 74 ff.) and W. G. Dixon (pp. 156 ff.). I thank John Stachel for correcting several errors in the first draft of this Section and for calling my attention to Havas's work.
2. I follow Einstein's view of the matter. See, for example, Einstein and Grommer (1927), pp. 2f.; Einstein and Infeld (1949), p. 210. But the above analysis of the relation between field equations and equations of motion in classical electrodynamics ought to be qualified in some respects. As we saw on p. 113, the Lorentz force law can be inferred from the Relativity Principle and Maxwell's equations. Moreover, as is well known, the equation of continuity (charge conservation) provides conditions of integrability of the Maxwell equations, whence not any arbitrary charge distribution but only one which satisfies the equation of continuity is compatible with them.
3. According to J. Stachel (1969*a*, 1972) no such constraints need arise in the case of a gravitational field with an external source. In that case one describes the source by a given energy tensor T and seeks to integrate the inhomogeneous Einstein equations (5.7.17) or (5.5.10). Since these equations entail that the contravariant energy tensor components T^{ij} relative to any chart must satisfy the conservation equations

$$T^{ij}{}_{;j} = 0 \qquad (5.5.4a)$$

and the covariant divergence in (5.5.4*a*) involves the very metric components that are the unknowns for which one is trying to solve the field equations, it seems at first sight that the energy tensor cannot be arbitrarily assigned. However, a theorem proved by C. Pereira (in his unpublished doctoral thesis of 1967) apparently removes all restrictions on the choice of the energy tensor *T*. According to Pereira's theorem, if $(\mathfrak{T}^{ij})$ is a 4×4 symmetric matrix of scalar fields on a spacetime region U and if (g_{ij}) are the components of an arbitrary Lorentz metric **g** relative to a chart x defined on U, one can always find on a neighbourhood of each point $P \in U$ a chart y^P such that, if we treat the $\mathfrak{T}^{ij}$ as the components relative to y^P of a (2, 0) tensor density $\mathfrak{T}$, it turns out that the covariant divergence of $\mathfrak{T}$ vanishes. (The covariant divergence must of course be computed using the *transformed* metric components $(\partial x^h/\partial y^i)\,(\partial x^k/\partial y^j)\,g_{hk}$.) Thus any arbitrary energy tensor could apparently be reconciled with the need of satisfying eqns. (5.5.4*a*). The reader will not fail to observe that, while the scalar fields $\mathfrak{T}^{ij}$ are completely arbitrary, their exact physical interpretation as energy tensor components depends on the structure of the charts y^P and will probably be quite different in their several domains. In particular, there is no guarantee that any particle worldlines defined by those interpretations will be geodesic or will have any other preestablished shape. Not to mention the fact, appositely stressed by Stachel (in a conversation with the author), that such an arbitrary array of scalar fields interpreted as energy tensor components relative to prospective charts will usually turn out to be physically inane.

4. Einstein and Grossmann (1913), p. 10.
5. W. Rindler (1977), pp. 182 f. See also A. S. Eddington (1924), pp. 126 f.
6. H. Weyl (1923), § 38. See also § 36 of the fourth edition on which the English translation is based.
7. A. Einstein and J. Grommer (1927); see also A. Einstein (1927).
8. A. Einstein and J. Grommer (1927), p. 4. See the elegant reconstruction of Einstein and Grommer's basic argument in J. Ehlers (1980*b*).
9. A. Einstein and N. Rosen (1935), p. 73.
10. Singularities in General Relativity are the subject of Section 6.4. As we shall see there, several proposals have been made for supplementing relativistic spacetime with a set of ideal points comprising the so-called singularities. However, in my opinion, such ideal points cannot be regarded as possible locations for events (e.g. events in the history of a particle) without drastically modifying the theory. In order to view them as such one would have indeed to equate the physical world with the Schmidt completion of relativistic spacetime (p. 217) or some other no less abstruse mathematical contraption, as step which, as far as I know, nobody has yet ventured to take and which appears to be quite unwarranted.
11. Cf. Einstein and Infeld (1949), p. 209: "All attempts to represent matter by an energy-momentum tensor are unsatisfactory and we wish to free our theory from any particular choice of such a tensor. Therefore we shall deal here only with gravitational equations in empty space, and matter will be represented by singularities of the gravitational field." Einstein, Infeld and Hoffmann (1938), p. 65, reject the assumption of a specific energy tensor for matter because such tensors "must be regarded as purely temporary and more or less phenomenological devices for representing the structure of matter, and their entry into the equations makes it impossible to determine how far the results obtained are independent of the particular assumption made concerning the constitution of matter". Less formally, Infeld (1957), p. 398, remarks that Einstein always thought that to use the field equations (5.5.10) instead of the vacuum field equations in which $T^{ij} = 0$ "is somehow in bad taste, because we do not know in [5.5.10] what $[T^{ij}]$ is, and we mix a geometrical tensor on the left side with a physical tensor on the right side".
12. L. Infeld and A. Schild (1949), p. 410. To avoid misunderstandings I have substituted Latin for Greek indices in agreement with our usual convention.
13. Fock's theory is expounded at length in Chapter VI of his (1959). Its first formulation was simultaneously published in Russian and French in 1939. N. Petrova (1949) carried Fock's original approximation one step further. A. Papapetrou (1951*a*) gave a different derivation of the equations of motion of finite masses from the field equations, based on an approach similar to Fock's. In his recent text-book A. Papapetrou (1974), pp. 152–7, discusses the motion of test particles from this point of view.
14. To obtain the general relativistic form of the energy tensor of a perfect fluid substitute the general metric components g^{ij} for the Minkowski components η^{ij} in equations (4.5.17*).
15. On spinning test particles see A. Papapetrou (1951*b*), E. Corinaldesi and A. Papapetrou (1951).
16. In his encyclopedia article on General Relativity, P. G. Bergmann (1962*b*), p. 245, comments on the inadequacies of both Einstein's and Fock's work on the subject.
17. P. Havas (1979), p. 121, after repeating the argument of his earlier paper with J. N. Goldberg, emphasizes that the Geodesic Law is not "the full law of motion for an extended body" or "for particles which carry both monopole and dipole moments", but only an approximation to such full laws.
18. P. Havas (1979), p. 121.
19. W. G. Dixon (1970*a*, *b*, 1974/75, 1979), J. Ehlers and E. Rudolph (1977), R. Schattner (1979*a*, *b*). I follow J. Ehlers (1980*b*).
20. J. Ehlers (1980*a*), p. 283.
21. J. Ehlers (1980*a*), p. 283.
22. J. Ehlers (1980*b*).
23. The domain of dependence $D(S)$ of a subset S of a causal space was defined on p. 124. The concepts of causal geometry can be extended to arbitrary relativistic spacetimes as explained on page 192. Cf. also pp. 247 f., 252 f.
24. On the Cauchy problem and the dynamical formulation of General Relativity see the review

articles by Y. Choquet-Bruhat and J. W. York (1980), A. E. Fischer and J. E. Marsden (1979) and C. Teitelboim (1980). The articles by Bruhat (1962) and Arnowitt, Deser and Misner (1962) in Witten's *Gravitation* are still recommended as an introduction to more recent research.

25. Einstein (1913*a*), p. 1266.
26. Einstein (1918*a*), p. 154.
27. Einstein (1918*a*). Cf. Einstein (1916*d*), p. 692.
28. H. Bondi, F. A. E. Pirani and I. Robinson (1959).
29. Weinberg (1972), p. 251. The mathematical difficulties of the subject arise from the non-linearity of Einstein's field equations. The main conceptual difficulty was mentioned on p. 319, note 6, where I referred the reader to Pirani's trailblazing article of 1957. Due to the extreme weakness of gravitational interaction (roughly 10^{-40} times as strong as electromagnetism), gravitational radiation is very hard to detect. At the time of writing, the strongest evidence for its existence was that presented by J. H. Taylor *et al.* (1979): The binary pulsar 1913 + 16, presumably formed by two neutron stars orbiting about each other in 7.8 hours, appears to be losing energy and angular momentum at the rate expected from gravitational radiation due to the orbital motion of its constituents. Gravitational radiation is explained informally in a pleasant book by P. C. W. Davies (1980). For more rigorous presentations the reader may turn to any good graduate textbook of General Relativity, or to Pirani (1965). Grishchuk and Polnarev (1980) is a recent survey, with up-to-date references. On the current state of experimental research, see Douglass and Braginsky (1979).
30. The best recent surveys on the subject are to be found in Ehlers, ed. (1979).
31. See Geroch (1977). On pp. 216–8 we touch on the subject of ideal points at infinity, in connection with spacetime singularities.
32. The conformal treatment of null infinity was introduced by Roger Penrose; see Penrose (1964, 1968). To define a manifold-with-boundary we consider first the n-dimensional half-space $H^n \subset \mathbf{R}^n$. This consists of all real number n-tuples with non-negative first element. The boundary of H^n (in $\mathbf{R}^n$) is the $(n-1)$-manifold $\{0\} \times \mathbf{R}^{n-1}$. We view H^n as a topological space with the topology induced by the standard (Euclidean) topology of $\mathbf{R}^n$. An *n-manifold-with-boundary M* is defined like an n-manifold (p. 257), except that the charts of M are bijective mappings of subsets of M onto open sets of the half-space H^n. A point $P \in M$ belongs to the *boundary* $\partial M \subset M$ if and only if P is mapped by a chart on a point of the boundary of H^n. (The reader ought to satisfy himself that this property of being charted into the boundary of H^n is invariant under coordinate transformations.)
33. By "Schwarzschild mass" we understand primarily the integration constant that turns up in Schwarzschild's exact solution to Einstein's field equations *in vacuo* (designated by m in the Schwarzschild line element, as given on p. 335, note 2). The concept can be extended to other static solutions.
34. One must bear in mind, however, that because electricity can be both attractive and repulsive, whereas gravity, in the short range, is purely attractive, electromagnetic radiation starts with dipole terms, while gravitational radiation in the linearized approximation starts with quadrupole terms.
35. The linearized or weak-field approximation to General Relativity is obtained by setting the metric components g_{ij} equal to $\eta_{ij} + \gamma_{ij}$, with γ_{ij} small; thus assuming that spacetime is very nearly flat. See Einstein (1916*d*), or Rindler (1977), pp. 188 ff.
36. See the review article by W. L. Burke (1979) and the references given there.

Chapter 6

Section 6.1

1. There are several survey articles on special questions, such as the brilliant 100-page monograph on "Singularities and Horizons" by Tipler, Clarke and Ellis (1980), and I know of two long histories of 20th-century cosmology, by J. N. North (1965) and J. Merleau-Ponty (1965)—both of them published by a curious coincidence in the same year in which the discovery of the

background thermal radiation by Penzias and Wilson tipped the scales in favour of relativistic cosmology and triggered a new spurt of interest in it. But there is as yet no historical work on General Relativity that can be compared, say, with Max Jammer's books on Quantum Mechanics.

2. Let v, w be vectors at $P, Q \in \mathscr{M}$, respectively. $Q = tP$ for some $t \in \mathbf{R}^4$. w is *parallel* to v if and only if $w = t_{*P}(v)$.

3. If $\mathbf{R}^4$ acts transitively on $\mathscr{M}$ as a Lie group of translations a global chart $x: \mathscr{M} \to \mathbf{R}^4$ is defined by setting $x(tP) = t$ for a fixed "origin" $P \in \mathscr{M}$. If $x: \mathscr{M} \to \mathbf{R}^4$ is a global chart, a smooth transitive and effective action of the additive group of $\mathbf{R}^4$ on $\mathscr{M}$ is defined by setting $Q = tP$ whenever $x(Q) - x(P) = t$.

4. Levi-Cività, who had coauthored with Ricci the main source for the study of the absolute differential calculus invented by the latter (Ricci and Levi-Cività, 1901), had kept an eye on Einstein's use of it in his first theory of gravity. In the Spring of 1915 both men corresponded actively about what Levi-Cività saw as mathematical errors in Einstein (1914*b*). (Princeton Einstein Archive, Microfilm Reel 16. On April 2, 1915 Einstein wrote to Levi-Cività: "Eine so interessante Korrespondenz habe ich noch nie erlebt".) Levi-Cività's research on parallelism was, he says, directly encouraged by Einstein's theory of gravity, being guided by a desire to simplify the formal apparatus ("ridurre alquanto l'apparato formale") used for the definition of the Riemann tensor, in view of its essential role in "the mathematical development of Einstein's grandiose conception" (Levi-Cività (1917), p. 173). G. Hessenberg (1917), who independently achieved a non-metric characterization of geodesics, curvature and covariant differentiation which is essentially the same as Levi-Cività's and involves in effect a conception of vector parallelism (though the expression is not used), was also motivated by the wish to simplify "the vast and heavy apparatus of formulae" ("der umfängliche und schwerfällige Formelapparatus") of the theory of quadratic differential forms, in view of its newly acquired significance for Relativity. A decade earlier, L. E. J. Brouwer had explicitly introduced the concept of vector parallelism along a curve in a discussion of Riemannian manifolds of constant negative curvature (Brouwer (1906), pp. 5, 16, 18).

5. Levi-Cività (1977), pp. 100 ff. The first Italian edition of this book appeared in 1923.

6. To see that the enveloping surface S' is indeed developable consider a sequence σ_1 of equidistant values of t, $(t_0, \ldots, t_r)$, with $\gamma(t_0) = P$ and $\gamma(t_r) = Q$. The family of planes $\pi_0, \ldots, \pi_r$ is enveloped by a polyhedral surface S'_1 that coincides with π_i on the region containing $\gamma(t_i)$ which is bounded by the intersections of π_i with π_{i-1} and with π_{i+1} $(1 \le i \le r)$. S'_1 can obviously be developed onto a plane by rotating its sides about its edges. Now form new sequences σ_2, $\sigma_3, \ldots$ of values of t, by adding at each turn the midpoint between every two consecutive values in the preceding sequence. We thus generate an infinite sequence $S'_1, S'_2, S'_3, \ldots$ of developable surfaces which converges to S'.

7. An *embedding* f of an n-manifold M into an m-manifold M' is a diffeomorphism of M onto a submanifold of M' (see p. 261). f is isometric with respect to the metric $\mathbf{g}$ of M and the metric $\mathbf{g}'$ of M' if $\mathbf{g} = f^*\mathbf{g}'$. $(M, \mathbf{g})$ is *globally isometrically embeddable* (g.i.e.) in $(M', \mathbf{g}')$ if there is an isometric embedding of the former into the latter. $(M, \mathbf{g})$ is *locally isometrically embeddable* (l.i.e.) in $(M', \mathbf{g}')$ if each point of M has a neighbourhood U such that $(U, \mathbf{g})$ is g.i.e. in $(M', \mathbf{g}')$. The following results on embeddability have been proved after the publication of Levi-Cività's paper: (i) Any n-manifold can be embedded in $\mathbf{R}^{2n+1}$ (Whitney, 1936) and also in $\mathbf{R}^{2n}$ (Whitney, 1944*a*), but it may not be embeddable in $\mathbf{R}^{2n-1}$ (Whitney, 1944*b*). (ii) An n-manifold M with C^∞-metric $\mathbf{g}$ is l.i.e. in a similar m-manifold $(M', \mathbf{g}')$ if $m \ge (n/2)(n+3)$ and the number of positive and negative eigenvalues of $\mathbf{g}$ are respectively less than those of $\mathbf{g}'$ (Greene, 1970); moreover, if the manifolds and their metrics are analytic (p. 288, note 12), m can be as low as $(n/2)(n+1)$, exactly as Levi-Cività (1917, p. 176) claimed (Friedman, 1961). (iii) Requirements of g.i.e are of course more stringent; thus, while an arbitrary 4-manifold with a Lorentz C^∞-metric is l.i.e. in $\mathbf{R}^{14}$, endowed with a metric of suitable signature, it is g.i.e., as far as we know, in a flat $\mathbf{R}^{36}$ if it is compact (Greene, 1970), and in a flat $\mathbf{R}^{89}$ if it is non-compact (Clarke, 1969; Clarke's result holds for a C^3-metric). Further results and explanations on the subject, and a list of 144 references, are given by H. F. Goenner (1980*a*).

8. The tangent to γ at $\gamma(t)$ is $\gamma_{*t}\mathrm{d}/\mathrm{d}t = (\mathrm{d}x^i \cdot \gamma/\mathrm{d}t)\, \partial/\partial x^i|_{\gamma t}$. Substituting this expression for V in (6.1.1) we obtain the geodesic equations (D.8) on p. 281.

9. Weyl (1918*a*), p. 385; cf. Weyl (1919*b*), p. 112. Weyl's unified theory was first developed in his (1918*b*), submitted by Einstein to the Prussian Academy. (See also Weyl (1919*b*), (1921*c*) and the various editions of his book *Raum Zeit Materie*.) Upon receiving Weyl's manuscript, Einstein

wrote him that the main idea was "*a stroke of genius of the first rank*" (Einstein to Weyl, 6th April 1918, Princeton Einstein Archive, Microfilm Reel 24; Einstein's emphasis). However, a few days later, he remarked: "So schön Ihr Gedanke ist, muss ich doch offen sagen dass es nach meiner Ansicht ausgeschlossen ist, dass die Theorie der Natur entspricht." The objection mentioned in note 10 follows. (Einstein to Weyl, 15th April 1918, *ibid.*)

10. This gave rise to the following objection to Weyl's theory, put forward by Einstein in his letter of 15 April 1918 and printed at the end of Weyl (1918*b*): if the characteristic frequencies of atomic clocks depend not only on their actual spacetime locations but also on their past worldlines, spectral lines would be blurred.

11. See p. 307, note 5. Weyl defines the electromagnetic field tensor as in that note and obtains the first set of Maxwell equations (4.5.5) as a mathematical corollary.

12. Eddington (1921); (1924), pp. 213 ff.

13. Componentwise, $\mathrm{R}_{ij} = \frac{1}{2}(\mathrm{R}_{ij} + \mathrm{R}_{ji}) + \frac{1}{2}(\mathrm{R}_{ij} - \mathrm{R}_{ji})$.

14. Einstein (1923*a*) proposed a system of field equations, derivable from a variational principle, for Eddington's theory. Einstein (1923*b, c*) add further developments. The content of Einstein's three papers is reported in English in Eddington (1924), Supplementary Note 14, pp. 257 ff.

15. See Einstein (1930*a, b*), which supersede the earlier presentations in Einstein (1928*a, b*; 1929*a, b*). See also Cartan (1930) and Cartan/Einstein (1979), which contains the important correspondence that went on between them from 1929 to 1932.

16. Cartan to Einstein, 8th May 1929, in Cartan/Einstein (1979), p. 6. Observe that the implicit Lorentz metric is compatible, in the sense of page 192, with the non-symmetric connection that governs "parallelism at a distance", though the latter is obviously not its Levi-Cività connection. The field equations proposed by Einstein (1930*a*, p. 693) involve both the metric and the torsion components as the data from which the connection components are to be obtained by integration.

17. Einstein (1956), Appendix II, pp. 133–66.

18. Trautman (1965), p. 107, presents the matter as follows. Let Γ be an arbitrary symmetric linear connection on $\mathscr{M}$. Let the scalar field t and the symmetric (2, 0)-tensor field **g** be coupled to each other as above. Taking components with respect to a global chart x, let us postulate that: (i) $\mathrm{R}^{i}_{ijk} = 0$ (where R^{i}_{jkh} is defined by (D.2)); (ii) $\mathrm{R}^{i}_{jhm}\,(\partial t/\partial x^{k}) = \mathrm{R}^{i}_{khm}\,(\partial t/\partial x^{j})$; (iii) $g^{ij}\mathrm{R}^{h}_{kjm} = g^{ih}\mathrm{R}^{j}{}_{mhk}$; and (iv) $g^{ij}{}_{;k} = 0$. Given these postulates, Newton's theory of gravity consists of the vacuum field equations $\mathrm{R}_{ij} = 0$ and the equations of motion (5.4.7*)—as in General Relativity, the Ricci tensor vanishes *in vacuo* and freely falling particles obey the geodesic law. See also G. Dautcourt (1964).

19. The angle between two vectors v and w at a point P of a proper Riemannian manifold $(M, \mathbf{g})$ is, by definition, equal to *arc cos* $(\mathbf{g}_P(v, w)/(\mathbf{g}_P(v, v)\,\mathbf{g}_P(w, w))^{1/2})$.

20. If v is a vector at a point P and λ_P is an arbitrary real number, $\mathbf{g}_P(v, v)$ is greater than (equal to, less than) zero if and only if the same holds for $\lambda_P^2\,\mathbf{g}_P(v, v)$.

21. In the notation of note 19, the vector v is orthogonal to w if and only if $\mathbf{g}_P(v, w) = 0$. This criterion also makes sense if **g** is indefinite, in which case, of course, all null vectors must be pronounced self-orthogonal (cf. p. 92).

22. Weyl (1921*b*), p. 100. The theorem does imply this insofar as the free fall of particles and the propagation of light fix, respectively, the projective and the conformal structure of spacetime. However, this premiss must be taken with a grain of salt, for the timelike and null geodesics of General Relativity, on which indeed the said structures do depend, are not supposed to be the *actual* worldlines of real particles and light pulses, but the worldlines that such objects *would* describe under certain ideal circumstances.

23. Ehlers *et al.* (1972), p. 65. The authors refer to Helmholtz (1868)—a paper I have discussed in Torretti (1978*a*), pp. 155 ff.

24. I strongly advice the reader to study the original paper by Ehlers, Pirani and Schild (1972) or the subsequent exposition by Ehlers (1973), pp. 22–37. Ehlers and Schild (1973) prove important results on projective structures (in Weyl's sense).

25. Ehlers, Pirani and Schild (1972) postulate that the radar charts must constitute a C^3-atlas. Cf. my remarks on p. 348, note 4 to part A.

26. Another very interesting construction of the manifold structure of spacetime along similar lines is that given by N. M. J. Woodhouse (1973).

Section 6.2

1. See the references to Newton's water-bucket and two-globe experiments on p. 285, note 11.
2. H. Weyl (1923), p. 221.
3. Einstein/Born (1969), p. 258. See also Einstein to Besso, 10 August 1954, in Einstein/Besso (1972), p. 525.
4. Einstein to Mie, 8th February 1918, page 1; Princeton Einstein Archive, Microfilm Reel 17, Einstein's "relativistic standpoint" of 1918 ultimately rests on the observationist dogma he professed at that time: "Das Kausalgesetz hat nur dann den Sinn einer Aussage über die Erfahrungswelt, wenn als Ursachen und Wirkungen letzten Endes nur *beobachtbare Tatsachen* auftreten" (Einstein (1916*a*), p. 771—in English in Kilmister (1973), p. 143; I advise reading the whole paragraph).
5. Einstein to Mie, 22nd February 1918, page 1; Princeton Einstein Archive, Microfilm Reel 17. I have substituted Latin for Greek indices; see p. 319, n. 2.
6. Einstein (1918*b*), pp. 241 f.
7. Einstein (1918*b*), p. 241 n. Note that the German designation *Machsches Prinzip* means Machian Principle and does not entail that the principle was authored or even subscribed to by Mach himself.
8. His stance changed in the course of the following decade. In a lecture delivered in 1927, on the bicentennial of Newton's death, Einstein mentioned as a sign of "Newton's wisdom" the recognition by the latter that "the observable geometrical quantities (distances between the material particles) and their evolution in time do not completely characterize the motions from a physical point of view" and that "space must possess some sort of physical reality, [. . .] a reality of the same kind as the material particles and their distances" (Einstein (1934), p. 202).
9. Einstein (1916*b*), p. 103. The passages quoted by Einstein occur in Mach (1963), II.6, pp. 217, 224, 226. They include some well-known remarks on Newton's water-bucket experiment. This, Mach observed, can only teach us that no noticeable centrifugal forces act on the water when it moves relatively to the bucket's thin walls, whereas such forces do show up as soon as the water is moving together with the bucket relatively to the mass of the earth and the other sidereal bodies. However "nobody can say what the quantitative and qualitative outcome of the experiment would be if the bucket walls became thicker and more massive; eventually, several miles thick".
10. Einstein (1913*b*), p. 290.
11. Einstein (1912*a*, *b*). In this theory, the inertial mass that occurs in the formulae for impulse and energy is m/c, where m is a constant characteristic of the body under consideration, and c is the speed of light *in vacuo*, which—as we noted on p. 138—is proportional to the gravitational potential. As masses are piled up near the body, the gravitational potential decreases and m/c increases. "This agrees with Mach's bold idea that inertia originates in an interaction of the massive particle under consideration with all the rest" (Einstein and Grossmann (1913), p. 6).
12. Einstein (1913*a*), p. 1260.
13. Einstein to Mach, 25th June 1913; printed in fascimile in Hönl (1966), p. 27. Compare Mach's remark on a thick-walled rotating bucket in note 9.
14. Einstein (1956), pp. 101 f. On the linear approximation to General Relativity see p. 325, note 35.
15. Einstein (1956), p. 103.
16. Einstein (1917), p. 145.
17. Einstein (1915*c*); K. Schwarzschild (1916).
18. "Somit würde die Tragheit durch die (im Endlichen vorhandene) Materie zwar *beeinflusst* aber nicht *bedingt*" (Einstein (1917), p. 147).
19. He assumes for simplicity's sake that the metric is spatially isotropic at each point. Relative to a suitable chart x the line element is then given by $ds^2 = -A dx^\alpha dx^\alpha + B(dx^0)^2$. In a first approximation the inertial mass of a test particle is then equal to $mAB^{-1/2}$ (where m stands for the particle's proper mass). This expression will vanish at spatial infinity if A tends to 0 or B tends to ∞.
20. Obviously Einstein had in mind the velocities of stars belonging to the Galaxy. At the time of his writing a substantial segment of the astronomical profession still maintained that the Galaxy comprises all that can be observed in the sky. The fact that most nebulae are star systems or galaxies by themselves, lying at enormous distances from our own, was not established beyond reasonable doubt until the twenties, when the 100-inch Hooker telescope at Mt. Wilson was

trained on them. However, by 1917 V. M. Slipher had already measured the radial velocities of some nebulae and had found that NGC 1068 and NGC 4594 were receding from us at about 1000 km/s.

21. Riemann (1867), p. 22.

22. A topological space S is said to be *compact* if every open cover of S includes a finite subcover. Suppose that S is a compact spacelike hypersurface of a relativistic spacetime. S can surely be covered by a family of subsets of unit volume, open in S. Since S is compact there is a finite collection of at most m such subsets which also covers. S. Hence, the volume of S is not greater than m.

23. $\mathbf{S}^3$ denotes the set of all real number quadruples (a, b, c, d) such that $a^2+b^2+c^2+d^2 = 1$. We look on $\mathbf{S}^3$ as endowed with the topological, differential and metric structures induced in it by inclusion in $\mathbf{R}^4$. The metric structure in question is the proper Riemannian metric of constant positive curvature known as *spherical* or "Riemann" geometry.

24. We can choose a chart with coordinate functions t, φ, ψ, θ, such that $\partial/\partial t$ is the worldvelocity field of matter and φ, ψ and θ define a set of hyperspherical coordinates on each hypersurface orthogonal to $\partial/\partial t$. Relative to this chart, matter is "at rest" and all components of the energy tensor vanish except T^{00}, which is constant and equal to the uniform energy density ρ (cf. eqns (5.7.14) and pp. 118 ff.). Recalling that $T^h_h = \rho$ (p. 172) and noting that, relative to the chosen chart, $g_{0\alpha} = 0$ (for $\alpha = 1, 2, 3$), we can substitute the above values of the energy tensor components into (D. 6) to obtain:

$$\begin{aligned} R_{00} - g_{00}(\lambda + \tfrac{1}{2}\kappa\rho) &= -\kappa\rho(g_{00})^2 \\ R_{ij} - g_{ij}(\lambda + \tfrac{1}{2}\kappa\rho) &= 0 \quad \text{if either } i \text{ or } j \neq 0 \end{aligned} \tag{1}$$

As W. de Sitter (1917*c*) pointed out, equations (1) are satisfied by the g_{ij} implied by the line element:

$$ds^2 = dt^2 - R^2(d\varphi^2 + \sin^2\varphi(d\psi^2 + \sin^2\psi d\theta^2)) \tag{2}$$

where $R^2 = 1/\lambda$. The "cosmological constant" λ thus turns out to be equal to the curvature of each spacelike section of the world. De Sitter (1917*b*, p. 231; 1917*c*, pp. 7 f.) also noted that solution (2) can be realized in two topologically inequivalent spacetimes: one in which the spherical sections orthogonal to $\partial/\partial t$ are the spherical 3-spaces isometric to $\mathbf{S}^3$ countenanced by Einstein, and another in which they are elliptical 3-spaces, in the sense of Klein (1871). Elliptical 3-space is homeomorphic with the real projective 3-space $\mathbf{P}^3$ (whose points are the sets $\{\alpha\mathbf{r} \mid 0 \neq \alpha \in \mathbf{R}\}$, for each non-zero $\mathbf{r} \in \mathbf{R}^4$), which is harder to figure out than $\mathbf{S}^3$, but has the intuitive advantage that any two geodesics in it meet exactly once, whereas in spherical space all geodesics issuing from a point meet again at an "antipodal" point, However Einstein, in a postcard sent on 16th April 1919 to Klein himself, expressed his preference for the spherical over the elliptical model.

25. Einstein (1917), p. 149. Current data indicate that galaxies form clusters which in turn are grouped into superclusters up to 6×10^7 lightyears wide, but that the distribution of such superclusters is indeed homogeneous (E. J. Groth *et al.* (1977)). Cf. the fascinating discussion of "Homogeneity and Clustering" in P. J. E. Peebles (1980), Chapter I.

26. Einstein (1921*a*), p. 129: "The distribution of the visible stars is extremely irregular, so that we on no account may venture to set down the mean density of starmatter in the universe as equal, say, to the mean density in the Milky Way."

27. De Sitter showed that the eqns (1) in note 24, with $\rho = 0$, are satisfied by the g_{ij} implicit in the line element:

$$ds^2 = \cos^2\varphi dt^2 - R^2(d\varphi^2 + \sin^2\varphi(d\psi^2 + \sin^2\psi d\theta^2)) \tag{3}$$

where $R^2 = 3/\lambda$. He contrasted this "system *B*" with Einstein's "system *A*", defined by (2) in note 24. If $\mathbf{R}^5$ is endowed with the flat metric of signature $(+ - - - -)$ de Sitter's system *B* can be isometrically mapped onto the hypersurface $\{(t, a, b, c, d) \mid a^2 + b^2 + c^2 + d^2 - t^2 = const.\}$. On this "hyperboloid" (as on its readily visualizable analogue in $\mathbf{R}^3$, the ordinary hyperboloid $a^2 + b^2 - t^2 = const.$) geodesics orthogonal to the section $t = 0$ recede from each other and the sections $t = const.$ become larger and larger as the absolute value of t increases. (The geometry of de Sitter's system *B* was spelled out by Felix Klein (1918*b*); cf. E. Schrödinger (1956), Ch. I; a very clear explanation, with a diagram, is given by W. Rindler (1977), pp. 183–7.) Hence, de Sitter's world is static only because it is empty, but a configuration of test particles launched into it will not conserve its shape or size as t varies. De Sitter (1917*b*), p. 236, after quoting Slipher's values for the radial velocities of NGC. 1068 and NGC. 4594 (see note 20), observed that "about a *systematic* displacement towards the red of the spectral lines of nebulae we can, however, as yet say nothing with certainty. If in the future it should be proved that very distant objects have systematically positive apparent radial velocities this would be an indication that the system B, and not A, would correspond to the truth". This is, to my knowledge, the earliest suggestion that we may be living in an expanding universe.

28. Cf. W. de Sitter (1917*a*), pp. 1221 f.: "From the purely physical point of view, for the description of phenomena in our neighbourhood this question [namely, the choice between Einstein's solution (2) in note 24, de Sitter's solution (3) in note 27, and the flat Minkowski metric—R.T.] has no importance. In our neighbourhood the g_{ij} have in all cases within the limits of accuracy of our observations the values [η_{ij}], and the field-equations [D.6] are not different from [5.7.17]. The question thus really is: how are we to extrapolate outside our neighbourhood? The choice can thus not be decided by physical arguments, but must depend on metaphysical or philosophical considerations, in which of course also personal judgment or predilections will have some influence."

29. Einstein's remark, in his (1918*b*), p. 243, to the effect that, as far as he knew, the modified field equations did not admit a singularity-free empty solution must be read in the light of his (1918*c*), submitted to the Prussian Academy on March 7th, 1918 (i.e. one day after his (1918*b*) was received for publication in *Annalen der Physik*). He claims there that de Sitter's solution (formula (3) in note 27) contains an unavoidable singularity at $\varphi = \pi/2$. A fortnight later, however, he acknowledged to Felix Klein, who had taken exception to his claim: "You are perfectly right. De Sitter's world is in itself singularity-free and all its worldpoints are equivalent. A singularity arises through the substitution that furnishes the static form of the line element. [. . .] My critical remark with regard to de Sitter's solution requires correction. There is in fact a singularity-free solution of the gravitational equations without matter. But this world should by no means be considered as a physical possibility. For it is not possible to establish in it a time t such that the three-dimensional sections $t = const.$ do not intersect each other and that these sections are equal (metrically)" (Einstein to Klein, postcard, 20th March 1918; Princeton Einstein Archive, Microfilm Reel 14). In other words, Einstein now rejects de Sitter's world as a valid counterexample to Mach's Principle because it cannot be partitioned by a universal time into isometric spacelike slices. Why Nature should be expected to meet this Procrustean demand is anybody's guess.

30. For a more detailed analysis with numerous references to the literature, see H. Goenner (1970), (1980*b*).

31. A. H. Taub (1951); E. Newman, L. Tamburino and T. Unti (1963); C. W. Misner and A. H. Taub (1969). For a brief survey of these solutions see, for instance, M. P. Ryan and L. C. Shepley (1975), pp. 132–46. Cf. the interesting comments on Taub–NUT space by C. W. Misner (1967).

32. S. Weinberg (1972), p. 87. The same point was made independently in the title of an article by J. F. Woodward and W. Yourgrau (1972), on "The incompatibility of Mach's Principle and the Principle of Equivalence in Current Gravitation Theory"; however, the authors, who propose to discard the strong Equivalence Principle and hence General Relativity in order to save Mach's Principle, do not succeed in explaining—at any rate to someone as obtuse as myself—wherein the alleged incompatibility consists.

33. See J. Callaway (1954). The same objection can be raised against the more exact treatment of the Lense–Thirring problem as a perturbation of the Schwarzschild solution by D. R. Brill and J. M. Cohen (1966) and J. M. Cohen and D. R. Brill (1968).

34. If γ is not a geodesic and v_1, v_2 and v_3 are three spacelike vectors at a point P of γ which form an orthonormal tetrad with the tangent $\dot{\gamma}_P$, their parallels along γ at the infinitesimally distant point P' do not form an orthonormal tetrad with the tangent $\dot{\gamma}_{P'}$ (for the latter is not parallel to $\dot{\gamma}_P$). However, we do obtain such a tetrad if we take the projections of the said parallels onto the orthogonal complement of $\dot{\gamma}_{P'}$. If this operation is repeated again and again as we advance by infinitesimal displacements along the range of γ we generate a succession of spacelike triads which satisfies our intuitive notion of a system of non-rotating spatial axes (Weyl (1923*a*), p. 268). This idea can be formally expressed as follows. Let U be a field of timelike unit vectors in the relativistic spacetime ($\mathscr{W}$, **g**). The *Fermi derivative* $F_{\mathrm{U}}\mathrm{X}$ of a vector field X along U is equal to $\nabla_{\mathrm{U}}\mathrm{X} - \mathbf{g}(\nabla_{\mathrm{U}}\mathrm{U}, \mathrm{X})\mathrm{U} - \mathbf{g}(\mathrm{U}, \mathrm{X})\nabla_{\mathrm{U}}\mathrm{U}$ (where $\nabla_{\mathrm{U}}\mathrm{X}$ denotes the covariant derivative of X along U). The following properties the Fermi derivative are not hard to prove: (i) $F_{\mathrm{U}}\mathrm{U} = 0$; (ii) if X is

orthogonal to U—i.e. if $\mathbf{g}(X, U) = 0$—the value of $F_U X$ at each point P is the projection of the local value of $\nabla_U X$ on the space of vectors orthogonal to U_P; (iii) if X and Y are vector fields such that $F_U X = F_U Y = 0$, $\mathbf{g}(X, Y)$ is constant along each integral curve of U. Let γ be such an integral curve. A vector field X is *Fermi transported* or *Fermi* propagated along γ if, on the range of γ, $F_U X = 0$. A *compass of inertia* along γ is an ordered triple of mutually orthogonal vector fields of unit length, defined on the range of γ, each of which is orthogonal to U and Fermi propagated along γ. If γ is a geodesic, then, on its range, $\nabla_U U = 0$ and $F_U X = \nabla_U X$, so that X is Fermi propagated along γ if and only if it is also parallel along γ. Thus our definition of compass of inertia along a timelike geodesic is just a special case of the more general definition given here.

35. The *vorticity* (also called the *rotation*, or *spin*) of a fluid can be defined in several ways. The following concept of a vorticity tensor was introduced by J. L. Synge (1937). We represent all geometrical objects by their components relative to a given chart. Thus U_i and U^i stand for the covariant and contravariant forms of the worldvelocity field U, and $U_{i;k}$ denotes the covariant derivative of the former. The fluid's departure from geodesic motion is measured by the acceleration field $\dot{U}_i = U_{i;j}U^j$. The *vorticity tensor* (in effect, a 2-form) is given by:

$$\omega_{ij} = \tfrac{1}{2}(U_{i;j} - U_{j;i} + \dot{U}_i U_j - \dot{U}_j U_i)$$

See J. L. Synge (1960), pp. 169–73.

36. See K. Gödel (1949*a*, 1952), I. Ozswáth and E. L. Schücking (1969). Gödel (1952), p. 176, proposed the following "directly observable necessary and sufficient reason for the rotation of an expanding spatially homogeneous and finite universe": "*For sufficiently great distances there must be more galaxies in one half of the sky than on the other half.*"

37. Einstein (1922*a*), p. 19.

38. Einstein (1922*a*), p. 18; my italics.

39. Einstein to C. B. Weinberg, 1st December 1937; Princeton Einstein Archive, Microfilm Reel 17. Einstein summed up his own position as follows: "Alles Begriffliche kann seine Berechtigung nur auf seine *Beziehbarkeit auf* das, aber keineswegs auf eine *Ableitbarkeit aus* dem sinnlich Wahrnehmbaren gründen."

40. Thus, for example, the general relativistic expression for the components of the energy tensor of a perfect fluid is

$$T^{ij} = (\rho + p)U^i U^j + p g^{ij}$$

where U is the worldvelocity field, ρ the energy density and p the pressure. Cf. the general definition of the energy tensor in S. Weinberg (1972), Chapter 12, or J. V. Narlikar (1979), Lecture 6.

41. Einstein to Pirani, February 2nd, 1954; Princeton Einstein Archive, Microfilm Reel 17. This very important letter was written in response to a manuscript by Pirani, "On Mach's Principle and preferred directions in General Relativity". The words "alle nicht auf sie zurückführbaren Begriffe", whose translation I have placed in square brackets, are crossed out in the original MS; they bear witness to the association of Mach's Principle with positivistic reductionism in Einstein's mind. Cf. note 39.

42. D. W. Sciama (1953); C. Brans and R. H. Dicke (1961); F. Hoyle and J. V. Narlikar (1964, 1966). The subtle analyses by R. H. Dicke (1962, 1964*b*, *c*), though partly outdated, are still very instructive from a methodological point of view.

43. H. Hönl and H. Dehnen (1963, 1964) interpret Mach's Principle as the requirement that the inertial and gravitational forces that show up in each coordinate system be fully and uniquely determined by the *sum* of the energy-momentum (i.e. world-momentum) density of matter *and* the energy-momentum density of the gravitational field. In the light of this requirement the authors classify the solutions of the Einstein field equations into *strictly Machian* (Einstein's static universe, de Sitter's empty universe, the Friedmann worlds), *loosely Machian* (Ozswáth and Schücking's finite rotating universe) and anti-Machian (Minkowski spacetime, the Schwarzschild solution, the Lense-Thirring solution for a rotating central body). Gödel's infinite rotating universe, dismissed at first blush as anti-Machian (1963, p. 210), turns out, on a closer look, to be loosely Machian (1964, p. 288).

44. J. A. Wheeler (1964) conceives Mach's Principle as "a boundary condition to separate allowable solutions of Einstein's equations from physically inadmissible solutions" (p. 306). He states it thus: "The specification of a sufficiently regular closed three-dimensional geometry at two

immediately succeeding instants, and of the density and flow of matter-energy, is to determine the geometry of space-time, past, present, and future, and thereby the inertial properties of every infinitesimal test particle" (p. 333). Wheeler's formulation is in itself very interesting and is directly linked to some highly innovative research done on General Relativity in the last quarter of a century (e.g. by Arnowitt, Deser and Misner on GR as a dynamical theory; by Yvonne Choquet-Bruhat and others on the initial-value problem; the new vindication of the virtual uniqueness of Einstein's field equations by Hojman, Kuchař and Teitelboim, etc.) However, it seems to me that by allowing and indeed prescribing the *independent* specification of the global 3-geometry and its time rate of change on a spacelike slice of the world Wheeler exceeds the bounds of what may fairly be called Machian. Besides, in order to yield the required determination of the spacetime geometry, "past, present; and future", the spacelike slice in question must be a global Cauchy surface, i.e. a hypersurface that is met once (and once only) by *every* maximally extended null or timelike curve through *every* worldpoint outside it. But there is no shred of evidence that there is a global Cauchy surface in the real world.

45. According to Raine's criteria a solution is Machian if (i) the metric is uniquely determined by the Riemann tensor, and (ii) the Weyl tensor is a linear functional of the matter current. Raine's formal statement of (i) has full generality, but his—rather formidable—exact version of (ii) applies only to spacetimes admitting a Cauchy surface. Raine shows that Minkowski spacetime and all spacetimes that approach the Minkowski metric at infinity violate condition (i). Condition (ii) is violated by all empty non-flat spacetimes. Among the relativistic spacetimes admitting a global time coordinate that partitions them into geometrically homogeneous spacelike sections ("homogeneous worlds") only the isotropic ones, characterized by the Robertson–Walker line element (6.3.2), pass Raine's criteria, but all rotating or anisotropically expanding spatially homogeneous spacetimes are excluded as non-Machian.

Section 6.3

1. Unfortunately I can barely touch on the fascinating subject of relativistic cosmology. For a readable and reliable introduction I recommend the relevant chapters of W. Rindler (1977), pp. 192–244 or J. V. Narlikar (1979), pp. 173–230. At the time of writing the best up-to-date general surveys of the subject were, in my opinion, A. K. Raychaudhuri's *Theoretical Cosmology* (1979) and D. J. Raine's *The Isotropic Universe* (1981). M. P. Ryan and L. C. Shepley (1975) give a fairly clear presentation of the mathematical theory. A thorough student, who longs for a detailed, step-by-step argument will take delight in S. Weinberg (1972), pp. 375–633.

2. Kepler's paradox of 1610, subsequently discussed by E. Halley (1720*a*, *b*), among others now goes under the name of Olbers, the German astronomer who revived it in 1823. For an up-to-date treatment see E. R. Harrison (1964, 1965), or Harrison's article in the *American Journal of Physics* of 1977. Let me note only that Einstein's 1917 world model is in fact no less liable to the paradox than Newton's, for the light emitted by a star will endlessly circumnavigate the static spherical space until obstructed by another star. Harrison's brilliant solution rests on the fact that the energy supply of the stars must run out in a time much shorter than would be required to fill the sky with starlight.

3. Hubble's estimates were based, of course, on the apparent luminosity of the objects concerned. Due to a confusion between two kinds of *star* populations—distinguished by Baade in the 1940s—Hubble underestimated the distance to the brightest galaxies and hence to all others. M 31 is now believed to be some 2,000,000 lightyears away from us and to be twice as massive as our galaxy.

4. Eddington (1924), p. 162. The average velocity of the sample was + 572 km/s. The greatest negative (i.e. approaching) velocity in Eddington's list was − 300 km/s (NGC 221 and 224).

5. E. Hubble (1929), p. 109, Table 1. The rest of Eddington's list goes into Hubble's Table 2, of "Nebulae whose distance are estimated from radial velocities" (by means of Hubble's linear relation).

6. See the quotation on p. 330, note 27.

7. Eddington (1924), pp. 161 ff.; Weyl (1923*a*), pp. 294 f., 322 f.; Weyl (1923*c*); Hubble (1929), p. 173. Cf. notes 24 and 27 to Section 6.2.

8. Ryan and Shepley (1975), p. 57. The authors refer specifically to the *closed* Friedmann

solutions, i.e. those in which, as in Einstein's static universe, the spacelike hypersurfaces orthogonal to the worldlines of matter are *compact* (viz. proper Riemannian 3-manifolds of constant *positive* curvature). Giordano Bruno's world model is the "common sense" universe of stars without end in Euclidean space.

9. Einstein (1922*b*). Einstein (1923*d*) admitted that his objection to Friedmann was based on an error of calculation.

10. G. Lemaître (1927)—published in English translation in the MNRAS as Lemaître (1931*a*). Lemaître's assumptions are somewhat more general than Friedmann's inasmuch as he treats matter as a perfect fluid—with time-dependent, spatially isotropic pressure—not as pressureless dust. (He even suggests that the pressure might be responsible for the expansion of the universe!) However, the current habit of distinguishing between "Friedmann universes" in which $\lambda = 0$, and "Lemaître universes" in which λ can take any value, is historically unjustified, for Friedmann (1922), p. 378, explicitly made allowance for an arbitrary λ—"which we may also set equal to zero"—and studied the consequences of its satisfying several alternative inequalities.

11. Robertson was well acquainted with Friedmann's work but he learned about Lemaître's after his first paper was ready. Like Lemaître he made allowance for spatially isotropic positive pressure. The meaning of his postulate of spatial isotropy will be made clear below.

12. Einstein (1931), p. 236. Einstein assumes that the cosmic fluid has negligible pressure.

13. Einstein (1956), p. 127.

14. Friedmann (1922) only considers the case $K > 0$. The case $K < 0$ is the subject of Friedmann (1924). The possibility that $K = 0$, as in the Friedmann solution countenanced by Einstein and de Sitter (1932), is ignored by Friedmann, probably because he did not realize that it also yields—in fact for every value of λ—non-static solutions expanding from a point in the finitely distant past (or contracting to one in the finitely distant future).

15. This opinion can be traced back to F. Ueberweg (1851) and H. v. Helmholtz (1868). It was very influential until the advent of General Relativity, with which, of course, it is incompatible. Only the (complete) proper Riemannian n-manifolds of constant curvature have the said property. If $n = 3$, the manifold must admit a 6-parameter group of motions, which can only be the Euclidean group, the Bolyai–Lobachevsky group, or the group of motions of spherical or elliptical geometry.

16. Friedmann (1922), p. 379.

17. See, for instance, Ryan and Shepley (1975), Chapters 8, 9 and 10.

18. H. P. Robertson (1929), p. 823.

19. An extension of Fubini's result to Riemannian n-manifolds admitting an $(m+1)m/2$-parameter group of motions ($m < n$) is proved in detail by Weinberg (1972), pp. 396 ff., who also provides the requisite mathematical background.

20. Robertson (1929), p. 827, n. 9, criticized Friedmann for having overlooked that his assumptions of class II implied that $\partial/\partial x^0$ generates a congruence of geodesics. Friedmann had expressed the line element of de Sitter's solution—regarded as a special case of his own—in the same form used by de Sitter himself and reproduced above on p. 329, note 27. This line element is referred to a chart in which the parametric lines of the time coordinate function are not geodesics.

21. If the integral lines of $\partial/\partial x^0$ are timelike geodesics, x^0 must be related to the proper time τ by the affine transformation $x^0 = \alpha\tau + \beta$ ($\alpha, \beta \in \mathbf{R}$, $\alpha \neq 0$) along each such line. By (F2), the scale factor α must be the same everywhere.

22. As a matter of fact, (II.1*b*) was the *only* consequence that Robertson (1929) drew from his demand, for he had postulated (II.2) from the outset. See also Robertson (1933), p. 65 and Note C. However, (II.2) follows at once from the theorem by Fubini on which Robertson's argument rests.

23. This was Lie's solution of the "problem of space" formulated by Helmholtz. (S. Lie (1890); S. Lie and F. Engel (*TTG*), vol III, Abt. V.) The following consideration may satisfy the reader that every hypersurface $x^0 = const.$ must have a constant curvature. Let H be such a hypersurface. All sectional curvatures at a point $P \in H$ must be equal, say, to K_P, because H is isotropic at P. But if the scalar field $P \mapsto K_P$ varied as P ranges over H, its gradient would single out a particular direction at each point of H.

24. Robertson (1935) derives in fact all three assumptions of class II from what he calls, after E. A. Milne, the *cosmological principle*. This he states as follows: The "fundamental particle-observers" A, A′, . . ., are "each equipped with a clock, a theodolite, and apparatus for sending and receiving light-signals"; "the description of the whole system, as given by A in terms of his

immediate measurements, is to be identical with the description given by any other fundamental observer A′ in terms of his own measurements" (*loc. cit.*, p. 285). As I am not sure that I understand what all this means, I have preferred to glean from Robertson's proof the premises that he actually uses in the construction of a global time coordinate. Let me mention also that A. C. Walker (1935, 1937), who independently achieved the same results as Robertson, based them on (i) "the *cosmological principle*", according to which "the totality of observations that any observer can make on the system is identical with the totality of observations that any other observer can make on the system"; (ii) "the principle of symmetry", to the effect that "each fundamental observer sees himself to be a centre of spherical symmetry"; and (iii) Fermat's principle of least action, governing the light-paths (A. C. Walker (1937), pp. 94, 103, 91).

25. Relative to any chart whose time coordinate function is x^0 the metric components g_{ij} satisfy the following conditions: (i) $g_{0\alpha} = g_{\alpha 0} = 0$ (by (F1)); (ii) $g_{\alpha\beta}$ is equal to minus the product of a function $\bar{g}_{\alpha\beta}$ depending on the space coordinates only, and the square of a function S depending on x^0 alone (by F3); (iii) g_{00} depends on x^0 only (by F2).

26. See, for instance, Weinberg (1972), p. 403.

27. Note that in order to evaluate z one must know both v_1 and v_2. In practice, one assumes that the light received from distant galaxies was emitted in the remote past with the same frequency with which light of the same type—i.e. of the same atomic origin—is now radiated in our terrestrial laboratories. The type of light of a particular frequency is recognized by the pattern of neighbouring spectral lines.

28. The calculations leading to (6.3.8) show that the only non-zero components of the linear connection determined by the Robertson–Walker line element (6.3.2) are $\Gamma^{\alpha}_{0\beta} = \dot{S}S^{-1}\delta^{\alpha}_{\beta}$, $\Gamma^{0}_{\alpha\beta} = \dot{S}S\bar{g}_{\alpha\beta}$, and $\Gamma^{\alpha}_{\beta\gamma} = \bar{\Gamma}^{\alpha}_{\beta\gamma}$; where a dash over a symbol indicates that it denotes a component of the "spatial" metric $d\sigma$ or its linear connection. Substituting these values for the Γ^{i}_{jk} in the geodesic equations (D.8) we verify at once that the parametric lines of τ are geodesics.

29. See, for instance, Weinberg (1972), pp. 381–3.

30. Apparently Friedmann overlooked this, for he describes ρ as "an unknown function of the [four] world-coordinates" (Friedmann (1922), p. 380 n. 1), thereby suggesting that his assumptions are much less stringent than they are in fact.

31. Eliminating $\ddot{S}$ from (6.3.9) we obtain "Friedmann's equation":

$$\dot{S}^2 = \frac{1}{S}\left(\frac{\kappa\rho + \lambda}{3}S^3 - kS\right) \qquad (*)$$

The expression in parenthesis on the right-hand side of (*) is a polynomial in S that we shall denote by $f(S)$. Obviously, τ_B can be obtained by evaluating the elliptic integral $\int(\sqrt{u/f(u)})du$ from $u = S_0$ to $u = 0$. By plotting S as a function of τ we see at once that $0 < |\tau_B| < S_0/\dot{S}_0$. Note that if we had chosen S to be the negative square root of S^2, (6.3.10) would entail that $\ddot{S} > 0$ when $\lambda \leq 0$; in that case, S could not have a maximum, and would tend to 0 as τ approaches τ_B.

32. The different alternatives are carefully discussed by Rindler (1977), pp. 230 ff. Allowing a *positive* isotropic pressure makes no essential difference.

33. Cf. the discussion of the Robertson–Walker line element in Rindler (1977), pp. 207 ff.; especially the figure on p. 209.

34. Suppose that our Friedmann world ($\mathscr{W}$, **g**) contains a point $P_\infty = \lim P_n$. Then the metric **g**, and hence the Riemann tensor R and the curvature scalar R must exist and be smooth at P_∞. But then $R(P_\infty) = R(\lim P_n) = \lim R(P_n)$, and we know that the latter limit does not exist.

35. Cf. Friedmann (1922), p. 384 n.: "The time from the creation of the world is the time that has elapsed from the moment when space was a point to the present moment."

Section 6.4

1. A complex-valued function f is said to *have a singularity* at a point z of the complex plane if the following two conditions are met in *every* neighbourhood U of z: (i) f is analytic on an open subset of U; (ii) f cannot be extended to an analytic function defined on all U. If f has a singularity at z, z is said to be a *singular point* of f.

2. A star a few times more massive than the Sun collapses under its own weight when its

thermonuclear energy sources become exhausted (Oppenheimer and Snyder (1939)). If the star is and remains spherically symmetric, has zero angular momentum and is far removed from all other large bodies, the field surrounding it can be represented by the Schwarzschild solution to Einstein equations *in vacuo*, $\mathbf{R}_{ij} = 0$ (Schwarzschild (1916), Birkhoff (1923)). Let the Schwarzschild line element be given by

$$ds^2 = \left(1 - \frac{2Gm}{r}\right)dt^2 - \left(1 - \frac{2Gm}{r}\right)^{-1} dr^2 - r^2 (d\theta^2 + \sin^2\theta \, d\varphi^2)$$

where m is the star's mass and the gravitational constant $G = \kappa/8\pi$. When the contracting star attains a radius less than $2Gm$ ($\approx$ 6 km if m is twice the solar mass), nothing can arrest its quick compression to infinite density at $r = 0$. A test particle or a light pulse inside the sphere $r < 2Gm$ is irresistibly sucked into the singularity. In the more realistic case of a rotating star, the field at the final stages of collapse would most probably correspond to one of the Kerr solutions (Kerr (1963); Newman *et al.* (1965); Carter (1968, 1970, 1977)). The situation here is more complex but ultimately the same as in the idealized Schwarzschild case. For a clearer view of this matter see S. W. Hawking's luminous article on "Black Holes" in Lerner and Trigg (1981), the textbook presentations in Rindler (1977), pp. 149–56, and Ohanian (1976), Chapter 9, and the review of recent research by J. C. Miller and D. W. Sciama (1980).

3. Such ingredients include, it is true, the complete metric space $\mathbf{R}$. But even this can be built by "bootstrapping", as in the original version of the above ploy, first suggested by Cauchy and carried out by Weierstrass. Let $\mathbf{Q}$ be the field of rational numbers. A sequence $P_1, P_2, \ldots$ of rational numbers is called a Cauchy sequence if for every positive $\delta \in \mathbf{Q}$ there is an integer N such that $|P_r - P_s| < \delta$ whenever $N < r, s$. As is well known, not every Cauchy sequence of rationals converges to a limit in $\mathbf{Q}$. We shall say that two Cauchy sequences of rationals (P_n), (Q_n) are equivalent if their "mix" $P_1, Q_1, P_2, Q_2, \ldots$, is a Cauchy sequence. We define $\mathbf{R}$ as the collection of equivalence classes of Cauchy sequences of rationals. The function d: $\mathbf{R} \times \mathbf{R} \to \mathbf{R}$ assigns to any pair of "points" $(P_n), (Q_n) \in \mathbf{R}$, the Cauchy sequence of rationals $(|P_n - Q_n|)$. $(\mathbf{R}, d)$ is a complete metric space entirely built from the incomplete field $\mathbf{Q}$.

4. Kobayashi and Nomizu, *FDG*, vol. I, p. 172.

5. C. W. Misner (1963).

6. "Space–time is represented as a connected paracompact four-dimensional C^∞ manifold with pseudo-Riemannian metric tensor g_{ij} of signature minus two. [..] Space-time will be said to be singularity-free if the metric is a field of class C^2 and $\mathscr{M}$ is geodesically complete with respect to the metric" (S. W. Hawking (1966*a*), p. 512). "Space-time is said to be singularity-free if it is time complete (all timelike geodesics can be extended to arbitrary length) and if the metric is C^2" (S. W. Hawking (1966*b*), p. 491). "A space is said to be singular if it contains an inextendable timelike or null geodesic of finite affine length" (R. P. Geroch (1966), p. 446). Geroch refers to Misner (1963), but no such definition is to be found there. Having equated the absence of singularities with metric completeness in the case of proper Riemannian manifolds, Misner (1963), p. 925, cautiously observes that "although we lack a definition of 'non-singular' for Lorentz-signature Riemannian manifolds, we have two sufficient conditions: compactness and affine [i.e. geodesic—R.T.] completeness." In the same paper he shows that one of these conditions can be had without the other (see page 215). Misner discussed again the interpretation of spacetime singularities in the paper he coauthored with A. H. Taub (1968), Section II; no definition is proposed there.

7. Hawking and Ellis (1973), p. 258.

8. See, for example, p. 330, note 29.

9. I assume that τ_B is the g.l.b. of the range of the time coordinate function τ. It may seem that I thereby wantonly ignore the so-called periodic Friedmann solution, which holds if $\lambda < 0$, or if $\lambda = 0$ and $k = 1$, and can also obtain if $k = 1$ and $0 < \lambda < \lambda_E$ (where $\lambda_E = (2/\kappa\rho S^3)^2$ is the value of λ corresponding to the static Einstein universe). In the case of the periodic solution the graph of S as a function of τ is a cycloid, and the range of τ can be taken to be the entire real number field *minus* the periodically recurring zeroes of S. The interval between two successive zeroes of S is the time of one cosmic period during which the Friedmann world expands and recontracts. The manifold $\mathscr{W}^*$ mapped by τ onto the punctured real line is obviously disconnected, with a different component for each cosmic period. Consequently, $\mathscr{W}^*$ does not underly *one*, but *many* relativistic spacetimes, in the sense agreed on p. 5.22. In fact, the periodic Friedmann solution confirms our wisdom in demanding that a spacetime manifold should be path-connected. For let $P, Q \in \mathscr{W}^*$ such that

$\tau(P) < \tau_B < \tau(Q), S(\tau_B) = 0$. Then no information whatsoever can be transmitted from P to Q. P does not causally or chronologically precede Q in the sense of the theory of causal spaces (Section 4.6) and the only reason we have for saying that P comes before Q in time is the purely formal one that the interval assigned by the function τ to the cosmic period of the former, precedes in **R** the interval assigned by the same function to the cosmic period of the latter. But the cosmic periods of P and Q have very little to do with each other and shall be viewed as successive stages of one and the same world only if one is anxious to save by all means the eternity of Mother Nature (cf. note 26).

10. We are not compelled to parametrize each timelike geodesic γ of $(\mathscr{W}, \mathbf{g})$ by the proper time τ. But an affine transformation $\tau \mapsto \alpha\tau + \beta$ merely replaces τ_B by another finite real number. We may indeed parametrize all matter worldlines by $t = \log(\tau - \tau_B)$, whereby their domains become unbounded below (and equal to **R** if, from the outset, they are unbounded above). But t is not an affine parameter (p. 278) and its adoption cannot alter the geodesic incompleteness of $(\mathscr{W}, \mathbf{g})$. In this connection, I would like to state a theorem proved by C. J. S. Clarke (1971) which throws much light on the nature of spacetime singularities: If $(\mathscr{W}, \mathbf{g})$ is a strongly causal spacetime (in the sense defined on p. 125, i.e. if, as it is presumably the case in the world we live in, there are in $(\mathscr{W}, \mathbf{g})$ no closed or almost closed timelike curves), and if it contains incomplete *null* geodesics, there is a scalar field ω on $\mathscr{W}$ such that every null geodesic in $(\mathscr{W}, \omega^2\mathbf{g})$ is complete. Since the metrics $\mathbf{g}$ and $\omega^2\mathbf{g}$ are conformal (p. 192) they both define the same causal structure on $\mathscr{W}$ and the *paths* of their respective null geodesics are identical. However, while the affine parameters of some of the null geodesics of $\mathbf{g}$ are bounded above or below, the null geodesics of $\omega^2\mathbf{g}$—though they include parametrizations of the very lightpaths which appear to "run into a singularity" in $(\mathscr{W}, \mathbf{g})$—all have unbounded affine parameters, ranging from $-\infty$ to ∞.

11. See, for instance, Einstein (1956), p. 129.

12. Thus W. de Sitter (1933), p. 631, commenting on the apparent inconsistency between Friedmann's "time from the creation of the universe", as calculated from Hubble's Law, and the much longer period required for the formation of a star according to the then "modern" theories of stellar evolution, observed that the perfect spatial homogeneity and isotropy presupposed by the line element (6.3.2) "is an idealization of the problem, taking account of inertia only and neglecting gravitation, or, in other words, replacing the mutual gravitational action between the separate galaxies by the averaged action of all galaxies in the universe combined. This approximation is, of course, entirely sufficient so long as the mutual distances are large, but ceases to be an approximation when these distances become very small." As we approach $\tau = \tau_B$ "the mutual perturbations [. . .] will be large and different for different bodies, and [. . .] the exact equality of all velocities will consequently be destroyed". Hence, de Sitter concluded, "the conception of a universe shrinking to a mathematical point at one particular moment of time $[\tau = \tau_B]$ must thus be replaced by that of a near approach of all galaxies during a short interval of time near $[\tau = \tau_B]$".

13. More formally, $(\mathscr{W}, \mathbf{g})$ is *generic* if it meets the following "generic condition": every timelike or null geodesic γ in $\mathscr{W}$ has a point P at which the Riemann tensor R and the tangent $v = \dot{\gamma}_P$ satisfy the inequality

$$v^r v^s (v_i v_h \mathrm{R}_{jrsk} - v_j v_h \mathrm{R}_{irsk} - v_i v_k \mathrm{R}_{jrsh} + v_j v_h \mathrm{R}_{irsh}) \neq 0,$$

where $v_i = g_{ij} v^j$.

14. The weak convergence condition follows from the field equations (D.7) and the hypothesis that $\mathrm{T}_{ij}\mathrm{V}^i\mathrm{V}^j \geq 0$ for every timelike vector V. The latter hypothesis says that the local energy density of matter, as measured by an observer with worldvelocity V, is nonnegative. The strong convergence condition presupposes that $(\mathrm{T}_{ij} - \frac{1}{2}g_{ij}\mathrm{T}^h_h + \kappa^{-1}\lambda)\mathrm{V}^i\mathrm{V}^j \geq 0$ for every timelike V. This holds good, for example, if $\lambda = 0$, for a perfect fluid with positive density, unless it has a large negative pressure. See Hawking and Ellis (1973), pp. 94 f.

15. Proofs are given in the papers mentioned at the end of each theorem and also in the magnificent book by Hawking and Ellis (1973), where the mathematical and physical background required for understanding them is explained in detail. A less formal but highly instructive discussion of the singularity theorems will be found in Geroch and Horowitz (1979), pp. 258–66. F. J. Tipler *et al.* (1980), pp. 147–52 report on recent efforts to weaken the convergence and causality conditions presupposed by the theorems.

16. Condition (iii) of Theorem III can be formally stated as follows. For any given future-directed timelike geodesic γ through P we denote its field of unit tangents by V. The function $\theta = \mathrm{V}^i_{\ ;i}$ is called the divergence of V. Let w be an arbitrary unit timelike vector pointing to the

future at P. We set $f_\gamma = g_{ij}(P)w^i V_P^j$. Condition (iii) says that there are numbers b and δ such that, for each such geodesic γ, θ becomes less than $-\delta f_\gamma$ within a length b/f_γ measured from P along γ. Condition (iv-c) of Theorem IV says simply that the divergence θ, in the above sense, on each *null* geodesic through P becomes negative *at some point* to the future of P. (As noted above, we may substitute "past" for "future" throughout the foregoing statements.)

17. See, for instance, Geroch (1971*b*). A reasonable topology for the set $\mu(\mathscr{W})$ of Lorentz metrics on a 4-manifold $\mathscr{W}$ is, say, the following. Each $\mathbf{g} \in \mu(\mathscr{W})$ is a cross-section of the bundle $T_2(\mathscr{W})$ of (0, 2)-tensors over $\mathscr{W}$. Each open set $A \subset T_2(\mathscr{W})$ determines a subset $B_A \subset \mu(\mathscr{W})$ formed by all the Lorentz metrics which map $\mathscr{W}$ into A. The collection of all such sets B_A constitutes a basis for our topology.

18. Ryan and Shepley (1975), p. 81. Misner's example is the torus with Lorentz line element $ds^2 = -\cos x\, dx^2 + 2 \sin x\, dx\, dy + \cos x\, dy^2$. See Ryan and Shepley (1975), p. 95.

19. See, for instance, Misner and Taub (1968).

20. Awareness of this example, privately communicated by Geroch, led Hawking (1967) to abandon the definition of a singularity-free spacetime quoted in note 6 and to view "timelike and null geodesic completeness as *a minimum condition* for regarding space-time as singularity-free" (loc. cit., p. 169, my italics. See also Hawking and Ellis (1973), p. 258).

21. Hawking and Ellis (1973), p. 276.

22. Since Schmidt's construction depends exclusively on the canonic and connection forms of the principal bundle $F\mathscr{W}$, $\mathscr{W}$ could be just any n-manifold endowed with a linear connection Γ. If Γ is the Levi-Cività connection of a proper Riemannian metric $\mathbf{g}$, $\bar{\mathscr{W}}$ is homeomorphic to the metric completion of $(\mathscr{W}, \mathbf{g})$. This shows that the Schmidt completion $\bar{\mathscr{W}}$ of a manifold $\mathscr{W}$ with a linear connection is a natural generalization of the standard Cauchy completion of a manifold with a proper Riemannian metric.

23. The condition of being past and future distinguishing is one of a series of increasingly stronger "causality conditions" that may be imposed on $(\mathscr{W}, \mathbf{g})$. The weakest is the *chronology condition*: there are no closed timelike curves in $\mathscr{W}$. Then comes the *causality condition*: there are no closed causal (i.e. nowhere spacelike) curves. These two conditions are equivalent in a generic relativistic spacetime that satisfies the weak convergence condition (Hawking and Ellis (1973), p. 192, Prop. 6.4.5). To be *past and future distinguishing* is a stronger requirement, which is in turn strictly weaker than the property of being a strong causal space, defined on p. 125. A maximally strong condition of this kind is the condition of *stable causality*, which will be satisfied if the metric $\mathbf{g}$ has a neighbourhood U in the topological space of Lorentz metrics on $\mathscr{W}$ defined in note 17, such that for every $\mathbf{h} \in U$, $(\mathscr{W}, \mathbf{h})$ satisfies the chronology condition. This amounts to saying that $(\mathscr{W}, \mathbf{g})$ is stably causal if a small perturbation of the metric $\mathbf{g}$ will not generate closed timelike curves in $\mathscr{W}$.

24. The following proof, drawn from Geroch, Kronheimer and Penrose (1972), p. 550, will give an idea of the style of argument in this fascinating field of contemporary mathematical physics. Let γ be a timelike curve in $\mathscr{W}$. $I^-(\gamma)$ is clearly a past set. If $I^-(\gamma)$ is not an IP it can be represented as the union of two non-empty past sets, p and q, none of which is a part of the other. Choose points $P \in p - q$ and $Q \in q - p$. P and Q belong to $I^-(\gamma)$; there is a pair of points P', Q' on the range of γ such that $P < P'$ and $Q < Q'$. Without loss of generality we may assume that $P' < Q' \in I^-(\gamma)$. Since p and q are past sets whose union is $I^-(\gamma)$, whichever of them contains Q' contains both P and Q, contrary to our assumption. Consequently, $I^-(\gamma)$ is an IP. To prove the converse, let p be an IP of $\mathscr{W}$. We must try to find a timelike curve γ such that $p = I^-(\gamma)$. Choose a point $P \in p$. Set $a = I^-(p \cap I^+(P))$ and $b = I^-(p - I^+(P))$. Since p is a past set, $p = a \cup b$. Since p is an IP, either $p = a$ or $p = b$. Since $P \in b$, $p = q$. If P' is another point of p, P' belongs to the past of the intersection of p with the future of P and there must therefore be a third point $P'' \in p$ in the future of both P and P'. Choose a countable family of points $P_0, P_1, \ldots$, dense in p and a point $Q_1 \in p$ in the future of P_0. Then, for each integer $j > 1$ we can find a point $Q_j \in p$ in the future of both P_{j-1} and Q_{j-1}. Since each point of the sequence $Q_1, Q_2, \ldots$ lies in the past of his successor, there is a timelike curve γ in p that joins them all. For each $P \in p$, the set $U_P = p \cap I^+(P)$ is non-empty and open in p. Since the family (P_n) is dense in p, U_P contains one of its members, say, P_k. But $P_k < Q_{k+1}$, whence $P_k \in I^-(\gamma)$. Consequently, $p \subset I^-(\gamma)$. Since p is a past set and the range of γ is contained in it, $p = I^-(\gamma)$. Q.E.D.

25. See Ellis and King (1974), Ellis and Schmidt (1977); and the survey article by Tipler, Clarke and Ellis (1980), pp. 155–67.

26. A theorem recently proved by F. J. Tipler (1980) in a way explains why Friedmann's expanding and recontracting world cannot continue through the initial or the final singularity in

any genuine physical sense. For if it did so continue, it would go time and again through exactly the same process of evolution and involution. But according to Tipler's theorem a relativistic spacetime which (i) is generic, (ii) meets the strong convergence condition, (iii) contains compact Cauchy surfaces, and (iv) is uniquely determined from initial data on any of its Cauchy surfaces (cf. p. 332, note 44), cannot be time periodic; and the expanding and recontracting Friedmann world satisfies conditions (i)–(iv).

Chapter 7

Section 7.1

1. Aristotle, *Physica*, IV, 10, 218^{a}9; IV, 11, 220^{a}5; IV, 13, 222^{a}10–14.
2. To see that it is so, suppose that there is such a positive lower bound, and that the time lag between a cause on the sun and its effect on earth, or between a cause on earth and its effect on the sun, can be no less than 8 minutes. Then, a sudden change on the solar surface which I observe at 12 noon GMT is simultaneous, by the negative causal criterion, with all states of my body from 11.44 A.M. to 12 noon GMT, although they are certainly not simultaneous with one another.
3. Reichenbach (1928), p. 161. The stated criterion applies only to "time order at the same point". The expression "the same point" is *prima facie* meaningless in Relativity, but Reichenbach (1924), § 5, explains that by a "real point" he means the worldline of a material particle.
4. Reichenbach (1928), p. 170.
5. Reichenbach (1928), p. 170. The epithet "topological" is not a happy one, for—at least in a Newtonian or a relativistic universe—the relation defined above is not invariant under arbitrary homeomorphisms of spacetime onto itself and therefore is not a topological relation.
6. Reichenbach (1928), p. 171. See also the remarks in Reichenbach (1924), after Theorem 9.
7. Reichenbach (1928), p. 172; (1958), p. 146.
8. Grünbaum (1973), p. 29. Grünbaum's use of "metrical" in this context is no less idiosyncratic than Reichenbach's use of "topological" (see note 5). The paradigm of "metrical simultaneity" is Einstein's distant simultaneity in an inertial frame (Section 3.2), which is notoriously non-invariant under spacetime isometries (Lorentz point transformations) and therefore is not a metrical relation between "events" (worldpoints).
9. Reichenbach (1924), Definition 2.
10. Let t' denote the arrival time of the first bouncing signal sent from A to B at t. In a Newtonian world, $t = t'$. In a relativistic world, $t \neq t'$ unless A and B coalesce or collide at t. ($t < t'$ if, as one normally assumes, the clock at A runs "forward".) This is what we mean by saying that there is "a finite limiting velocity for causal propagation".
11. Note that in eqn. (7.1.1), if $t_A = t_0$, the choice of ε is irrelevant to the value of t_B (which is then always equal to t_0). This confirms that, as Adolf Grünbaum has repeatedly emphasized, Reichenbach's rule for the conventional definition of simultaneity is effective only if $t_A \neq t_0$, i.e. if there is "a finite limiting velocity for causal propagation" (note 10). If no such finite limiting velocity exists, the negative causal criterion is by itself sufficient to establish a transitive relation of simultaneity between events.
12. Reichenbach (1924), Definition 8; (1928), p. 151.
13. Cf. Arzeliès (1966), pp. 217–25.
14. Reichenbach in effect never attempts to apply the rule beyond this narrow scope. However, he did not introduce inertial motion and inertial frames right at the beginning of his axiomatics because he was trying to show how close one can get to defining them by means of light signals and clocks alone, without resorting to metre rods.
15. Eddington's proof is not even mentioned by Reichenbach (1928), pp. 158 ff., where he discusses—and dismisses—synchronization by clock transport in Special Relativity.
16. See Eddington (1923), §§4, 11. Philosophers took an interest in synchronization by infinitely slow clock transport after Bridgman explained it in his Relativity Primer (Bridgman (1963), p. 65) and Ellis and Bowman (1967) based on it their case for the non-conventionality of standard synchronism in Special Relativity. See, for instance, Grünbaum (1969), Salmon (1969), Van Fraassen (1969*a*), Janis (1969), Ellis (1971), Bowman (1977).

17. In his discussion of ε-Lorentz transformations, Winnie (1970) ignores two spatial dimensions. This can be done if the two charts involved in the transformation are matched (in the sense of p. 57), and the Reichenbach functions associated with each of them has its maximum or minimum value in the direction of the only spatial axis considered (which is also the direction in which the frame of one chart may be moving in the frame of the other).

18. Giannoni (1978*a*), p. 26, expands the set of Winnie transformations to form a group—which we may call the WG-group—by eliminating the causality condition (7.1.2) on Reichenbach functions. In this way, the rationale of Reichenbach's rule, which was to display the whole spectrum of transitive simultaneity relations which are compatible with natural "topological" simultaneity (coupled, eventually, with the Law of Inertia), drops out of sight. I do not mean this as a criticism of Giannoni, who explicitly disassociates his version of conventionalism from Reichenbach's (*ibid.*, p. 18).

19. Reichenbach (1928), p. 151, suggests as much when he says that standard synchronism is "essential" (*wesentlich*) for Special Relativity. But the theory, as we know, can very well do without standard or any other particular kind of synchronism if it is stated in a generally covariant form.

20. P. Mittelstaedt (1977) forcefully argues this point.

21. See on this point M. Friedman (1977); also the illuminating article by J. Earman (1972*a*).

22. Hawking and Ellis (1973), pp. 198–201.

23. Robb (1914) avoided the term "simultaneity" for philosophical reasons. As he says on p. 6: "According to the view here put forward, the only events which are really simultaneous are events which occur at the same place." Mehlberg (1937), on the other hand, demands a definition of simultaneity capable of providing "a necessary and sufficient condition of this relation, not just a convenient (*commode*) criterion" (p. 171). The definition (D3) given on pp. 169 f. is similar to the one in Mehlberg (1966) which I reproduce in the main text.

24. *Added in proof*: An English translation of Mehlberg (1935/37) is now available in volume I of Mehlberg, *Time, Causality, and the Quantum Theory* (Dordrecht: Reidel; ISBN 90–277–0721–9).

25. Before closing the present section I believe I should mention the unceasing stream of proposals for establishing "true" simultaneity by means of signals of known speed. The proponents reason that if a signal is emitted at time t from a point P and it is known to travel with speed v to a point Q, at a distance d from P, the signal will arrive in Q at time $t + d/v$. They forget however that in order to specify a speed we must refer it to a definite space and time. Thus, for example, the vacuum speed of light c whose latest best value in metres per second is printed in every handbook of physics is referred to the relative space of an arbitrary inertial frame and to the time defined in it by standard synchronism. (Relatively to a time defined by non-standard synchronism in an inertial frame the speed of light in that frame is obviously a non-constant function of direction.) The manner how the definition of time is implicit in an experimental design for the measurement of light velocity can be quite subtle. Thus, the fairly sophisticated method of Fung and Hsieh (1980) rests on the double assumption that "a wave front is everywhere perpendicular to the direction of propagation" and that "for a parallel beam all wave fronts propagate in a single direction at the same speed" (p. 655), two premises whose truth-value is evidently not invariant under arbitrary transformations of the time coordinate function (or even under transformations restricted to inertial time scales). Other recent proposals of this type are aptly criticized by P. Øhrstrom (1980). W. Salmon (1977) surveys the established methods of measuring the speed of light, showing that when they do not presuppose a definition of synchronism, they actually measure the so-called "two-way" speed of light—i.e. the quotient between the distance travelled by a bouncing light signal and the travel time measured by a clock at the source. I am not altogether satisfied with Salmon's analysis of Roemer's method. Roemer's idea can be summarized as follows. A distant observer at rest relatively to Jupiter will see any given Jovial moon hide behind the planet at fixed intervals, with constant frequency ν. For a terrestrial observer, the frequency increases (decreases) by $\Delta\nu$ when Jupiter is approaching (receding from) the earth. The value of $\Delta\nu$ depends, according to the classical Doppler formula (which Roemer of course did not know but is nevertheless implicit in his work), on the actual period $1/\nu$ of the Jovial moon, the relative velocity u of Jupiter and the earth, and the (one-way) speed of light c. Setting $1/\nu$ equal to the annual mean of the observed period we can readily compute c from the equation $c/(c+u) = (\nu + \Delta\nu)/\nu$, provided that we know u. Salmon says that "to ascertain the one-way speed of light across the earth's orbit" Roemer had to assume "that the clocks used to time the eclipses of Jupiter's satellite continue to tell the same time as they travel from one end of the orbit to another" (p. 264). But the truth is that

in order to obtain the required data Roemer had to time only a series of 40 consecutive eclipses of the Jovial moon—lasting, all in all, some 10 weeks—on two different occasions when the distance between Jupiter and the earth was, respectively, increasing and decreasing. The clock he used for this purpose might, for aught we know, have stopped and had its annual cleaning between the two series of observations. (Indeed, with better instruments it would suffice to measure on each occasion the period of approximately 42.5 hours from one eclipse to the next.) Standard synchronism enters into Roemer's calculation through the value assigned to the velocity u with which Jupiter recedes from the earth.

Section 7.2

1. "Freie Erfindungen des menschlichen Geistes"—Einstein (1934), p. 180. Einstein proceeds, on p. 182: "The scientists of [the 18th and 19th centuries] were usually possessed with the idea that the fundamental concepts and the fundamental laws of physics were not, in a logical sense, free inventions of the human spirit, but could be derived from experiments by 'abstraction'—i.e. in a logical way. A clear recognition of the erroneousness of this notion came only with the general theory of relativity, which showed that one could account for the relevant set of empirical facts in a more satisfactory and complete manner on a foundation (*Fundament*) quite different from Newton's. But quite apart from the question of superiority, the fictive character of the foundations (*Grundlagen*) is perfectly evident from the fact that two essentially different foundations can be produced that largely agree with experience. This proves, at any rate, that every attempt at a logical derivation (*einer logischen Ableitung*) of the fundamental concepts and the fundamental laws of mechanics from elementary experiences is doomed to failure." (Unless Einstein was fighting straw men, "logische Ableitung" in the last sentence can only mean "induction", not "deduction", as the English translator S. Bargmann rendered it.) The preceding passage occurs in the same lecture of June 10th, 1933, from which the Einstein motto of the present book is taken. The latter was Einstein's answer to the question: "If, then, it is true that the axiomatic foundation of theoretical physics is not inferred (*erschlossen*) from experience, but must be freely invented, may we hope to find the right way at all?" (*ibid*., p. 182).
2. Kant (1787), p. xii.
3. Kant (1781), p. 231; (1787), p. 283.
4. Kant (1787), pp. 145 f.; (1783), p. 111 (Ak., IV, 318).
5. Kant (1790), p. 124 (Ak., VIII, 249).
6. The five items listed above correspond or are entailed by the Kantian principles of the Analogies of Experience (1, 2, and 3), the Anticipations of Perception (4) and the Axioms of Intuition (5). (1) and (3) were upset by Special Relativity, (5) by General Relativity, (4) by the Old Quantum Theory, and (2) by Quantum Mechanics.
7. Cf. Kant (1783), p. 101 (Ak., IV, 312).
8. On the origins and evolution of conventionalism, see J. A. Coffa (1980). On Reichenbach's philosophy of geometry, see Kamlah (1977), Beauregard (1977), Nerlich (1976), B. Ellis (1963). We shall concentrate on Reichenbach's version of geometric conventionalism, because it is generally believed that it flows from a careful study of Relativity. A more complete treatment of the subject would have to consider also the early works of Carnap (1922, 1924, 1925), and, of course, the conventionalist philosophy of geometry formulated, while Einstein was still a child, by Henri Poincaré (1887, 1891). The latter provides the backdrop for Einstein's discussion of conventionalism in the well-known lecture, "Geometry and Experience" (1921*a*). I do not wish to repeat what I have said about Poincaré in my (1978). However, the following remarks are new, and will illustrate and clarify the main point I have been trying to make above. Poincaré's conventionalism is rooted in Felix Klein's work "On the so-called non-Euclidean geometry" (1871, 1873), in which the classical non-Euclidean geometry of Bolyai and Lobachevsky—called by Klein "hyperbolic"—Euclidean or "parabolic" geometry, and a new type of non-Euclidean geometry, discovered by Klein and which he called "elliptic", were united in a single one-parameter family of "projective metrics", i.e. metric structures for projective space. Let P stand for complex projective 3-space. We denote by A the set of all real points of P; by B, the set A minus an arbitrary plane (the "plane at infinity"); by C, a connected subset of B bounded by a real quadric Q. As is well known, the automorphisms (projective permutations) of P map quadrics onto quadrics. The automorphisms

of P that map C and its boundary Q respectively onto themselves constitute a subgroup G of the full group of automorphisms G_{full}. Klein showed, following a suggestion of Cayley (1859), that the G-invariant properties and relations of the points of C satisfy, under a suitable interpretation, the propositions of hyperbolic geometry. Likewise, if Q' denotes a purely imaginary quadric, the set of automorphisms of P that map Q' onto itself is a subgroup G' of G_{full}, whose elements also map A onto itself. The G'-invariant properties and relations of the points of A satisfy elliptic geometry. Finally, if Q'' denotes a certain specific degenerate imaginary quadric, the set of automorphisms of P that map Q'' onto itself is again a sub-group G'' of G_{full}, whose elements also map B onto itself. The G''-invariant properties and relations of the points of B satisfy parabolic, i.e. Euclidean geometry. Suppose that, as was normally assumed in Poincaré's time, all physical experiments and observations are contained in a finite open subset S of a Euclidean space. S can then be embedded in P by an embedding function i that maps straights into straights and satisfies the relation $i(S) \subset C \subset B \subset A \subset P$. Consequently, we can alternatively regard S as a part of a model of hyperbolic, Euclidean or elliptic space. Since, as Poincaré (1887), p. 216, duly noted, "the existence of a group is not incompatible with that of another group", all those alternative views of physical space can coexist in peace and none is truer than the other. In my opinion, the foregoing argument conclusively proves that the choice of one of the said geometries as physical geometry is a purely conventional matter, *provided that physical space is identified with S*. This identification presupposes that (i) the locus of bodies has the topological and affine properties taken for granted in Newtonian kinematics, and (ii) its global structure, beyond the range of actual experiment and observation, is physically irrelevant. Prior commitment to (i) and (ii) sets the stage for the conventional choice of a physical geometry in accordance with the Cayley–Klein theory of projective metrics. But the prerequisites of such a choice are no longer there if, as in General Relativity, other foundational commitments displace (i) and (ii). Poincaré claimed later than even the topology of physical space is conventionally fixed. Since he never successfully identified the unstructured carrier of alternative physical topologies nor stated the terms of reference for choosing between them, I conclude that Poincaré's *topological* conventionalism—in contrast to his *metrical* conventionalism—involved a confusion between "worldmaking" commitments and intramundane choices.

9. M. Schlick (1918), p. 352.
10. Reichenbach (1920), p. 48.
11. Reichenbach (1920), p. 38. Reichenbach adds, *ibid.*, pp. 38 f.: "It is useless to try to consider the individual perceptions as the defined elements of reality. The content of each perception is far too complex to serve as an element of coordination. [. . .] Before we coordinate a perception we must introduce order in it, by 'distinguishing what is relevant from what is irrelevant'. But such distinction already involves a coordination presupposing certain equations or the laws expressed by them."
12. Reichenbach (1920), p. 40.
13. *Ibid.*, p. 50. Kant's words are from Kant (1781), p. 93/(1787), p. 126.
14. Reichenbach (1920), p. 40.
15. *Ibid.*, p. 41.
16. *Ibid.*, p. 85.
17. *Ibid.*, p. 84. The change of constitutive principles in the course of history gives rise to an interesting philosophical problem that is boldly tackled by Reichenbach. *How can experience refute the constitutive principles on which it rests?* Moreover, *how can experience built on discarded principles be of any use in the search of new constitutive principles?* The first question is fairly easy to answer. Rejection of constitutive principles does not ensue from a conflict between them and the experience built in accordance with them, but follows upon the failure of such experience to materialize. If the description of perceived realities in terms of the chosen principles of coordination turns out to be inconsistent, we ought not to say that the principles clash with experience, but rather that experience itself is falling apart. Such situations can often be redressed by redescribing part of the available sense evidence as hallucinatory (or as a forgery). If this way out is closed, the principles of coordination must be revised, not because they are contradicted by experience, but because they have proved unable to continually constitute experience. It is much harder to find a satisfactory reply to the second question. Its inherent difficulty has led some contemporary philosophers to view scientific revolutions and conceptual change as essentially irrational processes, which are set in motion by the shortcomings of earlier science but cannot be

guided by former experience. Reichenbach's solution is based, he says, on the method used by Einstein for revising the conceptual system of Newtonian science after its experimental breakdown. If the coordination between scientific concepts and perceived realities could ever be exact it would certainly be absurd to test new principles in the light of experiences interpreted according to the old principles which the new ones are expected to replace. But the coordination of concepts and reality is always merely approximate. As technical improvements increase attainable precision or widen the scope of observation, familiar principles that have long provided a satisfactory framework for the description of nature begin generating inconsistencies. On the other hand, the same principles that are shown to be incompatible with one another when a high degree of approximation is demanded or when the whole range of values of the relevant physical variables is taken into account, can still yield a fairly consistent description of phenomena if a greater margin of imprecision is allowed or a narrower range of such values is admitted for consideration. Progress in the formulation of ever more suitable principles of coordination can therefore be achieved by successive approximations. Reichenbach's solution of the problem of conceptual change is satisfactory as far as it goes, but one may wonder whether it can encompass truly radical departures from established scientific tradition. In what sense can we say that Aristotelian physics approximates Newtonian dynamics? Or that Keplerian ellipses converge to Ptolemaic epicycles in the limit of naked eye astronomy? Indeed, if scientific change could occur only by successive approximations, science would be bound forever to the arbitrary—perhaps genetically conditioned—choice of an initial set of constitutive principles and no genuine scientific revolutions could take place.

18. Reichenbach (1924), p. 5.

19. "Theorie des täglichen Lebens"—Reichenbach (1924), p. 6.

20. Strictly speaking, Definition 9 cannot even be said to coordinate the class of light rays with the concept of a straight line, for it says only that the former is *contained* in the extension of the latter, not that it is *equal* to it. It is therefore hard to understand why the statement is termed a "definition". But then, logical neatness is not the main strength of Reichenbach's axiomatics.

21. Reichenbach (1928), p. 23.

22. Reichenbach (1928), p. 23.

23. But is it really a coordinative definition? The reference to temperature should make us wary. "At 0°C" can indeed be explained in terms of melting ice at sea level, but reference must then be made to the purity of the water and the normality of the atmospheric pressure. A good deal of theory went also into the design of the Paris rod. The current SI metre, equal to 1,650,763.73 times the wavelength *in vacuo* of the radiation corresponding to the transition between the $^2p_{10}$ and the $^5d_5$ levels of an atom of krypton 86, is of course firmly anchored to contemporary theoretical physics.

24. "Unter Kraft verstehen wir ein Etwas, das wir für eine *geometrische Veränderung* verantwortlich machen"—Reichenbach (1928), p. 38.

25. Reichenbach (1928), p. 21. Universal forces were introduced as "forces d'espèce X" in Reichenbach (1922), pp. 34 ff. See also Reichenbach (1924), p. 68, where he introduces the terms "metrical force" (for forces of type X) and "physical force" (for all other forces).

26. Reichenbach (1928), p. 32. Note that this definition does not simply coordinate the concept "rigid body" with "this thing here". We never come across a solid which is not subject to differential forces or from which the influence of such forces has been already—i.e. before we approach it with our theory-laden views—"corrected away".

27. Reichenbach (1928), p. 49.

28. Reichenbach (1928), p. 293: "In der Gravitation [liegt] der Typus einer *universellen* Kraft [vor]." Cf. Reichenbach (1922), p. 38: "La gravitation, qui jusqu'ici avait été concue comme toutes les autres forces, est précisément, si l'on y regarde de plus près, une force d'espèce X." Reichenbach (1924), p. 68: "Zu den metrischen Kräften gehört auch die Gravitation; darin besteht der Sinn der Einsteinschen Gravitationstheorie."

29. Reichenbach (1928), pp. 293 f.

30. Einstein (1956), p. 79.

31. Reichenbach must have had some inkling that General Relativity does not simply do away with the force of gravity, for right after the passage quoted on p. 236 (ref. 29) he proclaims nevertheless that "we must ascribe to the gravitational field the physical reality of a force field", namely, "the force field which we regard not as the cause of the *perturbation* of geometrical

relations, but as the cause of *geometry as such*". For "even if we do not introduce a force to explain a deviation of a measuring instrument from some standard geometry, we do still invoke a force as a cause for the fact that there is a uniform concordance of all measuring instruments" (Reichenbach (1928), p. 294). I dare say that in this text Reichenbach's use of the word "force" has lost the last remnant of precision.

32. As far as I can tell, the idea is due to Poincaré (1895), who bid us imagine a world enclosed in a great sphere of radius R, in which all solids have the same coefficient of expansion, thermal equilibrium is instantaneously reached, and the absolute temperature at distance r from the centre is proportional to $R^2 - r^2$.

33. Russell (1959), p. 39. Russell referred specifically to his own earlier *Essay on the Foundations of Geometry* (1897), in which he had asserted that the free mobility of rigid bodies was a condition of the possibility of physical experience. The pervasive influence of Helmholtz's view was indeed so great that 13 years after the discovery of the field equations, Einstein himself characterized geometry as "la science des possibilités de deplacement des corps solides", and added that in General Relativity the metric tensor defines "le mouvement des solides librement déplaçables, en l'absence d'effets électromagnétiques" (Einstein (1928*c*), p. 164; I thank Adolf Grünbaum for calling my attention to this passage).

34. Hugo Dingler had repeatedly pointed out that the manufacturers of precision instruments *assume* the validity of Euclidean geometry and *control* their products by its rules. Likewise, surveyors *correct* their observations by Euclid's theorem about the sum of the three interior angles of a triangle. According to Dingler, preference for Euclidean geometry was mandated by its unique position among the proper Riemannian geometries of constant curvature (the others involve an arbitrary parameter). On Dingler's philosophy of geometry, see Torretti (1978*b*) and the references listed there.

35. Einstein to Barnett, June 19th, 1948; quoted by J. Stachel (1977), p. ix. Einstein (1928*c*), pp. 164 f. had already made the same point.

36. On empiricism, apriorism and conventionalism in 19th century philosophy of geometry see, for instance, Torretti (1978*a*), Chapter 4. Conventionalism had at the time a clear advantage over the other two. Apriorism, when not shielded by ignorance, had to make such concessions to the plurality of geometries, that the "pure form of externality" became increasingly vague. Empiricists, on the other hand, were unable to explain the leap from blurred data to neat mathematical structures, or to justify the collection of empirical data on geometry by means of geometrically characterized instruments and processes.

37. Reichenbach (1928), p. 42.

38. Cf. F. A. E. Pirani (1956). Pirani's method for measuring the curvature components rests on a theorem in differential geometry known as *Jacobi's equation* or the *equation of geodesic deviation*. Take any n-manifold M ($n \geq 2$), with a symmetric linear connection that determines the covariant differential operator ∇ and the curvature R. Consider a family of geodesics in M, indexed by a parameter v, which forms a congruence in a submanifold $M' \subset M$, diffeomorphic with $\mathbf{R}^2$. We assume that the indexing by the parameter v is so contrived that there is a chart defined on M', with coordinate functions u and v, such that each geodesic of the family is a parametric line of u (with u varying along it as an affine parameter), while v assigns to each point of M' the index of the geodesic through that point. Set $\mathrm{U} = \partial/\partial u$, $\mathrm{V} = \partial/\partial v$. Since the connection is symmetric (i.e. torsion free), $\nabla_{\mathrm{U}}\mathrm{V} - \nabla_{\mathrm{V}}\mathrm{U} = [\mathrm{U}, \mathrm{V}]$, by (C.6.6.). But $[\mathrm{U}, \mathrm{V}] = 0$ (p. 262). Consequently, by (C.6.7),

$$\nabla_{\mathrm{U}}\nabla_{\mathrm{U}}\mathrm{V} = \nabla_{\mathrm{U}}\nabla_{\mathrm{V}}\mathrm{U} = \nabla_{\mathrm{V}}\nabla_{\mathrm{U}}\mathrm{U} + \mathrm{R}(\mathrm{U}, \mathrm{V})\mathrm{U}$$

By the definition of geodesics, (C.7.1), $\nabla_{\mathrm{U}}\mathrm{U} = 0$. We thus have that

$$\nabla_{\mathrm{U}}\nabla_{\mathrm{U}}\mathrm{V} = \mathrm{R}(\mathrm{U}, \mathrm{V})\mathrm{U} \qquad (*)$$

which is Jacobi's equation. In terms of the components of U, V and R relative to an arbitrary chart of M, Jacobi's equation reads:

$$\partial^2 \mathrm{V}^i/\partial u^2 = \mathrm{R}^i_{jkh}\mathrm{U}^j\mathrm{U}^k\mathrm{V}^h \qquad (**)$$

If we are dealing with timelike geodesics in a relativistic spacetime we can choose proper time as the affine parameter reflected by u. Let two particles A and B describe two nearby geodesics of the family, and refer equation (**) to a local Lorentz chart x based at A. Then, $x^0 \approx u$, $\mathrm{V}^0 = \partial x^0/\partial v \approx \partial u/\partial v = 0$, and, if B is very near A, the space components $\mathrm{V}^\alpha = \partial x^\alpha/\partial v$ of V along

A's worldline are proportional to the space coordinates of B in the local inertial frame. Hence, the quantities $\partial^2 V^\alpha/\partial u^2$ measure the acceleration of B with respect to A. The following, mutually complementary explanations of geodesic deviation and its physical significance can be useful as an introduction to Pirani's paper: Clarke (1979), pp. 74–76; Dodson and Poston (1977), pp. 453 f.; Weinberg (1972), pp. 453 f.; Misner *et al.* (1973), pp. 265–75 (nicely illustrated); Synge (1960), pp. 19–25, 57–64.

39. The normal scientific practice of testing a hypothesis by methods that partly presuppose it is studied at length, under the suggestive name of "bootstrapping", by Clark Glymour (1980).

40. The main primary sources for the study of Grünbaum's philosophy of geometry are collected in Grünbaum (1968), (1973). Grünbaum (1950, 1952, 1953) give us some insight into the philosopher's original motivation. The important essay, Grünbaum (1977), contains much that is completely new. The reader may also wish to consult Grünbaum (1967), a book on Zeno of Elea. Among the numerous discussions of the subject, I find the following to be particularly illuminating: Putnam (1963), Earman (1970), Glymour (1972*a*), Stein (1977*b*). See also Massey (1969), Van Fraassen (1969*b*), Fine (1971), Friedman (1972), Horwich (1975), Quinn (1976), Nerlich (1976), pp. 155–85, Angel (1980), pp. 236–44.

41. Grünbaum (1970), pp. 488–94; (1973), pp. 468–74. What distinguishes R-metrics from other cross-sections of the bundle of symmetric (0, 2)-tensors over a manifold S is that their range cannot include "degenerate" tensors. (A symmetric (0, 2)-tensor t at a point $P \in S$ is said to be "degenerate" if there is a non-zero vector $v \in S_P$ such that $t(v, w) = 0$ for every $w \in S_P$. Note that according to this definition a "non-degenerate" symmetric (0, 2)-tensor at $P \in S$ is an inner product on S_P; see p. 303, note 1.

42. Grünbaum (1962), p. 413; (1963), p. 10; (1968), p. 13; (1973), p. 10. The following passage precedes the above, and may contribute to clarify it: "There is no *intrinsic* attribute of the space between the end points of a line-segment AB, or any relation between these two points themselves, in virtue of which the interval AB could be said to contain the same amount of space as the space between the termini of another interval CD not coinciding with AB. Corresponding remarks apply to the time continuum." See also note 44.

43. Grünbaum (1970), p. 518; (1973), p. 498.

44. Grünbaum has often linked his reasons for denying that space and time, in and by themselves, are metric structures, to Riemann's distinction between discrete and continuous manifolds. Riemann said that the quantitative comparison between different parts of a manifold (distinguished from the rest of it by a characteristic or a boundary) "is made by counting in the case of discrete quantities, and in the case of continuous quantities by measurement". He further noted that "measurement consists in a superposition of the quantities to be compared, and therefore requires some means of transporting (*forttragen*) one quantity as a measure of the other" (Riemann (1867), p. 9). According to Riemann, this implies that "while in a discrete manifold the principle of metric relations is already contained in the concept of the manifold, in a continuous one it must come from somewhere else" (*ibid.*, p. 23). Commenting on these ideas, Grünbaum (1968), pp. 216 f., bids us consider "an interval AB in the mathematically continuous physical space of, say, a given blackboard, and also an interval $T_0 T_1$ in the continuum of instants constituted by, say, the movement of a classical particle". Contrary to what would be the case in a discrete manifold, "neither the cardinality of AB nor any other property *built into* the interval provides a measure of its particular spatial extension, and similarly for the temporal extension of $T_0 T_1$. For AB has the same cardinality as any of its *proper* subintervals and also as any other nondegenerate interval interval CD. Corresponding remarks apply to the time interval $T_0 T_1$. If the intervals of physical space or time *did* possess a built-in measure or '*intrinsic metric*', then relations of *congruence* (and also of incongruence) would obtain among disjoint space intervals AB and CD on the strength of that intrinsic metric. And in that hypothetical case, neither the existence of congruence relations among disjoint intervals nor their epistemic ascertainment would logically involve the iterative application and transport of any length standard. But the intervals of mathematically continuous physical space and time are devoid of a built-in metric. And in the absence of such an intrinsic metric, the basis for the extensional measure of an interval of physical space or time must be furnished by a relation of the interval to a body or process which is applied to it from outside itself and is thereby '*extrinsic*' to the interval. Hence *the very existence* and not merely the epistemic ascertainment of relations of congruence (and of incongruence) among disjoint space intervals AB and CD of continuous physical space will depend on the respective

relations sustained by such intervals to an extrinsic metric standard which is applied to them. Thus, *whether two disjoint intervals are congruent at all or not will depend on the particular coincidence behavior of the extrinsic metric standard under transport* and not only on the particular intervals AB and CD. Similarly for the case of disjoint time intervals $T_0 T_1$ and $T_2 T_3$, and the role of clocks." These considerations, which, if true, would evidently apply also *mutatis mutandis*, to the continuous spacetime manifold of Relativity, are sufficient, according to Grünbaum, to establish the conventionality of geometry on solid grounds, for "an extrinsic metric standard is self-congruent under transport *as a matter of convention* and not as a matter of spatial fact, although any *concordance* between its congruence findings and those of another such standard is indeed a matter of *fact*" (Grünbaum (1968), p. 217).

45. Grünbaum (1968), p. 209.

46. Grünbaum (1968), p. 209. Where I insert "[GGC]" Grünbaum writes "Riemann's doctrine". Cf. note 44.

47. Let M be an n-manifold. By a *foliation* F of M, with codimension q $(1 \leq q < n)$, I mean a family of path-connected subsets of M called *leaves*, such that (i) each point $P \in M$ lies in one and only one leaf of F, and (ii) if a given $L \in F$ contains a particular $P \in M$, there is a chart x of M defined on a neighbourhood U of P whose last q coordinate functions are constant on $L \cap U$. It follows that each leaf of F is an $(n-q)$-submanifold of M. A foliation of spacetime into hypersurfaces is therefore a foliation with codimension 1; a congruence of curves in spacetime is a foliation of spacetime with codimension 3.

48. Note that Robertson–Walker spacetimes are not usually handled in this way. Instead of a single cosmic space of worldlines (the parametric lines of τ) with line element $ds_3 = d\sigma$, we consider each fibre of τ as a distinct space (of events), with its own time-dependent line element $ds_3 = S(\tau)\,d\sigma$. The choice between these two approaches is indeed conventional for, as Eddington (1933), p. 90, remarked, "the theory of the 'expanding universe' might also be called the theory of the 'shrinking atom'." But no matter which approach we prefer, the space metrics ds_3 are not arbitrarily postulated, but result from the spacetime metric.

49. Nerlich (1976), p. 184; cf. *ibid.*, p. 129. In the "Reply to Hilary Putnam" from which the last two indented quotations (refs. 45 and 46) are taken, Grünbaum considers what we may call "local" or "restricted" space metrics, dependent on a particular spacetime chart x, defined on a region U. A "space" is carved out of U by identifying all worldpoints whose space coordinates by x^1, x^2 and x^3 are identical. If the spacetime metric is expressed, relative to x, by $(ds^4)^2 = g_{ij}\,dx^i dx^j$, it is arguably appropriate to set $(ds_3)^2 = \gamma_{\alpha\beta}\,dx^\alpha\,dx^\beta$, where $\gamma_{\alpha\beta} = g_{\alpha\beta} - (g_{0\alpha}g_{0\beta}/g_{00})$. (See, for instance, Møller (1952), pp. 237 ff.; Arzeliès (1961), pp. 70 ff.; Landau and Lifschitz (1962), pp. 272 ff.) The physical significance of such a chart-dependent line element is of course very doubtful and it is altogether misleading to describe $\gamma_{\alpha\beta}$ as "the spatial metric tensor". However, what really matters for our evaluation of GGC is that ds_3 is uniquely determined by ds_4 for any given x, i.e. for any given space carved out of space-time in this way.

50. Grünbaum (1977), p. 342. Here and in the following quotations I substitute Latin letters for indices ranging over $\{0, 1, 2, 3\}$.

51. Grünbaum (1977), pp. 342, 343.

52. Grünbaum (1977), pp. 344 f. Like Stein (1977*b*), p. 380, I am unable to understand how an *open-ended* set can be *canonical.*

53. Grünbaum (1977), p. 345.

54. Grünbaum (1977), p. 346.

55. Grünbaum (1977), p. 356.

56. Grünbaum (1977), p. 351, notes that his thesis "that the g_{ij} fields of NE-GQ worlds acquire geometrical significance through the mediation of external entities like the FCS" enables us to "accomodate philosophically" some physical theories that endow spacetime with two metrics, realized by two *different* FCS. Thus, in a modification of General Relativity proposed by P. A. M. Dirac with a view to reconciling it with Quantum Mechanics, "macroscopic and atomic metric standards generate *incompatible* metric ratios among space-time intervals", so that "only the macroscopic standards physically realize the scalar $|g_{ij}\,dx^i\,dx^j|^{1/2}$ corresponding to the g_{ij} field of Einstein's equations". But if a theory of this kind does actually postulate two distinct spacetime metrics, which are not to be regarded merely as a coarser and a finer approximation to one and the same field, but as two different fields, defined on every spacetime region, no matter how small, I do not see how one would be philosophically better served by distinguishing between the two

"physical" fields, which are discernible in principle at each worldpoint, and their respective "geometrical" counterparts, which exist only where they are "constituted" as such by the appropriate FCS. Since both "physical" fields have by hypothesis the "formal mathematical properties" of a metric, no philosophical problem raised by their actual coexistence at every worldpoint can be solved by the expedient of denying them the epithet "geometrical".

Section 7.3

1. "Si quid intelligitur temporis, quod in nullas iam vel minutissimas momentorum partes dividi possit, id solum est quod praesens dicatur; quod tamen ita raptim a futuro in praeteritum transvolat, ut nulla morula extenditur. Nam si extenditur, dividitur in praeteritum et futurum: praesens nullum habet spatium" (St. Augustine, *Confessiones*, XI.15.20; cf. XI.21.27).

2. I read in the *Cambridge Encyclopedia of Astronomy* (1977), p. 24, that Rigel (β Orionis) lies at a distance of 250 parsec $\approx$ 815 lightyears. Frederick's fourth expedition to Italy began in 1166.

3. Denote the said Lorentz chart by y and suppose that its frame moves past us, in the direction opposite to Rigel, with a speed $v = 0.99999$ ($c = 1$). If Rigel's space coordinates in our own rest frame are $x^1 = -815$, $x^2 = 0$, $x^3 = 0$, and the time at which the light we see was emitted is $x^0 = -815$, we readily calculate, using (3.4.18), that $y^0 = -1.82$. Referred to y, the light we see left the star less than two years ago (and Rigel itself is of course only 1.82 lightyears away).

4. It was held indeed by Aristotle himself, a century before the Stoics. See *De Interpretatione*, 9. By the way, I do not believe that the ancient Stoics were truly guilty of logical determinism, but rather that they held, against Aristotle, that every proposition is either true or false, because they subscribed to strict determinism on physical grounds.

5. Writing Lp for "it is necessary that p", we have that $L(p \vee \sim p)$ does not logically imply that $Lp \vee \sim Lp$.

6. Let A' and B' be the images of A and B by a Lorentz chart x whose time axis goes through A and B. Draw λ through A' perpendicular to $A'B'$. All points on λ are the images by x of events simultaneous with A in the frame of x. Choose any point C' on λ such that $C'B'$ is the image by x of a spacelike geodesic. C' is the image by x of an event C. Draw μ through A' so that μ is not perpendicular to $C'B'$ and makes with λ an angle complementary to $\measuredangle A'C'B'$. μ is the image by x of the time axis of a Lorentz chart y in whose frame C is simultaneous with B.

7. Putnam (1979), p. 198. A different sort of ambiguity turns up on p. 200. Putnam discusses the following allegedly binary relation: Rxy if and only if x is simultaneous with y in the coordinate system of x. However, the definiens is not binary but ternary, for the first occurrence of x ranges over events, whereas the second must range over some other domain. (To speak of "the coordinate system of a given event" is a category mistake.)

8. Rietdijk (1966), p. 342.

9. This notion of smooth agreement can be made precise as follows. We say that Q is a neighbouring point of $P \in \mathscr{W}$ if Q lies on the neighbourhood of P on which Exp_P^{-1} (the inverse of the local exponential mapping) is a diffeomorphism. There is then one and (up to reparametrization) only one geodesic γ from P to Q which never leaves that neighbourhood. The local time orientations at P and Q agree smoothly if any future-directed vector at P is sent by parallel transport along γ to future-directed vectors at each point in γ's range up to and including Q.

10. It is clear that a time-orientable spacetime cannot contain a closed curve that is not time preserving. On the other hand, if our spacetime $(\mathscr{W}, \mathbf{g})$ is not time-orientable there is, for each choice of a local time orientation at every point of $\mathscr{W}$, a pair of neighbouring points P and Q at which the time orientations do not agree smoothly in the sense of note 9. But then, for every arbitrary fixed $O \in \mathscr{W}$, a loop $OPQO$ which is geodesic from P to Q cannot be time preserving.

11. Remember that two curves γ, γ', which map the same interval $I \subset \mathbf{R}$ into the same manifold $\mathscr{W}$, are said to be *homotopic* if one of them is continually deformable into the other, i.e. if there is a continuous mapping f of $I \times [0, 1]$ into $\mathscr{W}$, such that, for each $t \in I$, $f(t, 0) = \gamma(t)$ and $f(t, 1) = \gamma'(t)$.

12. Let S be a path-connected topological space. A connected topological space $\bar{S}$ is said to be a *covering space* over S with *covering map* f, if there is a continuous mapping f of $\bar{S}$ onto S, such that every $P \in S$ has a connected open neighbourhood U onto which each connected component of $f^{-1}(U)$ is mapped homeomorphically by f. $\bar{S}$ is a *universal covering space* over S if it is simply connected. Thus, for example, the plane $\mathbf{R}^2$ is a universal covering space over the cylinder $\mathbf{R} \times \mathbf{S}$,

with covering map $(x, y) \mapsto (x, (\cos 2\pi y, \sin 2\pi y))$; note that each fibre of this covering map has infinitely many points. If S is a differentiable manifold, every covering space over S has a unique differentiable structure such that the covering map is smooth. A connected manifold M has a unique (up to isomorphism) universal covering manifold $\overline{M}$. $\overline{M}$ is a principal fibre bundle over M whose structural group is the fundamental group of M. See Steenrod (1951), pp. 67 ff.

13. The preceding discussion of time-orientability closely follows the excellent article by Geroch and Horowitz (1979), which also contains a careful yet not too technical discussion of the causal structure of relativistic spacetimes that I heartily recommend to the reader. The notion of topologically distinct yet observationally indistinguishable spacetimes is the subject of a deep mathematico-philosophical inquiry by C. Glymour (1972*b*, 1977). D. Malament (1977*b*) provides a luminous and insightful introduction to Glymour's work.

14. We referred to the Gödel solutions on p. 331, note 36; H. Stein (1970*a*) discusses them with special regard to their peculiar causal structure. On the Kerr solutions, see the references mentioned on p. 335, note 2, and (for a philosophical discussion) B. Kanitscheider (1979*a*), pp. 168–72. The extraordinary properties of the Taub–NUT solution are discussed by Misner (1967). The recent treatise by D. Kramer *et al.* (1980) systematically studies and classifies the exact solutions to the Einstein field equations known to date, including those just mentioned.

15. S. Hawking (1971), p. 396.

16. S. Hawking and G. F. R. Ellis (1973), p. 189.

17. S. Hawking and G. F. R. Ellis (1973), p. 197.

18. R. Geroch and G. T. Horowitz (1979), p. 238.

19. Note that the common practice of referring a particle horizon to a particle at a worldpoint, and not just to a worldpoint, is redundant and may be misleading.

20. Roger Penrose has given a different definition of particle horizons, which is not tied to the existence of a congruence of timelike curves. A Penrose particle horizon is, like an event horizon, the boundary of a set of worldpoints defined with respect to a particle or its worldline. The *Penrose particle horizon* of a timelike worldline γ is the boundary of γ's chronological future $I^+(\gamma)$. Note that the event horizon of γ is in effect the boundary of γ's chronological past $I^-(\gamma)$ (because γ's causal and chronological past have the same closure and hence the same boundary). The event horizon of γ "separates those events which are observable by an observer whose worldline is γ from those events unobservable by him", while the Penrose particle horizon of γ "separates those events from which a particle with worldline γ can be observed (in principle), from those events from which the particle cannot be so observed" (Penrose (1968), p. 189). A short reflection should make clear that a non-empty Penrose particle horizon will occur in every case in which the Rindler particle horizon is non-empty. (For the latter is a boundary of the particles which must remain concealed to a given observation, whereas the former is the boundary of the potential observations to which a given particle must remain concealed.) However , in spite of the alluring simplicity and greater generality of Penrose's concept, Rindler's has prevailed in the literature.

21. See, for instance, Hawking (1972); Hawking and Ellis (1973), Chapter 9. A particle outside a black hole cannot be reached by causal curves originating inside it. Thus, all events within the black hole lie beyond the horizon of every particle that does not fall into it. These statements do not hold if the energy contents of a black hole can "tunnel through" its boundary according to the laws of Quantum Mechanics, as in Hawking (1975)—cf. the popular exposition by Hawking (1977).

22. MacCallum (1971).

23. Cf. Misner (1969*b*).

24. Misner (1980), p. 225. Misner points out that A. H. Taub's work on anisotropic Big Bang worlds shows that the "horizon paradox" we have described above is not merely a consequence of the ideally perfect isotropy of the Friedmann worlds, which could be avoided in a more realistic cosmological model.

Appendix

A. Differentiable Manifolds

1. Gauss (1827), Riemann (1867). I have discussed Gauss's methods and Riemann's theory in Torretti (1978*a*), pp. 67–109.

2. A mapping f of a topological space onto another is a homeomorphism if (i) f is bijective (i.e. one-one and onto), and (ii) both f and its inverse f^{-1} map open sets onto open sets. A homeomorphism is thus an isomorphism of topological spaces. Two such spaces are said to be homeomorphic if there is a homeomorphism of one onto the other.

3. The adjective *real* refers to the fact that the charts have their ranges in $\mathbf{R}^n$, and so have real-valued coordinate functions. A *complex* manifold has complex-valued coordinate functions. More generally, one may replace $\mathbf{R}^n$ in the above definitions by an arbitrary n-dimensional Banach space (i.e. a normed vector space in which all Cauchy sequences converge). It is worth noting that a C^k-differentiable structure can be defined on an arbitrary set S as follows: An *n-chart* is a bijective mapping of a part of S onto an open subset of an n-dimensional Banach space; a C^k-atlas A of S and the respective maximal atlas $A_{\max}$ are defined as above, with 'n-chart' substituted for 'chart'; S is given the weakest topology that makes every n-chart in $A_{\max}$ into a homeomorphism.

4. The exact order of differentiability of a mapping is usually of no consequence in physics, because any C^r-differentiable mapping of a C^k-differentiable manifold into another can be approximated to any desired degree by a C^k-differentiable mapping, if $1 \leq r < k \leq \infty$. (Munkres (1966), p. 41.) There is no significant loss of generality in confining our discussions to C^∞-differentiable manifolds, because any maximal C^1-atlas of a manifold S includes a C^∞-subatlas. (Munkres (1966, p. 46.) On the other hand, the postulated order of differentiability of some unphysical auxiliary functions—e.g. a metric at infinity—can have significant physical implications; see Ehlers (1980*a*), p. 284. And of course the difference between C^1-differentiability and mere continuity can be dramatic; this matters because even within the scope of a straight differential-geometric theory there are physical situations which one might wish to deal with by means of unsmooth continuous or merely bounded functions, and it is a moot question whether one can actually do so without dismantling the theory. See Taub (1979); Tipler, Clarke and Ellis (1980), pp. 156 f.

5. That is:

$$\bar{f}_{,i}(u) = \lim_{h \to 0} (1/h)(f \cdot x^{-1}(u_1, \ldots, u_i + h, \ldots, u_n) - f \cdot x^{-1}(u_1, \ldots, u_i, \ldots, u_n)),$$

where $u = (u_1, \ldots, u_n)$ is any point in the range of x.

6. Proof: Let $v = \Sigma_i a_i \partial/\partial x^i$. Then $vx^j = a_j$. Hence, v is identically 0 only if $a_j = 0$ for all indices j.

7. This definition of *vector field* is equivalent to the one given on p. 260. Note that if V is a vector field on S, $f \mapsto \mathrm{V}f$ is a linear mapping of $\mathscr{F}(S)$ into itself, which is also called V in the literature. The homonymy is justified insofar as the mapping of the linear space of linear endomorphisms of $\mathscr{F}(S)$ into the module of vector fields on S that sends each such V to its namesake is a linear isomorphism.

8. See, for instance, MacLane and Birkhoff, *Algebra*, p. 238.

9. See, for instance, Y. Choquet-Bruhat *et al.* (*AMP*), pp. 117–21.

B. Fibre Bundles

1. Lie groups were defined on p. 288, note 12. The action "on the right" of a Lie group on a manifold was defined on p. 288, note 13.

2. In other words, if $a, b \in E$, there is a $g \in G$ such that $ag = b$ if and only if $\pi(a) = \pi(b)$.

3. For a proof, see Kobayashi and Nomizu, *FDG*, vol. I, p. 52.

4. Denote this mapping by f. If g, $h \in G$, $a \in E$, $s \in F$, $f(gh, (a, s)) = (agh, h^{-1}g^{-1}s) = f(h, f(g, (a, s)))$.

5. If $A \in GL(n, \mathbf{R})$ is a linear automorphism of $\mathbf{R}^n$, A can be extended to a unique linear automorphism of the tensor algebra on $\mathbf{R}^n$, which maps $\mathbf{t}^q_r$ onto itself (for each value of q and r). Since $\mathbf{t}^q_r$ (for each q and r) is constructed in a definite way from $\mathbf{R}$, $\mathbf{R}^n$ and the dual space $(\mathbf{R}^n)^* = \mathbf{t}_1$ (the space of real-valued linear functions on $\mathbf{R}^n$), A extended is completely determined if we know its restrictions to $\mathbf{R}^n$ and to $(\mathbf{R}^n)^*$. The former is of course A. We set the latter equal to $(A^T)^{-1}$, i.e. the inverse of the transpose of A.

C. Linear Connections

1. Other alternatives are explained and compared by Y. Choquet-Bruhat *et al.* (1977), pp. 287 ff.

2. Remember that the *kernel* ker f of a linear mapping f is the subset of its domain which f maps on zero.

3. The p-linear mapping ω_P of S_P into V is determined by its value at each list of p vectors drawn from a basis of S_P. If $n < p$, all such lists are repetitious.

4. The Lie derivative of a vector field along another was defined on p. 262. .

5. Kobayashi and Nomizu, *FDG*, vol. I, pp. 36 f., Prop. 3.11.

6. If $\mathbf{i}$ denotes the canonic inclusion of $\pi^{-1}(\pi q)$ into E, $\pi \cdot \mathbf{i}$ is a constant mapping and $(\pi \cdot \mathbf{i})_* = \pi_* \cdot \mathbf{i}_* = 0$. Hence $\mathbf{i}_{*q}(V_q) \subset \ker \pi_{*q}$. Since $\ker \pi_{*q}$ is m-dimensional it must be identical with V_q.

7. As lifts are so to speak the key to our subject, let me propose another view of them. Since each space of horizontal vectors H_q is n-dimensional we can think of it as the image of a linear injection σ_q of $B_{\pi q}$ into E_q. σ_q is uniquely defined by the requirement that $\pi_{*q} \cdot \sigma_q$ be the identity on $B_{\pi q}$. Evidently, σ_q varies smoothly with q and is consistent with the right action $\overline{R}$ (so that for any $g \in G$, $\sigma_{qg} = \overline{R}_{g*q} \cdot \sigma_q$). σ_q may be said to *lift* the tangent space $B_{\pi q}$ to E_q. The lift of the vector field X on B is the vector field X* on E given by $q \mapsto \sigma_q(\mathrm{X}_{\pi q})$. The uniqueness of X* is ensured by its definition.

8. A rigorous proof of this fact will be found in Kobayashi and Nomizu, *FDG*, vol. I, pp. 69 f., Prop. 3.1. We can make it plausible as follows: Let X be a vector field on a neighbourhood of γ's range, which assigns to each point of the latter the tangent to γ at that point. If X* is the lift of X and t denotes the mapping $u \mapsto u + t$ of $\mathbf{R}$ onto itself, $\gamma \cdot t$ is the restriction to $t(I)$ of the maximal integral curve of X* through p.

9. Kobayashi and Nomizu, *FDG*, vol. I, p. 142, Prop. 7.3.

10. Kobayashi and Nomizu, *FDG*, vol. I, p. 143, Prop. 7.4.

11. See Kobayashi and Nomizu, *FDG*, vol. I, p. 76, example 5.2, and the remarks on Ω and Θ on pp. 77 and 120. A study of the said example 5.2 will disclose the deep necessity underlying our seemingly artificial definitions (C.6.1) and (C.6.2). The "easily verified" fact mentioned by Kobayashi and Nomizu on p. 76, line 13, rests on the following: each $p \in \pi^{-1}(P) \subset FS$ is equal to qg for some g in the structure group; q^{-1} sends each $r \in \mathbf{R}^n$ to the equivalence class $[q, r] \in S_P \subset TS$, which, if $p = qg$, is identical with $[p, g^{-1}r]$, the image of $g^{-1}r$ by p.

12. Kobayashi and Nomizu, *FDG*, vol. I, p. 135, Prop. 5.3.

13. Since this result is quite fundamental for General Relativity we may as well prove it. For any n-tuple $(r^1, \ldots, r^n) \in \mathbf{R}^n$, the equations $x^i \cdot \gamma(t) = r^i t$ define a geodesic γ such that $\gamma(0) = P$. Since $\mathrm{d}x^i \cdot \gamma/\mathrm{d}t = r^i$ and $\mathrm{d}^2 x^i \cdot \gamma/\mathrm{d}t^2 = 0$, (C.7.2) entails that $\Gamma^i_{jk}(P) r^j r^k = 0$. As this holds for all n-tuples $(r^1, \ldots, r^n)$, $\Gamma^i_{jk}(P) + \Gamma^i_{kj}(P) = 0$. If $\Gamma^i_{kj} = \Gamma^i_{jk}$, it follows that $\Gamma^i_{jk}(P) = -\Gamma^i_{jk}(P) = 0$.

14. For a proof, see Kobayashi and Nomizu, *FDG*, vol. I, pp. 79 ff., Prop. 6.1.

15. Although I have hitherto taken for granted that the reader is acquainted with linear algebra, a definition of contraction might not be out of place here. If f is a q-linear function on a linear space V and g is an r-linear function on a linear space W, the *tensor product* $f \otimes g$ is the $(q+r)$-linear function defined on $V^q \times W^r$ by

$$(f \otimes g)(v_1, \ldots, v_q, w_1, \ldots, w_r) = f(v_1, \ldots, v_q) g(w_1, \ldots, w_r) \qquad (v_i \in V,\ w_i \in W).$$

It is clear that, for suitable multilinear functions f, g, h, $f \otimes (g \otimes h) = (f \otimes g) \otimes h$. As we know, a (q, r) tensor field on an n-manifold S can be regarded as a $(q+r)$-linear mapping of the module $(\mathscr{V}_1(S))^q \times (\mathscr{V}^1(S))^r$ into its ring of scalars $\mathscr{F}(S)$. We shall say that a (q, r)-tensor field on S is *simple* if it is the tensor product of q vector fields and r covector fields on S. Obviously the module $\mathscr{V}^q_r(S)$ of (q, r)-tensor fields on S is spanned by its simple elements. Hence a linear mapping of $\mathscr{V}^q_r(S)$ need only be defined on the latter. The *contraction* $\mathbf{Ctr}^i_j$ of (q, r)-tensor fields on S with respect to their i-th contravariant and j-th covariant indices $(1 \le i \le q, 1 \le j \le r)$ is the linear mapping of $\mathscr{V}^q_r(S)$ into $\mathscr{V}^{q-1}_{r-1}(S)$ defined as follows: for any vector fields on S, $\mathrm{X}_1, \ldots, \mathrm{X}_q$ and any covector fields on S, $\omega_1, \ldots, \omega_r$,

$$\mathrm{Ctr}^i_j(\mathrm{X}_1 \otimes \ldots \otimes \mathrm{X}_q \otimes \omega_1 \otimes \ldots \otimes \omega_r) = \omega_j(\mathrm{X}_i)(\mathrm{X}_1 \otimes \ldots \otimes \mathrm{X}_{i-1} \otimes \mathrm{X}_{i+1} \otimes \ldots \otimes \mathrm{X}_q \otimes \omega_1 \otimes \ldots \otimes \omega_{j-1} \otimes \omega_{j+1} \otimes \ldots \otimes \omega_r).$$

It follows immediately from this definition that if A is, say, a (2, 2)-tensor field given, on the domain of a chart x, by $A = A^{ij}{}_{kh}\,(\partial/\partial x^i \otimes \partial/\partial x^j \otimes dx^k \otimes dx^h)$, the contraction of A with respect to the first contravariant and the first covariant indices is given on the domain of x by $\mathrm{Ctr}^1_1(A) = A^{ij}{}_{kh}(\partial x^k/\partial x^i)(\partial/\partial x^j \otimes dx^h) = A^{ij}{}_{ih}(\partial/\partial x^j \otimes dx^h)$. The (^j_h)-th component of $\mathrm{Ctr}^1_1(A)$ relative to x is therefore equal to $A^{ij}{}_{ih}$ (summation intended over the repeated index i!). The contraction of tensors at a point is defined analogously.

ADDENDA to page 302, note 34. On the other hand, Elie Zahar (1982), with whose reading of Poincaré's conventionalism I substantially agree, maintains that Poincaré discovered Special Relativity independently in 1905 and that his philosophical insights contributed decisively to this achievement. A similar opinion is voiced in this series by Jerzy Giedymin (1982). In my review of Giedymin's book (*Diálogos*, 40, November 1982), I reaffirm and further elaborate the views I put forward above, in Section 3.8.

References

The following list purports to contain all writings mentioned in the book in abbreviated form. It also includes some books and articles I have not mentioned, but to which I feel obliged.

The abbreviations designating periodicals will be readily understandable to most readers and to any competent librarian. Some of them, which are perhaps too concise, are deciphered below:

AJP	*American Journal of Physics.*
Arch. Néerl.	*Archives néerlandaises des sciences exactes et naturelles.*
AWWS	*Akademie der Wissenschaften in Wien. Math.-naturw. Kl. Sitzungsberichte.*
BJHS	*British Journal for the History of Science.*
BJPS	*British Journal for the Philosophy of Science.*
CR	*Comptes Rendus de l'Académie des Sciences (Paris).*
FP	*Foundations of Physics.*
Gött. Abh.	*Abhandlungen der K. Gesellschaft der Wissenschaften zu Göttingen.*
Gött. Nachr.	*Nachrichten der K. Gesellschaft der Wissenschaften zu Göttingen. Math.-phys. Kl.*
GRG	*General Relativity and Gravitation.*
HSPS	*Historical Studies in the Physical Sciences.*
JDMV	*Jahresbericht der Deutschen Mathematiker-Vereinigung.*
KNAWPr.	*Koninklijke Nederlandse Akademie van Wetenschappen. Proceedings.*
KNAWVersl.	*Koninklijke Nederlandse Akademie van Wetenschappen. Verslag van de gewone vergaderingen der wis- en natuurkundige afdeeling.*
KS	*Kantstudien.*
Leipz. Ber.	*K. Sächsische Akademie der Wissenschaften zu Leipzig. Berichte über die Verhandlungen der Math.-phys. Kl.*
MNRAS	*Monthly notices of the Royal Astronomical Society.*
MR	*Mathematical Reviews.*
PS	*Philosophy of Science.*
RSLPr.	*Royal Society (London) Proceedings.*
RLSPT	*Royal Society (London) Philosophical Transactions.*
SHPS	*Studies in History and Philosophy of Science.*
ZAW	*Zeitschrift für allgemeine Wissenschaftstheorie.*

Abraham, M. (1909*a*), "Zur Elektrodynamik bewegter Körper". *Rend. Circ. Mat. Palermo*, 28: 1–28.

Abraham, M. (1909*b*), "Zur elektromagnetischer Mechanik". *Phys. Ztschr.* 10: 737–41.

Abraham, M. (1910), "Sull'elettrodinamica di Minkowski". *Rend. Circ. Mat. Palermo*, 30: 33–46.

Abraham, M. (1912*a*), "Zur Theorie der Gravitation", *Phys. Ztschr.* 13: 1–5, 176.

Abraham, M. (1912*b*), "Der freie Fall". *Phys. Ztschr.* 13: 310–11.

Abraham, M. (1912*c*), "Die Erhaltung der Energie und der Materie im Schwerkraftfelde". *Phys. Ztschr.* 13: 311–14.

Abraham, M. (1912*d*), "Relativität und Gravitation. Erwiderung auf eine Bemerkung des Hrs. A. Einsteins". *Ann. Physik* (4) 38: 1056–8.

Abraham, M. (1912*e*), "Nochmals Relativität und Gravitation. Bemerkungen zu A. Einsteins Erwiderung". *Ann. Physik* (4) 39: 444–8.

Abraham, M. (1912*f*), "Das Gravitationsfeld". *Phys. Ztschr.* 13: 793–7.

Abraham, R. and J. E. Marsden (1978), *Foundations of Mechanics*, Second Edition, Reading, Mass.: Benjamin/Cummings.

Airy, G. B. (1871), "On a supposed alteration in the amount of astronomical aberration of light, produced by the passage of the light through a considerable thickness of refracting medium". *RSLPr*, A 20: 35–9.

Alexandrov, A. D. (1967), "A contribution to chronogeometry". *Can. J. Math.* 19: 1119–28.

Alexandrov, A. D. and V. V. Ovchinnikova (1953), "Notes on the Foundations of Relativity". *Vestnik Leningr. Univ.* 11: 95 ff. (In Russian. I have not seen this paper; I owe the reference to A. D. Alexandrov, 1967.)

Alley, C. O., L. S. Cutler, R. Reisse, R. Williams, C. Steggerda, J. Rayner, J. Mullendore and S. Davis (1977), "Atomic-clock measurements of the General Relativistic time differences produced by aircraft flights using both direct and laser-pulse remote time comparison". Preprint.

Alväger, T., A. Nilsson and J. Kjellman (1964), "On the independence of the velocity of light on the motion of the light source". *Ark. Fys.* 26: 209–21.

Ampère, A. M. (1827), "Mémoire sur la théorie mathématique des phénomènes électrodynamiques, uniquement déduite de l'expérience, dans lequel se trouvent réunis les Mémoires que M. Ampère a communiqués à l'Académie royale des Sciences, dans les séances des 4 et 26 décembre 1820, 10 juin 1822, 22 décembre 1823, 12 septembre et 28 novembre 1825". *Mémoires de l'Académie Royale des Sciences de l'Institut de France*, Paris: Chez Firmin Didot, Père et Fils. Tome VI. Année 1823. (Note that this and the five preceding volumes of this second series of Paris Academy *mémoires* bear the publication date 1827.)

Anderson, J. L. (1964), "Relativity principles and the role of coordinates in physics". In Chiu and Hoffmann (1964), pp. 175–94.

Anderson, J. L. (1967), *Principles of Relativity Physics*. New York: Academic Press.

Angel, R. B. (1980), *Relativity: The Theory and its Philosophy*. Oxford: Pergamon.

Arago, F. (1853), "Mémoire sur la vitesse de la lumière", *CR*, 36: 38–49.

Archibald, R. C. (1914), "Time as a fourth dimension". *Am. Math. Soc. Bull.* 20: 409–12.

Aristotelis Opera edidit Academia Regia Borussica. Aristoteles graece ex recognitione Immanuelis Bekkeri. Berlin: G. Reimer, 1831. 2 vols.

Aristotle's Physics. A revised text with introduction and commentary by W. D. Ross. Oxford: Clarendon, 1936. (Quotations from Aristotle's *Physica* are from this text.)

Arnowitt, R., S. Deser and C. W. Misner (1962), "The dynamics of general relativity". In Witten (1962), pp. 227–265.

Arzeliès, H. (1961), *Rélativité généralisée. Gravitation.* Fasc. I. Paris: Gauthier-Villars.

Arzeliès, H. (1966), *Relativistic Kinematics.* Oxford: Pergamon.

Augustine, Saint. *Obras de San Agustín.* Texto bilingüe. Tomo Segundo. Las Confesiones. Edición crítica y anotada por A. C. Vega. Madrid: B.A.C., 1946.

Babcock, G. C. and T. G. Bergman (1964), "Determination of the constancy of the speed of light". *J. Opt. Soc. Am.* 54: 147–51.

Bacry, H. and J. M. Lévy-Leblond (1968), "Possible kinematics". *J. Math. Phys.* 9: 1605–14.

Bailey, J., K. Borer, F. Combley, H. Drumm, F. Krienen, F. Lange, E. Picasso, W. von Ruden, F. J. M. Farley, J. H. Field, W. Flegel and P. M. Hattersley (1977), "Measurements of relativistic time dilatation for positive and negative muons in a circular orbit". *Nature*, 268: 301–5.

Bateman, H. (1910), "The transformation of the electrodynamical equations", *London Math. Soc. Proc.* (2) 8: 223–64.

Bauer, H. (1918), "Ueber die Energiekomponenten des Gravitationsfeldes". *Phys. Ztschr.* 19: 163–5.

Beauregard, L. A. (1977), "Reichenbach on conventionalism". *Synthèse*, 34: 265–80.

Benz, W. (1967), "Zur Linearität relativistischer Transformationen". *JDMV*, 70: 100–8.

Bergmann, P. G. (1942), *Introduction to the Theory of Relativity.* Englewood Cliffs: Prentice-Hall.

Bergmann, P. G. (1962*a*), "The special theory of relativity". In Flügge (1962), pp. 109–202.

Bergmann, P. G. (1962*b*), "The general theory of relativity". In Flügge (1962), pp. 203–72.

Berkson, W. (1974), *Fields of Force. The Development of a World View from Faraday to Einstein.* New York: Wiley.

Bertotti, B. (1962), "The theory of measurement in general relativity". In Møller, *EGT*, pp. 174–201.

Berzi, V. and V. Gorini (1969), "Reciprocity principle and the Lorentz transformations", *J. Math. Phys.* 10: 1518–24.

Bilaniuk, O. M. P., V. K. Deshpande and E. C. G. Sudarshan (1962), " »Meta« relativity". *AJP*, 30: 718–723.
Birkhoff, G. D. (1923), *Relativity and Modern Physics*. Cambridge, Mass.: Harvard University Press.
Bondi, H. (1961), *Cosmology*. Second Edition. Cambridge: The University Press.
Bondi, H. (1964), *Relativity and Common Sense. A new approach to Einstein*. New York: Doubleday.
Bondi, H., F. A. E. Pirani and I. Robinson (1959), "Gravitational waves in general relativity. III. Exact plane waves", *RSLPr.* A 251: 519–33.
Bork, A. M. (1964), "The fourth dimension in 19th century physics". *Isis*, 55:326–38.
Born, M. (1909*a*), "Die träge Masse und das Relativitätsprinzip". *Ann. Physik* (4) 28: 571–84.
Born, M. (1909*b*), "Die Theorie des starren Elektrons in der Kinematik des Relativitätsprinzips". *Ann. Physik* (4) 30: 1–56, 840.
Born, M. (1909*c*), "Ueber die Dynamik des Elektrons in der Kinematik des Relativitätsprinzips". *Phys. Ztschr.* 10: 814–17.
Born, M. (1910), "Ueber die Definition des starren Körpers in der Kinematik der Relativitätstheorie". *Phys. Ztschr.* 11: 233–4.
Born, M. (1956), *Physics of my generation*. Oxford: Pergamon.
Boscovich, R. J. (1759), *Philosophiae naturalis theoria redacta ad unicam legem virium in natura existentium*. Vienna: A. Bernardi.
Bosshardt, B. (1976), "On the *b*-boundary of the closed Friedmann model". *Comm. Math. Phys.* 46: 263–8.
Bowman, P. A. (1977), "On conventionality and simultaneity—another reply". In Earman *et al.* (1977), pp. 433–47.
Brace, D. B. (1904), "On double refraction in matter moving through the aether". *Philos. Mag.* (6) 7: 317–29.
Bradley, J. (1728), "A Letter from the Reverend Mr. James Bradley Savilian Professor of Astronomy at Oxford and F.R.S. to Dr. Edmund Halley Astronom. Roy. etc. giving an Account of a new discovered Motion of the Fix'd Stars", *RSLPT*, 35: 637–61.
Braginsky, V. B. and V. I. Panov (1972), "Verification of the equivalence of inertial and gravitational mass". *Sov. Phys. JETP*, 34: 463–66. (Russian orig. in *Zh. Eksp. Teor. Fiz.* 61: 873–9; 1971.)
Brans, C. H. (1962), "Mach's Principle and the locally measured gravitational constant in general relativity". *Phys. Rev.* 125: 388–96.
Brans, C. H. and R. H. Dicke (1961), "Mach's Principle and a relativistic theory of gravitation". *Phys. Rev.* 124: 925–35.
Brecher, K. (1977), "Is the velocity of light independent of its source?" *Phys. Rev. Lett.* 39: 1051–4.
Bridgman, P. W. (1963), *A sophisticate's primer of Relativity*. London: Routledge.
Brill, D. R. and J. M. Cohen (1966), "Rotating masses and their effect on inertial frames". *Phys. Rev.* 143: 1011–15.
Brouwer, L. E. J. (1906), "The force field of the non-Euclidean spaces with negative curvature". *KNAWPr.* 9: 116–33.
Bruhat, Y. See Choquet-Bruhat, Y.
Brush, S. F. (1967), "Note on the history of the Fitzgerald–Lorentz contraction". *Isis*, 58: 230–2.
Buck, R. C. and R. S. Cohen, eds. (1971), *PSA 1970. In memory of Rudolf Carnap*. Dordrecht: Reidel.
Bunge, M. (1964), "Phenomenological theories". In Bunge, ed., *The Critical Approach to Science and Philosophy*. New York: The Free Press, pp. 234–54.
Bunge, M. (1966), "Mach's critique of Newtonian mechanics". *AJP*, 34: 585–96.
Bunge, M. (1967), *Foundations of Physics*. Berlin: Springer.
Bunge, M. (1968), "Physical time: the objective and relational theory". *PS*, 35: 355–88.
Bunge, M. and A. García Máynez (1976), "A relational theory of physical space". *Int. J. Theor. Phys* 15: 961–72.
Burke, W. L. (1979), "The slow-motion approximation in radiation problems". In Ehlers (1979), pp. 220–48.
Callaway, J. (1954), "Mach's Principle and the unified field theory". *Phys. Rev.* 96: 778–80.

Čapek, Milič, ed. (1976), *The Concepts of Space and Time. Their structure and their development.* Dordrecht: Reidel.

Carathéodory, C. (1924), "Zur Axiomatik der speziellen Relativitätstheorie". *Preuss. Ak. Wiss. Phys.-math. Kl. Sitzungsber.*, pp. 12–27.

Carmeli, M., S. I. Fickler and L. Witten, eds. (1970), *Relativity. Proceedings of the Relativity Conference in the Midwest, held at Cincinnati, Ohio, June 2–6, 1969.* New York: Plenum.

Carnap, R. (1922), *Der Raum. Ein Beitrag zur Wissenschaftslehre.* (Kantstudien Ergänzungsheft 56.) Berlin: Reuther & Reichard.

Carnap, R. (1924), "Dreidimensionalität des Raumes und Kausalität. Eine Untersuchung über den logischen Zusammenhang zweier Fiktionen". *Ann. Philos.* 4: 105–30.

Carnap, R. (1925), "Ueber die Abhängigkeit der Eigenschaften des Raums von denen der Zeit". *KS*, 30: 331–45.

Cartan, E. (1922), "Sur les équations de la gravitation d'Einstein". *J. Math. Pures Appl.* (9) 1: 141–203.

Cartan, E. (1923/24/25), "Sur les variétés à connexion affine et la théorie de la relativité généralisée". *Ann. Ec. Norm. Sup.* 40: 325–412; 41: 1–25: 42: 17–88.

Cartan, E. (1930), "Notice historique sur la notion de parallélisme absolu". *Math. Ann.* 102: 698–706.

Cartan, E./Einstein, A. (1979), *Letters on Absolute Parallelism 1929–1932.* Original text, English transl. by J. Leroy and J. Ritter. Ed. by R. Debever. Princeton: Princeton University Press.

Cartan, H. (1967), *Calcul différentiel.* Paris: Hermann.

Carter, B. (1968), "Global structure of the Kerr family of gravitational fields". *Phys. Rev.* 174: 1559–71.

Carter, B. (1970), "An axisymmetric black hole has only two degrees of freedom". *Phys. Rev. Lett.* 26: 331–3.

Carter, B. (1971), "Causal structure in space-time". *GRG*, 1: 349–91.

Carter, B. (1977), "The vacuum black hole uniqueness theorem". In Ruffini (1977), pp. 243–54.

Castagnino, M. A. (1971), "The Riemannian structure of space-time as a consequence of a measurement method". *J. Math. Phys.* 12: 2203–11.

Chevalley, C. (1946) *Theory of Lie Groups, I.* Princeton: Princeton University Press.

Chiu, Hong-Yee and W. F. Hoffmann, eds. (1964), *Gravitation and Relativity.* New York: Benjamin.

Choquet-Bruhat, Y. (1962), "The Cauchy problem". In Witten (1962), pp. 130–68.

Choquet-Bruhat, Y., C. DeWitt-Morette and M. Dillard-Bleick (*AMP*), *Analysis, Manifolds and Physics.* Amsterdam: North-Holland, 1977.

Choquet-Bruhat, Y. and J. W. York (1980), "The Cauchy problem". In Held (1980), vol. I, pp. 99–172.

Christoffel, E. B. (1869*a*), "Ueber die Transformation der homogenen Differential-ausdrücke zweiten Grades". *J. r. angew. Math.* 70: 46–70.

Christoffel, E. B. (1869*b*), "Ueber ein die Transformation homogenen Differential-ausdrücke zweiten Grades betreffendes Theorem". *J. r. angew. Math.* 70: 241–5.

Clarke, C. J. S. (1969), "On the global isometric embedding of pseudo-Riemannian manifolds". *RSLPr.* 314 A (1969) 417–28.

Clarke, C. J. S. (1971), "On the geodesic completeness of causal space-times", *Camb. Phil. Soc. Proc.* 69: 319–24.

Clarke, C. J. S. (1977), "Time in general relativity". In Earman *et al.* (1977), pp. 94–108.

Clarke, C. J. S. (1979), *Elementary General Relativity.* London: Arnold.

Clemence, G. M. (1947), "The relativity effect in planetary motions". *Rev. Mod. Phys.* 19: 361–64.

Coffa, J. A. (1980), "From geometry to tolerance. Sources of conventionalism in 19th century geometry". Preprint. (Forthcoming in vol. 7 of the Pittsburgh Series in the Philosophy of Science.)

Cohen, J. M. and D. R. Brill (1968), "Further examples of » Machian « effects on rotating bodies in general relativity". *Nuovo Cimento* (10) 56B: 209–19.

Collins, C. B. and S. W. Hawking (1973), "Why is the universe isotropic?" *Astroph. J.* 180: 317–34.

Corinaldesi, E. and A. Papapetrou (1951), "Spinning test particles in general relativity, II". *RSLPr.* A 209:29–68.
Cullwick, E. G. (1981), "Einstein and special relativity. Some inconsistencies in his electrodynamics". *BJPS*, 32: 167–76.
Cunningham, E. (1910), "The principle of relativity in electrodynamics and an extension thereof". *London Math. Soc. Proc.* (2) 8: 77–98.
D'Agostino, S. (1975), "Hertz's researches on electromagnetic waves". *HSPS*, 6: 261–323.
Dautcourt, G. (1964), "Die Newtonische Gravitationstheorie als strenger Grenzfall der allgemeinen Relativitätstheorie". *Acta Phys. Polon.* 25: 637–47.
Davies, P. C. W. (1980), *The Search for Gravity Waves.* Cambridge: The University Press.
Descartes, R. (*AT*), *Oeuvres, publiées par Charles Adam et Paul Tannéry sous les auspices du Ministère de l'Instruction Publique.* Paris: Cerf, 1897–1910. 12 vols.
Descartes, R. (1637), *Discours de la methode pour bien conduire sa raison, et chercher la verité dans les sciences. Plus la dioptrique. Les meteores. Et la geometrie. Qui sont des essais de cete methode.* Leyde: I. Maire. (Published anonymously.)
DeWitt, C. and B. DeWitt, eds. (1964), *Relativity, Groups and Topology.* New York: Gordon & Breach.
DeWitt, C. and J. A. Wheeler, eds. (1968), *Battelle Rencontres. 1967 Lectures in Mathematics and Physics.* New York: Benjamin.
Dicke, R. H. (1959), "New research on old gravitation". *Science*, 129: 621–4.
Dicke, R. H. (1962), "Mach's Principle and equivalence". In Møller, *EGT*, pp. 1–50.
Dicke, R. H. (1964*a*), "Remarks on the observational basis of general relativity". In Chiu and Hoffmann (1964), pp. 1–16.
Dicke, R. H. (1964*b*), "The many faces of Mach". In Chiu and Hoffmann (1964), pp. 121–41.
Dicke, R. H. (1964*c*), "The significance for the solar system of time-varying gravitation". In Chiu and Hoffmann (1964), pp. 142–74.
Dicke, R. H. (1964*d*), "Experimental relativity". In DeWitt and DeWitt (1964), pp. 165–313.
Dicke, R. H. and H. M. Goldenberg (1974), "The oblateness of the sun". *Astroph. J. Suppl.* 27: 131–82.
Dingler, H. (1921), *Physik und Hypothese. Versuch einer induktiven Wissenschaftslehre nebst einer kritischen Analyse der Fundamente der Relativitätstheorie.* Berlin u. Leipzig: Vereinigung Wissenschaftlicher Verleger.
Dirac, P. A. M. (1938), "Classical theory of radiating electrons". *RSLPr.* A 167: 148–69.
Dixon, W. G. (1970*a*), "Dynamics of extended bodies in general relativity. I. Momentum and angular momentum". *RSLPr.* A 314: 499–527.
Dixon, W. G. (1970*b*), "Dynamics of extended bodies in general relativity. II. Moments of the charge-current vector". *RSLPr.* A 319: 509–47.
Dixon, W. G. (1974/75); "Dynamics of extended bodies in general relativity. III. Equations of motion". *RSLPT*, A 277: 59–119.
Dixon, W. G. (1978), *Special Relativity. The Foundations of Macroscopic Physics.* Cambridge: The University Press.
Dixon, W. G. (1979), "Extended bodies in general relativity". In Ehlers (1979), pp. 156–219.
Dodson, C. T. J. and T. Poston (1977), *Tensor Geometry. The geometric viewpoint and its uses.* London: Pitman.
Dorling, J. (1976), "Special relativity out of Euclidean geometry". Preprint, 22 May 1976.
Dorling, J. (1977), "The eliminability of masses and forces in Newtonian particle mechanics: Suppes reconsidered". *BJPS*, 28: 55–57.
Dorling, J. (1978), "Did Einstein need general relativity to solve the problem of absolute space? Or had the problem already been solved by special relativity?" *BJPS*, 29: 311–23.
Douglass, D. H. and V. B. Braginsky (1979), "Gravitational-radiation experiments". In Hawking and Israel (1979), pp. 90–137. (References on pp. 846–50.)
Droste, J. (1916), "The field of a single centre in Einstein's theory of gravitation, and the motion of a particle in that field". *KNAWPr.* 19: 197–215.
Dutta, M., T. K. Mukherjee and M. K. Sen (1970), "On linearity in the special theory of relativity". *Int. J. Theor. Phys.* 3: 85–91.
Dyson, F. W., A. S. Eddington and C. Davidson (1920), "A determination of the deflection of

light by the Sun's gravitational field, from observations made at the total eclipse of May 29, 1919". *RSLPT*, A 220: 291–333.

Eardley, D. M. and L. Smarr (1979), "Time functions in numerical relativity: marginally bound dust collapse". *Phys. Rev.* D 19: 2239–59.

Earman, J. (1970), "Are spatial and temporal congruence conventional?" *GRG*, 1: 143–57.

Earman, J. (1972*a*), "Notes on the causal theory of time". *Synthèse*, 24: 74–86.

Earman, J. (1972*b*), "Implications of causal propagation outside the null cone". *Austral. J. Philos.* 50: 222–37.

Earman, J. (1974), "Covariance, invariance and the equivalence of frames". *FP*, 4: 267–89.

Earman, J. (1977), "How to talk about the topology of time". *Noús*, 11: 211–26.

Earman, J. and M. Friedman (1973), "The meaning and status of Newton's law of inertia and the nature of gravitational forces". *PS*, 40: 329–59.

Earman, J. and C. Glymour (1978*a*), "Lost in the tensors: Einstein's struggles with covariance principles 1912–1916". *SHPS*, 9: 251–78.

Earman, J. and C. Glymour (1978*b*), "Einstein and Hilbert: Two months in the history of general relativity". *Arch. Hist. Exact Sci.* 19: 291–307.

Earman, J. and C. Glymour (1980*a*), "Relativity and eclipses: The British eclipse expeditions of 1919 and their predecessors". *HSPS*, 11: 49–85.

Earman, J. and C. Glymour (1980*b*), "The gravitational red shift as a test of general relativity: History and analysis". *SHPS*, 11: 175–214.

Earman, J. and C. Glymour, "Abraham and Einstein: Two theories of gravitation". Preprint, undated.

Earman, J., C. Glymour and J. Stachel, eds. (1977), *Foundations of Space-Time Theories.* (Minnesota Studies in the Philosophy of Science, 8.) Minneapolis: University of Minnesota Press.

Eddington, A. S. (1920), *Space Time and Gravitation. An outline of General Relativity Theory.* Cambridge: The University Press.

Eddington, A. S. (1921), "A generalization of Weyl's theory of the electromagnetic and gravitational fields". *RSLPr*, A 99: 104–22.

Eddington, A. S. (1923), *The Mathematical Theory of Relativity.* Cambridge: The University Press.

Eddington, A. S. (1924), *The Mathematical Theory of Relativity.* Second Edition. Cambridge: The University Press.

Eddington, A. S. (1930), "On the instability of Einstein's spherical world". *MNRAS*, 90: 668–78.

Eddington, A. S. (1933), *The Expanding Universe.* Cambridge: The University Press.

Edelen, D. (1962), *The Structure of Field Space. An axiomatic formulation of field physics.* Berkeley: University of California Press.

Edwards, W. F. (1963), "Special relativity in anisotropic space". *AJP*, 31: 482–9.

Ehlers, J. (1973), "Survey of general relativity theory". In Israel (1973), pp. 1–125.

Ehlers, J. ed. (1979), *Isolated Gravitating Systems in General Relativity.* (Intern. Sch. of Physics » Enrico Fermi « , Course LXVII.) Amsterdam: North-Holland.

Ehlers, J. (1980*a*), "Isolated systems in general relativity". *Ann. N. Y. Acad. Sci.* 336: 279–94.

Ehlers, J. (1980*b*), "Problems of relativistic celestial mechanics". Preprint. (Lecture at Acc. dei Lincei, Rome.)

Ehlers, J., F. A. E. Pirani and A. Schild (1972), "The geometry of free fall and light propagation". In O'Raifeartaigh (1972), pp. 63–84.

Ehlers, J. and E. Rudolph (1977), "Dynamics of extended bodies in general relativity. Center-of-mass description and quasirigidity". *GRG*, 8: 197–217.

Ehlers, J. and A. Schild (1973), "Geometry in a manifold with projective structure". *Comm. Math. Phys.* 32: 119–46.

Ehrenfest, P. (1909), "Gleichförmige Rotation starrer Körper und Relativitätstheorie". *Phys. Ztschr.* 10: 918.

Ehresmann, Ch. (1957), "Les connexions infinitésimales dans un espace fibré différentiable". In *Colloque de Topologie (Espaces Fibrés), Bruxelles 1950.* Paris: Masson, pp. 29–55.

Eichenwald, A. (1903), "Ueber die magnetischen Wirkungen bewegter Körper im elektrostatischen Felde". *Ann. Physik* (4) 11: 1–30; 421–41.

Einstein, A. (1905*a*), *Eine neue Bestimmung der Moleküldimensionen*. Bern: Wyss. (Reprinted with a 'Nachtrag' in *Ann. Physik* (4) 19: 289–306; 1906.)
Einstein, A. (1905*b*), "Ueber einen die Erzeugung und Verwandlung des Lichtes betreffenden heuristischen Gesichtspunkt". *Ann. Physik* (4) 17: 132–48.
Einstein, A. (1905*c*), "Die von der molekulärkinetischen Theorie der Wärme geforderte Bewegung von in ruhenden Flüssigkeiten suspendierten Teilchen". *Ann. Physik* (4) 17: 549–60.
Einstein, A. (1905*d*), "Zur Elektrodynamik bewegter Körper". *Ann. Physik* (4) 17: 891–921.
Einstein, A. (1905*e*), "Ist die Trägheit eines Körpers von seinem Energiegehalt abhängig?". *Ann. Physik* (4) 18: 639–41.
Einstein, A. (1906), "Das Prinzip von der Erhaltung der Schwerpunktsbewegung und die Trägheit der Energie". *Ann. Physik* (4) 20: 627–33.
Einstein, A. (1907*a*), "Die vom Relativitätsprinzip geforderte Trägheit der Energie". *Ann. Physik* (4) 23: 371–84.
Einstein, A. (1907*b*), "Ueber das Relativitätsprinzip und die aus demselben gezogenen Folgerungen". *Jahrb. Radioakt. Elektr.* 4: 411–462. ("Berichtigungen", *ibid.* 5: 98–9; 1908.)
Einstein, A. (1911*a*), "Ueber den Einfluss der Schwerkraft auf die Ausbreitung des Lichtes". *Ann. Physik* (4) 35: 808–909.
Einstein, A. (1911*b*), "Relativitätstheorie". *Vierteljahrschrift Naturforschende Gesellschaft Zürich*, 56: 1–14.
Einstein, A. (1912*a*), "Lichtgeschwindigkeit und Statik des Gravitationsfeldes". *Ann. Physik* (4) 38: 355–69.
Einstein, A. (1912*b*), "Zur Theorie des statischen Gravitationsfeldes", *Ann. Physik* (4) 38: 443–58.
Einstein, A. (1912*c*), "Gibt es eine Gravitationswirkung, die der elektrodynamischen Induktionswirkung analog ist?" *Vierteljahrschrift gerichtl. Medizin.* 44: 37–40.
Einstein, A. (1912*d*), "Relativität und Gravitation. Erwiderung auf eine Bemerkung von M. Abraham". *Ann. Physik* (4) 38: 1059–64.
Einstein, A. (1912*e*), "Bemerkung zu Abrahams vorangehender Auseinadersetzung 'Nochmals Relativität und Gravitation'". *Ann. Physik* (4) 39: 704.
Einstein, A. (1913*a*), "Zum gegenwärtigen Stand des Gravitationsproblems". *Phys. Ztschr.* 14: 1249–66.
Einstein, A. (1913*b*), "Physikalische Grundlagen einer Gravitationstheorie". *Vierteljahrschrift Naturforschende Gesellschaft Zürich*, 58: 284–90.
Einstein, A. (1914*a*), "Principielles zur verallgemeinerten Relativitätstheorie und Gravitationstheorie". *Phys. Ztschr.* 15: 176–80.
Einstein, A. (1914*b*), "Der formale Grundlage der allgemeinen Relativitätstheorie". *Preuss. Ak. Wiss. Sitzungsber.*, pp. 1030–85.
Einstein, A. (1915*a*), "Zur allgemeinen Relativitätstheorie". *Preuss. Ak. Wiss. Sitzungsber.*, pp. 778–86.
Einstein, A. (1915*b*),"Zur allgemeinen Relativitätstheorie (Nachtrag)". *Preuss. Ak. Wiss. Sitzungsber.*, pp. 799–801.
Einstein, A. (1915*c*), "Erklärung der Perihelbewegung des Merkur aus der allgemeinen Relativitätstheorie". *Preuss. Ak. Wiss. Sitzungsber.*, pp. 831–9.
Einstein, A. (1915*d*), "Die Feldgleichungen der Gravitation". *Preuss. Ak. Wiss. Sitzungsber.*, pp. 844–7.
Einstein, A. (1916*a*), "Die Grundlagen der allgemeinen Relativitätstheorie". *Ann. Physik* (4) 49: 769–822.
Einstein, A. (1916*b*), "Ernst Mach". *Phys. Ztschr.* 17: 101–4.
Einstein, A. (1916*c*), "Eine neue formale Deutung der Maxwellschen Feldgleichungen der Elektrodynamik". *Preuss. Ak. Wiss. Sitzungsber.*, pp. 184–8.
Einstein, A. (1916*d*), "Näherungsweise Integration der Feldgleichungen der Gravitation". *Preuss. Ak. Wiss. Sitzungsber.*, pp. 688–96.
Einstein, A. (1916*e*), "Ueber Friedrich Kottlers Abhandlung 'Ueber Einsteins Aequivalenzhypothese und die Gravitation'." *Ann. Physik* (4) 51: 639–42.
Einstein, A. (1916*f*), "Hamiltonsches Prinzip und allgemeine Relativitätstheorie", *Preuss. Ak. Wiss. Sitzungsber.*, pp. 1111–16.

Einstein, A. (1917), "Kosmologische Betrachtungen zur allgemeinen Relativitätstheorie". *Preuss. Ak. Wiss. Sitzungsber.*, pp. 142–52.
Einstein, A. (1918*a*), "Ueber Gravitationswellen". *Preuss. Ak. Wiss. Sitzungsber.*, pp. 154–67.
Einstein, A. (1918*b*), "Prinzipielles zur allgemeinen Relativitätstheorie", *Ann. Physik* (4) 55: 241–4.
Einstein, A. (1918*c*), "Kritisches zu einer von Hrn. De Sitter gegebenen Lösung der Gravitationsgleichungen". *Preuss. Ak. Wiss. Sitzungsber.*, pp. 270–2.
Einstein, A. (1918*d*), "Der Energiesatz in der allgemeinen Relativitätstheorie". *Preuss. Ak. Wiss. Sitzungsber.*, pp. 448–59.
Einstein, A. (1918*e*), "Bemerkung zu Herrn Schrödingers Notiz 'Ueber ein Lösungssystem der allgemein kovarianten Gravitationsgleichungen'." *Phys. Ztschr.* 19: 165–6.
Einstein, A. (1918*f*), "Dialog über Einwände gegen die Relativitätstheorie". *Naturwiss.* 6: 697–702.
Einstein, A. (1920), *Aether und Relativitätstheorie.* Berlin: Springer.
Einstein, A. (1921*a*), "Geometrie und Erfahrung". *Preuss. Ak. Wiss. Sitzungsber.*, pp. 123–30.
Einstein, A. (1921*a**), *Geometrie und Erfahrung. Erweiterte Fassung eines Festvortrages.* Berlin: Springer. (Expanded version of (1921*a*).)
Einstein, A. (1921*b*), "Ueber eine naheliegende Ergänzung des Fundamentes der allgemeinen Relativitätstheorie". *Preuss. Ak. Wiss. Sitzungsber.*, pp. 261–4.
Einstein, A. (1922*a*), *Sidelights on Relativity*, transl. by G. B. Jeffery and W. Perrett. London: Methuen. (Translation of (1920) and (1921*a**).)
Einstein, A. (1922*b*), "Bemerkung zu der Arbeit von A. Friedmann, 'Ueber die Krümmung des Raumes'." *Ztschr. f. Physik*, 11: 326.
Einstein,A. (1923*a*), "Zur allgemeinen Relativitätstheorie". *Preuss. Ak. Wiss. Phys.-math. Kl. Sitzungsber.*, pp. 32–8.
Einstein, A. (1923*b*), "Bemerkung zu meiner Arbeit 'Zur allgemeinen Relativitätstheorie'." *Preuss. Ak. Wiss. Phys.-math. Kl. Sitzungsber.*, pp. 76–7.
Einstein, A. (1923*c*), "Zur affinen Feldtheorie". *Preuss. Ak. Wiss. Phys.-math. Kl. Sitzungsber.*, pp. 137–40.
Einstein, A. (1923*d*), "Notiz zu der Arbeit von A. Friedmann, 'Ueber die Krümmung des Raumes'." *Ztschr. f. Physik*, 16: 228.
Einstein, A. (1924), "Ueber den Aether". *Schweizerische Naturforsch. Gesellsch. Verhandl.* 105.2: 85–93.
Einstein, A. (1927), "Allgemeine Relativitätstheorie und Bewegungsgesetz". *Preuss. Ak. Wiss. Phys.-math. Kl. Sitzungsber.*, pp. 235–45.
Einstein, A. (1928*a*), "Riemann-Geometrie mit Aufrechterhaltung des Begriffes des Fernparallelismus". *Preuss. Ak. Wiss. Phys.-math. Kl. Sitzungsber.*, pp. 217–21.
Einstein, A. (1928*b*), "Neue Möglichkeit fur eine einheitliche Feldtheorie von Gravitation und Elektrizität". *Preuss. Ak. Wiss. Phys.-math. Kl. Sitzungsber.*, pp. 224–7.
Einstein, A. (1928*c*), "À propos de 'La Déduction Relativiste' de M. Émile Meyerson". *Rev. Philos.* 105: 161–6.
Einstein, A. (1928*d*), Review of H. Reichenbach, *Philosophie der Raum-Zeit-Lehre. Deutsche Literaturzeitung*, Heft 1, Sp. 19–20.
Einstein, A. (1929*a*), "Einheitliche Feldtheorie". *Preuss. Ak. Wiss. Phys.-math. Kl. Sitzungsber.*, pp. 2–7.
Einstein, A. (1929*b*), "Einheitliche Feldtheorie und Hamiltonsches Prinzip". *Preuss. Ak. Wiss. Phys.-math. Kl. Sitzungsber.*, pp. 156–9.
Einstein, A. (1930*a*), "Auf die Riemann-Metrik und den Fern-Parallelismus gegründete einheitliche Feldtheorie". *Math. Ann.* 102: 685–97.
Einstein, A. (1930*b*), "Théorie unitaire du champ physique". *Ann. Inst. Henri Poincaré*, 1: 1–24.
Einstein, A. (1931), "Zum kosmologischen Problem der allgemeinen Relativitätstheorie". *Preuss. Ak. Wiss. Phys.-math. Kl. Sitzungsber.*, pp. 235–7.
Einstein, A. (1934), *Mein Weltbild.* Amsterdam: Querido Verlag.
Einstein, A. (1935), "Elementary derivation of the equivalence of mass and energy". *Am. Math. Soc. Bull.* 41: 223–30.
Einstein, A. (1940), "Considerations concerning the Fundaments of Theoretical Physics". *Science*, 91: 487–92.

Einstein, A. (1949), "Autobiographisches". In Schilpp (1949), pp. 2–94.
Einstein, A. (1954), *Relativity. The Special and the General Theory. A popular exposition.* 15th ed. Authorised transl. by R. W. Lawson. London: Methuen. (The first German edition appeared in 1916.)
Einstein, A. (1956), *The Meaning of Relativity. (The Stafford Little Lectures of Princeton University, May 1921.)* Fifth edition, including the relativistic theory of the non-symmetric field. Princeton: Princeton University Press.
Einstein, A. (1982), "How I created the theory of relativity". *Physics Today*, 35/8: 45–47. (Kyoto lecture of 14 December 1922.)
Einstein, A./Besso, M. (1972), *Correspondance 1903–1955.* Traduction, notes et introduction de P. Speziali. Paris: Hermann. (Note that the 1979 edition, which bears the same title page, does not include the original German text of the letters.)
Einstein, A./Born, H. and M. (1969), *Briefwechsel 1916–1955.* München: Nymphenburger Verlagshandlung.
Einstein, A. and A. D. Fokker (1914), "Die Nordströmsche Gravitationstheorie vom Standpunkt des absoluten Differentialkalküls". *Ann. Physik* (4) 44: 321–8.
Einstein, A. and J. Grommer (1927), "Allgemeine Relativitätstheorie und Bewegungsgesetz". *Preuss. Ak. Wiss. Phys.-math. Kl. Sitzungsber.*, pp. 2–13.
Einstein, A. and M. Grossmann (1913), *Entwurf einer verallgemeinerten Relativitätstheorie und einer Theorie der Gravitation.* Leipzig: Teubner. (Though described as a "Separatabdruck" of the next item, it appeared several months earlier and does not include the "Bemerkungen" at the end.)
Einstein, A. and M. Grossmann (1913*), "Entwurf einer verallgemeinerten Relativitätstheorie und einer Theorie der Gravitation". *Ztschr. Math. u. Phys.* 62: 225–61.
Einstein, A. and M. Grossmann (1914), "Kovarianzeigenschaften der Feldgleichungen der auf die verallgemeinerte Relativitätstheorie gegründeten Gravitationstheorie". *Ztschr. Math. u. Phys.* 63: 215–25.
Einstein, A. and L. Infeld (1940), "The gravitational equations and the problem of motion, II", *Ann. Maths.* 41: 455–64.
Einstein, A. and L. Infeld (1949), "On the motion of particles in general relativity theory". *Can. J. Math.* 1: 209–41.
Einstein, A., L. Infeld and B. Hoffmann (1938), "The gravitational equations and the problem of motion". *Ann. Maths.* 39: 65–100.
Einstein, A. and J. Laub (1908*a*), "Ueber die elektromagnetischen Grundgleichungen für bewegte Körper". *Ann. Physik* (4) 26: 532–40.
Einstein, A. and J. Laub (1908*b*), "Ueber die im elektromagnetischen Felde auf ruhende Körper ausgeübten ponderomotorischen Kräfte". *Ann. Physik* (4) 26: 541–50.
Einstein, A. and N. Rosen (1935), "The particle problem in general relativity theory". *Phys. Rev.* 48: 73–7.
Einstein, A. and W. de Sitter (1932), "On the relation between the expansion and the mean density of the universe". *Nat. Ac. Sci. (U.S.) Proc.* 18: 213–14.
Einstein, A. and A. Sommerfeld (1968), *Briefwechsel. Sechzig Briefe aus dem goldenen Zeitalter der modernen Physik.* Hrsg. u. komm. v. A. Hermann. Basel: Schwabe.
Eisenhart, L. P. (1950), *Riemannian Geometry.* Princeton: Princeton University Press.
Ellis, B. (1963), "Universal and differential forces". *BJPS*, 14: 177–94.
Ellis, B. (1971), "On conventionality and simultaneity—a reply". *Austral. J. Philos.* 49: 177–203.
Ellis, B. and P. Bowman (1967), "Conventionality in distant simultaneity". *PS*, 34: 116–36.
Ellis, G. F. R. and A. R. King (1974), "Was the big bang a whimper?" *Comm. Math. Phys.* 38: 119–56.
Ellis, G. F. R. and B. G. Schmidt (1977), "Singular spacetimes". *GRG*, 8: 915–54.
Eötvös, R. (*GA*), *Gesammelte Arbeiten.* Budapest: Akadémiai Kiadó, 1953.
Eötvös, R. (1889), "Ueber die Anziehung der Erde auf verschiedene Substanzen". In Eötvös, *GA*, pp. 17–20. (Read at academy meeting of 20.1.1889; pub. in *Math. u. Naturwiss. Ber. Ung.* 8: 65–8.)
Eötvös, R., D. Pekár and E. Fekete (1922), "Beiträge zum Gesetze der Proportionalität von Trägheit und Gravität". *Ann. Physik* (4) 68: 11–66. (Results submitted in 1909. Also in Eötvös, *GA*, pp. 309–72.)

Epstein, P. S. (1911), "Ueber relativische Statik". *Ann. Physik* (4) 36: 779–95.

Esposito, F. P. and L. Witten, eds. (1977), *Asymptotic Structure of Space-Time*. New York: Plenum.

Faraday, M. (*ERE*), *Experimental Researches on Electricity*. London: R. & J. E. Taylor, 1839–1855. 3 vols.

Feenberg, E. (1974), "Conventionality in distant simultaneity". *FP*, 4: 121–6.

Feinberg, G. (1967), "Possibility of faster-than-light particles". *Phys. Rev.* 159: 1089–1105.

Fermi, E. (1922), "Sopra i fenomeni che avvengono in vicinanza di una linea oraria". *R. Acc. Lincei, Rend. Cl. Sci. Fis. Mat. Nat.* 311: 184–7, 306–9.

Field, H. (1980), *Science without numbers. A defence of nominalism*. Princeton: Princeton University Press.

Fine, A. (1971), "Reflections on a relational theory of space". *Synthèse*, 22: 448–81.

Fischer, A. E. and J. E. Marsden (1979), "The initial value problem and the dynamical formulation of general relativity". In Hawking and Israel (1979), pp. 138–211. (References on pp. 850–9.)

Fitzgerald, G. F. (1889), "The ether and the earth's atmosphere". *Science*, 13: 390.

Fitzgerald, P. (1970), "Tachyons, backwards causation and freedom". In Buck and Cohen (1971), pp. 415–36.

Fizeau, A. H. L. (1851), "Sur les hypothèses relatives à l'éther lumineux, et sur une expérience qui paraît démontrer que le mouvement des corps change la vitesse avec laquelle la lumière se propage dans leur intérieur". *CR*, 33: 349–55.

Flügge, S., ed. (1962), *Encyclopedia of Physics (Handbuch der Physik). Vol. IV. Principles of Electrodynamics and Relativity*. Berlin: Springer.

Fock, V. (1939), "Sur le mouvement des masses finies d'après la théorie de gravitation einsteinienne". *Acad. Sci. URSS J. Phys.* 1: 81–116. (Russian orign. in *Zh. Eksp. Teor. Fiz.* 9: 375–410.)

Fock, V. (1957), "Three lectures on relativity theory". *Rev. Mod. Phys.* 29: 325–33.

Fock, V. (1959), *The Theory of Space, Time and Gravitation*. Transl. from the Russian by N. Kemmer. London: Pergamon.

Fokker, A. D. (1965), *Time and Space, Weight and Inertia. A chronogeometrical introduction to Einstein's theory*. Translation by D. Bijl, edited by D. Field. Oxford: Pergamon.

Fomalont, E. B. and R. A. Sramek (1976), "Measurements of the solar gravitational deflection of radio waves in agreement with general relativity". *Phys. Rev. Lett.* 36: 1475–8.

Fox, J. G. (1962), "Experimental evidence for the second postulate of special relativity". *AJP*, 30: 297–300.

Fox, J. G. (1965), "Evidence against emission theories". *AJP*, 33: 1–17.

Fox, R., C. G. Kuper and S. G. Lipson (1970), "Faster-than-light group velocities and causality violation". *RSLPr.* A 316: 515–24.

Francis, R. (1981), "Convention-free simultaneity in the causal theory of time". Preprint, January 1981.

Frank, P. (1909), "Die Stellung des Relativitätsprinzips im System der Mechanik und der Elektrodynamik". *AWWS*, 118: 373–446.

Frank, P. and H. Rothe (1911), "Ueber die Transformation der Raumzeitkoordinaten von ruhenden auf bewegte Systeme". *Ann. Physik* (4) 34: 825–53.

Frank, P. and H. Rothe (1912), "Zur Herleitung der Lorentz Transformation". *Phys. Ztschr.* 13: 750–3.

Frege, G. (1891), "Ueber das Trägheitsgesetz". *Ztschr. f. Philos. u. phil. Kritik*, 98: 145–61.

French, A. P. (1968), *Special Relativity*. New York: Norton.

Fresnel, A. J. (1818), "Lettre d'Augustin Fresnel à François Arago sur l'influence du mouvement terrestre dans quelques phénomènes d'optique". *Ann. Chim. Phys.* 9: 57 ff. (Also in Fresnel, *Oeuvres complètes*, Paris: Imprimerie Impériale, 1868, vol. II, pp. 627–36.)

Freudenthal, H. (1968), "L'algèbre topologique, en particulier les groupes topologiques et de Lie". *Revue de Synthèse* (3) 49/52: 223–43.

Friedman, A. (1961), "Local isometric imbedding of Riemannian manifolds with indefinite metrics". *J. Math. & Mech.* 10: 625–49.

Friedman, M. (1972), "Grünbaum on the conventionality of geometry". *Synthèse*, 24: 219–35.

Friedman, M. (1973), "Relativity principles, absolute objects and symmetry groups". In Suppes (1973), pp. 296–320.

Friedman, M. (1977), "Simultaneity in Newtonian mechanics and special relativity". In Earman *et al.* (1977), pp. 403–32.

Friedmann, A. (1922), "Ueber die Krümmung des Raumes", *Ztschr. f. Physik*, 10: 377–386.

Friedmann, A. (1924), "Ueber die Möglichkeit einer Welt mit konstanter negativer Krümmung des Raumes". *Ztschr. f. Physik*, 21: 326–32.

Friedrichs, K. (1927), "Eine invariante Formulierung des Newtonschen Gravitationsgesetzes und des Grenzüberganges vom Einsteinschen zum Newtonschen Gesetz". *Math. Ann.* 98: 566–75.

Fubini, G. (1904), "Sugli spazii a quattro dimensioni che ammettono un gruppo continuo di movimenti". *Annali Mat. Pura e Appl.* (3) 9: 33–90.

Fung, Shing-Fai and K. C. Hsieh (1980), "Is the isotropy of the speed of light a convention?" *AJP*, 48: 654–57.

Galileo Galilei (*EN*), *Le opere*. Nuova ristampa della Edizione Nazionale. Firenze: G. Barbera, 1964–1966. 20 vols.

Galileo Galilei (1967), *Dialogue concerning the Two Chief World Systems—Ptolemaic and Copernican.* Transl. by S. Drake. Berkeley: University of California Press.

Galison, P. L. (1979), "Minkowski's space-time. From visual thinking to the absolute world". *HSPS*, 10: 85–121.

Gauss, C. F. (1827), "Disquisitiones generales circa superficies curvas". *Comm. soc. reg. sci. Gott. cl. math.* 6: 99–146.

Geroch, R. P. (1966), "Singularities in closed universes". *Phys. Rev. Lett.* 17: 445–7.

Geroch, R. P. (1967), "Topology in general relativity". *J. Math. Phys.* 8: 782–6.

Geroch, R. P. (1968*a*), "What is a singularity in general relativity?" *Ann. Physics*, 48: 526–40.

Geroch, R. P. (1968*b*), "Local characterization of singularities in general relativity". *J. Math. Phys.* 9: 450–65.

Geroch, R. P. (1970), "Domain of dependence". *J. Math. Phys.* 11: 437–49.

Geroch, R. P. (1971*a*), "Space-time structure from a global viewpoint". In R. K. Sachs (1971), pp. 71–103.

Geroch, R. P. (1971*b*), "General relativity in the large". *GRG*, 2: 61–74.

Geroch, R. P. (1977), "Asymptotic structure of space-time". In Esposito and Witten (1977), pp. 1–106.

Geroch, R. P. and G. T. Horowitz (1979), "Global structure of spacetimes". In Hawking and Israel (1979), pp. 212–93.

Geroch, R., E. H. Kronheimer and R. Penrose (1972), "Ideal points in space-time". *RSLPr.* A 327: 545–67.

Giannoni, C. B. (1971), "Einstein and the Lorentz-Poincaré theory of relativity". In Buck and Cohen (1971), pp. 575–89.

Giannoni, C. B. (1973), "Special relativity in accelerated systems". *PS*, 40: 382–92.

Giannoni, C. B. (1978*a*), "Relativistic mechanics and electrodynamics without one-way velocity assumptions". *PS*, 45: 17–46.

Giannoni, C. B. (1978*b*), "A universal axiomatization of kinematical theories". In P. D. Asquith and I. Hacking, *PSA 1978*, East Lansing: Philosophy of Science Association, pp. 60–70.

Giannoni, C. B. (1979), "Comment on 'Relative simultaneity in the special theory of relativity' ". *PS*, 46: 306–9.

Giedymin, J. (1982), *Science and Convention.* Oxford: Pergamon.

Gillespie, C. C., ed. (*DSB*), *Dictionary of Scientific Biography*, New York: Scribner's, 1970 ff.

Glymour, C. (1972*a*), "Physics by convention". *PS*, 39: 322–40.

Glymour, C. (1972*b*), "Topology, cosmology and convention". *Synthèse*, 24: 195–218.

Glymour, C. (1977), "Indistinguishable space-times and the fundamental group". In Earman *et al.* (1977), pp. 50–9.

Glymour, C. (1980), *Theory and evidence.* Princeton: Princeton University Press.

Gödel, K. (1949*a*), "An example of a new type of cosmological solutions of Einstein's field equations of gravitation". *Rev. Mod. Phys.* 21: 447–50.

Gödel, K. (1949*b*), "A remark about the relationship between relativity theory and idealistic philosophy". In Schilpp (1949), pp. 557–62.

Gödel, K. (1952), "Rotating universes in general relativity theory". *Proc. Internat. Congr. of Mathematicians, Cambridge, Mass. 1950*, Providence: American Mathematical Society, vol. I, pp. 175–81.

Goenner, H. F. (1970), "Mach's Principle and Einstein's theory of gravitation". In R. S. Cohen and R. J. Seeger, eds., *Ernst Mach: Physicist and Philosopher*, Dordrecht: Reidel, pp. 200–15.

Goenner, H. F. (1980*a*), "Local isometric embedding of Riemannian manifolds and Einstein's theory of gravitation". In Held (1980), vol. 1, pp. 441–68.

Goenner, H. F. (1980*b*), "Machsches Prinzip und Theorien der Gravitation". Preprint, March 1980.

Goldberg, S. (1967), "Henri Poincaré and Einstein's theory of relativity". *AJP*, 35: 934–44.

Goldberg, S. (1969), "The Lorentz theory of the electron and Einstein's theory of relativity". *AJP*, 37: 982–94.

Goldberg, S. (1970), "Poincaré's silence and Einstein's relativity. The role of theory and experiment in Poincaré's physics". *BJHS*, 5: 73–84.

Gooding, D. (1981), "Final steps to the field theory: Faraday's study of magnetic phenomena, 1845–1850", *HSPS*, 11: 231–75.

Green, G. (1839), "On the laws of the reflexion and refraction of light at the common surface of two non-crystallized media". *Camb. Philos. Soc. Trans.* 7: 1–24; 113–20.

Greene, R. E. (1970), *Isometric embeddings of Riemannian and pseudo-Riemannian manifolds.* (Memoirs of the American Mathematical Society, No. 97.) Providence: American Mathematical Society.

Grishchuk, L. P. and A. G. Polnarev (1980), "Gravitational waves and their interaction with matter and fields". In Held (1980), vol. 2, pp. 393–434.

Grøn, Ø. (1975), "Relativistic description of a rotating disk". *AJP*, 43: 869–76.

Grossmann, M. (1913), "Mathematische Bemerkungen zur Gravitationstheorie". *Vierteljahrschrift Naturforschende Gesellschaft Zürich*, 58: 291–7.

Groth, E. J., J. E. Peebles, M. Seldner and R. M. Soneira (1977), "The clustering of galaxies". *Sci. Amer.* 237, 5: 76–98.

Grünbaum, A. (1950), "Relativity and the atomicity of becoming". *Rev. of Metaph.* 4: 143–86.

Grünbaum, A. (1952), "A consistent conception of the extended linear continuum as an aggregate of unextended elements". *PS*, 19: 288–306.

Grünbaum, A. (1953), "Whitehead's method of extensive abstraction". *BJPS*, 4: 215–26.

Grünbaum, A. (1962), "Geometry, chronometry and empiricism". In H. Feigl and G. Maxwell, eds., *Scientific Explanation, Space and Time* (Minnesota Studies in the Philosophy of Science, 3.) Minneapolis: University of Minnesota Press, pp. 405–526.

Grünbaum, A. (1963), *Philosophical Problems of Space and Time.* New York: Knopf.

Grünbaum, A. (1967), *Modern Science and Zeno's Paradoxes.* Middletown, Conn.: Wesleyan University Press.

Grünbaum, A. (1968), *Geometry and Chronometry in Philosophical Perspective.* Minneapolis: University of Minnesota Press.

Grünbaum, A. (1969), "Simultaneity by slow clock transport in the special theory of relativity". *PS*, 36: 5–43. (Reprinted in Grünbaum (1973), pp. 670–708.)

Grünbaum, A. (1970), "Space, time and falsifiability, Part I". *PS*, 37: 469–588. (Reprinted in Grünbaum (1973). pp. 449–568.)

Grünbaum, A. (1973), *Philosophical Problems of Space and Time.* Second, enlarged edition. Dordrecht: Reidel.

Grünbaum, A. (1977), "Absolute and relational theories of space and space-time". In Earman *et al.* (1977), pp. 303–74.

Grünbaum, A. and A. I. Janis (1977), "The geometry of the rotating disk in the special theory of relativity". *Synthèse*, 34: 281–99.

Guth, E. (1970), "Contribution to the history of Einstein's geometry as a branch of physics". In Carmeli *et al.* (1970), p. 161–207.

Hafele, J. C. and R. E. Keating (1972), "Around-the-world atomic clocks". *Science*, 177: 166–70.

Hall, A. R. and M. B. Hall, eds. (1978), *Unpublished Scientific Papers of Isaac Newton. A Selection from the Portsmouth Collection in the University Library, Cambridge.* Cambridge: The University Press.

Halley, E. (1720*a*), "Of the infinity of the sphere of fix'd stars". *RSLPT*, 31: 22–4.

Halley, E. (1720*b*), "Of the number, order, and light of the fix'd stars". *RSLPT*, 31: 24–6.

Harrison, E. R. (1964), "Olbers' Paradox". *Nature*, 204: 271–2.
Harrison, E. R. (1965), "Olbers' paradox and the background radiation density in an isotropic homogeneous universe". *MNRAS*, 131: 1–12.
Harrison, E. R. (1977), "The dark night sky paradox". *AJP*, 45: 119–24.
Harvey, A. L. (1965), "Brief review of Lorentz-covariant scalar theories of gravitation". *AJP*, 33: 449–60.
Havas, P. (1964), "Four-dimensional formulations of Newtonian mechanics and their relation to the special and the general theory of relativity". *Rev. Mod. Phys.* 36: 938–65.
Havas, P. (1967), "Foundation problems in general relativity". In M. Bunge, ed. *Delaware Seminar in the Foundations of Physics.* Berlin: Springer, pp. 124–48.
Havas, P. (1968), "Causality requirements and the theory of relativity". *Synthèse*, 18: 75–102.
Havas, P. (1979), "Equations of motion and radiation reaction in the special and general theory of relativity". In Ehlers (1979), pp. 74–155.
Havas, P. and J. N. Goldberg (1962), "Lorentz–invariant equations of motion of point masses in the general theory of relativity". *Phys. Rev.* 128: 398–414.
Hawking, S. W. (1965), "Occurrence of singularities in open universes". *Phys. Rev. Lett.* 15: 689–90.
Hawking, S. W. (1966*a*), "The occurrence of singularities in cosmology". *RSLPr.* A 294: 511–21.
Hawking, S. W. (1966*b*), "The occurrence of singularities in cosmology, II". *RSLPr.* A 295: 490–3.
Hawking, S. W. (1966*c*), "Singularities in the universe". *Phys. Rev. Lett.* 17: 444–5.
Hawking, S. W. (1966*d*), "Perturbations of an expanding universe". *Astroph. J.* 145: 544–54.
Hawking, S. W. (1967), "The occurrence of singularities in cosmology, III, Causality and singularities". *RSLPr.* A 300: 187–201.
Hawking, S. W. (1969), "The existence of cosmic time functions". *RSLPr.* A 308: 433–5.
Hawking, S. W. (1971), "Stable and generic properties in general relativity". *GRG*, 1: 393–400.
Hawking, S. W. (1973), "The event horizon". In C. DeWitt and B. DeWitt, eds., *Black Holes: Les Astres Occlus. Les Houches Summer School 1972.* New York: Gordon & Breach, pp. 1–56.
Hawking, S. W. (1975), "Particle creation by black holes". *Comm. Math. Phys.* 43: 199–220.
Hawking, S. W. (1977), "The quantum mechanics of black holes", *Sci. Amer.* 236, 1: 34–40.
Hawking, S. W. and G. F. R. Ellis (1965), "Singularities in homogeneous world models". *Phys. Letters*, 17: 246–7.
Hawking, S. W. and G. F. R. Ellis (1968), "The cosmic black-body radiation and the existence of singularities in our universe". *Astroph. J.* 152: 25–36.
Hawking, S. W. and G. F. R. Ellis (1973), *The Large Scale Structure of Space-Time.* Cambridge: The University Press.
Hawking, S. W. and W. Israel, eds. (1979), *General Relativity. An Einstein centenary survey.* Cambridge: The University Press.
Hawking, S. W., A. King and P. McCarthy (1976), "A new topology for curved space-time which incorporates the causal, differential and conformal structures". *J. Math. Phys.* 17: 174–81.
Hawking, S. W. and R. Penrose (1970), "The singularities of gravitational collapse and cosmology". *RSLPr.* A 314: 529–48.
Heintzmann, H. and P. Mittelstaedt (1968), "Physikalische Gesetze in beschleunigten Bezugssystemen". *Springer Tracts in Modern Physics*, 47: 185–225.
Held, A., ed. (1980), *General Relativity and Gravitation. One Hundred Years after the Birth of Albert Einstein.* New York: Plenum. 2 vols.
Helmholtz, H. von (1868), "Ueber die Tatsachen, die der Geometrie zum Grunde liegen". *Gött. Nachr.*, pp. 193–221.
Herivel, J. (1965), *The Background to Newton's* Principia. *A study of Newton's dynamical researches in the years 1664–84.* Oxford: Clarendon.
Herneck, F. (1966), "Die Beziehungen zwischen Einstein und Mach". *Wiss. Ztschr. Univ. Jena Math.-naturw. Reihe*, 15: 1 ff.
Hertz, H. (*EW*), *Electric Waves, being researches on the propagation of electric action with finite velocity through space.* Transl. by D. E. Jones. New York: Dover.
Hertz, H. (1890*a*), "Ueber die Grundgleichungen der Elektrodynamik für ruhende Körper". *Ann. Physik* (3) 40: 577–624.

Hertz, H. (1890*b*), "Ueber die Grundgleichungen der Elektrodynamik für bewegte Körper". *Ann. Physik* (3) 41: 369–99.
Hessenberg, G. (1971), "Vektorielle Begründung der Differentialgeometrie". *Math. Ann.* 78: 187–217.
Hilbert, D. (*GG*), *Grundlagen der Geometrie*. 10. Auflage. Stuttgart: Teubner, 1968.
Hilbert, D. (1910), "Hermann Minkowski". *Math. Ann.* 68: 445–71.
Hilbert, D. (1915), "Die Grundlagen der Physik. (Erste Mitteilung.)" *Gött. Nachr.*, pp. 395–407.
Hilbert, D. (1917), "Die Grundlagen der Physik. (Zweite Mitteilung.)" *Gött. Nachr.*, pp. 53–76.
Hill, H. A. and R. T. Stebbins (1975), "The intrinsic visual oblateness of the sun". *Astroph. J.* 200: 471–83.
Hirosige, T. (1968), "Theory of relativity and the ether". *Jap. Stud. Hist. Sci.* 7: 37–58.
Hirosige, T. (1969), "Origins of Lorentz' theory of electrons and the concept of the electromagnetic field". *HSPS*, 1: 151–269.
Hirosige, T. (1976), "The ether problem, the mechanistic worldview, and the origins of the theory of relativity". *HSPS*, 7: 3–82.
Hoek, M. (1868), "Détermination de la vitesse avec laquelle est entraînée une onde lumineuse traversant un milieu en mouvement". *Arch. néerl.* 3: 180–5.
Hoek, M. (1869), "Détermination de la vitesse avec laquelle est entraîné un rayon lumineux traversant un milieu en mouvement". *Arch. néerl.* 4: 443–50.
Hoffmann, B., ed. (1966), *Perspectives in Geometry and Relativity. Essays in Honor of Vlácav Hlavatý*. Bloomington: Indiana University Press.
Hoffmann, B. (1972), "Einstein and tensors". *Tensor*, 26: 157–72.
Hojman, S. A., K. Kuchař and C. Teitelboim (1976), "Geometrodynamics regained". *Ann. Physics* 96: 88–135.
Holton, G. (1960), "On the origins of the special theory of relativity". *AJP*, 28: 627–36.
Holton, G. (1964), "Notes on the thematic analysis of science: The case of Poincaré and relativity". In *Mélanges Koyré*, Paris: Hermann, pp. 257–68.
Holton, G. (1968/69), "Influences on Einstein's early work in relativity theory". *Amer. Scholar*, 37: 59–79.
Holton, G. (1969), "Einstein, Michelson and the »crucial« experiment". *Isis*, 60: 132–97.
Holton, G. (1973), *Thematic Origins of Scientific Thought. Kepler to Einstein.* Cambridge, Mass.: Harvard University Press. (The four preceding items by Holton are reproduced in this volume.)
Hönl, H. (1966), "Zur Geschichte des Machschen Prinzips". *Wiss. Ztschr. Univ. Jena Math.-naturw. Reihe*, 15: 25–36.
Hönl, H. and H. Dehnen (1963, 1964), "Ueber Machsche und anti-Machsche Lösungen der Feldgleichungen der Gravitation". *Ann. Physik* (7) 11: 201–15; 14: 271–95.
Horwich, P. (1975), "Grünbaum on the metric of space and time". *BJPS*, 26: 199–211.
Hoyle, F. and J. V. Narlikar (1964), "A new theory of gravitation". *RSLPr.* A 282: 191–207.
Hoyle, F. and J. V. Narlikar (1966), "A conformal theory of gravitation". *RSLPr.* A 294: 138–48.
Hubble, E. (1929), "A relation between distance and radial velocity among extra-galactic nebulae". *Nat. Ac. Sci. (U.S.) Proc.* 15: 168–73.
Humason, M. L. (1931), "Apparent velocity-shifts in the spectra of faint nebulae". *Astroph. J.* 74: 35–42.
Ignatowsky, W. von (1910), "Einige allgemeine Bemerkungen zum Relativitätsprinzip". *Verh. Deutsch. Phys. Ges.* 12: 788–96.
Ignatowsky, W. von (1910*), "Einige allgemeine Bemerkungen zum Relativitätsprinzip". *Phys. Ztschr.* 11: 972–6.
Ignatowsky, W. von (1911*a*), "Das Relativitätsprinzip". *Arch. f. Math. u. Phys.* (3) 17: 1–24; 18: 17–41.
Ignatowsky, W. von (1911*b*), "Eine Bemerkung zu meiner Arbeit 'Einige allgemeine Bemerkungen zum Relativitätsprinzip'". *Phyz. Ztschr.* 12: 779.
Infeld, L. and J. Plebański (1960), *Motion and Relativity*. Warsaw: Państwowe Wydawnictwo Naukowe.
Infeld, L. and A. Schild (1949), "The motion of test particles in general relativity". *Rev. Mod. Phys.* 21: 408–13.
Israel, W., ed. (1973), *Relativity, Astrophysics and Cosmology*. Dordrecht: Reidel.
Ives, H. E. (1952), "Derivation of the mass-energy relation". *J. Opt. Soc. Am.* 42: 540–3.

Iyanaga, S. and Y. Kawada, eds. (1977), *Encyclopedic Dictionary of Mathematics by the Mathematical Society of Japan*. Transl. rev. by K. O. May. Cambridge, Mass.: The MIT Press, 1977. 2 vols.

Jackson, F. and R. Pargetter (1977), "Relative simultaneity in the special relativity". *PS*, 44: 464–74.

Jackson, F. and R. Pargetter (1979), "A reply to Torretti [1979] and Giannoni [1979]". *PS*, 46: 310–15.

Jammer, M. (1954), *Concepts of Space. The History of the Theories of Space in Physics*. Cambridge, Mass.: Harvard University Press.

Jammer, M. (1961), *Concepts of Mass in Classical and Modern Physics*. Cambridge, Mass.: Harvard University Press.

Jammer, M. (1979), "Some foundational problems in the special theory of relativity". In Toraldo di Francia (1979), pp. 203–36.

Janis, A. I. (1969), "Synchronism by slow transport of clocks in noninertial frames of reference". *PS*, 36: 74–81.

Johnson, R. (1977), "The bundle boundary in some special cases". *J. Math. Phys.* 18: 898–902.

Kamlah, A. (1977), "Hans Reichenbach's relativity of geometry". *Synthèse*, 34: 249–63.

Kanitscheider, B. (1971), *Geometrie und Wirklichkeit*. Berlin: Duncker & Humblot.

Kanitscheider, B. (1972), "Die Rolle der Geometrie innerhalb physikalischer Theorien". *Ztschr. philos. Forsch.* 26: 42–55.

Kanitscheider, B. (1973), "Geochronometrie und Geometrodynamik". *ZAW*, 4: 261–303.

Kanitscheider, B. (1974), *Philosophisch-historische Grundlagen der physikalischen Kosmologie*. Stuttgart: Kohlhammer.

Kanitscheider, B. (1976*a*), *Vom absoluten Raum zur dynamischen Geometrie*. Mannheim: Bibliographisches Institut.

Kanitscheider, B. (1976*b*), "Das Zeitproblem bei geodätischer Unvollständigkeit". In Kanitscheider, ed., *Sprache und Erkenntnis, Festschrift für Gerhard Frey zum 60. Geburtstag*. Innsbruck: Innsbrucker Beiträge zur Kulturwissenschaft, Bd. 19, pp. 315–31.

Kanitscheider, B. (1977), "Singularitäten, Horizonte und das Ende der Zeit". *Philos. Naturalis*, 16: 480–511.

Kanitscheider, B. (1979*a*), *Philosophie und moderne Physik: Systeme, Strukturen, Synthesen*. Darmstadt: Wissenschaftliche Buchgesellschaft.

Kanitscheider, B. (1979*b*), "Die philosophische Relevanz der Kosmologie". In Nelkowski *et al.* (1979), pp. 336–57.

Kant, I. (Ak.), *Kant's gesammelte Schriften, herausgegeben von der K. Preussischen Akademie der Wissenschaften*. Berlin, 1902 ff.

Kant, I. (1756), *Metaphysicae cum geometria iunctae usus in philosophia naturali, cuius specimen I continet monadologiam physicam*.

Kant, I. (1770), *De mundi sensibilis atque intelligibilis forma et principiis*. Regiomonti: I. I. Kanter.

Kant, I. (1781), *Critik der reinen Vernunft*. Riga: J. F. Hartknoch.

Kant, I. (1783), *Prolegomena zu einer jeden künftigen Metaphysik die als Wissenschaft wird auftreten können*. Riga: J. F. Hartknoch.

Kant, I. (1786), *Metaphysische Anfangsgründe der Naturwissenschaft*. Riga: J. F. Hartknoch.

Kant, I. (1787), *Critik der reinen Vernunft*. Zweyte hin und wieder verbesserte Auflage. Riga: J. F. Hartknoch.

Kant, I. (1790), *Ueber eine Entdeckung nach der alle neue Critik der reinen Vernunft durch eine ältere entbehrlich gemacht werden soll*. Königsberg: F. Nicolovius.

Kerr, R. P. (1963), "Gravitational field of a spinning mass as an example of algebraically special metrics". *Phys. Rev. Lett.* 11: 237–8.

Keswani, G. H. (1964/5, 1965/6), "Origin and concept of relativity". *BJPS*, 15: 286–306; 16: 19–32, 273–94.

Kilmister, C. W. (1970), *Special Theory of Relativity*. Oxford: Pergamon.

Kilmister, C. W. (1973), *General Theory of Relativity*. Oxford: Pergamon.

Klein, F. (1871, 1873), "Ueber die sogenannte Nicht-Euklidische Geometrie". *Math. Ann.* 4: 573–625; 6: 112–45.

Klein, F. (1872), *Vergleichende Betrachtungen über neuere geometrische Forschungen*. Erlangen: A. Düchert.

Klein, F. (1893), "Vergleichende Betrachtungen über neuere geometrische Forschungen". *Math. Ann.* 43: 63–100. (Revised version of Klein (1872).)

Klein, F. (1911), "Ueber die geometrischen Grundlagen der Lorentz-Gruppe". *Phys. Ztschr.* 12: 17–27.

Klein, F. (1917), "Zu Hilberts erster Note über die Grundlagen der Physik". *Gött. Nachr.*, pp. 469–82.

Klein, F. (1918*a*), "Ueber die Differentialgesetze für die Erhaltung von Impuls und Energie in der Einsteinschen Gravitationstheorie". *Gött. Nachr.*, pp. 171–89.

Klein, F. (1918*b*), "Ueber die Integralform der Erhaltungssätze und die Theorie der räumlich geschlossenen Welt". *Gött. Nachr.*, pp. 394–423.

Klein, F. (1967), *Vorlesungen über die Entwicklung der Mathematik im 19. Jahrhundert*. New York: Chelsea. Two vols. in one. (Reprint of Berlin edition of 1926–1927.)

Kobayashi, S. and K. Nomizu (*FDG*), *Foundations of Differential Geometry*. New York: Wiley, 1963/1969. 2 vols.

Koslow, A., ed. (1967), *The Changeless Order. The Physics of Space, Time and Motion*. New York: Braziller.

Kottler, F. (1912), "Ueber die Raumzeitlinien der Minkowskischen Welt". *AWWS* (II*a*), 121: 1659–1759.

Koyré, A. (1957), *From the closed world to the infinite universe*. Baltimore: The Johns Hopkins Press.

Kramer, D., H. Stephani, E. Herlt and M. MacCallum (1980), *Exact Solutions of Einstein's Field Equations*. Cambridge: The University Press.

Kretschmann, E. (1917), "Ueber den physikalischen Sinn der Relativitätspostulaten". *Ann. Physik* (4) 53: 575–614.

Kronheimer, E. and R. Penrose (1967), "On the structure of causal spaces". *Camb. Philos. Soc. Proc.* 63: 481–501.

Kundt, W. and B. Hoffmann (1962), "Determination of gravitational standard time". In *Recent Developments in General Relativity (Infeld Festschrift)*. New York: Pergamon, pp. 303–6.

Lagrange, J. L. (1797), *Théorie des fonctions analytiques*. Paris: Imprimerie de la République, Prairial, An V.

Lalan, V. (1937), "Sur les postulats qui sont à la base des cinématiques". *Bull. Soc. Math. France*, 67: 83–99. (Abstract in *CR*, 203: 1491; 1936.)

Lanczos, C. (1972), "Einstein's path from special to general relativity". In O'Raifeartaigh (1972), pp. 5–19.

Landau, L. D. and E. M. Lifshitz (1960), *Mechanics*. Transl. by J. B. Sykes and J. S. Bell. Oxford: Pergamon.

Landau, L. D. and E. M. Lifshitz (1962), *The Classical Theory of Fields*. Revised Second Edition. Transl. by M. Hamermesh. Oxford: Pergamon.

Lange, L. (1885*a*), "Ueber die wissenschaftliche Fassung des Galilei'schen Beharrungsgesetzes". *Philos. Stud.* 2: 266–97.

Lange, L. (1885*b*), "Ueber das Beharrungsgesetz". *Leipz. Ber.* 37: 333–51.

Lange, L. (1886), *Die geschichtliche Entwicklung des Bewegungsbegriffes und ihr voraussichtliches Endergebnis. Ein Beitrag zur historischen Kritik der mechanischen Principien*. Leipzig: W. Engelmann.

Lange, L. (1902), "Das Inertialsystem vor dem Forum der Naturforschung. Kritisches und Antikritisches". In *Festschrift Wilhelm Wundt zum 70. Geburtstag*. Leipzig: W. Engelmann, vol. II, pp. 1–71.

Larmor, J. (1897), "A dynamical theory of the electric and luminiferous medium. Part III. Relations with material media". *RSLPT*, 190: 205–300.

Larmor, J. (1900), *Aether and Matter. A development of the dynamical relations of the aether to material systems on the basis of the atomic constitution of matter*. Cambridge: The University Press.

Latzer, R. (1972), "Nondirected light signals and the structure of time". *Synthèse*, 24: 236–80.

Laue, M. von (1907), "Die Mitführung des Lichtes durch bewegte Körper nach dem Relativitätsprinzip". *Ann. Physik* (4) 23: 989–90.

Laue, M. von (1911), "Zur Dynamik der Relativitätstheorie". *Ann. Physik* (4) 35: 524–42.

Laue, M. von (1911*b*), "Zur Diskussion über den starren Körper in der Relativitätstheorie". *Phys. Ztschr.* 12: 85–7.

Laue, M. von (1920), "Theoretisches über neuere optische Beobachtungen zur Relativitätstheorie". *Phys. Ztschr.* 21: 659–62.

Laue, M. von (1955). *Die Relativitätstheorie. 1. Band. Die spezielle Relativitätstheorie.* 6. durchgesehene Aufl. Braunschweig: Vieweg. (The first edn appeared in 1911 and was the first textbook on Relativity.)

Laue, M. von (1965), *Die Relativitätstheorie. 2. Band. Die allgemeine Relativitätstheorie.* 5. durchgesehene und erweiterte Aufl. hrsg. v. F. Beck. Braunschweig: Vieweg. (The first edn appeared in 1921.)

Lee, A. R. and T. M. Kalotas (1975), "Lorentz transformations from the first postulate". *AJP*, 43: 434–7. ("Comments" by several authors in *AJP*, 44: 988–1002; 1976.)

Leibniz, G. W. (*GP*), *Die philosophischen Schriften von Gottfried Wilhelm Leibniz herausgegeben von C. J. Gerhardt.* Hildesheim: Olms, 1965. 7 vols. (Reprint of Berlin edition of 1875–1890.)

Lemaître, G. (1927), "Un univers homogène de masse constante et de rayon croissant, rendant compte de la vitesse radiale des nébuleuses extra-galactiques". *Ann. Soc. Sci. Bruxelles*, A 47: 49–59.

Lemaître, G. (1931*a*), "A homogeneous universe of constant mass and increasing radius accounting for the radial velocity of extragalactic nebulae". *MNRAS*, 91: 483–90. (Eng. transl. of preceding item by Lemaître.)

Lemaître, G. (1931*b*), "The expanding universe". *MNRAS*, 91: 490–501.

Lense, J. and H. Thirring (1918), "Ueber den Einfluss der Eigenrotation der Zentralkörper auf die Bewegung der Planeten und Monde nach der Einsteinschen Gravitationstheorie". *Phys. Ztschr.* 19: 156–63.

Lerner, R. G. and G. L. Trigg, eds. (1981), *Encyclopedia of Physics.* Reading, Mass.: Addison-Wesley.

Levi-Cività, T. (1917), "Nozione di parallelismo in una varietà qualunque". *Rend. Circ. Mat. Palermo*, 42: 173–205.

Levi-Cività, T. (1927), "Sur l'écart géodésique". *Math. Ann.* 97: 291–320.

Levi-Cività, T. (1977), *The Absolute Differential Calculus (Calculus of Tensors).* Transl. by M. Long. Ed. by E. Persico. New York: Dover.

Lévy-Leblond, J.-M. (1976), "One more derivation of the Lorentz transformation". *AJP*, 44: 271–7.

Lewis, G. N. (1908), "A revision of the fundamental laws of matter and energy". *Philos. Mag.* (6) 16: 707–17.

Lewis, G. N. and R. C. Tolman (1909), "The principle of relativity and non-Newtonian mechanics". *Philos. Mag.* (6) 18: 510–23.

Lichnerowicz, A. (1962), *Elements of Tensor Calculus.* Transl. by J. W. Leech and D. J. Newman. London: Methuen.

Lie, S. (1890), "Ueber die Grundlagen der Geometrie". *Leipz. Ber.* 42: 284–321,355–418.

Lie, S. (1893), *Vorlesungen über continuierlichen Gruppen mit geometrischen und anderen Anwendungen.* Bearbeitet und hrsg. von G. Scheffers. Leipzig: Teubner.

Lie, S. and F. Engel. (*TTG*), *Theorie der Transformationsgruppen.* New York: Chelsea, 1970, 3 vols. (Reprint of Leipzig edition of 1888–1893.)

Lindsay, R. B. and H. Margenau (1936), *Foundations of Physics.* New York: Wiley.

Longair, M. S., ed. (1974), *Confrontation of Cosmological Theories with Observational Data.* (IAU Symposium No. 63.) Dordrecht: Reidel.

Lorentz, H. A. (*CP*). *Collected Papers.* The Hague: Nijhoff. 9 vols.

Lorentz, H. A. (1886), "De l'influence du mouvement de la terre sur les phénomènes lumineux". *KNAWVersl.* 2: 297–372 (in Dutch; the French version appeared first in *Arch. Néerl.* 21: 103–176, 1887; rep. in Lorentz, *CP*, 4: 153–214).

Lorentz, H. A. (1892*a*), "La théorie électromagnetique de Maxwell et son application aux corps mouvants". *Arch. Néerl.* 25: 363 ff. (Also in Lorentz, *CP*, 2: 164–343.)

Lorentz, H. A. (1892*b*), "The relative motion of the earth and the ether". *KNAWVersl.* 1: 74–79 (in Dutch; the English version in Lorentz, *CP*, 4: 219–23).

Lorentz, H. A. (1895), *Versuch einer Theorie der electrischen und optischen Erscheinungen in bewegten Körpern.* Leiden: Brill. (Also in Lorentz, *CP*, 5: 1–137.)

Lorentz, H. A. (1898), "Die Fragen, welche die translatorische Bewegung des Lichtäthers betreffen". *Verh. Ges. deutsch. Naturf. u. Aerzte*, 70: 56–65. (Also in Lorentz, *CP*, 7: 101–15.)

Lorentz, H. A. (1899), "Théorie simplifiée des phénomènes électriques et optiques dans les corps

en mouvement". *KNAWVersl.* 7: 507–522 (in Dutch; the French version in Lorentz, *CP*, 5: 139–55).
Lorentz, H. A. (1899*), "Simplified theory of electrical and optical phenomena in moving systems". *KNAWPr.* 1: 427 ff. (Revised English version of Lorentz (1899); rep. in Schaffner (1972), pp. 254–73.)
Lorentz, H. A. (1900), "Considérations sur la pesanteur". *KNAWVersl.* 8: 603 ff. (in Dutch; the French version appeared first in *Arch. Néerl.* 7: 325 ff., 1902; rep. in Lorentz, *CP*, 5: 198–215).
Lorentz, H. A. (1904), "Electromagnetic phenomena in a system moving with any velocity smaller than that of light". *KNAWPr.* 6: 809–31. (Also in Lorentz, *CP*, 5: 172–197. In Dutch in *KNAWVersl.* 12: 986–1009, 1904.)
Lorentz, H. A. (1909), *The Theory of Electrons and its applications to the phenomena of radiant heat. A Course of Lectures delivered in Columbia University, New York*, in March and April 1906. Leipzig: Teubner.
Lorentz, H. A. (1916), *The Theory of Electrons . . .*[as in preceding item]. Second Edition. Leipzig: Teubner.
Lorentz, H. A., A. Einstein and H. Minkowski (*RP*), *Das Relativitätsprinzip. Eine Sammlung von Abhandlungen.* Darmstadt: Wissenschaftliche Buchgesellschaft. 1958. (Reprint of 5th edn of 1923.)
Lorentz, H. A., A. Einstein, H. Minkowski and H. Weyl (*PR*), *The Principle of Relativity. A Collection of Original Memoirs on the Special and General Theory of Relativity.* Transl. by W. Perrett and G. B. Jeffery. London: Methuen, 1923.
Lovelock, D. (1971), "The Einstein tensor and its generalizations". *J. Math. Phys.* 12: 498–501.
Lovelock, D. (1972), "The four-dimensionality of space and the Einstein tensor". *J. Math. Phys.* 13: 874–6.
Lovelock, D. and H. Rund (1975), *Tensors, Differential Forms and Variational Principles.* New York: Wiley.
MacCallum, M. A. H. (1971), "Mixmaster universe problem". *Nature. Phys. Sci.* 230: 112–13.
MacCullagh, J. (1848), "An essay towards a dynamical theory of crystalline reflexion and refraction". *R. Irish Acad. Trans.* 21: 17–50. (Read December 9th, 1839.)
Mach, E. (1883), *Die Mechanik in ihrer Entwicklung. Historisch-kritisch dargestellt.* Leipzig: Brockhaus.
Mach, E. (1963), *Die Mechanik historisch-kritisch dargestellt.* Darmstadt: Wissenschaftliche Buchgesellschaft. (Reprint of 9th edn of preceding item by Mach, Leipzig 1933.)
Machamer, P. K. and R. G. Turnbull, eds. (1976), *Motion and Time, Space and Matter. Interrelations in the History of Philosophy and Science.* Columbus: Ohio State University Press.
MacLane, S. and G. Birkhoff (1967). *Algebra.* New York: Macmillan.
Malament, D. (1977*a*), "Causal theories of time and the conventionality of simultaneity". *Noûs*, 11: 293–300.
Malament, D. (1977*b*), "Observationally indistinguishable spacetimes: Comments on Glymour's paper". In Earman *et al.* (1977), pp. 61–80.
Malament, D. (1977*c*), "The class of continuous timelike curves determines the topology of spacetime". *J. Math. Phys.* 18: 1399–1404.
Marder, L. (1971), *Time and the space-traveller.* London: Allen & Unwin.
Markus, L. (1955), "Line element fields and Lorentz structures on differential manifolds". *Ann. Maths.* 62: 411–17.
Marzke, R. F. and J. A. Wheeler (1964), "Gravitation as geometry. I. The geometry of space-time and the geometrodynamical standard meter". In Chiu and Hoffmann (1964), pp. 40–64.
Mascart, E. E. N. (1872/74), "Sur les modifications qu'éprouve la lumière par suite du mouvement de la source lumineuse et du mouvement de l'observateur". *Ann. Ec. Norm.* 1: 157–214; 3: 363–420.
Massey, G. J. (1969), "Towards a clarification of Grünbaum's concept of intrinsic metric". *PS*, 36: 331–45.
Maxwell, J. C. (*SP*), *The Scientific Papers of James Clerk Maxwell*, ed. by W. D. Niven. Cambridge: The University Press, 1890. 2 vols.
Maxwell, J. C. (*TEM*), *A Treatise on Electricity and Magnetism.* Third edition. New York: Dover. 2 vols. (Reprint of Oxford edition of 1891.)
Maxwell, J. C. (1861/62), "On physical lines of force". *Philos. Mag.* (4) 21: 162–175, 281–291,

338–348; 23: 12–24, 85–95. (Reprinted in Maxwell, *SP*, 1: 415–513.)
Maxwell, J. C. (1865), "A dynamical theory of the electromagnetic field". *RSLPT*, 155: 459–512. (Reproduced with omission of Parts V and VII in Tricker (1966), pp. 226–86.)
Maxwell, J. C. (1879/80), "On a possible mode of detecting a motion of the solar system through the luminiferous ether". By the late Professor J. Clerk Maxwell, F. R. S. In a letter to Mr. D. P. Todd of the Nautical Almanac Office, Washington, U. S. Communicated by Professor Stokes, Sec. R. S. Received January 7, 1880. *RSLPr*. 30: 108–110. (The text of Maxwell's letter is reproduced in Hawking and Israel (1979), pp. xvii f.)
McCormmach, R. (1970), "H. A. Lorentz and the electromagnetic view of nature". *Isis*, 61: 459–97.
McCormmach, R. (1976), "Editor's foreword". *HSPS*, 7: xii–xxxv.
McMullin, E. (1978), *Newton on matter and activity*. Notre Dame: University of Notre Dame Press.
Mehlberg, H. (1935), "Essai sur la théorie causale du temps. I. La théorie causale du temps chez ses principaux représentants". *Stud. philos.* (Lemberg), 1: 119–260. (See p. 339, n. 24.)
Mehlberg, H. (1937), "Essai sur la théorie causale du temps. II. Durée et causalité". *Stud. philos.* (Lemberg), 2: 111–231. (See p. 339, n. 24.)
Mehlberg, H. (1966), "Relativity and the atom". In P. K. Feyerabend and G. Maxwell, eds., *Mind, Matter and Method.* Minneapolis: University of Minnesota Press, pp. 449–91.
Mehra, J. (1974*a*), "Origins of the modern theory of gravitation. The historical origins of the theory of general relativity from 1907 to 1919". *Bull. Soc. R. Sci. Liège*, 43: 190–213.
Mehra, J. (1974*b*), *Einstein, Hilbert and the Theory of Gravitation. Historical origins of General Relativity Theory*. Dordrecht: Reidel.
Merleau-Ponty, J. (1965), *Cosmologie du XXe Siècle*. Paris: Gallimard.
Merleau-Ponty, J. (1974), *Leçons sur la genèse des théories physiques. Galilée, Ampère, Einstein.* Paris: Vrin.
Michelson, A. A. (1881), "The relative motion of the earth and the luminiferous ether". *Amer. J. Sci.* (3) 22: 120–9.
Michelson, A. A. and E. W. Morley (1886), "Influence of motion of the medium on the velocity of light". *Amer. J. Sci.* (3) 31: 377–86.
Michelson, A. A. and E. W. Morley (1887), "On the relative motion of the earth and the luminiferous aether". *Philos. Mag.* (5) 24: 449–63. (Also in *Amer. J. Sci.* (3) 34: 333–41.)
Mie, G. (1912*a*, *b*; 1913), "Grundlagen einer Theorie der Materie". *Ann. Physik* (4) 37: 511–34; 39: 1–40; 40: 1–66.
Mie, G. (1914), "Bemerkungen zur Einsteinschen Gravitationstheorie". *Phys. Ztschr.* 15: 115–22; 169–76.
Miller, A. I. (1973), " A study of Henri Poincaré's 'Sur la dynamique de l'électron' ". *Arch. Hist. Exact Sci.* 10: 207–328.
Miller, A. I. (1980), "On some other approaches to electrodynamics in 1905". In Woolf (1980), pp. 66–91.
Miller, A. I. (1981), *Albert Einstein's Special Theory of Relativity. Emergence (1905) and Early Interpretation (1905–1911)*. Reading, Mass.: Addison-Wesley.
Miller, J. C. and D. W. Sciama (1980), "Gravitational collapse to the black hole state". In Held (1980), vol. 2, pp. 359–91.
Minkowski, H. (1908), "Die Grundgleichungen für die elektromagnetischen Vorgänge in bewegten Körper". *Gött. Nachr.*, pp. 53–111.
Minkowski, H. (1909), "Raum und Zeit". *Phys. Ztschr.* 10: 104–111.
Minkowski, H. (1910), "Die Grundgleichungen für die elektromagnetischen Vorgänge in bewegten Körper". *Math. Ann.* 68: 472–525. (Same as Minkowski (1908).)
Minkowski, H. (1915), "Das Relativitätsprinzip". *JDMV*, 24: 372–382. (Lecture of 1907; also in *Ann. Physik* (4) 47: 927 ff.)
Minkowski, H. and M. Born (1910), "Eine Ableitung der Grundgleichungen für die elektromagnetischen Vorgänge in bewegten Körpern vom Standpunkte der Elektronentheorie". *Math. Ann.* 68: 525–51.
Misner, C. W. (1963), "The flatter regions of Newman, Unti and Tamburino's generalised Schwarzschild space". *J. Math. Phys.* 4: 924–37.
Misner, C. W. (1967), "Taub-NUT space as a counterexample to almost anything". In J. Ehlers,

ed., *Relativity Theory and Astrophysics. I. Relativity and Cosmology*. (Am. Math. Soc. Lectures in Applied Maths. Vol. 8.) Providence: American Mathematical Society, pp. 160–9.
Misner, C. W. (1969*a*), "Absolute zero of time". *Phys. Rev.* 186: 1328–33.
Misner, C. W. (1969*b*), "Mixmaster universe". *Phys. Rev. Lett.* 22: 1071–4.
Misner, C. W. (1980), "Values and arguments in homogeneous spaces". In Tipler, *EGR*, pp. 221–31.
Misner, C. W. and A. H. Taub (1969), "A singularity-free empty universe". *Sov. Phys. JETP*, 28: 122–33. (Russian orig. in *Zh. Eksp. Teor. Fiz.* 55: 233–55; 1968.)
Misner, C. W., K. S. Thorne and J. A. Wheeler (1973), *Gravitation*. San Francisco: Freeman.
Mittelstaedt, P. (1976), *Der Zeitbegriff in der Physik. Physikalische und philosophische Untersuchungen zum Zeitbegriff in der klassischen und in der relativistischen Physik*. Mannheim: Bibliographisches Institut.
Mittelstaedt, P. (1977), "Conventionalism in special relativity". *FP*, 7: 573–83.
Mitton, S., ed. (1977), *The Cambridge Encyclopedia of Astronomy*. New York: Crown.
Møller, C. ed. (*EGT*), *Evidence for Gravitational Theories*. (Intern. Sch. of Physics »Enrico Fermi«, Course XX.) New York: Academic Press, 1962.
Møller, C. (1952), *The Theory of Relativity*. Oxford: Clarendon.
Møller, C. (1962), "Tetrad fields and conservation laws in general relativity". In Møller, *EGT*, pp. 252–64.
Mortensen, C. and G. Nerlich (1978), "Physical topology". *J. Philos. Logic*, 7: 209–23.
Munkres, J. R. (1966), *Elementary Differential Topology*. Princeton: Princeton University Press.
Narlikar, J. V. (1979), *Lectures on General Relativity and Cosmology*. London: Macmillan.
Nelkowski, H., A. Hermann, H. Poser, R. Schrader and R. Seiler, eds. (1979), *Einstein Symposion Berlin aus Anlass der 100. Wiederkehr seines Geburtstages*. (Lecture Notes in Physics, vol. 100.) Berlin: Springer.
Nerlich, G. (1976), *The Shape of Space*. Cambridge: The University Press.
Nerlich, G. (1979*a*), "Is curvature intrinsic to physical space?" *PS*, 46: 439–458.
Nerlich, G. (1979*b*), "What can geometry explain?" *BJPS*, 30: 69–83.
Neumann, C. (1870), *Ueber die Principien der Galilei-Newton'schen Theorie. Akademische Antrittsvorlesung gehalten in der Aula der Universität Leipzig am 3. November 1869*. Leipzig: Teubner.
Newman, E. T., E. Couch, K. Chinnapared, A. Exton, A. Prakash and R. Torrence (1965), "Metric of a rotating charged mass". *J. Math. Phys.* 6: 918–19.
Newman, E. T., L. Tamburino and T. J. Unti (1963), "Empty space generalization of the Schwarzschild metric". *J. Math. Phys.* 4: 915–23.
Newton, I. (*Cajori*), *Sir Isaac Newton's Mathematical Principles of Natural Philosophy and his System of the World*. Translated into English by Andrew Motte in 1729. The translations revised, and supplied with an historical and explanatory appendix by Florian Cajori. Berkeley: University of California Press, 1934.
Newton, I. (*PNPM*), *Philosophiae Naturalis Principia Mathematica*. The Third Edition (1726) with variant readings assembled and edited by A. Koyré and I. B. Cohen with the Assistance of A. Whitman. Cambridge, Mass.: Harvard University Press, 1972. 2 vols.
Newton, I. (1756), *Four Letters from Sir Isaac Newton to Doctor Bentley containing some arguments in proof of a Deity*. London: R. & J. Dodsley.
Noether, E. (1918), "Invariante Variationsprobleme". *Gött. Nachr.*, pp. 235–57.
Nomizu, K. (1956), *Lie Groups and Differential Geometry*. Tokyo: The Mathematical Society of Japan.
Nomizu, K. (1979), *Fundamentals of Linear Algebra*. New York: Chelsea.
Nordström, G. (1912*a*), "Relativitätsprinzip und Gravitation". *Phys. Ztschr.* 13: 1126–29.
Nordström, G. (1912*b*), "Träge und schwere Masse in der Relativitätsmechanik". *Ann. Physik* (4) 40: 856–78.
Nordström, G. (1913), "Zur Theorie der Gravitation vom Standpunkt des Relativitätsprinzips". *Ann. Physik* (4) 42: 533–54.
Nordström, G. (1914), "Die Fallgesetze und Planetenbewegungen in der Relativitätstheorie". *Ann. Physik* (4) 43: 1101–10.
Nordström, G. (1918), "On the energy of the gravitation field in Einstein's theory". *KNAWPr.* 20: 1238–45.

North, J. D. (1965), *The Measure of the Universe. A History of Modern Cosmology*. Oxford: Clarendon.

Ohanian, H. C. (1976), *Gravitation and Spacetime*. New York: Norton.

Øhrstrom, P. (1980), "Conventionality in distant simultaneity". *FP*, 10: 333–43.

Oppenheimer, J. R. and H. Snyder (1939), "On continued gravitational contraction". *Phys. Rev.* 56: 455–9.

O'Raifeartaigh, L. (1958), "Fermi coordinates". *R. Irish Acad. Proc.* 59 A: 15–24.

O'Raifeartaigh, L., ed. (1972), *General Relativity. Papers in honour of J. L. Synge*. Oxford: Clarendon.

Ozsváth, I. and E. Schücking (1969), "The finite rotating universe". *Ann. Physics*, 55: 166–204.

Palatini, A. (1919), "Deduzione invariantiva delle equazioni gravitazionali dal Principio di Hamilton". *Rend. Circ. Mat. Palermo*. 43: 203–12.

Papapetrou, A. (1951*a*), "Equations of motion in general relativity". *Proc. Phys. Soc. (London)*, A 64: 61–75, 302–10.

Papapetrou, A. (1951*b*), "Spinning test particles in general relativity, I". *RSLPr.* A 209: 248–58.

Papapetrou, A. (1974), *Lectures on General Relativity*. Dordrecht: Reidel.

Pars, L. A. (1921), "The Lorentz transformation". *Philos. Mag.* (6) 42: 249–58.

Pauli, W. (1958), *Theory of Relativity*. Transl. by G. Field. With supplementary notes by the author. London: Pergamon. (This is a translation of Pauli's article "Relativitätstheorie" in *Encyclopädie der mathematischen Wissenschaften*, Leipzig: Teubner, 1921, vol. V, 2, pp. 539–775.)

Peebles, P. J. E. (1971), *Physical Cosmology*. Princeton: Princeton University Press.

Peebles, P. J. E. (1980), *The Large-Scale Structure of the Universe*. Princeton: Princeton University Press.

Penrose, R. (1964), "Conformal treatment of infinity". In DeWitt and DeWitt (1964), pp. 565–86.

Penrose, R. (1965), "Gravitational collapse and space-time singularities". *Phys. Rev. Lett.* 14: 57–9.

Penrose, R. (1968), "Structure of space-time". In DeWitt and Wheeler (1968), pp. 121–235.

Penrose, R. (1974), "Singularities in cosmology". In Longair (1974), pp. 263–72.

Penrose, R. (1977), "Space-time singularities". In Ruffini (1977), pp. 173–81.

Penrose, R. (1979), "Singularities and time-asymmetry": In Hawking and Israel (1979), pp. 581–638.

Penrose, R. (1980), "On Schwarzschild causality—a problem for »Lorentz covariant« general relativity". In Tipler, *EGR*, pp. 1–12.

Penrose, R. and W. Rindler (1965), "Energy conservation as the basis of relativistic mechanics". *AJP*, 33: 55–9.

Penzias, A. A. and R. W. Wilson (1965), "A measurement of excess antenna temperature at 4080 Mc/s". *Astroph. J.* 142: 419–21.

Perrin, R. (1979), "Twin paradox: a complete treatment from the point of view of each twin". *AJP*, 47: 317–19.

Petrov, A. Z. (1969), *Einstein Spaces*. Transl. by R. F. Kelleher. Oxford: Pergamon.

Petrova, N. M. (1949), "On the equations of motion and the mass tensor for systems of finite mass in the general theory of relativity". *Zh. Eksp. Teor. Fiz.* 19: 989–999. (In Russian; reviewed by A. J. Coleman in *MR*, 11: 467–8, 1950.)

Pirani, F. A. E. (1956), "On the physical significance of the Riemann tensor". *Acta Phys. Polon.* 15: 389–405.

Pirani, F. A. E. (1957), "Invariant formulation of gravitational radiation theory". *Phys. Rev.* 105: 1089–99.

Pirani, F. A. E. (1965), "Introduction to gravitational radiation theory". In Trautman *et al.* (1965), pp. 249–73.

Pirani, F. A. E. (1970), "Noncausal behaviour of classical tachyons". *Phys. Rev.* D 1: 3224–5.

Pirani, F. A. E. and A. Schild (1966), "Conformal geometry and the interpretation of the Weyl tensor". In Hoffmann (1966), pp. 291–309.

Planck, M. (1906), "Das Prinzip der Relativität und die Grundgleichungen der Mechanik". *Verh. Deutsch. Phys. Ges.* 4: 136–41.

Planck, M. (1907), "Zur Dynamik bewegter Systeme". *Preuss. Ak. Wiss. Sitzungsber.*, pp. 542–70.
Poincaré, H. *Oeuvres*. Publiées sous les auspices de l'Académie des Sciences. Paris: Gauthier-Villars. 1916–1956. 11 vols.
Poincaré, H. (*DP*), *Dernières Pensées*. Paris: Flammarion, 1963.
Poincaré, H. (*SH*), *La Science et l'Hypothèse*. Paris: Flammarion, 1968.
Poincaré, H. (*VS*), *La Valeur de la Science*. Paris: Flammarion, 1970.
Poincaré, H. (1887), "Sur les hypothèses fondamentales de la géométrie". *Bull. Soc. Math. France*, 15: 203–16. (Also in Poincaré, *Oeuvres*, 9: 79–91.)
Poincaré, H. (1891), "Les géométries non-euclidiennes". *Rev. gén. des sciences* 2: 769–74.
Poincaré, H. (1893), "Le continu mathématique". *Rev. Métaph. Mor.* 1: 26–34.
Poincaré, H. (1895), "L'èspace et la géométrie". *Rev. Métaph. Mor.* 3: 631–46.
Poincaré, H. (1898), "La mesure du temps". *Rev. Métaph. Mor.* 6: 1–13.
Poincaré, H. (1900), "La théorie de Lorentz et le principe de la réaction". *Arch. Néerl.* 5: 252–278. (Also in Poincaré, *Oeuvres*, 9: 464–88.)
Poincaré, H. (1905), "Sur la dynamique de l'électron". *CR*, 140: 1504–8.
Poincaré, H. (1906), "Sur la dynamique de l'électron". *Rend. Circ. Mat. Palermo*, 21: 129–75.
Poincaré, H. (1954), *Electricité et Optique. La Lumière et les Théories Electrodynamiques. Leçons professées à la Sorbonne en 1888, 1890 et 1899.* 2e. ed. rev. et compl. par J. Blondin et E. Néculiéa. Paris: Gauthier-Villars. (In fact a reprint of the much earlier second edition.)
Pound, R. V. and J. L. Snider (1965), "Effect of gravity on gamma radiation". *Phys. Rev.* 140B: 788–803.
Prokhovnik, S. J. (1967), *The Logic of Special Relativity*. Cambridge: The University Press.
Putnam, H. (1963), "An examination of Grünbaum's philosophy of geometry". In B. Baumrin, ed. *Philosophy of Science, The Delaware Seminar*, vol. 2. New York: Interscience, pp. 205–255. (Also in Putnam (1979), pp. 93–129.)
Putnam, H. (1967), "Time and physical geometry". *J. Philos.* 64: 240–7. (Also in Putnam (1979), pp. 198–205.)
Putnam, H. (1979), *Mathematics, Matter and Method. Philosophical Papers, Volume I.* Second Edition. Cambridge: The University Press.
Pyenson, L. (1977), "Hermann Minkowski and Einstein's special theory of relativity". *Arch. Hist. Exact Sci.* 17: 71–95.
Pyenson, L. (1979), "Physics in the shadow of mathematics. The Göttingen electron-theory seminar of 1905". *Arch. Hist. Exact Sci.* 21: 55–89.
Pyenson, L. (1980), "Einstein's education: Mathematics and the laws of nature". *Isis*, 71: 399–425.
Quinn, P. L. (1976), "Intrinsic metrics on continuous spatial manifolds". *PS*, 43: 396–414.
Raine, D. J. (1975), "Mach's Principle in general relativity". *MNRAS*, 171: 507–28.
Raine, D. J. (1981), *The Isotropic Universe. An Introduction to Cosmology*. Bristol: Adam Hilger.
Raychaudhuri, A. K. (1955), "Relativistic cosmology, I". *Phys. Rev.* 98: 1123–6.
Raychaudhuri, A. K. (1979), *Theoretical Cosmology*. Oxford: Clarendon.
Rayleigh, Lord (1902), "Does motion through the aether cause double refraction?" *Philos. Mag.* (6) 4: 678–83.
Recami, E. and R. Mignani (1974), "Classical theory of tachyons (special relativity extended to superluminal frames and objects)". *Riv. Nuovo Cimento*, 4: 209–90.
Regge, T. (1961), "General relativity without coordinates". *Nuovo Cimento*, 19: 558–71.
Reichenbach, H. (1920), *Relativitätstheorie und Erkenntnis Apriori*. Berlin: Springer.
Reichenbach, H. (1922), "La signification philosophique de la théorie de la relativité". *Revue Philos.* 94: 5–61.
Reichenbach, H. (1924), *Axiomatik der relativistischen Raum-Zeit-Lehre*. Braunschweig: Vieweg.
Reichenbach, H. (1928), *Philosophie der Raum-Zeit-Lehre*. Berlin: W. de Gruyter.
Reichenbach, H. (1958), *The Philosophy of Space and Time*. Transl. by M. Reichenbach and J. Freund. New York: Dover. (English translation of Reichenbach (1928).)
Reinhardt, M. (1973), "Mach's Principle—a critical review". *Ztschr. Naturf.* 28*a*: 529–37.
Reissner, H. (1916), "Ueber die Eigengravitation des elektrischen Feldes nach der Einsteinschen Theorie". *Ann. Physik* (4) 50: 106–20.

Ricci, G. and T. Levi-Cività (1901), "Méthodes de calcul différentiel absolu et leurs applications". *Math. Ann.* 54: 125–201.
Riemann, G. F. B. (1867), "Ueber die Hypothesen, welche der Geometrie zugrunde liegen". *Gött. Abh.* 13: 133–52. (My references are to an offprint, with pages numbered from 7 to 26.)
Riemann, G. F. B. (1902), *Gesammelte mathematische Werke und wissenschaftlicher Nachlass.* Hrsg. unter Mitwirkung von R. Dedekind u. H. Weber. Nachträge hrsg. von M. Noether u. W. Wirtinger. Leipzig: Teubner.
Rietdijk, C. W. (1966), "A rigorous proof of determinism derived from the special theory of relativity". *PS*, 33: 341–4.
Rietdijk, C. W. (1976), "Special relativity and determinism". *PS*, 43: 598–609.
Rindler, W. (1956), "Visual horizons in world models". *MNRAS*, 116: 662–77.
Rindler, W. (1960), *Special Relativity.* Edinburgh: Oliver & Boyd.
Rindler, W. (1977), *Essential Relativity, Special, General and Cosmological.* Second Edition. Berlin: Springer.
Ritz, W. (1908*a*), "Recherches critiques sur l'électrodynamique générale". *Ann. Chim. Phys.* (8) 13: 145–275.
Ritz, W. (1908*b*), "Recherches critiques sur les théories électrodynamiques de Cl. Maxwell et de H. A. Lorentz". *Arch. Sci. Phys. et Nat.* (4) 26: 209–36.
Ritz, W. (1908*c*), "Du rôle de l'ether en physique". *Scîentia*, 3: 260–74.
Robb, A. A. (1914), *A Theory of Time and Space.* Cambridge: The University Press. (The "Introduction" to this book was published *under the same title* in Cambridge by Heffer & Sons in 1913.)
Robb, A. A. (1921), *The Absolute Relations of Time and Space.* Cambridge: The University Press.
Robb, A. A. (1936), *Geometry of Time and Space.* Cambridge: The University Press. (This is a revised version of A. A. Robb (1914).)
Robertson, H. P. (1929), "On the foundations of relativistic cosmology". *Nat. Ac. Sci. (U.S.) Proc.* 15: 822–9.
Robertson, H. P. (1933), "Relativistic cosmology", *Rev. Mod. Phys.* 5: 62–90.
Robertson, H. P. (1935/36), "Kinematics and world-structure". *Astroph. J.* 82: 284–301; 83: 187–201, 257–71.
Robertson, H. P. (1949*a*), "Postulate *versus* observation in the special theory of relativity". *Rev. Mod. Phys.* 21: 378–82.
Robertson, H. P. (1949*b*), "Geometry as a branch of physics". In Schilpp (1949), pp. 315–32.
Robertson, H. P. and T. W. Noonan (1968), *Relativity and Cosmology.* Philadelphia: Saunders.
Roentgen, W. C. (1885), "Versuche über die elektromagnetische Wirkung der dielektrischen Polarisation". *Preuss. Ak. Wiss. Sitzungsber.*, pp. 195–8.
Roentgen, W. C. (1888), "Ueber die durch Bewegung eines im homogenen electrischen Felde befindlichen Dielectricums hervorgerufene electrodynamische Kraft". *Ann. Physik* (3) 35: 264–270.
Roentgen, W. C. (1890), "Beschreibung des Apparates, mit welchem die Versuche über die electrodynamische Wirkung bewegter Dielectrica ausgeführt wurden". *Ann. Physik* (3) 40: 93–108.
Roll, P. G., R. Krotkov and R. H. Dicke (1964), "The equivalence of inertial and passive gravitational mass". *Ann. Physics*, 26: 442–517.
Rosenfeld, L. (1956), "The velocity of light and the evolution of electrodynamics". *Suppl. Nuovo Cimento*, 4: 1630–1669.
Rossi, B. and D. B. Hall (1941), "Variation of the rate of decay of mesotrons with momentum". *Phys. Rev.* 59: 223–8.
Ruffini, R., ed. (1977), *Proceedings of the First Marcel Grossmann Meeting on General Relativity.* Amsterdam: North-Holland.
Rund, H. and D. Lovelock (1972), "Variational principles in the general theory of relativity". *JDMV*, 74: 1–65.
Russell, B. (1897), *An Essay on the Foundations of Geometry.* Cambridge: The University Press.
Russell, B. (1959), *My Philosophical Development.* London: Allen & Unwin.
Ryan, M. P. and L. C. Shepley (1975), *Homogeneous Relativistic Cosmologies.* Princeton: Princeton University Press.
Sachs, R. K., ed. (1971), *General Relativity and Cosmology* (Intern. Sch. of Physics »Enrico

Fermi«, Course XLVII). New York: Academic Press.
Salmon, W. C. (1969), "The conventionality of simultaneity". *PS*, 36: 44–63.
Salmon, W. C. (1977), "The philosophical significance of the one-way speed of light". *Noûs*,11: 253–92.
Sandage, A. (1961), "The ability of the 200-inch telescope to discriminate between selected world models". *Astroph. J.* 133: 355–92.
Sandage, A. (1968), "Observational cosmology". *Observatory*, 88: 91–106.
Schaffner, K. F. (1969), "The Lorentz electron theory and relativity". *AJP*, 37: 498–513.
Schaffner, K. F. (1972), *Nineteenth-Century Aether Theories.* Oxford: Pergamon.
Schaffner, K. F. (1976), "Space and time in Lorentz, Poincaré, and Einstein: Divergent approaches to the discovery and development of the special theory of relativity". In Machamer and Turnbull (1976), pp. 465–507.
Schattner, R. (1979*a*), "The center of mass in general relativity". *GRG*, 10: 377–93.
Schattner, R. (1979*b*), "The uniqueness of the center of mass in general relativity". *GRG*, 10: 395–9.
Schild, A. (1962*a*), "Gravitational theories of the Whitehead type and the principle of equivalence". In Møller, *EGT*, pp. 69–115.
Schild, A. (1962*b*), "The principle of equivalence". *The Monist*, 47: 20–39.
Schilpp, P. A., ed. (1949), *Albert Einstein: Philosopher-Scientist.* Evanston, Ill.:The Library of Living Philosophers.
Schlick, M. (1918), *Allgemeine Erkenntnislehre.* Berlin: Springer.
Schmidt, B. G. (1971), "A new definition of singular points in general relativity". *GRG*, 1: 269–80.
Schmidt, H. (1958), "A causal theory for waves propagating with a velocity greater than that of light". *Ztschr. f. Physik*, 151: 408–20.
Schouten, J. A. and E. R. van Kampen (1930), "Zur Einbettungs- und Krümmungstheorie nichtholonomer Gebilde". *Math. Ann.* 103: 752–83.
Schrödinger, E. (1918*a*), "Die Energiekomponente des Gravitationsfeldes". *Phys. Ztschr.* 19: 4–7.
Schrödinger, E. (1918*b*), "Ueber ein Lösungssystem der allgemein kovarianten Gravitationsgleichungen". *Phys. Ztschr.* 19: 20–2.
Schrödinger, E. (1954), *Space-Time Structure.* Cambridge: The University Press.
Schrödinger, E. (1956), *Expanding Universes.* Cambridge: The University Press.
Schur, F. (1886), "Ueber den Zusammenhang der Räume constanten Riemannschen Krümmungsmaasses mit den projectiven Räumen". *Math. Ann.* 27: 537–67.
Schutz, B. F. (1980), *Geometrical Methods of Mathematical Physics.* Cambridge: The University Press.
Schutz, J. W. (1973), *Foundations of Special Relativity. Kinematic Axioms for Minkowski Space-Time.* (Lecture Notes in Mathematics, vol. 361). Berlin: Springer.
Schwartz, S. P., ed. (1977), *Naming, Necessity and Natural Kinds.* Ithaca: Cornell University Press.
Schwarzschild, K. (1916), "Ueber das Gravitationsfeld eines Massenpunktes nach der Einsteinschen Theorie", *Preuss. Ak. Wiss. Sitzungsber.*, pp. 189–96.
Sciama, D. W. (1953), "On the origin of inertia". *MNRAS*, 113: 34–42.
Scribner, C. (1964), "Henri Poincaré and the principle of relativity". *AJP*, 32: 672–8.
Seelig, C. (1954), *Albert Einstein: Eine dokumentarische Biographie.* Zürich: Europa-Verlag.
Sexl, R. U. and H. K. Urbantke (1976), *Relativität, Gruppen, Teilchen. Spezielle Relativitätstheorie als Grundlage der Feld- und Teilchenphysik.* Wien: Springer.
Shankland, R. S. (1963, 1973), "Conversations with Albert Einstein". *AJP*, 31: 47–57; 41: 895–901.
Shankland, R. S. (1964), "Michelson-Morley experiment". *AJP*, 32: 16–35.
Shapiro, I. I. (1980*a*), "Experimental tests of the general theory of relativity". In Held (1980), vol. 2, pp. 469–89.
Shapiro, I. I. (1980*b*), "Experimental challenges posed by the general theory of relativity". In Woolf (1980), pp. 115–36.
Silberstein, L. (1924). *The Theory of Relativity.* Second edition, enlarged. London: Macmillan.
Sitter, W. de (1911), "On the bearing of the principle of relativity on gravitational astronomy". *MNRAS*, 71: 388–415.

Sitter, W. de (1913*a*), "A proof of the constancy of the velocity of light". *KNAWPr.* 15: 1297–8.
Sitter, W. de (1913*b*), "On the constancy of the velocity of light". *KNAWPr.*, 16: 395–6.
Sitter, W. de (1913*c*), "Ein astronomischer Beweis für die Konstanz der Lichtgeschwindigkeit". *Phys. Ztschr.* 14: 429.
Sitter, W. de (1917*a*), "On the relativity of inertia. Remarks concerning Einstein's latest hypothesis". *KNAWPr.* 19: 1217–25.
Sitter, W. de (1917*b*), "On the curvature of space". *KNAWPr.* 20: 229–43.
Sitter, W. de (1917*c*), "On Einstein's theory of gravitation, and its astronomical consequences, III". *MNRAS*, 78: 3–28.
Sitter, W. de (1933), "On the expanding universe and the time-scale". *MNRAS*, 93: 628–34.
Sklar, L. (1974), *Space, Time and Spacetime*. Berkeley: University of California Press.
Smart, J. J. C., ed. (1964), *Problems of Space and Time*, New York: Macmillan.
Sommerfeld, A. (1910), "Zur Relativitätstheorie. I. Vierdimensionale Vektoralgebra. II. Vierdimensionales Vektoranalysis". *Ann. Physik* (4) 32: 749–76; 33: 649–89.
Sommerfeld, A. (1964), *Lectures on Theoretical Physics. III. Electrodynamics.* Transl. by E. G. Ramberg. New York: Academic Press.
Spivak, M. (1965), *Calculus on Manifolds.* New York: Benjamin.
Spivak, M. (*CIDG*), *A Comprehensive Introduction to Differential Geometry.* Second Edition. Berkeley: Publish or Perish, 1979. 5 vols.
Stachel, J. (1969*a*), "Specifying sources in general relativity". *Phys. Rev.* 180: 1256–61.
Stachel, J. (1969*b*), "Comments on 'Causality requirements and the theory of relativity' [Havas, 1968]". In R. S. Cohen and M. Wartofsky, eds., *Boston Studies in the Philosophy of Science*, vol. 5, pp. 179–98 (Dordrecht: Reidel).
Stachel, J. (1972), "External sources in general relativity". *GRG*, 3: 257–67.
Stachel, J. (1974), "The rise and fall of geometrodynamics". In K. F. Schaffner and R. S. Cohen, eds., *PSA 1972*. Dordrecht: Reidel, pp. 31–54.
Stachel, J. (1977), "Notes on the Andover Conference". In Earman *et al.* (1977) pp. vii–xiii.
Stachel, J. (1979), "The genesis of general relativity". In Nelkowski *et al.* (1979), pp. 428–42.
Stachel, J. (1980*a*), "Einstein and the rigidly rotating disk". In Held (1980), vol. 1, pp. 1–15.
Stachel, J. (1980*b*), "Einstein's search for general covariance, 1912–1915. Talk given at [the] Ninth International Conference on General Relativity and Gravitation. Jena, DDR, July 17, 1980". Preprint.
Stachel, J. (1980*c*), *Guide to the Duplicate Einstein Archive and Control Index.* Princeton. Mimeographed.
Stachel, J. and R. Torretti (1982), "Einstein's first derivation of mass–energy equivalence". *AJP*, 50: 760–763.
Steenrod, N. (1951), *The Topology of Fibre Bundles.* Princeton: Princeton University Press.
Stein, H. (1967), "Newtonian space-time". *Texas Quart.* 10: 174–200.
Stein, H. (1968), "On Einstein–Minkowski space-time". *J. Philos.* 65: 5–23.
Stein, H. (1970*a*), "On the paradoxical time-structures of Gödel". *PS*, 37: 589–601.
Stein, H. (1970*b*), "On the notion of field in Newton, Maxwell, and beyond". In R. H. Stuewer, ed., *Historical and Philosophical Perspectives in Science.* Minneapolis: University of Minnesota Press, pp. 264–287, 299–310.
Stein, H. (1977*a*),"Some philosophical prehistory of general relativity". In Earman *et al.* (1977), pp. 3–49.
Stein, H. (1977*b*), "On space-time ontology. Extracts from a letter to Adolf Grünbaum". In Earman *et al.* (1977), pp. 374–402.
Stoker, J. J. (1969), *Differential Geometry.* New York: Wiley.
Stokes, G. G. (1845), "On the aberration of light". *Philos. Mag.* (3) 27: 9–15.
Strauss, M. (1972), *Modern Physics and its Philosophy. Selected Papers in the Logic, History and Philosophy of Science.* Dordrecht: Reidel.
Sudarshan, E. C. G. and N. Mukunda (1974), *Classical Dynamics: A modern perspective.* New York: Wiley.
Suppes, P. (1959), "Axioms for relativistic kinematics with or without parity". In L. Henkin, P. Suppes and A. Tarski, eds., *The Axiomatic Method.* Amsterdam: North-Holland, pp. 291–307.
Suppes, P., ed. (1973), *Space, Time and Geometry.* Dordrecht: Reidel.

Süssmann, G. (1969), "Begründung der Lorentz-Gruppe allein mit Symmetrie- und Relativitäts-Annahmen". *Ztschr. f. Naturf.* 24*a*: 495–8.
Suvorov, S. G. (1979), "Einstein: the creation of the theory of relativity and some gnosiological lessons". *Sov. Phys. USPEKHI*, 22: 528–54.
Synge,J. L. (1937), "Relativistic hydrodynamics". *London Math. Soc. Proc.* 43: 376–416.
Synge, J. L. (1955), *Relativity: The Special Theory*. Amsterdam: North-Holland.
Synge, J. L. (1960), *Relativity: The General Theory*. Amsterdam: North-Holland.
Synge, J. L. and A. Schild (*TC*), *Tensor Calculus*. Toronto: University of Toronto Press, 1949. (Reprinted by Dover, New York.)
Szekeres, G. (1960), "On the singularities of a Riemannian manifold". *Publ. Math. Debrecen*, 7: 285–301.
Tanaka, S. (1960), "Theory of matter with super light velocity". *Progr. Theor. Phys. (Japan)*, 24: 171–200.
Taub, A. H. (1951), "Empty space-times admitting a three-parameter group of motions". *Ann. Maths.* 53: 472–90.
Taub, A. H. (1979), "Remarks on the symposium on singularities". *GRG*, 10: 1009–10.
Taylor, E. F. and J. A. Wheeler (1963), *Spacetime Physics*. San Francisco: Freeman.
Taylor, J. G. (1975), *Special Relativity*. Oxford: Clarendon.
Taylor, J. H., L. A. Fowler and P. M. McCulloch (1979), "Measurements of general relativistic effects in the binary pulsar PSR 1913 + 16". *Nature*, 277: 437–40.
Teitelboim, C. (1980), "The Hamiltonian structure of space-time". In Held (1980), vol. 1, pp. 195–225.
Ter Haar, D. (1967), *The Old Quantum Theory*. Oxford: Pergamon.
Terletskii, Y. P. (1968). *Paradoxes in the Theory of Relativity*. Transl. from Russian. New York: Plenum.
Thirring, H. (1918), "Ueber die Wirkung rotierender ferner Massen in der Einsteinschen Gravitationstheorie". *Phys. Ztschr.* 19: 33–9.
Thirring, H. (1921), "Berichtigung zu meiner Arbeit: 'Ueber die Wirkung rotierender ferner Massen in der Einsteinschen Gravitationstheorie'". *Phys. Ztschr.* 22: 29–30.
Thomson, J. (1884*a*), "On the law of inertia, the principle of chronometry and the principle of absolute clinural rest, and of absolute rotation". *R. Soc. Edinb. Proc.* 12: 568–78.
Thomson, J. (1884*b*), "A problem on point-motions for which a reference-frame can so exist as to have the motions of the points relative to it, rectilinear and mutually proportional". *R. Soc. Edinb. Proc.* 12: 730–42.
Tipler, F. J., ed. (*EGR*), *Essays in General Relativity. A Festschrift for Abraham Taub*. New York: Academic Press, 1980.
Tipler, F. J. (1976), "Causality violation in asymptotically flat space-times". *Phys. Rev. Lett.* 37: 879–82.
Tipler, F. J. (1977), "Singularities and causality violation". *Ann. Physics*, 108: 1–36.
Tipler, F. J. (1980), "General relativity and the eternal return". In Tipler, *EGR*, pp. 21–37.
Tipler, F. J., C. J. S. Clarke and G. F. R. Ellis (1980), "Singularities and horizons—a review article". In Held (1980), vol. 2, pp. 97–206.
Tolman, R. C. (1912), "Non-Newtonian mechanics, the mass of a moving body". *Philos. Mag.* (6) 23: 375–80.
Tolman, R. C. (1929), "On the possible line elements for the universe". *Nat. Ac. Sci. (U.S.) Proc.* 15: 297–304.
Tolman, R. C. (1934), *Relativity, Thermodynamics and Cosmology*. Oxford: Clarendon.
Tonnelat, M.-A. (1971), *Histoire du Principe de Rélativité*. Paris: Flammarion.
Toraldo di Francia, G., ed. (1979), *Problems in the Foundations of Physics*. (Intern. Sch. of Physics »Enrico Fermi«, Course LXXII.) Amsterdam: North-Holland.
Torretti, R. (1978*a*), *Philosophy of Geometry from Riemann to Poincaré*. Dordrecht: Reidel.
Torretti, R. (1978*b*), "Hugo Dingler's philosophy of geometry". *Diálogos*, 32: 85–128.
Torretti, R. (1979*a*), "Jackson and Pargetter's criterion of distant simultaneity". *PS*, 46: 302–5.
Torretti, R. (1979*b*), "Mathematical theories and philosophical insights in cosmology". In Nelkowski *et al.* (1979), pp. 320–35.
Trautman, A. (1962), "Conservation laws in general relativity". In Witten (1962) pp. 169–98.

Trautman, A. (1965), "Foundations and current problems of general relativity". In Trautman *et al.* (1965), pp. 1–248.

Trautman, A. (1966), "Comparison of Newtonian and relativistic theories of space-time". In Hoffmann (1966), pp. 413–25.

Trautman, A., F. A. E. Pirani and H. Bondi (1965), *Lectures on General Relativity.* (Brandeis Summer Institute in Theoretical Physics, 1964, vol. I.) Englewood Cliffs, N. J. :Prentice-Hall.

Tricker, R. A. R. (1965), *Early Electrodynamics. The First Law of Circulation.* Oxford: Pergamon.

Tricker, R. A. R. (1966), *The Contributions of Faraday and Maxwell to Electrical Science.* Oxford: Pergamon.

Trouton, F. T. and H. R. Noble (1904), "The mechanical forces acting on a charged condenser moving through space". *RSLPT*, A 202: 165–81.

Ueberweg, F. (1851), "Die Prinzipien der Geometrie wissenschaftlich dargestellt". In *Die Welt- und Lebensanschauung Friedrich Ueberwegs*, hrsg. v. M. Brasch. Leipzig: G. Engel, 1889, pp. 263–307. (Orig. publ. in *Arch. f. Philologie u. Pädagogik*, vol. 17.)

Umov, N. (1910), "Einheitliche Ableitung der Transformationen, die mit dem Relativitätsprinzip verträglich sind". *Phys. Ztschr.* 11: 905–8.

Van Fraassen, B. C. (1969*a*), "Conventionality in the axiomatic foundations of the special theory of relativity". *PS*, 36: 64–73.

Van Fraassen, B. C. (1969*b*), "On Massey's explication of Grünbaum's conception of metric". *PS*, 36: 346–53.

Van Fraassen, B. C. (1970), *An Introduction to the Philosophy of Space and Time.* New York: Random House.

Varićak, V. (1912), "Ueber die nicht-euklidische Interpretation der Relativtheorie". *JDMV*, 21: 103–27.

Veblen, O. and J. H. C. Whitehead (1932), *The Foundations of Differential Geometry.* Cambridge: The University Press.

Vermeil, H. (1917), "Notiz über das mittlere Krümmungsmass einer *n*-fach ausgedehnten Riemannschen Mannigfaltigkeit", *Gött. Nachr.*, pp. 334–44.

Vizgin, V. P. and Ya. A. Smorodinskiĭ (1979), "From the equivalence principle to the equations of gravitation". *Sov. Phys. USPEKHI*, 22: 489–513.

Voigt, W. (1887), "Ueber das Doppler'sche Prinzip". *Gött. Nachr.*, pp. 41–51.

Walker. A. G. (1935), "On the formal comparison of Milne's kinematical system with the systems of general relativity". *MNRAS*, 95: 263–9.

Walker, A. G. (1937), "On Milne's theory of world-structure". *London Math. Soc. Proc.* 42: 90–127.

Walker, A. G. (1944*a*), "Completely symmetric spaces". *J. London Math. Soc.* 19: 219–26.

Walker, A. G. (1944*b*), "Complete symmetry in flat space". *J. London Math. Soc.* 19: 227–9.

Weinberg, S. (1973), *Gravitation and Cosmology. Principles and Applications of the General Theory of Relativity.* New York: Wiley.

Weinstock, R. (1964), "Derivation of the Lorentz transformation equations without a linearity assumption". *AJP*, 32: 260–4.

Wertheimer, M. (1959), *Productive Thinking.* Enlarged edition, ed. by Michael Wertheimer. New York: Harper.

Weyl, H. (1918*a*), "Reine Infinitesimalgeometrie", *Math. Ztschr.* 2: 384–411.

Weyl, H. (1918*b*), "Gravitation und Elektrizität". *Preuss. Ak. Wiss. Sitzungsber.*, pp. 465–80.

Weyl, H. (1919*a*), *Raum-Zeit-Materie. Vorlesungen über allgemeine Relativitätstheorie.* Berlin: Springer.

Weyl, H. (1919*b*), "Eine neue Erweiterung der Relativitätstheorie". *Ann. Physik* (4) 59: 101–33.

Weyl, H. (1921*a*), *Raum-Zeit-Materie. Vorlesungen über allgemeine Relativitätstheorie.* Vierte, erweiterte Auflage. Berlin: Springer.

Weyl, H. (1921*b*), "Zur Infinitesimalgeometrie. Einordnung der projektiven und konformen Auffassung". *Gött. Nachr.*, pp. 99–112.

Weyl, H. (1921*c*), "Ueber die physikalischen Grundlagen der erweiterten Relativitätstheorie". *Phys. Ztschr.* 22: 473–80.

Weyl, H. (1923*a*), *Raum-Zeit-Materie. Vorlesungen über allgemeine Relativitätstheorie.* Fünfte, umgearbeitete Auflage. Berlin: Springer.

Weyl, H. (1923*b*), *Mathematische Analyse des Raumproblems. Vorlesungen gehalten in Barcelona und Madrid.* Berlin: Springer.
Weyl, H. (1923*c*), "Zur allgemeinen Relativitätstheorie". *Phys. Ztschr.* 24: 230–2.
Weyl, H. (1931), "Geometrie und Physik". *Naturwiss.* 19: 49–58.
Weyl, H. (1950), *Space-Time-Matter.* Transl. by H. L. Brose. New York: Dover. (This is a translation of Weyl (1921*a*).)
Wheeler, J. A. (1964), "Mach's Principle as a boundary condition for Einstein's equations". In Chiu and Hoffmann (1964), pp. 303–49.
Wheeler, J. A. (1968), *Einsteins Vision. Wie steht es heute mit Einsteins Vision, alles als Geometrie aufzufassen.* Berlin: Springer.
Whitehead, A. N. (1916), "La théorie relationniste de l'espace". *Rev. Métaph. Mor.* 23: 423–54.
Whitehead, A. N. (1919), *An Enquiry concerning the Principles of Natural Knowledge.* Cambridge: The University Press.
Whitehead, A. N. (1920), *The Concept of Nature.* Cambridge: The University Press.
Whitehead, A. N. (1922), *The Principle of Relativity, with applications to physical science.* Cambridge: The University Press.
Whitney, H. (1936), "Differentiable manifolds". *Ann. Maths.* 37: 645–80.
Whitney, H. (1944*a*), "The self-intersections of a smooth *n*-manifold in 2*n*-space". *Ann. Maths.* 45: 220–46.
Whitney, H. (1944*b*), "The singularities of a smooth *n*-manifold in a $(2n-1)$-space". *Ann. Maths.* 45: 247–93.
Whitrow, G. J. (1933), "A derivation of the Lorentz formulae". *Quart. J. Math.* 4: 161–72.
Whitrow, G. J. (1980), *The Natural Philosophy of Time.* Second Edition. Oxford: Clarendon.
Whitrow, G. J. and G. E. Morduch (1965), "Relativistic theories of gravitation. A comparative analysis with particular reference to astronomical tests". In A. Beer, ed., *Vistas in Astronomy*, vol. 6, pp. 1–67. (Oxford: Pergamon.)
Whittaker, E. T. (1928), "Note on the law that light-rays are the null-geodesics of a gravitational field". *Camb. Philos. Soc. Proc.*, 24: 32–4.
Whittaker, E. T. (1951/53), *A History of the Theories of Aether and Electricity.* London: Nelson. 2 vols.
Wien, W. (1900), "Ueber die Möglichkeit einer elektromagnetischen Begründung der Mechanik". *Arch. Néerl.* 5: 96–104.
Will, C. W. (1979), "The confrontation between gravitation theory and experiment". In Hawking and Israel (1979), pp. 24–89. (References on pp. 833–46.)
Williamson, R. B. (1977), "Logical economy in Einstein's 'On the electrodynamics of moving bodies'". *SHPS*, 8: 49–60.
Wilson, H. A. (1905), "On the electric effect of rotating a dielectric in a magnetic field". *RSLPT*, A 204: 121–37.
Winnie, J. A. (1970), "Special relativity without one-way velocity assumptions". *PS*, 37: 81–99, 223–38.
Winnie, J. A. (1977), "The causal theory of space-time". In Earman *et al.* (1977), pp. 134–205.
Witten, L., ed. (1962), *Gravitation: An Introduction to Current Research.* New York: Wiley.
Wood, A. B., G. A. Tomlison and L. Essex (1937), "The effect of the Fitzgerald–Lorentz contraction on the frequency of longitudinal vibration of a rod". *RSLPr.* A 158: 606–33.
Woodhouse, N. M. J. (1973), "The differentiable and causal structures of space-time". *J. Math. Phys.* 14: 495–501.
Woodruff, A. E. (1962), "Action at a distance in nineteenth century electrodynamics". *Isis*, 53: 439–59.
Woodruff, A. E. (1968), "The contributions of Hermann von Helmholtz to electrodynamics". *Isis*, 59: 300–11.
Woodward, J. F. and Yourgrau, W. (1972), "The imcompatibility of Mach's principle and the principle of equivalence in current gravitation theory". *BJPS*, 23: 111–16.
Woolf, H., ed. (1980), *Some Strangeness in the Proportion. A Centennial Symposium to Celebrate the Achievements of Albert Einstein.* Reading, Mass.: Addison-Wesley.
Zahar, E. (1973), "Why did Einstein's programme supersede Lorentz's?" *BJPS*, 24: 95–123, 223–62.
Zahar, E. (1978), "»Crucial« experiments: A case study". In G. Radnitzky and G. Anderson, eds.,

Progress and Rationality in Science. Dordrecht: Reidel, pp. 71–97.
Zahar, E. (1979), "The mathematical origins of general relativity and of unified field theory". In Nelkowski *et al.* (1979), pp. 370–96.
Zahar, E. (1980), "Einstein, Meyerson and the role of mathematics in physical discovery". *BJPS*, 31: 1–43.
Zahar, E. (1982), "Poincaré et la découverte du Principe de Relativité". Preprint, 6 February.
Zeeman, E. C. (1964), "Causality implies the Lorentz group". *J. Math. Phys.* 5: 490–3.
Zeeman, E. C. (1967), "The topology of Minkowski space". *Topology*, 6: 161–70.

Index

1. Symbols

2. Names

3. Terms and Subjects